Graphene
Synthesis and Applications

Nanomaterials and Their Applications

Series Editor: M. Meyyappan

Carbon Nanotubes: Reinforced Metal Matrix Composites
Arvind Agarwal, Srinivasa Rao Bakshi, Debrupa Lahiri

Inorganic Nanoparticles: Synthesis, Applications, and Perspectives
Edited by Claudia Altavilla, Enrico Ciliberto

Graphene: Synthesis and Applications
Edited by Wonbong Choi, Jo-won Lee

Nanorobotics: An Introduction
Lixin Dong, Bradley J. Nelson

Plasma Processing of Nanomaterials
Edited by R. Mohan Sankaran

Graphene
Synthesis and Applications

Edited by
Wonbong Choi
Jo-won Lee

CRC Press
Taylor & Francis Group
Boca Raton London New York

CRC Press is an imprint of the
Taylor & Francis Group, an **informa** business

CRC Press
Taylor & Francis Group
6000 Broken Sound Parkway NW, Suite 300
Boca Raton, FL 33487-2742

First issued in paperback 2020

Version Date: 20110608

ISBN-13: 978-0-367-57686-8 (pbk)
ISBN-13: 978-1-4398-6187-5 (hbk)

Library of Congress Cataloging-in-Publication Data

Graphene : synthesis and applications / [edited by] Wonbong Choi, Jo-won Lee.
p. cm. -- (Nanomaterials and their applications ; 3)
Includes bibliographical references and index.
ISBN 978-1-4398-6187-5 (hardback)
1. Graphene. 2. Graphene--Industrial applications. I. Choi, Wonbong, 1963- II. Lee, Jo-won, 1952-

QD341.H9G695 2012
620'.5--dc23 2011017149

Visit the Taylor & Francis Web site at
http://www.taylorandfrancis.com

and the CRC Press Web site at
http://www.crcpress.com

Contents

Preface

This book aims to present an overview of recent advancements of research in graphene, in the areas of synthesis, properties, and applications, such as electronics, heat transport, field emission, sensors, composites, and energy. Researchers from various sectors including physics, chemistry, materials, and electrical engineering have prepared their contributed chapters based on their research expertise in these fields. Although graphene created enormous research activities in recent years due to its excellent electrical, optical, and mechanical properties, most of the applications are still in their infancy. Therefore, it is an appropriate time to compile a book presenting a comprehensive review of the current status of graphene, specially focused on synthesis and future applications.

Graphene, a monolayer of sp2 bonded carbon atoms in a honeycomb lattice, created a surge in research activities during the last 6-7 years owing to its high current density, ballistic transport, chemical inertness, high thermal conductivity, optical transmittance, and super hydrophobicity at nanometer scale. Graphene is considered to be one of the miracle materials in the twenty-first century. The first report (by A. K. Geim and his co-workers) of employing a simple technique called micromechanical cleavage to extract graphene has attracted worldwide attention and earned them the Nobel Prize in Physics in the year 2010. Though graphene was known well before that time, Geim and the group's research in 2004 created huge interest in the field.

In a sense, graphene is more attractive than its allotrope, carbon nanotubes (CNTs) since the 2-dimensional form of graphene is much better, from the fabrication and application point of view, than 1-dimensional CNTs. Utilizing its extremely high mobility (200,000 cm^2/Vs at RT in comparison with 1,400 cm^2/Vs for Si and 8500 cm^2/Vs for GaAs), stand-alone high frequency transistor with a cut-off frequency as high as 300 GHz could be designed. On the other hand, electromigration problems in interconnects could be avoided with high current capacity (10^8 A/cm^2) and low resistance (1 μΩ-cm: 35% less than Cu) of graphene. Graphene heat pad has shown good promise owing to its high thermal conductivity (5 kW/mK—10 times larger than Cu and Al). Graphene can also be a strong candidate for replacement of ITO as a transparent electrode (which is a necessity as earth crust has very low reserve of Indium). Graphene could have over 90% of transparency and 30 Ohm/sq of sheet resistivity, making it most suitable for transparent conducting electrode applications. Graphene has an extraordinary mechanical strength/weight ratio exceeding that of any known material. Graphene also has the highest surface to volume ratio, utilizing two-side surfaces. Thus, graphene-based chemical sensors can be used to detect explosives in luggage and volatile organic compounds in air by converting chemical reactions into electrical signals. Graphene might revolutionize battery technologies, where it can be used as a super-conductive membrane between a battery's poles. This battery could supply a huge energy for a short period of time. Graphene has a larger spin diffusion length, too—so we expect potential high injection efficiency of spins for spintronics.

Nevertheless, the main worry at this juncture is that graphene can become a hype, unless the issues are resolved within a short time period.

Graphene was originally visualized as a material that could replace Si in digital logic circuits since it exhibits excellent electrical properties, including very high electron mobility. However, one problem is that graphene does not have a band gap, which is a prerequisite for a semiconductor. Furthermore, a graphene transistor is very difficult to turn off, with an on/off ratio as high as 1,000 at room temperature. To open a stable band gap of ~1 eV, it is necessary to make graphene ribbons smaller than 2 nm wide with single atom precision. Variation in width of a graphene sheet results in deviation of band gap energy. Mobility of graphene is severely degraded if the ribbon edges are rough and if the substrate underneath is not flat. In turn, reproducibility issues are challenging for the success of future graphene nanoelectonics.

Graphene grown by the chemical vapor method is not a single crystal, due to unavoidable occurrence of nucleation and growth during the process. This leads to degradation and variation in properties of graphene-based electronic devices. Graphene transistors have not yet revealed better analog properties over other single crystalline high mobility semiconductors (such as III-V compound semiconductor) since the transistors have already approached the range of almost 1 THz cut-off frequency. So, it seems that graphene's promise in next generation electronics is not easily achievable. Its future may lie elsewhere such as in passive devices and/or components less sensitive on variation of its energy band gap.

Silicon was discovered in 1824, but the first transistor was made by Bell Labs scientists in 1947. It took 123 years to create the first transistor, but with Ge. In contrast, CNT was discovered in 1991 and the first CNT transistor was demonstrated in 1998. Graphene was manufactured in the 2-D stable form through an easy and reproducible process in 2004—just seven years ago. Thus, we need some patience for the materialization of our dream in the application of nanocarbon materials—CNTs and graphene. Lots of research efforts should be made until some big breakthroughs happen with CNTs and graphene.

We would like to conclude with the old saying "Where there's a will there's a way." "Though your beginning was small, yet your latter end should greatly increase." We truly hope that the 11 chapters in this book will be beneficial for the scientific community including students, teachers, researchers and project managers.

Wonbong Choi and Jo-won Lee

Introduction

Graphene, a one-atom-thick planar sheet of carbon atoms densely packed in a honeycomb crystal lattice, has revolutionized the scientific frontiers in nanoscience and condensed matter physics due to its exceptional electrical, physical, and chemical properties. Expected as a possible replacement for silicon in electronics and applications in many other advanced technologies, graphene has sparked enormous interest in many research groups around the world, and has resulted in an abrupt increase in publications on the subject and recently in Geim and Novoselov's Nobel Prize in Physics. The reported properties and applications of graphene have opened up new opportunities for future devices and systems. Although graphene is known as one of the best electronic materials, synthesizing a single sheet of graphene for industrial applications has been less explored. This book aims to present an overview of the advancement of research in graphene in the areas of synthesis, properties, and applications, such as electronics, heat dissipation, field emission, sensors, composites, and energy. Eleven chapters are presented by experts from each research area. Wherever applicable, the limitations of present knowledge base and future research directions have also been highlighted.

CHAPTER 1: TAILORING THE PHYSICAL PROPERTIES OF GRAPHENE

C. G. Rocha, M. H. Rümmeli, I. Ibrahim, H. Sevincli, F. Börrnert, J. Kunstmann, A. Bachmatiuk, M. Pötschke, W. Li, S. A. M. Makharza, S. Roche, B. Büchner, and G. Cuniberti

The basic properties of graphene are described, with emphasis on its potential in electronics, mechanical devices, and photonics. In addition, the state of the art regarding other important physical aspects of graphene beyond its electrical properties is also reviewed including its mechanical, magnetic, and thermal properties.

CHAPTER 2: GRAPHENE SYNTHESIS

Santanu Das and Wonbong Choi

Major graphene synthesis methods are described with detailed information regarding process parameters and graphene characteristics. Among various synthesis techniques, emphasis is given on mechanical exfoliation, chemical synthesis, chemical vapor deposition, and epitaxial growth as the most popular graphene synthesis methods among scientists and researchers. Other important issues of graphene synthesis, such as fabrication of functionalized graphene, large-scale graphene growth and graphene transfer onto other substrates, have also been summarized.

CHAPTER 3: QUANTUM TRANSPORT IN GRAPHENE-BASED MATERIALS AND DEVICES: FROM PSEUDOSPIN EFFECTS TO A NEW SWITCHING PRINCIPLE

Stephan Roche, Frank Ortmann, Alessandro Cresti, Blanca Biel, and David Jimenez

This theoretical chapter presents an overview of some electronic and transport features of clean and chemically modified graphene-based materials and devices, either described with simple models or tight-binding models elaborated from first principles simulations.

CHAPTER 4: ELECTRONIC AND PHOTONIC APPLICATIONS FOR ULTRAHIGH-FREQUENCY GRAPHENE-BASED DEVICES

Taiichi Otsuji, Tetsuya Suemitsu, Akira Satou, Maki Suemitsu, Eiichi Sano, Maxim Ryzhii, and Victor Ryzhii

This chapter provides the recent advances in theoretical and experimental studies of graphene-based materials in electronic and photonic device applications. Due to unique carrier transport and optical properties, including massless and gapless energy spectra, graphene will break through many technological limits on conventional electronic and photonic devices. One of the most promising applications for the electronic devices is the introduction of graphene as the channel material in field-effect transistors (FETs). The optical properties of graphene can also provide many advantages in optoelectronic applications such as new types of terahertz lasers as well as high-sensitive, ultrafast photodetector and phototransistor operation of graphene on junction and graphene channel FETs. A detailed discussion and experimental study are described.

CHAPTER 5: GRAPHENE THIN FILMS FOR UNUSUAL FORMAT ELECTRONICS

Chao Yan, Houk Jang, Youngbin Lee, and Jong-Hyun Ahn

This chapter provides an introduction to graphene films for electronic application, focusing on growth and transfer techniques that can be used to synthesize and fabricate films on unusual substrates such as flexible, stretchable substrates. The content is organized into six main sections: introduction to graphene for high-performance electronics, fabrication of graphene thin films using the chemical vapor deposition method and printing approach for large area electronics, applications in radio frequency transistors and flexible electronic systems, integration of graphene for touch screen panels, organic solar cells, and light-emitting diodes, graphene-based gas barrier films, and a summary.

CHAPTER 6: NANOSIZED GRAPHENE: CHEMICAL SYNTHESIS AND APPLICATIONS IN MATERIALS SCIENCE

Chongjun Jiao and Jishan Wu

Currently, nanosized graphene is among the most widely studied families of organic compounds. Synthesis and applications of nanosized graphene are described in Chapter 6. This chapter summarizes modern processes to synthesize nanosized graphenes with different sizes, shapes, and edge structures, and their basic physical properties. In addition, the fundamental structure and physical property relationship is introduced, and then applications of nanosized graphene in materials science are discussed.

CHAPTER 7: GRAPHENE-REINFORCED CERAMIC AND METAL MATRIX COMPOSITES

Debrupa Lahiri and Arvind Agarwal

This chapter deals with graphene-reinforced metals and ceramic nanocomposites, their synthesis techniques, and potential applications. A classification of composite preparation techniques is offered, based on the mechanism. Future scope and potential of these nanocomposites are also discussed.

CHAPTER 8: GRAPHENE-BASED BIOSENSORS AND GAS SENSORS

Subbiah Alwarappan, Shreekumar Pillai, Shree R. Singh, and Ashok Kumar

This chapter describes some of the important electrochemical biosensing applications of graphene. More specifically, the electrochemistry of graphene, the direct electrochemistry of enzymes on graphene, graphene's electrocatalytic activity toward biomolecules, graphene-based enzyme biosensors, graphene-based DNA sensors, and environmental sensors are discussed.

CHAPTER 9: FIELD EMISSION AND GRAPHENE: AN OVERVIEW OF THE CURRENT STATUS

Indranil Lahiri and Wonbong Choi

This chapter focuses on application of graphene-based field emission devices. In recent years, graphene, the two-dimensional allotrope of carbon, has shown good promise for application in field emission devices. Field emitters are widely used in high-powered microwave devices, miniature x-ray sources, displays, sensors, and the electron gun in electron microscopes. Application of graphene is predicted to open a new door to flexible and transparent field emission devices. The basic theory of field emission phenomenon and other materials used in field emission devices are mentioned briefly.

CHAPTER 10: GRAPHENE AND GRAPHENE-BASED MATERIALS IN SOLAR CELL APPLICATION

Indranil Lahiri and Wonbong Choi

This chapter focuses on the application of graphene and other graphene-based materials in solar cells. Before summarizing the current status of research on graphene-based solar cells, important properties of graphene that are relevant to solar cell application are discussed.

CHAPTER 11: GRAPHENE: THERMAL AND THERMOELECTRIC PROPERTIES

Suchismita Ghosh and Alexander A. Balandin

This chapter summarizes the thermal properties of graphene and graphene multilayers with an emphasis on thermal conduction and thermoelectric power. The experimental and theoretical investigation of heat conduction in suspended graphene layers is reviewed. The superior thermal properties of graphene are beneficial for all proposed electronic device applications and make graphene a promising material for thermal management.

Contributors

Arvind Agarwal
Department of Mechanical and Materials Engineering
Florida International University
Miami, Florida

Jong-Hyun Ahn
School of Advanced Materials Science and Engineering
Sungkyunkwan University
Suwon, South Korea

Subbiah Alwarappan
Department of Mechanical Engineering
Nanotechnology Research and Education Center
University of South Florida
Tampa, Florida

A. Bachmatiuk
IFW Dresden-Leibniz Institute for Solid State and Materials Research Dresden
Dresden, Germany

Alexander A. Balandin
Department of Electrical Engineering
Materials Science and Engineering Program
University of California Riverside
Riverside, California

Blanca Biel
Departamento de Electrónica y Tecnologia de Computadores
Universidad de Granada
Facultad de Ciencias
Granada, Spain

F. Börrnert
IFW Dresden-Leibniz Institute for Solid State and Materials Research Dresden
Dresden, Germany

B. Büchner
IFW Dresden-Leibniz Institute for Solid State and Materials Research Dresden
Dresden, Germany

Wonbong Choi
Nanomaterials and Device Laboratory
Department of Mechanical and Materials Engineering
Florida International University
Miami, Florida

Gianaurelio Cuniberti
Institute for Materials Science and Max Bergmann Center of Biomaterials
Dresden University of Technology
Dresden, Germany

Allessandro Cresti
IMEP-LAHC, UMR 5130
Université de Savoie
Minatec
Grenoble, France

Santanu Das
Nanomaterials and Device Laboratory
Department of Mechanical and Materials Engineering
Florida International University
Miami, Florida

Suchismita Ghosh
STTD, Intel Corporation
Hillsboro, Oregon

I. Ibrahim
Institute for Materials Science and
Max Bergmann Center of Biomaterials
Dresden University of Technology
Dresden, Germany

Houk Jang
School of Advanced Materials Science and Engineering
Sungkyunkwan University
Suwon, Korea

Chongjun Jiao
Department of Chemistry
National University of Singapore
Singapore

David Jimenéz
Departament d'Enginyeria Electrò
Escola Tècnica Superior d'Enginyeria
Universitat Autònoma de Barcelona
Bellaterra, Spain

Wu Jishan
Department of Chemistry
National University of Singapore
Singapore

Ashok Kumar
Department of Mechanical Engineering
Nanotechnology Research and Education Center
University of South Florida
Tampa, Florida

J. Kunstmann
Institute for Materials Science and
Max Bergmann Center of Biomaterials
Dresden University of Technology
Dresden, Germany

Debrupa Lahiri
Department of Mechanical and Materials Engineering
Florida International University
Miami, Florida

Indranil Lahiri
Nanomaterials and Device Lab
Department of Mechanical and Materials Engineering,
Florida International University
Miami, Florida

Jo-won Lee
The National Program of Tera-level Nano Devices
Hanyang University, Korea

Youngbin Lee
School of Advanced Materials Science and Engineering
Sungkyunkwan University
Suwon, Korea

W. Li
Institute for Materials Science and
Max Bergmann Center of Biomaterials
Dresden University of Technology
Dresden, Germany

S. A. M. Makharza
IFW Dresden-Leibniz Institute for Solid State and Materials Research Dresden
Dresden, Germany

Frank Ortmann
CEA, INAC, SPRAM, GT
Grenoble Cedex, France

Taiichi Otsuji
Research Institute of Electrical Communication
Tohoku University
Sendai, Japan

Shreekumar Pillai
Department of Nanobiotechnology
Alabama State University
Montgomery, Alabama

M. Pötschke
Institute for Materials Science and
Max Bergmann Center of Biomaterials
Dresden University of Technology
Dresden, Germany

C. G. Rocha
Institute for Materials Science and
Max Bergmann Center of Biomaterials
Dresden University of Technology
Dresden, Germany

Stephan Roche
Institute for Materials Science and
Max Bergmann Center of Biomaterials
Dresden University of Technology
Dresden, Germany
and
Catalan Institute of Nanotechnology
Barcelona, Spain
Institucio Catalana de Recerca i Estudis Avancats (ICREA)
Barcelona, Spain

M. H. Rümmeli
IFW Dresden-Leibniz Institute for Solid State and Materials Research Dresden
Dresden, Germany

Maxim Ryzhii
Computer Nano-Electronics Laboratory
University of Aizu
Aizu-Wakamatsu, Japan

Victor Ryzhii
Computer Nano-Electronics Laboratory
University of Aizu
Aizu-Wakamatsu, Japan

Eiichi Sano
Research Center for Integrated Quantum Electronics
Hokkaido University
Sapporo, Japan

Akira Satou
Research Institute of Electrical Communication
Tohoku University
Sendai, Japan

H. Sevincli
Institute for Materials Science and
Max Bergmann Center of Biomaterials
Dresden University of Technology
Dresden, Germany

Shree R. Singh
Department of Nanobiotechnology
Alabama State University
Montgomery, Alabama

Maki Suemitsu
Research Institute of Electrical Communication
Tohoku University
Sendai, Japan

Tetsuya Suemitsu
Research Institute of Electrical Communication
Tohoku University
Sendai, Japan

Chao Yan
School of Advanced Materials Science and Engineering
Sungkyunkwan University
Suwon, Korea

1 Tailoring the Physical Properties of Graphene

C. G. Rocha, M. H. Rümmeli, I. Ibrahim, H. Sevincli, F. Börrnert, J. Kunstmann, A. Bachmatiuk, M. Pötschke, W. Li, S. A. M. Makharza, S. Roche, B. Büchner, and G. Cuniberti

CONTENTS

1.1 INTRODUCTION

For many years graphene was deemed an "academic" material where its perfect honeycomb monolayer structure of carbon atoms was treated solely as a theoretical model for describing the properties of various carbon-based materials such as graphite, fullerenes, and carbon nanotubes. Older theoretical predictions [1,2,3], studying pristine two-dimensional (2D) crystals, presumed graphene would be unstable in reality due to thermal fluctuations that prevent long-range crystalline order at finite temperatures. This presumption was strongly supported by various experimental investigations with thin films in which the samples became unstable as their thickness was reduced. Now, early in the twenty-first century, graphene has emerged as a real sample [4,5]. The initial works by Geim and Novoselov showed the isolation of astonishingly thin carbon films and eventually monolayer graphene by simply using scotch tape. Since its discovery, the variety of physical phenomena explored using graphene has expanded at a remarkably fast pace inspiring a wide variety of novel technological applications. Spurred on by potential future applications like single-electron transistors [6], flexible displays [7,8]. and solar cells [9], a lot of research effort is being devoted to understanding the main physical properties of graphene. For this reason, the following subsections of this review aim to introduce the reader to the basic features of graphene, in particular, its unique electronic structure and related electrical transport properties. Later, the state of the art regarding other important physical aspects of graphene beyond its electrical properties is also reviewed including its mechanical, magnetic, and thermal properties.

1.2 BASIC ELECTRONIC PROPERTIES OF GRAPHENE-BASED STRUCTURES

Graphene is defined as a single layer of carbon atoms arranged in a hexagonal lattice, as illustrated in Figure 1.1a. Its atomic structure can also be used as a basic building block to construct other carbon-based materials: (1) it can be folded into fullerenes, (2) rolled up into nanotubes, or (3) stacked into graphite. The primitive cell of graphene is composed of two non-equivalent atoms, A and B, and these two sublattices are translated from each other by a carbon–carbon distance $a_{c-c} = 1.44$ Å.

A single carbon atom has four valence electrons with a ground-state electronic shell configuration of [He] $2s^2\,2p^2$. In the case of graphene, the carbon–carbon chemical bonds are due to hybridized orbitals generated by the superposition of 2s with $2p_x$ and $2p_y$ orbitals. The planar orbitals form the energetically stable and localized σ-bonds with the three nearest-neighbor carbon atoms in the honeycomb lattice, and they are responsible for most of the binding energy and for the elastic properties of the graphene sheet. The remaining free $2p_z$ orbitals present π symmetry orientation and the overlap of these orbital states between neighboring atoms plays a major role in the electronic properties of graphene. For this reason, a good approximation for describing the electronic structure of graphene is to adopt an orthogonal nearest-neighbor tight-binding approximation assuming that its electronic states can

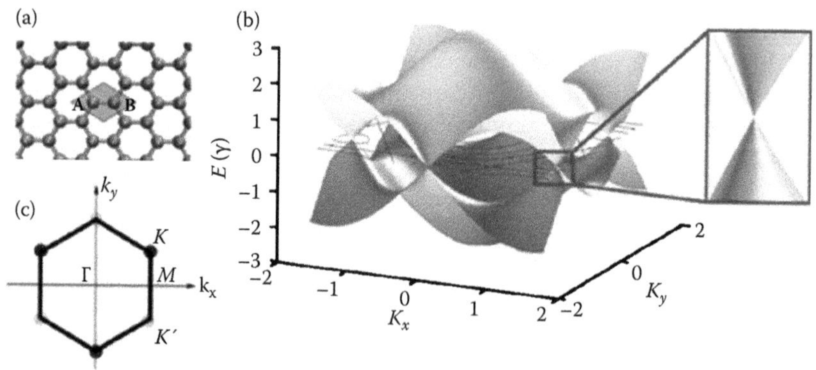

FIGURE 1.1 (*See color insert.*) (a) Honeycomb lattice of graphene. The shadowed area delineates the unit cell of graphene with its two nonequivalent atoms labeled by A and B. (b) Band energy dispersion obtained via tight binding approximation. The inset highlights the conical-shape dispersion around the charge neutrality point. (c) First Brillouin zone.

be simply represented by a linear combination of $2p_z$ orbitals. Solving the Schrödinger equation, which reduces into a matrix secular expression, one can obtain the energy dispersion relation of π (bonding) and π* (antibonding) bands [10,11,12].

$$E(k_x,k_y)=\pm\gamma\sqrt{1+4\cos\left(\frac{\sqrt{3}k_xa}{2}\right)\cos\left(\frac{k_ya}{2}\right)+4\left[\cos\left(\frac{k_ya}{2}\right)\right]^2} \tag{1.1}$$

where k_x and k_y are the components of the k vector that are folded onto the first hexagonal Brillouin zone (shown in Figure 1.1c) and $\gamma = 2.75$ eV is the hopping energy. The electronic structure of graphene can also be represented by closed-form expressions obtained analytically for the single-electron propagators written on a real-space basis [13]. In Figure 1.1b, one can see the band structure of graphene obtained from such a simple tight-binding model, which yields symmetric conduction and valence bands with respect to the Fermi energy (also called the *charge neutrality point* or *Dirac point*) set at 0 eV. Graphene valence and conduction bands are degenerate at 6 points located on the corners of the Brillouin zone, also called K and K′ valleys. The hexagonal region (Brillouin zone) has a side length of $4\pi/3a$ and delineates the Fermi surface of the graphene as shown in Figure 1.1c. Since the Fermi surface of graphene is compacted to a zero dimension zone composed of a finite set of 6 points on its Brillouin zone, graphene is usually termed a *semimetal* material with no overlap or zero-gap semiconductor. It is easy to see that the electronic properties of graphene are invariant by interchanging the K and K′ states, which means that the two valleys are related by time-reversal symmetry. Fascinating physical phenomena can be unveiled while attempting to break this effective time-reversal symmetry.

The low-energy dispersion near the valleys exhibits a circular conical shape, as displayed in the inset of Figure 1.1b, unlike the quadratic energy–momentum relation obeyed by electrons at the band edges in conventional semiconductors. Comparing this linear energy relation of graphene with the dispersion of massless relativistic particles obtained from the Dirac equation, one can see that graphene charge carriers can behave as Dirac fermions with an effective Fermi velocity that is around 300 times smaller than the speed of light [5]. This makes graphene a reliable system to study quantum electrodynamic phenomena, an area of investigation previously limited to particle physics and cosmology investigations. In this sense, several research groups have already addressed a variety of unusual phenomena that are revealed by graphene materials, which are characteristic of Dirac relativistic particles, for instance, the absence of localization effects even when disorder elements can take place [14,15], robust metallic conductivity even in the limit of nominally zero carrier concentration, and the half-integer quantum Hall effect.

Additional band features can be learned from the energy spectrum of graphene when the adopted model goes beyond the simple orthogonal tight-binding approach or Dirac formalism. More robust techniques, such as ab initio methods, predict that antibonding bands are located at a higher energy with respect to the bonding states if the overlapping integral matrix is nonorthogonal [16]. Sophisticated implementations for single-π band tight binding schemes considering up to the third-nearest neighbor interactions and overlap elements can result in an accurate description of the electronic properties in relation to first principle calculations [17].

The amazing electronic properties of graphene have greatly motivated the scientific community to pursue a better understanding of their main physical features with the bonus of converting them into real technological applications. However, the absence of an energy band gap greatly restricts its use on digital devices. Thus, alternative strategies capable of inducing a band gap in graphene are being sought. Several strategies have already been successfully adopted to modify the electronic structure of graphene and include chemical doping, interaction with substrates, and the application of mechanical forces or external electric/magnetic fields. Stacked graphene layers in the form of bilayers or graphite structures [18,19,20] also offer a promising route for band gap manipulation. Advanced lithographic techniques [21] employed to tailor wide graphene samples into nanoscale structures have shown that lateral confinement of charge carriers can work as an efficient energy gap–tuning parameter. Such narrow graphene structures are known as *graphene nanoribbons* (GNR) and it has been demonstrated that their energy gap scales inversely with the width. The following section is dedicated to a review of the main physical properties of such confined graphene systems.

1.2.1 Graphene Nanoribbons

Besides the idealizations of graphenelike 2D membranes, atomistic models of thin graphene strips were also addressed primarily to investigate the nature of edge dislocations and the appearance of defective dangling bonds in carbon networks [22]. Such narrow graphene strips, known as graphene nanoribbons, were also not expected to exist in nature. The discovery that graphene materials can be fabricated in the free state and combined with modern lithography techniques has confirmed that confined graphene structures are experimentally feasible. Currently, the synthesis of graphene nanoribbon samples has advanced considerably beyond that possible with conventional lithographic methods. For instance, "ribbons" with widths smaller than 10 nm have been synthesized via crystallographic etching [23,24], sonochemical techniques [25], and even through the unzipping of carbon nanotubes [26–28]. An original fabrication process for graphene nanoribbons with atomic-scale precision has recently been realized through the controlled assembly of molecular precursors consisting of polycyclic aromatic hydrocarbon compounds [29].

The physical properties of graphene nanoribbons are highly dependent on their width and the topology of the edge structures. There are two canonical types of graphene edges, referred to as *armchair* (AGNR) and *zigzag* (ZGNR) ribbons, and examples of their atomic structure can be seen in Figure 1.2. The atoms located on the edges are highlighted in green and W denotes the width of the ribbon. The width of an armchair ribbon can be defined in terms of the number of dimer lines: $W_a = (N_a - 1)a/2$ for armchair ribbons and $W_z = (N_z - 1)\sqrt{3}a/2$ for zigzag ribbons; N_a and N_z are their respective number of carbon chains.

The electronic structure of graphene nanoribbons can be represented in a simple manner following a single-π band tight binding description or Dirac approach where "particle-in-a-box" boundary conditions are applied to the ribbon's terminations. In

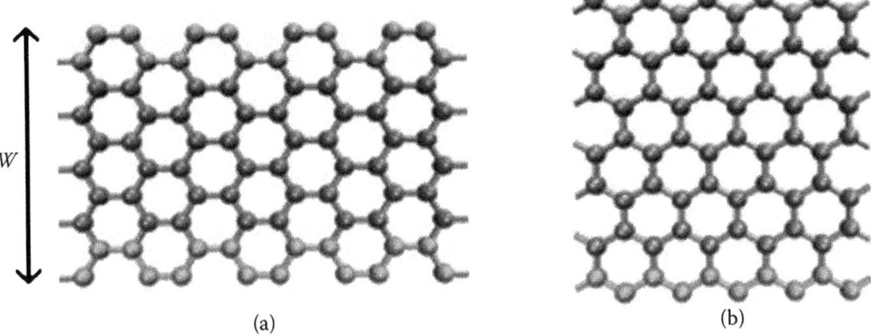

FIGURE 1.2 (*See color insert.*) Atomic structure of an (a) armchair- and a (b) zigzag- edge graphene nanoribbon. Green color atoms delineate the respective edge-shape and W denotes the width of the ribbon.

this case, the wave vector components lying in the width direction will be quantized, whereas those parallel to the axial direction remain continuous for infinite systems. In other words, limiting the width of a *bulk* graphene sheet means "slicing" the energy band structure of Figure 1.1b in well-defined directions; their projections can be seen on the Fermi surface as presented on the top panels of Figure 1.3. The quantization lines correspond to the allowed k states for three distinct graphene nanoribbons—AGNR(8), AGNR(9), and ZGNR(8)—placed over graphene's Brillouin zone. Whenever one of these states crosses one of the graphene's valleys, valence and conduction bands touch each other at the Fermi level and the ribbon exhibits metallic behavior, otherwise it is semiconducting [30].

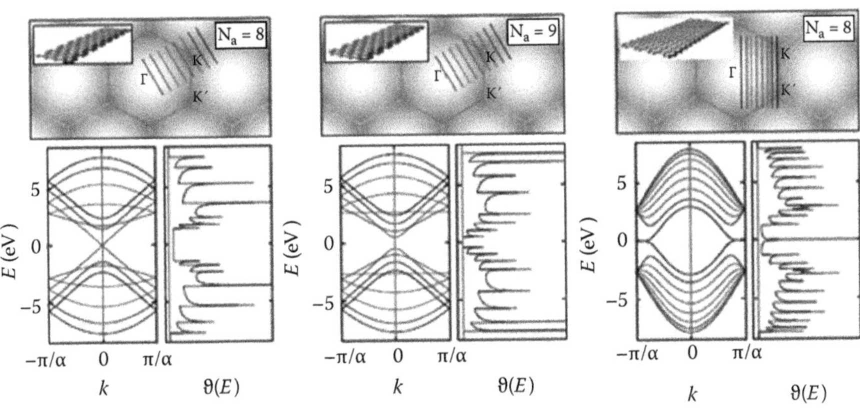

FIGURE 1.3 (*See color insert.*) (Top panels) Zone-folding diagram for three different graphene nanoribbons: left, AGNR(8); middle, AGNR(9); and right, ZGNR(8). The parallel lines in the Brillouin zone represent the allowed quantized states of the ribbon projected in momentum space. Their respective energy band structures and density of states curves are displayed on the lower panels. (Adapted from N. Nemec, Quantum transport in carbon-based nanostructures. Dr. rer. nat. (equiv. PhD) thesis, University of Regensburg, September 2007.)

Their respective energy dispersion relation and density of states curves calculated via nearest-neighbor tight-binding approximation are also shown in the lower panels. According to this simple description, one can predict that zigzag ribbons of any width show a singular edge state that decays exponentially into the center of the ribbon. Such edge states are twofold degenerate at the Fermi energy and reveal a nondispersive feature that lasts about 1/3 of the total size of the graphene Brillouin zone. As a consequence, the density of states of zigzag ribbons is characterized by a pronounced peak located at the charge neutrality point. Although there are still controversies concerning the associated energy eigenvalue of the edge state, the detection of such a peak has been accomplished through scanning tunneling microscopy measurements performed near zigzag edge sections of graphite [31]. In stark contrast, no such localized state appears in nanoribbons having an armchair edge configuration. Moreover, this simple model shows that armchair ribbons can change their electronic character depending on their width. An armchair nanoribbon can behave as a metal when the number of atoms along its width is equal to $3j + 2$, where j is an integer. This class of armchair ribbons exhibits semiconducting behavior when more sophisticated electronic structure models are applied or the edge atoms are parameterized to include the effects of hydrogen passivation. The remaining armchair ribbons in the $3j$ and $3j + 1$ categories are all semiconductors independent of the adopted model [22].

The challenge of inducing a band gap in graphene seems to be solved by cutting it into ribbons. On the other hand, the edges bring additional problems. Graphene nanoribbons indeed possess a band gap, but their edges have inherent edge disorder [32]. It turns out that their electronic properties are strongly reliant on the topological details of the atoms located on their extremities. Roughness, or even chemical groups bound to the edges, can also affect the electronic features of the ribbons. In this sense, studies focusing on disorder effects in graphene structures are of extreme relevance for envisioning the main mechanisms behind their electronic response.

1.3 EDGE DISORDER IN GRAPHENE STRUCTURES

The dominant scattering processes and resulting transport features of graphene are very dependent on the range of the disorder potential and the robustness or destruction of the underlying sublattice symmetries. A variety of physical behaviors can be unveiled when short-range interactions take place in graphene since all possibilities of intravalley and intervalley scattering events between K and K′ points are allowed. Short-range potentials formed, for instance, by atomically sharp defects such as vacancies [33,34]; Anderson disorder or edge deformations induce chirality breaking, leading to strong backscattering events and localization effects. In particular, graphene nanoribbons are naturally subjected to edge disorder due to the high reactivity of edges that can be subjected to chemical passivation, roughness, and structural reconstruction [35]. In addition, confinement effects are expected to maximize the sensitivity of the structures regarding the presence of disorder. Joule heating techniques capable of vaporizing carbon atoms from the edges has been carefully employed to pattern the morphology of nanoribbon extremities. The successful stabilization of sharp edge reconstructions mostly formed with either zigzag or armchair configurations has been observed [36]. Nonetheless, improving the quality of the edge-shapes in practice remains a difficult task. A multitude of different edge topologies have been characterized and so researchers must cope with a vast physical scenario involving prominent localization effects induced by edge disorder [25,37,38].

Depending on the edge shape of the graphene structure, different band gaps for similarly sized systems can be generated since its electronic structure is greatly influenced by disorder. Transport measurements realized in etched graphene samples have demonstrated that sharp resonances can appear inside the transport gap, providing evidence that the atomic details of tailored graphene systems play an important role in their conducting properties [39]. Atomistic models for edge disorder are often used to investigate the impact of topological edge roughness on the conducting properties of the ribbons. The boundaries are initially assumed to be perfect. Subsequently, the erosion of the edges can be simulated by randomly setting the hopping elements of neighboring atoms to zero or setting their onsite energies to very high values. The calculations indicate that even when very weak edge disorder effects are simulated, a prominent modification in the conductance profile of the nanoribbons is obtained [40,41]. In Figure 1.4, the impact on the conductance of a zigzag-edge ribbon considering different complexities for the edge disorder is

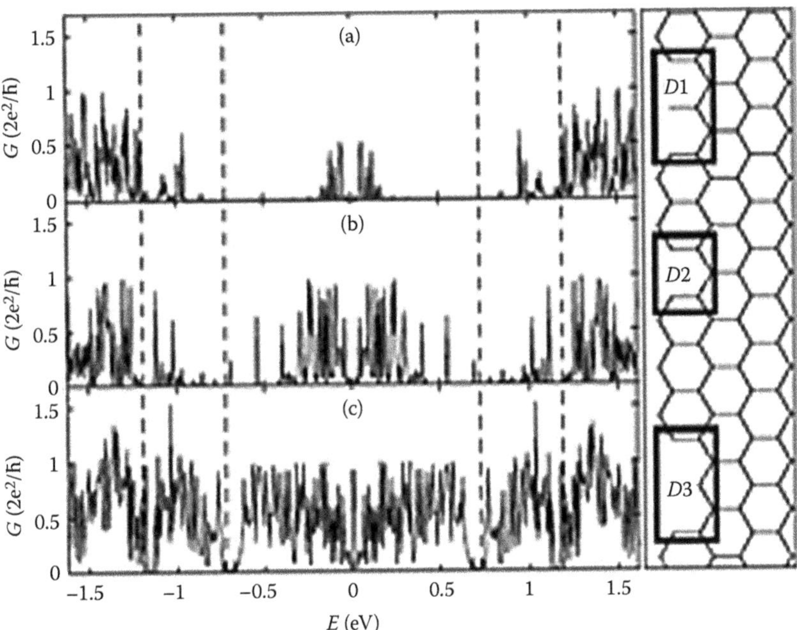

FIGURE 1.4 (Left panels) Conductance as a function of energy obtained from a disordered ZGNR(16) with length $L = 500$ nm and imposing a probability of 7.5% for removing carbon atoms from the edges. (Right panels) Schematic pictures representing the different types of defects: (D1) is an example of the Klein defect and (D2) and (D3) correspond to the defects where one and two hexagons, respectively, are missing. (Adapted from A. Cresti and S. Roche. 2009. Edge-disorder-dependent transport length scales in graphene nanoribbons: From Klein defects to the superlattice limit. *Physical Review B* 79: 233404-1-4.)

shown [35]. The D1 defect is known as the Klein defect and is composed of a zigzag edge with a single carbon bound on the edges [42,43]. The D2 and D3 imperfections consist of withdrawing one or two consecutive hexagons from the edges. The atoms were randomly removed with an equal probability of 7.5% and length of the scattering region is L = 500 nm. The conductance around the charge neutrality point is highly suppressed especially in the presence of Klein and D2 defects. A few resonance peaks survive within the energy range where one conductive channel was active regardless of whether the defects were absent or not. A robust conducting profile is observed when D3 defects exist, suggesting that the system can evolve from a quasiballistic to a localized regime depending on the details of the defects and as the length of the scattering region increases.

Similar studies performed in disordered armchair ribbons have shown that these later structures are relatively more sensitive to edge disorder in comparison to zigzag configurations. In other words, disordered armchair structures have a greater propensity to manifest localization effects. It was demonstrated that only 10% of edge defects are enough to wash out the electronic transmission in a wide range of energies of metallic armchair structures [44]. Specifically, for the case of semiconducting armchair ribbons, it has been shown that edge disorder is capable of transforming their electronic character into Anderson insulators as long as their width is kept relatively wide in order to minimize the impact of disordered edges. Insulating character is also achieved in ribbons with lengths large enough to avoid the direct tunneling of electrons along the channels [45,46].

The impact of the edges on the electronic structure of nanoribbons can be controlled by chemical passivation [47] where different species could, in principle, react with the carbon atoms situated on the extremities rather than the commonly used hydrogen saturation. In reality, not only the edges of the ribbons can be thought of as the most energetic location for dopants. The specific topology of nanoribbon edges enters as an additional control parameter for the segregation of impurities across their width and their distribution can be tuned by gate potentials [48]. Such studies involving doping processes in graphene nanoribbons open a wide field of applications in the industry of chemical and biosensor devices.

In addition to such amazing electronic properties, graphene structures are also exceptional materials for transferring heat [49,50]. The investigation involving heat conductivity can favor the elaboration of effective heat-dissipating devices in order to cool down electronic components. In addition, the variety of phenomena explored using graphene increased after its confirmation as the strongest and lightest material ever to be measured. The next sections are devoted to discussing the main achievements in the field of heat conductivity, thermal vibrations, and mechanical properties in graphene structures.

1.4 VIBRATIONAL PROPERTIES AND THERMAL PROPERTIES

At room temperature (RT), the thermal conductivity (κ) of single-layer graphene is mostly due to acoustic phonons [51]. The high value of κ is attributed to the absence of crystal defects, and suppression of Umklapp processes as the number of layers is reduced [52], that is, long mean-free paths of phonons. Such high values suggest that graphene can play a key role in future nanoelectronic devices [53].

The phonon branches of graphene can be grouped as in-plane (LA, TA, LO, and TO) and out-of-plane (flexural) modes (ZA and ZO). The acoustic flexural mode (ZA) in two-dimensional crystals has a quadratic dispersion in the vicinity of the Γ point of the Brillouin zone, and exhibits a singularity in the density of states at zero energy. At finite temperatures, thermal fluctuations are expected to give rise to atomic displacements as large as the interatomic distance; therefore low-dimensional crystals should be unstable [1,2]. However, macroscopic samples of graphene are shown to be stable and preserve their crystal quality, which is believed to be due to the existence of microscopic crumpling in the third dimension [54].

The ballistic thermal conductivity of graphene is isotropic [51]. In the temperature range below 20 K, the ZA mode is detrimental to conduction. Above 20 K, the LA and TA modes also contribute, however the ZA mode dominates the thermal conduction. The exact solution of the phonon Boltzmann equation shows that the ZA mode is the dominant heat carrier at higher temperatures as well [55]. It is also shown that anharmonic scattering is significantly restricted for the flexural modes due to selection rules, and this behavior is robust to inclusion of ripples and isotopic impurities. There is a variety of values reported for κ of single-layer suspended graphene at room temperature ranging from 600 to 5000 $Wm^{-1}K^{-1}$ [49,56,57]. The disagreement needs to be clarified.

On the other hand, when graphene lies on a supporting substrate, phonons leak across the interface and the flexural modes are scattered strongly. Nonetheless, κ of graphene on SiO_2 substrate has been measured as 600 $Wm^{-1}K^{-1}$ at RT, considerably higher than that of copper [50,58]. The reduction of κ due to a supporting substrate also points to the interplay between the phonon scattering mechanisms and the number of graphene layers. Measurements on few-layer graphene, with the number of layers ranging between 2 and 10, shows the dimensional crossover from two dimensions to bulklike behavior, and the crossover is assigned to the intralayer coupling of low-energy phonons and enhanced Umklapp scatterings [52]. As a result, κ drops from 2800 to 1300 $Wm^{-1}K^{-1}$ when the number of layers is increased from 2 to 4. In most device applications, graphene will be encased within dielectric materials, which will alter the thermal properties significantly. Measurements on single-layer graphene encased within SiO_2 show that the thermal conductivity is suppressed down to 160 $Wm^{-1}K^{-1}$, and for few-layer graphene it increases with the number of layers approaching the limit of in-plane κ for bulk graphite [59].

Similar to the electronic states in GNRs, only standing wave solutions are allowed perpendicular to the ribbon axis [60]. Therefore the wave vector is discrete in this direction, $q_{\perp,n} = n\pi/W$, where W is the ribbon width, and $n = 0, \ldots, N - 1$. The phonon branches of GNRs can be interpreted to consist of six fundamental modes that correspond to the modes of graphene, and their $6(N - 1)$ overtones [61], except for the fact that there exist 4 acoustic modes in quasi-one-dimensional crystals. In the ballistic regime, the thermal conductance of pristine GNRs is predicted to display a power law T at low temperatures, ranging from 1 to 1.5, from narrow to wide ribbons [51,62]. Since one has edges, it is unavoidable to have some irregularities in the edge shape and width of the GNRs. The effect of disorder in the edge shape of GNRs increases with decreasing ribbon width and suppresses thermal conductivity strongly for GNRs having widths of 20 nm and below [63,64]. The thermal conductivity of GNRs with sub-20-nm widths was measured as ~1000 $Wm^{-1}K^{-1}$ in agreement with theoretical calculations [65,66].

1.5 MECHANICAL PROPERTIES OF GRAPHENE

Graphene received the title of "strongest material ever" after the confirmation of its sustaining breaking strengths of 42 N/m with an intrinsic mechanical strain of ~ 25% and Young's modulus of Y ~ 1.0 TPa [67]. Its mechanical thickness can also be controlled as demonstrated through mechanical stress measurements performed on graphene sheets subjected to deformations induced by depositing different insulating capping layers [68]. The experimental findings regarding the main mechanical features of graphene have been confirmed by several theoretical works using different techniques. Among them, ab initio [69], tight binding [70], molecular dynamics simulations [71,72], and semiempirical models [73,74] have successfully estimated the Young's modulus and other intrinsic mechanical quantities of graphene.

The outstanding mechanical properties of graphene have also attracted interest from electronic applications due to the potential use that these light, stiff, and flexible materials can offer for designing building-block components in nanoelectromechanical systems (NEMS). In particular, the fabrication of low-cost NEMS devices requires a complete correspondence between mechanical and electrical responses of the conductive channel. In this sense, the operation mechanism of an efficient NEMS based on graphene relies strictly on the feasibility of performing band gap engineering with the aid of external mechanical forces. Detailed analysis of the physical properties of uniaxially strained "graphene-bulk" has been widely studied by Raman spectroscopy [68,75,76] and suggests that manipulation of the band gap is possible. Nevertheless, most of these experiments were conducted on samples placed on top of flexible substrates, which can gradually stretch or bend the sheets. The pure electromechanical response of suspended 2D graphene is still under debate. According to several theoretical works, the electronic structure of suspended graphene is extremely resistant against mechanical forces, being able to support reversible elastic deformations above 20% [69,70].

Band gap engineering of strained graphene materials is possible when tailored structures, such as nanoribbons, are mechanically perturbed. It has already been shown that the transport and electronic features of graphene nanoribbons can be efficiently tuned as a function of strain [77–81]. These studies highlight important aspects of the synthesis of graphene-based molecular electromechanical devices [82]. The conductance of uniaxially stretched graphene nanoribbons is shown to be strongly dependent on their edge shape, as can be seen in Figure 1.5. Ribbons with armchair edge symmetry can undergo a metal–semiconductor transition as mechanical strain increases, whereas zigzag ribbons exhibit a more robust transport behavior against stretching. Very small strain values are sufficient to open an energy gap in

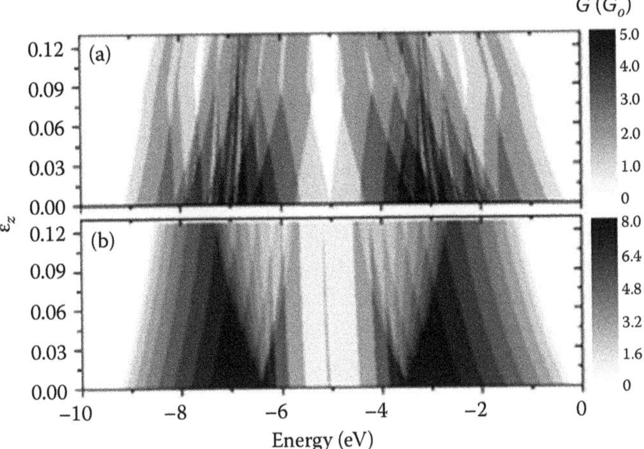

FIGURE 1.5 (*See color insert.*) Contour plots of conductance as a function of Fermi energy and mechanical strain for (top panel) an AGNR(11) and (lower panel) a ZGNR(10). (Adapted from M. Poetschke, C.G. Rocha, L.E.F. Foa Torres, S. Roche, and G. Cuniberti. Modeling graphene-based nanoelectromechanical devices. *Physical Review B* 81 (2010): 193404-1-4.)

AGNRs, confirming that their electronic character is sensitive to mechanical stress. In this sense, armchair edge ribbons are more suitable for engineering electromechanical devices in comparison to zigzag geometries.

1.6 GRAPHENE'S TRANSPORT PROPERTIES UNDER EXTERNAL FIELDS

The successful realization of graphene-based nanodevices depends mostly on patterning effective circuit architectures in which their electronic properties can be modified in a predetermined and reversible way. In fact, interesting quantum phenomena can be observed when the physical properties of low-dimensional systems are tuned by external fields such as electric or magnetic fields and gate voltages under dc conditions. Studies considering external fields in the stationary regime have been widely investigated both theoretically and experimentally. For instance, a graphene sheet experiencing the presence of a modulated electrical potential sustains strong modifications in its low-energy properties. The robust degeneracy at the Dirac point is split and the isotropic conelike structure of the energy relation is now composed of two distinct valley structures with highly anisotropic dispersions [83]. Theoretical investigations of graphene ribbons working as a transmission channel under transversal electric fields have demonstrated that the number of transmission modes can be controlled with the aid of an external voltage [84–86]. More importantly, the conductance of the system varies sharply by integer multiples of the quantum conductance with respect to the strength of the electric field. Additional transport features can be visualized when a rotating gate plate acts on the graphene ribbons. The transmission is shown to be dependent on the gate orientation and on the width of the ribbons [87]. External electric fields can also be used to effectively tune important physical quantities of graphene such as work function [88] and electron–phonon coupling [89]. An efficient alignment tool was idealized by the application of external electric fields where the graphene membranes can be, in principle, oriented in particular directions in space via electric polarization effects [90]. Moreover, the electronic properties of graphene were finely tuned through the adsorption of molecules with strong electric dipole moments, capable of inducing a local electric field on the structures. Band gap engineering in graphene hosts was theoretically addressed by considering that the intensity of the external electric field can be controlled by means of the density of ad molecules [91]. An even wider set of electronic responses can be obtained from graphene nanostructures when a combination of both electric and magnetic fields is applied. Energy-gap modulation can be achieved in graphene nanoribbon channels exposed to fields oriented in a type of Hall configuration [92] as shown in Figure 1.6. The lowest and highest energy states of an initially semiconducting AGNR are shown to collapse at the Dirac point at a critical electric field, guiding the system toward a semimetallic arrangement. The competition between localization and delocalization effects generated by the respective magnetic and electric fields gives rise to a rich set of electronic responses that can certainly be implemented into promising electronic devices. Essentially, the fields induce a broad set of refinements in graphene's energy spectrum such as k ↔ −k symmetry breaking, drastic modification of low-energy dispersions, subband spacings, and edge states [93,94]. Furthermore, at a critical electric and magnetic field ratio, it has demonstrated that the Landau spectrum contracts, viz. the Landau energy level spacing, gradually decreases [95]. At the same time, electric

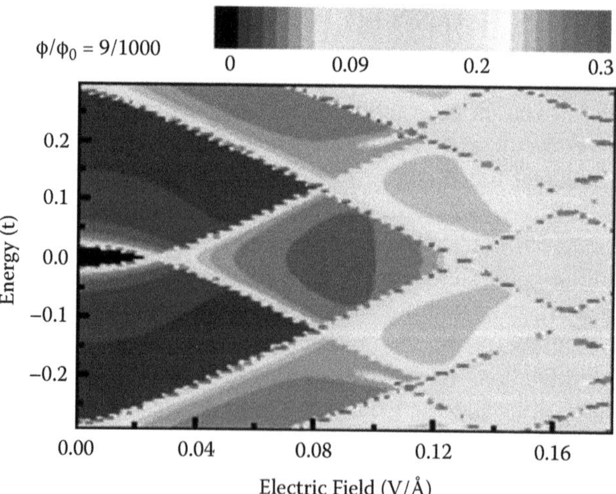

FIGURE 1.6 (*See color insert.*) Local density of states contour plot for a 24-AGNR as a function of the electric field intensity for a fixed magnetic flux of $\phi/\phi_0 = 9/1000$. Black corresponds to null density of states while the highest LDOS value is highlighted by red color. (Adapted from C. Ritter, S.S. Makler, and A. Latgé. Energy-gap modulations of graphene nanoribbons under external fields: A theoretical study. *Physical Review B* 77: 195443-1-5.; *Physical Review B* 82 (2008): 089903-1-2. With permission.)

excitations are found to disturb the magnetic susceptibility and the characteristic Haas-van Alphen oscillations observed in the magnetization curves calculated for graphene systems under magnetic fields [96]. Such anomalous phenomena are concluded to be associated with the relativistic flavor of the low-energy charge carriers in graphene.

Another possibility to control the electronic transmission of carbon-based nanomaterials is through the use of time-dependent excitations [97]. Recent studies targeting the use of ac fields in graphene materials [98–100] shed light on this growing research area, often overshadowed by studies considering external fields in the stationary regime. Under ac signals, several theoretical works have highlighted graphene's potential as a spectrometer device operating even at high-frequency noise. In particular, for the case in which a homogeneous ac gate can act on graphene channels, it has been shown that it is possible to achieve full control of the conductance patterns which, remarkably, resemble Fabry-Pérot interference patterns of light-wave cavities [98,101]. The results presented in Figure 1.7, obtained for an AGNR resonator, suggest several possibilities to tune the conductance profiles ranging from the standard dc regime (panel [a]), to suppression (panel [b]), phase change (panel [c]) of the oscillations, and robust behaviors (panel [d]) interpreted as a wagon-wheel effect held in the quantum domain. There is also an increasing interest in graphene's photovoltaic Hall effect since it was confirmed that photo-induced dc currents can

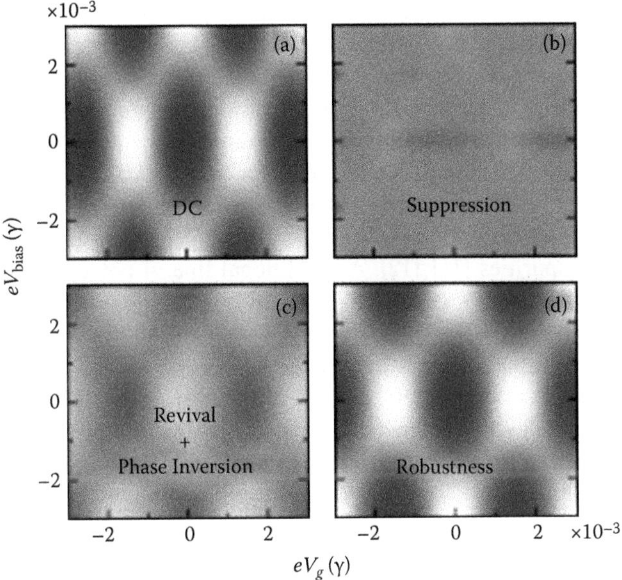

FIGURE 1.7 (*See color insert.*) Fabry-Pérot conductance interference patterns for an AGNR as a function of bias and gate voltages calculated for different driving frequencies and amplitudes associated with a time-dependent gate potential that follows a harmonic time dependency. White and dark blue colors correspond to maximum and minimum conductance values, respectively. (Adapted from C.G. Rocha, L.E.F. Foa Torres, and G. Cuniberti. Ac transport in graphene-based Fabry-Pérot devices. *Physical Review B* 81 (2010): 115435-1-8.)

be induced under intense light [102]. Such investigations underscore the potential for synthesizing organic solar cells based on graphene.

1.7 MAGNETIC PROPERTIES OF GRAPHENE

As discussed in Section 1.2.1, zigzag graphene nanoribbons have so-called edge states. According to numerous theoretical studies based on density functional theory or on the mean-field Hubbard model, it is believed that the associated peak in the density of states at the Fermi energy (E_F) gives rise to a magnetic instability, where the edge states become spin-polarized [22,103,104]. In the magnetic ground state, a band gap at E_F is opened, and the atoms are ferromagnetically ordered along one edge, and antiferromagnetically ordered between opposite edges. This antiferromagnetic ground state is consistent with the Lieb theorem for a biparticle lattice within a Hubbard model description [105]. The phenomenon of edge magnetism is not restricted to ideal ZGNRs but is believed to occur in any graphene system that has zigzag edge segments.

One of the first suggestions to use the magnetic edges for spintronics applications came from Son et al. [106]. They showed that the application of a transverse electric field causes the antiferromagnetic ZGNRs to become half-metallic. Without an electric field, the system is in the antiferromagnetic ground state with a band gap at E_F. The external electric field shifts the electronic states so that the band gap of one spin component is increased while the band gap of the other spin component is closed, such that the system becomes a metal with spin-polarized electrons, i.e., a spin valve. In essence, the effect of the transverse electric field is to break the symmetry between the left and the right edge. This symmetry breaking can be achieved without an electric field in different ways, for example, by saturating the left and the right edge of the ZGNR with different functional groups [107,108] or by edge selective defects [109,110]. Other graphene-based spin-valve devices were theoretically proposed using gate-driven spin currents originating from graphene arrays doped with magnetic impurities [111,112]. An additional line of potential application targets using the phenomenon of edge magnetism in order to build a magnetoresistor [113,114]. But up to now, there is little experimental evidence for graphene edge magnetism [115,116]. Magnetic edges have never been observed in local probe microscopy; only an indirect observation has been reported recently [117].

The inability to observe magnetic edge states is supported by recent theoretical works that show that the ideal zigzag edge morphology is not very likely to exist. In fact, other nonmagnetic edge morphologies are thermodynamically much more favorable [118–120]. Furthermore, it was shown that even in perfect ZGNRs, the antiferromagnetic ground state is stable only at very low temperatures [120]. So currently, there is some evidence accumulating that the phenomenon of edge magnetism is not applicable at room temperature.

A different type of intrinsic magnetism can be induced by certain types of point defects in bulk graphene. Point defects are, for example, lattice vacancies (missing carbon atoms) [121] and chemisorbed atoms [104,122]. Similar to edges, a point defect interrupts the ideal sp^2 lattice structure and induces electronic states that can be magnetic. Room temperature ferromagnetism was observed in proton-irradiated highly oriented pyrolytic graphite [123]. This observation was theoretically explained to come from magnetic point defects that are created by proton radiation [124].

Furthermore, magnetic properties can be induced by foreign magnetic atoms that are either adsorbed to graphene (adatoms) or replace carbon atoms in the honeycomb lattice (substitutional dopants) [124].

1.8 SYNTHESIS OF GRAPHENE

As discussed previously, the potential for graphene in materials and devices is massive. There are great expectations that this material will provide numerous socioeconomic benefits. However, it is important to be aware that many of the claims being made have been made for other well-known carbon nanostructures, namely, carbon nanotubes and fullerenes. Much of the promise for these nanostructures has yet to emerge in real applications. One of the biggest bottlenecks is what on the surface seems a simple technical issue—their synthesis and formation into atomically precise structures with high degrees of reproducibility. This is particularly important for molecular electronics applications. This same difficulty also applies to graphene, and hence one of the most important areas, if not the *most* important area to address is its synthesis and manipulation. The required knowledge and understanding to provide atomically precise fabrication of this material, in a reproducible manner, that is compatible with current semiconductor technology is still lacking. Nonetheless, great strides have been made in synthesizing graphene and this is discussed in greater detail in Chapter 2. Here we simply mention the techniques used to synthesize and functionalize graphene.

By far the most common route to synthesize graphene is chemical vapor deposition (CVD). There are many CVD variations. Thermal CVD is commonly applied to graphene formation over transition metals, including copper [126–129], nickel [130–136], iridium [137–139], and ruthenium [140–144].

Thermal CVD techniques can also be used for graphene synthesis over dielectrics, namely, sapphire [145] and various other oxides [146,147]. Free-standing carbon nanosheets and planar graphite films with a few graphene layers have been successfully synthesized by plasma enhanced CVD (PECVD) [130,148,149].

A widely used technique to synthesize graphene is the thermal decomposition of hexagonal α-SiC (*6H*-SiC and *4H*-SiC). It has the advantage that it is very clean because the epitaxially matching support crystal provides the carbon itself and no metal is involved. The technique dates back to the early 1960s when Badami found graphite on SiC by x-ray scattering after heating SiC to 2150°C in an ultrahigh vacuum (UHV) [150]. The parallel publication of the electrical response of graphene in 2004 by Novoselov et al. and Berger et al. (who used graphene grown from SiC) provided a new impetus to optimize the growth conditions of graphene on SiC [4,151].

An easy production and low-cost technique is the exfoliation of graphite. The more common exfoliation routes include mechanical exfoliation [4], ultrasound treatment in solution [152], and intercalation steps [153].

There is also an interest in producing functionalized graphene as this can open a band gap. Hydrogenation of graphene is one route. The hydrogenation changes the sp^2 structure of graphene to sp^3 hybridization. Graphene can be synthesized using a stream of hydrogen atoms [154], reactive ball milling between anthracite coal and cyclohexane [155], exchanging the fluorine in fluorinated graphite by hydrogen [156] or by dissolved metal reduction in liquid ammonia [157]. Fluorinated graphene (FG) is another proven functionalization route. There are different methods to produce FG, namely the extraction of single layers of FG from commercially available fluorinated graphite [5], the exposure of graphene to fluorine gas at ~500°C [158], placing graphene in a fluorine-based plasma [159], or exposure to xenon difluoride [160].

1.9 GRAPHENE AND SOME OF ITS APPLICATIONS

In this review, we surveyed a vast literature about graphene's unique physical properties and the main experimental techniques used to synthesize them. The multitude of topics addressed in this review attests to the prominent potential that this material has to transform the actual nanotechnology landscape with promising applications [161]. The interest in graphene has mobilized both academic and industry realms making it an ideal candidate for the design of modern nanoscale transistors, chemical and biosensors, flexible and organic light-emitting diodes (OLEDs) displays, solar and fuel cells, and other innovations. The restricted graphene mass-production and limited reproducibility in device performances are still important matters that researchers should consider in order to push graphene-based technology into a commercial status. However, the fast development of graphene research leaves no doubt that this material will revolutionize several markets such as electronics, medicine, and energy storing in the near future.

Medicine studies can also benefit from graphene's amazing properties. In particular, graphene possesses great sensorial response to external analytes, enabling the design of nanosensors to diagnose diseases [162]. Accurate biosensors can be created from DNA-functionalized graphene samples, which are capable of detecting external DNA genes associated with diseases [163]. Graphene can also have a huge impact in environmental monitoring applications bolstered by the design of graphene-based nanoscale gas sensors [164–166].

Another attractive innovation based on graphene materials reaches the electronics scope where researchers have been able to develop bendable transparent and conductive membranes composed of graphene for engineering flexible-panel displays [167]. Recent studies have revealed that graphene-based OLEDs can even top the performance of indium tin oxide (ITO) compounds, commonly used in transparent conductive electrodes [129]. Graphene is also considered to be the basis of future computing chips after the successful realization of high-speed graphene-based transistors operating at outstanding cutoff frequencies of 700–1400 GHz [168].

All these important innovations, which were generated after the first isolation of graphene layers, indicate that the use of these materials is not limited to providing simply a theoretical model that can describe the physical properties of several organic nanostructures. Graphene is occupying a centerpiece position in many scientific advances that can change our way of making and using technology. As mentioned by A.K. Geim, we are witnessing a scientific excitement similar to the one experienced around 100 years ago with the discovery of polymers that recently supplied our lives with plastics. We expect that the innovations resulting from graphene will prove even more exciting.

REFERENCES

1. Peierls, R. 1935. Quelques properties typiques des corpes solides. *Annales d' Institut Henri Poincare* 5: 177.
2. Landau, L. 1937. Zur Theorei der phasenumwandlugen II. *Physikalische Zeitschrift Sowjetunion* 11: 26.
3. Mermin, N. D. 1968. Crystalline order in two dimensions. *Physical Review* 176: 250.
4. Novoselov, K. S., Geim, A. K., Morozov, S. V., Jiang, D., Zhang, Y., Dubonos, S. V., Grigorieva, I. V., and Firsov., A. A. 2004. Electric field effect in atomically thin carbon films. *Science* 306: 666.
5. Novoselov, K. S., Geim, A. K., Morozov, S. V., Jiang, D., Katsnelson, M. I., Grigorieva, I.V., Dubonos, S. V., and Firsov, A. A. 2005. Two-dimensional gas of massless Dirac fermions in graphene. *Nature* 438: 197–200.
6. Lin, Y.-M., Dimitrakopoulos, C., Jenkins, K. A., Farmer, D. B., Chiu, H.-Y., Grill, A., and Avouris, Ph. 2010. 100-GHz transistors from water-scale epitaxial graphene. *Science* 327: 662.
7. Kim, K. S., Zhao, Y., Jang, H., Lee, S. Y., Kim, J. M., Kim, K. S., Ahn, J.-H., Kim, P., Choi, J.-Y., and Hong, B. H. 2009. Large-scale pattern growth of graphene films for stretchable transparent electrodes. *Nature* 457: 706–710.
8. De Arco, L. G., Zhang, Y., Schlenker, C. W., Ryu, K., Thompson, M. E., and Zhou, C. 2010. Continuous, highly flexible, and transparent graphene films by chemical vapor deposition for organic photovoltaics. *ACS Nano* 4: 2865.
9. Yang, N., Zhai, J., Wang, D., Chen, Y., and Jiang, L. 2010. Two-dimensional graphene bridges enhanced photo-induced charge transport in dye-sensitized solar cells. *ACS Nano* 4: 887.
10. Wallace, P. R. 1947. The band theory of graphene. *Physical Review* 71: 662.
11. Dubois, S. M.-M., Zanolli, Z., Declerck, X., and Charlier, J.-C. 2009. Electronic properties and quantum transport in graphene-based nanostructures. *The European Physical Journal B* doi: 10.1140/epjb/e2009-00327-8.
12. Hobson, J. B., and Nierenberg, W. A. 1953. The statistics of a two-dimensional, hexagonal net. *Physical Review* 89: 662.
13. Power, S. R., and Ferreira, M. S. 2010. Graphene electrons beyond the linear dispersion regime. arXiv: 1010.0908v2.
14. Morozov, S. V., Novoselov, K. S., Katsnelson, M. I., Schedin, F., Ponomarenko, L. A., Jiang, D., and Geim, A. K. 2006. Strong suppression of weak localization in graphene. *Physical Review Letters* 97: 016801.
15. Horsell, D. W., Tikhonenko, F. V., Gorbachev, R. V., and Savchenko, A. K. 2008. Weak localization in monolayer and bilayer graphene. *Philosophical Transactions of the Royal Society A* 366: 245.
16. Konstantinova, E., Dantas, S. O., and Barone, P. M. V. B. 2006. Electronic and elastic properties of two-dimensional carbon planes. *Physical Review B* 74: 035417.
17. Reich, S., Maultzsch, J., Ordejón, P., and Thomsen, C. 2002. Tight-binding description of graphene. *Physical Review B* 66: 035412.
18. Ohta, T., Bostwick, A., Seyller, T., Horn, K., and Rotenberg, E. 2006. Controlling the electronic structure of bilayer graphene. *Science* 313: 951.

19. Feldman, B. E., Martin, J., and Yacoby, A. 2009. Broken-symmetry states and divergent resistance in suspended bilayer graphene. *Nature Physics* 5: 889–983.
20. González, J. W., Santos, H., Pacheco, M., Chico, L., and Brey, L. 2010. Electronic transport through bilayer graphene. *Physical Review B* 81 (2010): 195406.
21. Melinda, Y. H., Barbaros, O., Zhang, Y., and Kim, P. 2007. Energy band-gap engineering of graphene nanoribbons. *Physical Review Letters* 98: 206805.
22. Nakada, K., Fujita, M., Dresselhaus, G., and Dresselhaus, M. S. 1996. Edge state in graphene ribbons: Nanometer size effect and edge shape dependence. *Physical Review B* 54: 17954.
23. Chen, Z. H., Lin, Y. M., Rooks, M. J., and Avouris, P. 2007. Graphene nano-ribbon electronics. *Physica E* 40: 228.
24. Han, M. Y., Ozyilmaz, B., Zhang, Y. B., and Kim, P. 2007. Energy band-gap engineering of graphene nanoribbons. *Physical Review Letters* 98: 206805.
25. Li, X., Wang, X., Zhang, L., Lee, S., and Dai, H. 2008. Chemically derived, ultrasmooth graphene nanoribbon semiconductors. *Science* 319: 1229–1233.
26. Kosynkin, D. V., Higginbotham, A. L., Sinitskii, A., Lomeda, J. R., Dimiev, A., Price, B. K., and Tour, J. M. 2009. Longitudinal unzipping of carbon nanotubes to form graphene nanoribbons. *Nature* 458: 872–877.
27. Elas, A. L., Botello-Mendez, A. R., Meneses-Rodriguez, D., Gonzalez, V. J., Ramirez-Gonzalez, D., Ci, L., Munoz-Sandoval, E., Ajayan, P. M., Terrones, H., and Terrones, M. 2009. Longitudinal cutting of pure and doped carbon nanotubes to form graphitic nanoribbons using metal clusters as nanoscalpels. *Nano Letters* 10: 366.
28. Santos, H., Chico, L., and Brey, L. 2009. Carbon nanoelectronics: Unzipping tubes into graphene ribbons. *Physical Review Letters* 103: 086801-1-4.
29. Cai, J., Ruffieux, P., Jaafar, R., Bieri, M., Braun, T., Blankenburg, S., Muoth, M., Seitsonen, A. P., Saleh, M., Feng, X., Müllen, K., and Fasel, R. 2010. Atomically precise bottom-up fabrication of graphene nanoribbons. *Nature* 466: 470.
30. Nemec, N. Quantum transport in carbon-based nanostructures. Dr. rer. nat. (equiv. PhD) thesis, University of Regensburg, September (2007).
31. Niimi, Y., Matsui, T., Kambara, H., Tagami, K., Tsukada, M., and Fukuyama, H. 2005. Scanning tunneling microscopy and spectroscopy studies of graphite edges. *Applied Surface Science* 241: 43.
32. Son, Y.-W., Cohen, M. L., and Louie, S. G. 2006. Energy gaps in graphene nanoribbons. *Physical Review Letters* 97: 216803.
33. Rosales, L., Pacheco, M., Barticevic, Z., Leon, A., Latge, A., and Orellana, P. A. 2009. Transport properties of antidote superlattices of graphene nanoribbons. *Physical Review B* 80: 073402-1-4.
34. Ritter, C., Pacheco, M., Orellana, P., and Latge, A. 2009. Electron transport in quantum antidots made of four-terminal graphene ribbons. *Journal of Applied Physics* 106: 104303-1-6.
35. Cresti, A., and Roche, S. 2009. Edge-disorder-dependent transport length scales in graphene nanoribbons: From Klein defects to the superlattice limit. *Physical Review B* 79: 233404-1-4.
36. Jia, X., Hofmann, M., Meunier, V., Sumpter, B. G., Campos-Delgado, J., Romo-Herrera, J. M., Son, H., Hsieh, Y.-P., Reina, A., Kong, J., Terrones, M., and Dresselaus, M. S. 2009. Controlled formation of sharp zigzag and armchair edges in graphitic nanoribbons. *Science* 323: 1701.
37. Kobayashi, Y., Fukui, K.-i., Enoki, T., Kusakabe, K., and Kaburagi, Y. 2005. Observation of zigzag and armchair edges using scanning tunneling microscopy and spectroscopy. *Physical Review B* 71: 193406.

38. Niimi, Y., Matsui, T., Kambara, H., Tagami, K., Tsukada, M., and Fukuyama, H. 2006. Scanning tunneling microscopy and spectroscopy of the electronic local density of states of graphite surfaces. *Physical Review B* 73: 085421.
39. Ihn, T., Cuettinger, J., Molitor, F., Schnez, S., Schurtenberger, E., Jacobsen, A., Hellmueller, S., Frey, T., Droescher, S., Stampfer, C., and Ensslin, K. 2010. Graphene single-electron transistors. *Materials Today* 13: 44.
40. Evaldsson, M., Zozoulenko, I. V., Xu, H., and Heinzel, T. 2008. Edge-disorder-induced Anderson localization and conduction gap in graphene nanoribbons. *Physical Review B* 78: 161407R.
41. Mucciolo, E. R., Castro Neto, A. H., and Lewenkopf, C. H. 2009. Conductance quantization and transport gaps in disordered graphene nanoribbons. *Physical Review B* 79: 075407.
42. Klein, D. J. 1994. Graphitic polymer strips with edge states. *Chemical Physics Letters* 217: 261.
43. Klein, D. J., and Bytautas, L. 1999. Graphitic edges and unpaired π-electron spins. *Journal of Physical Chemistry A* 103: 5196.
44. Areshkin, D., Gunlycke, D., and White, C. T. 2007. Ballistic transport in graphene nanostripes in the presence of disorder: Importance of edge effects. *Nano Letters* 7: 204.
45. Gunlycke, D., Areshkin, D., and White, C. T. 2007. Semiconducting graphene nanostripes with edge disorder. *Applied Physics Letters* 90: 142104.
46. Querlioz, D., Apertet, Y., Valentin, A., Huet, K., Bournel, A., Galdin-Retailleau, S., and Dollfus, P. 2008. Suppression of the orientation effects on bandgap graphene nanoribbons in the presence of edge disorder. *Applied Physics Letters* 92: 042108.
47. Rosales, L., Pacheco, M., Barticevic, Z., Latge, A., and Orellana, P. A. 2009. Conductance gaps in graphene nanoribbons designed by molecular aggregations. *Nanotechnology* 20: 095705-1-6.
48. Power, S. R., de Menezes, V. M., Fagan, S. B., and Ferreira, M. S. 2009. Model of impurity segregation in graphene nanoribbons. *Physical Review B* 80: 235424-1-5.
49. Balandin, A. A., Ghosh, S., Bao, W., Calizo, I., Teweldebrhan, D., Miao, F., and Lau, C. N. 2008. Superior thermal conductivity of single-layer graphene. *Nano Letters* 8: 902–907.
50. Seol, J. H., Jo, I., Moore, A. L., Lindsay, L., Aitken, Z. H., Pettes, M. T., Li, X., Yao, Z., Huang, R., Broido, D., Mingo, N., Ruoff, R. S., and Shi, L. 2010. Two-dimensional phonon transport in supported graphene. *Science* 328: 213–216.
51. Saito, K., Nakamura, J., and Natori, A. 2007. Ballistic thermal conductance of a graphene sheet. *Physical Review B* 76 (2007): 115409-1-4.
52. Ghosh, S., Bao, W., Nika, D. L., Subrina, S., Pokatilov, E. P., Lau, C. N., and Balandin, A. A. 2010. Dimensional crossover of thermal transport in few-layer graphene. *Nature Materials* 9: 555–558.
53. Ghosh, S., Calizo, I., Teweldebrhan, D., Pokatilov, E. P., Nika, D. L., Balandin, A. A., Bao, W., Miao, F., and Lau, C. N. 2008. Extremely high thermal conductivity of graphene: Prospects for thermal management applications in nanoelectronic circuits. *Applied Physics Letters* 92: 151911-1-3.
54. Meyer, J. C., Geim, A. K., Katsnelson, M. I., Novoselov, K. S., Booth, T. J., and Roth, S. 2007. The structure of suspended graphene sheets. *Nature* 446: 60–3.
55. Lindsay, L., Broido, D. A., and Mingo, N. 2010. Flexural phonons and thermal transport in graphene. *Physical Review B* 82: 115427-1-6.
56. Faugeras, C., Faugeras, B., Orlita, M., Potemski, M., Nair, R. R., and Geim, A. K. 2010. Thermal conductivity of graphene in corbino membrane geometry. *ACS Nano* 4: 1889–1892.

57. Cai, W., Moore, A. L., Zhu, Y., Li, X., Chen, S., Shi, L., and Ruoff, R. S. 2010. Thermal transport in suspended and supported monolayer graphene grown by chemical vapor deposition. *Nano Letters* 10: 1645–1651.
58. Seol, J. H., Moore, A. L., Shi, L., Jo, I., and Yao, Z. 2011. Thermal conductivity measurement of graphene exfoliated on silicon dioxide. *Journal of Heat Transfer* 133: 022403-1-7.
59. Jang, W., Chen, Z., Bao, W., Lau, C. N., and Dames, C. 2010. Thickness-dependent thermal conductivity of encased graphene and ultrathin graphite. *Nano Letters* 10: 3909–3913.
60. Yamamoto, T., Watanabe, K., and Mii, K. 2004. Empirical-potential study of phonon transport in graphitic ribbons. *Physical Review B* 70: 245402-1-7.
61. Gillen, R., Mohr, M., Thomsen, C., and Maultzsch, J. 2009. Vibrational properties of graphene nanoribbons by first-principles calculations. *Physical Review B* 80: 155418-1-9.
62. Munoz, E., Lu, J., and Yakobson, B. I. 2010. Ballistic thermal conductance of graphene ribbons. *Nano Letters* 10: 1652–1656.
63. Sevincli, H., and Cuniberti, G. 2010. Enhanced thermoelectric figure of merit in edge-disordered zigzag graphene nanoribbons. *Physical Review B* 81: 113401-1-4.
64. Li, W., Sevincli, H., Cuniberti, G., and Roche, S. 2010. Phonon transport in large-scale carbon-based disordered materials: Implementation of an efficient order-N and real-space Kubo methodology. *Physical Review B* 82: 041410-1-4.
65. Murali, R., Yang, Y., Brenner, K., Beck, T., and Meindl, J. D. 2009. Breakdown current density of graphene nanoribbons. *Applied Physics Letters* 94: 243114-1-3.
66. Wirtz, L., and Rubio, A. 2004. The phonon dispersion of graphite revisited. *Solid State Communications* 131: 141–152.
67. Lee, C., Wei, X., Kysar, J. W., and Hone, J. 2008. Measurement of elastic properties and intrinsic strength of monolayer graphene. *Science* 321: 385.
68. Ni, Z. H., Wang, H. M., Ma, Y., Kasim, J., Wu, Y. H., and Shen, Z. X. 2008. Tunable stress and controlled thickness modification in graphene by annealing. *ACS Nano* 2: 1033; Ni, Z. H., Yu, T., Lu, Y. H., Wang, Y. Y., Feng, Y. P., and Shen, Z. X. 2008. Uniaxial strain on graphene: Raman spectroscopy study and band-gap opening. *ACS Nano* 2: 2301.
69. Liu, F., Ming, P., and Li, J. 2007. Ab initio calculation of ideal strength and phonon instability of graphene under tension. *Physical Review B* 76: 064120.
70. Pereira, V. M., Peres, N. M. R., and Castro Neto, A. H. 2009. Tight-binding approach to uniaxial strain in graphene. *Physical Review B* 80: 045401; Pereira, V. M., and Castro Neto, A. H. 2009. Strain engineering of graphene's electronic structure. *Physical Review Letters* 103: 046801.
71. Gao, Y., and Hao, P. 2009. Mechanical properties of monolayer graphene under tensile and compressive loading. *Physica E* 41: 1561.
72. Xu, Z. 2009. Graphene nanoribbons under tension. *Journal of Computational and Theoretical Nanoscience* 6: 625.
73. Scarpa, F., Adhikari, S., and Srikantha Phani, A. 2009. Effective elastic mechanical properties of single layer graphene sheets. *Nanotechnology* 20: 065709.
74. Lu, Q., and Huang, R. 2010. Effect of edge structures on elastic modulus and fracture of graphene nanoribbons under uniaxial tension. arXiv: 1007.3298.
75. Mohiuddin, T. M. G., Lombardo, A., Nair, R. R., Bonetti, A., Savini, G., Jalil, R., Bonini, N., Basko, D. M., Galiotis, C., Marzari, N., Novoselov, K. S., Geim, A. K., and Ferrari, A. C. 2009. Uniaxial strain in graphene by Raman spectroscopy: G peak splitting, Grueneisen parameters, and sample orientation. *Physical Review B* 79: 205433-1-8.
76. Huang, M., Yan, H., Chen, C., Song, D., Heinz, T. F., and Hone, J. 2009. Phonon softening and crystallographic orientation of strained graphene studied by Raman spectroscopy. *Proceedings of the National Academy of Sciences USA* 106: 7304–7308.

77. Poetschke, M., Rocha, C. G., Foa Torres, L. E. F., Roche, S., and Cuniberti, G. 2010. Modeling graphene-based nanoelectromechanical devices. *Physical Review B* 81: 193404-1-4.
78. Erdogan, E., Popov, I., Rocha, C. G., Cuniberti, G., Roche, S., and Seifert, G. 2011. Engineering carbon chains from mechanically stretched graphene-based materials. *Physical Review B* 83: 041401(R).
79. Hod, O., and Scuseria, G. E. 2009. Electromechanical properties of suspended graphene nanoribbons. *Nano Letters* 9: 2619–2622.
80. Topsakal, M., and Ciraci, S. 2010. Elastic and plastic deformation of graphene, silicene, and boron nitride honeycomb nanoribbons under uniaxial tension: A first-principles density-functional theory study. *Physical Review B* 81: 024107-1-6.
81. Topsakal, M., Bagci, V. M. K., and Ciraci, S. 2010. Current-voltage (I-V) characteristics of armchair graphene nanoribbons under uniaxial strain. *Physical Review B* 81: 205437-1-5.
82. Isacsson, A. 2010. Nanomechanical displacement detection using coherent transport in ordered and disordered graphene nanoribbon resonators. arXiv: 1010.0508v1.
83. Ho, J. H., Chiu, Y. H., Tsai, S. J., and Lin, M. F. 2009. Semimetallic graphene in a modulated electric field. *Physical Review B* 79: 115427.
84. Novikov, D. S. 2007. Transverse field effect in graphene ribbons. *Physical Review Letters* 99: 056802.
85. Raza, H., and Kan, E. C. 2008. Armchair graphene nanoribbons: Electronic structure and electric-field modulation. *Physical Review B* 77: 245434.
86. Kan, E.-J., Li, Z., Yang, J., and Hou, J. G. 2007. Will zigzag graphene nanoribbon turn half metal under electric field? *Applied Physics Letters* 91: 243116.
87. Kinder, J. M., Dorando, J. J., Wang, H., and Chan, G. K.-L. 2009. Perfect reflection of chiral fermions in gated graphene nanoribbons. *Nano Letters* 9: 1980–1983.
88. Yu, Y.-J., Zhao, Y., Louis, S. R., Kwang, E. B., Kim, S., and Kim, P., 2009. Tuning the graphene work function by electric fields. *Nano Letters* 9: 3430–3434.
89. Yan, J., Zhang, Y., Kim, P., and Pinczuk, A. 2007. Electric field effect tuning electron-phonon coupling in graphene. *Physical Review Letters* 98: 166802.
90. Wang, Z. 2009. Alignment of graphene nanoribbons by an electric field. *Carbon* 47: 3050–3053.
91. Dalosto, S. D., and Levine, Z. H. 2008. Controlling the band gap in zigzag graphene nanoribbons with an electric field induced by a polar molecule. *Journal of Physical Chemistry C* 112: 8196–8199.
92. Ritter, C., Makler, S. S., and Latgé, A. 2008. Energy-gap modulations of graphene nanoribbons under external fields: A theoretical study. *Physical Review B* 77: 195443-1-5.; *Physical Review B* 82: 089903-1-2.
93. Chen, S. C., Wang, T. S., Lee, C. H., and Lin, M. F. 2008. Magneto-electronic properties of graphene nanoribbons in the spatially modulated electric field. *Physics Letters A* 372: 5999–6002.
94. Ho, J. H., Lai, Y. H., Lu, C. L., Hwang, J. S., Chang, C. P., and Lin, M. F. 2006. Electronic structure of a monolayer graphite layer in a modulated electric field. *Physics Letters A* 359: 70–75.
95. Lukose, V., Shankar, R., and Baskaran, G. 2007. Novel electric field effects on Landau levels in graphene. *Physical Review Letters* 98: 116802-1-4.
96. Zhang, S., Ma, N., and Zhang, E. 2010. The modulation of the Haas-van Alphen effect in graphene by electric field. *Journal of Physics: Condensed Matter* 22: 115302-1-8.
97. Kohler, S., Lehmann, J., and Haenggi, P. 2005. Driven quantum transport on the nanoscale. *Physics Reports* 406: 379–443.
98. Foa Torres, L. E. F., and Cuniberti, G. 2009. Controlling the conductance and noise of driven carbon-based Fabry-Perot devices. *Applied Physics Letters* 94: 222103-1-3.

99. Zhu, R., and Chen, H. 2009. Quantum pumping with adiabatically modulated barriers in graphene. *Applied Physics Letters* 95: 122111-1-3.
100. Prada, E., San-Jose, P., and Schomerus, H. 2009. Quantum pumping in graphene. *Physical Review B* 80: 245414-1-5.
101. Rocha, C. G., Foa Torres, L. E. F., and Cuniberti, G. 2010. Ac transport in graphene-based Fabry-Pérot devices. *Physical Review B* 81: 115435-1-8.
102. Oka, T., and Aoki, H. 2009. Photovoltaic Hall effect in graphene. *Physical Review* B 79: 081406-1-4.
103. Fujita, M., Wakabayashi, K., Nakada, K., and Kusakabe, K. 1996. Peculiar localized state at zigzag graphite edge. *Journal of the Physical Society of Japan* 65:1920-1-4.
104. Yazyev, O. V. 2010. Emergence of magnetism in graphene materials and nanostructures. *Reports on Progress in Physics* 73:056501-1-16.
105. Lieb, E. H. 1989. Two theorems on the Hubbard model. *Physical Review Letters* 62:1201-1-4.
106. Son, Y.-W. Cohen, M. L., and Louie, S. G. 2006. Half-metallic graphene nanoribbons. *Nature* 444:347-1-3.
107. Kan, E.-J., Li, Z., Yang, J., and Hou, J. G. 2008. Half-metallicity in edge-modified zigzag graphene nanoribbons. *Journal of the American Chemical Society* 130:4224-1-2.
108. Cantele, G., Lee, Y.-S., Ninno, D., and Marzari, N. 2009. Spin channels in functionalized graphene nanoribbons. *Nano Letters* 9:3425-1-4.
109. Wimmer, M., Adagideli, I., Berber, S., Tománek, D., and Richter, K. 2008. Spin currents in rough graphene nanoribbons: Universal fluctuations and spin injection. *Physical Review Letters* 100:177207-1-4.
110. Lakshmi, S., Roche, S., and Cuniberti, G., 2009. Spin valve effect in zigzag graphene nanoribbons by defect engineering. *Physical Review B* 80:193404-1-4.
111. Guimaraes, F. S. M., Costa, A. T., Muniz, R. B., and Ferreira, M. S. 2010. Graphene-based spin-pumping transistor. *Physical Review B* 81: 233402-1-4.
112. Guimaraes, F. S. M., Costa, A. T., Muniz, R. B., and Ferreira, M. S. 2010. Graphene as a non-magnetic spin-current lens. arXiv: 1009.6228v1.
113. Kim, W. Y., and Kim, K. S. 2008. Prediction of very large values of magnetoresistance in a graphene nanoribbon device. *Nature Nanotechnology* 3:408-1-5.
114. Muñoz-Rojas, F., Fernández-Rossier, J., and Palacios, J. J. 2009. Giant magnetoresistance in ultrasmall graphene-based devices. *Physical Review Letters* 102:136810-1-4.
115. Sepioni, M., Nair, R. R., Rablen, S., Narayanan, J., Tuna, F., Winpenny, R., Geim, A. K., and Grigorieva, I. V. 2010. Limits on intrinsic magnetism in graphene. *Physical Review Letters* 105:207205-1-4.
116. Mermin, N. D., and Wagner, H. 1966. Absence of ferromagnetism or antiferromagnetism in one- or two-dimensional isotropic Heisenberg models. *Physical Review Letters* 17:1133-1-4.
117. Joly, V. L. J., Kiguchi, M., Hao, S.-J., Takai, K., Enoki, T., Sumii, R., Amemiya, K., Muramatsu, H., Hayashi, T., Kim, Y. A., Endo, M., Campos-Delgado, J., López-Urías, F., Botello-Méndez, A., Terrones, H., Terrones, M., and Dresselhaus, M. S. 2010. Observation of magnetic edge state in graphene nanoribbons. *Physical Review B* 81:245428-1-6.
118. Wassmann, T., Seitsonen, A. P., Saitta, A. M., Lazzeri, M., and Mauri, F. 2008. Structure, stability, edge states, and aromaticity of graphene ribbons. *Physical Review Letters* 101:096402-1-4.
119. Koskinen, P., Malola, S., and Häkkinen, H. 2008. Self-passivating edge reconstructions of graphene. *Physical Review Letters* 101:115502-1-4.
120. Kunstmann, J., Özdoğan, C., Quandt, A., and Fehske, H. 2011. Stability of edge states and edge magnetism in graphene nanoribbons. *Physical Review B* 83:045414-1-8.
121. Lehtinen, P. O., Foster, A. S., Ma, Y., Krasheninnikov, A., and Nieminen, R. M. 2004. Irradiation-induced magnetism in graphite: A density functional study. *Physical Review Letters* 93:1872021-1-4.

122. Duplock, E. J., Scheffler, M., and Lindan, P. J. D. 2004. Hallmark of perfect graphene. *Physical Review Letters* 92:225502-1-4.
123. Esquinazi, P., Spemann, D., Höhne, R., Setzer, A., Han, K. H., and Butz, T. 2003. Induced magnetic ordering by proton irradiation in graphite. *Physical Review Letters* 91:227201-1-4.
124. Yazyev, O. V. 2008. Magnetism in disordered graphene and irradiated graphite. *Physical Review Letters* 101:037203-1-4.
125. Sevinçli, H., Topsakal, M., Durgun, E., and Ciraci, S. Electronic and magnetic properties of 3D transition-metal atom adsorbed graphene and graphene nanoribbons. *Physical Review B* 77:195434.
126. Li, X., Cai, W., An, J., Kim, S., Nah, J., Yang, D., Piner, R., Velamakanni, A., Jung, I., Tutuc, E., Banerjee, S. K., Colombo, L., and Ruoff, R. S. 2009. Large-area synthesis of high-quality and uniform graphene films on copper foils. *Science* 324: 1312–1314.
127. Li, X., Magnuson, C. W., Venugopal, A., An, J., Suk, J. W., Han, B., Borysiak, M., Cai, W., Velamakanni, A., Zhu, Y., Fu, L., Vogel, E. M., Voelkl, E., Colombo, L. and Ruoff, R. S. 2010. Graphene films with large domain size by a two-step chemical vapor deposition process. *Nano Letters* 10(11]:4328–4334.
128. Hamilton, J. C., and Blakely, J. M. 1980. Carbon segregation to single crystal surfaces of Pt, Pd and Co. *Surface Science* 91(1]: 199–218.
129. Wu, W., Liu, Z., Jauregui, L. A., Yua, Q., Pillai, R., Cao, H., Bao, J., Chen, Y. P., and Pei, S.-S. 2010. Wafer-scale synthesis of graphene by chemical vapor deposition and its application in hydrogen sensing. *Sensors and Actuators B* 150: 296–300.
130. Obraztsova, A. N., Obraztsova, E. A., Tyurnina, A. V., and Zolotukhin, A. A. 2007. Chemical vapor deposition of thin graphite films of nanometer thickness. *Carbon* 45: 2017–2021.
131. Yu, Q., Lian, J., Siriponglert, S., Li, H., Chen, Y. P., and Pei, S.-S. 2008. Graphene segregated on Ni surfaces and transferred to insulators. *Applied Physics Letters* 93: 113103-1-3.
132. Dedkov, Y. S., Fonin, M., Rüdiger, U., and C. Laubschat. 2008. Rashba effect in the graphene/Ni(111] system. *Physical Review Letters* 100: 107602-1-4.
133. Fuentes-Cabrera, M., Baskes, M. I., Melechko, A. V., and Simpson, M. L. 2008. Bridge structure for the graphene/Ni(111] system: A first principles study. *Physical Review B* 77:035405-1-5.
134. Dedkov, Y.S., Fonin, M., Rüdiger, U., and Laubschat, C. 2008. Graphene-protected iron layer on Ni(111]. *Applied Physics Letters* 93: 022509-1-3.
135. Usachov, D., Dobrotvorskii, A.M., Varykhalov, A., Rader, O., Gudat, W., Shikin, A. M., and Adamchuk, V. K. 2008. Experimental and theoretical study of the morphology of commensurate and incommensurate graphene layers on Ni single-crystal surfaces. *Physical Review B* 78: 085403-1-8.
136. Varykhalov, A., Sánchez-Barriga, J., Shikin, A. M., Biswas, C., Vescovo, E., Rybkin, A., Marchenko, D., and Rader, O. 2008. Electronic and magnetic properties of quasi-freestanding graphene on Ni. *Physical Review Letters* 101:157601-1-4.
137. Coraux, J., N'Diaye, A. T., Busse, C., and Michely, T. 2008. Structural coherency of graphene on Ir(111]. *Nano Letters* 8 (2): 565–570.
138. Preobrajenski, A. B., Ng, M. L., Vinogradov, A. S., and N. Mårtensson, N. 2008. Controlling graphene corrugation on lattice-mismatched substrates. *Physical Review B* 78: 073401-1-4.
139. Pletikosić1, I., Kralj, M., Pervan, P., Brako, R., Coraux, J., N'Diaye, A. T., Busse, C., and Michely, T. 2009. Dirac cones and minigaps for graphene on Ir(111]. *Physical Review Letters* 102: 056808-1-4.
140. Yi, P., Dong-Xia, S., and Hong-Jun, G. 2007. Formation of graphene on Ru(0001] surface. *Chinese Physics* 16 (11]: 3151–3153.

141. Sutter, P. W., Flege, J.-I., and Sutter, E. A. 2008. Epitaxial graphene on ruthenium. *Nature Materials* 7:406–411.
142. Marchini, S., Günther, S., and Wintterlin, J. 2007. Scanning tunneling microscopy of graphene on Ru(0001]. *Physical Review B* 76:075429-1-9.
143. Vázquez de Parga, A. L., Calleja, F., Borca, B., Passeggi, M. C. J., Hinarejos, J. J., Guinea, F., and Miranda, R. 2008. Periodically rippled graphene: Growth and spatially resolved electronic structure. *Physical Review Letters* 100:056807-1-4.
144. Jiang, D.-E., Du, M.-H., and Dai, S. 2009. First principles study of the graphene/Ru(0001] interface. *Journal of Chemical Physics* 130:074705-1-5.
145. Hwang, J., Shields, V. B., Thomas, C. I., Shiraraman, S., Hao, D., Kim, M., Woll, A. R., Tompa, G. S., and Spencer, M. G. 2010. Epitaxial growth of graphitic carbon on C-face SiC and sapphire by chemical vapor deposition (CVD]. *Journal of Crystal Growth* 312:3219–3224.
146. Rümmeli, M. H., Kramberger, C., Grüneis, A., Ayala, P., Gemming, T., Büchner, B., and Pichler, T. 2007. On the graphitization nature of oxides for the formation of carbon nanostructures. *Chemistry of Materials* 19(17]:4105–4107.
147. Rümmeli, M. H., Bachmatiuk, A., Scott, A., Börrnert, F., Warner, J. H., Hoffman, V., Lin, J.-H., Cuniberti, G., and Büchner, B. 2010. Direct low-temperature nanographene CVD synthesis over a dielectric insulator. *ACS Nano* 4(7):4206–4210.
148. Wang, J. J., Zhu, M. Y., Outlaw, R. A., Zhao, X., Manos, D. M., Holloway, B. C., and Mammana, V. P. 2004. Free-standing subnanometer graphite sheets. *Applied Physics Letters* 85:1265–1267.
149. Somani, P. R., Somani, S. P., and Umeno, M. 2006. Planar nano-graphenes from camphor by CVD. *Chemical Physics Letters* 430: 56–59.
150. Badami, D. V. 1962. Graphitization of α-silicon carbide. *Nature* 193:569–570.
151. Berger, C., Song, Z., Li, T., Li, X., Ogbazghi, A. Y., Feng, R., Dai, Z., Marchenkov, A. N., Conrad, E. H., First, P. N., and de Heer, W. A. 2004. Ultrathin epitaxial graphite: 2D electron gas properties and a route toward graphene-based nanoelectronics. *Journal of Physical Chemistry B* 108:19912-6.
152. Hernandez, Y., Nicolosi, V., Lotya, M., Blighe, F. M., Sun, Z., De, S., McGovern, I. T., Holland, B., Byrne, M., Gun'Ko, Y. K., Boland, J. J., Niraj, P., Duesberg, G., Satheesh, K., Goodhue, R., Hutchison, J., Scardaci, V., Ferrari, A. C., and Coleman, J. N. 2008. High-yield production of graphene by liquid-phase exfoliation of graphite. *Nature Nanotechnology* 3:563–568.
153. Zhu, J. 2008. New solutions to a new problem. *Nature Nanotechnology* 3:528–529.
154. Elias, D. C., Nair, R. R., Mohiuddin, T. M., Morozov, S. V., Blake, P., Halsall, M. P., Ferrari, A. C., Boukhalov, D. W., Katsnelson, M. I., Geim, A. K., and Novoselov, K. S. 2009. Control of graphene's properties by reversible hydrogenation: Evidence for graphane. *Science* 323:610-613.
155. Lueking, A. D., Gutierrez, H. R., Foneseca, D. A., Narayanan, D. L., VanEssendelft, D., Jain, P., and Clifford, C. E. B. 2006. Combined hydrogen production and storage with subsequent carbon crystallization. *Journal of American Chemical Society* 128:7758-60.
156. Sato, Y., Watano, H., Hagiwara, R., and Ito, Y. 2006. Reaction of layered carbon fluorides CxF (x = 2.5–3.6) and hydrogen. *Carbon* 44: 664-670.
157. Pekker, S., Salvetat, J.-P., Jakab, E., Bonard, J.-M., and Forro, L. 2001. Hydrogenation of carbon nanotubes and graphite in liquid ammonia. *Journal of Physical Chemistry B* 105: 7938–7943.
158. Felten, A., Bittencourt, C., Pireaux, J. J., Van Lier, G., and Charlier, J. C. 2005. Radio-frequency plasma functionalization of carbon nanotubes surface O_2, NH_3, and CF_4 treatments. *Journal of Applied Physics* 98: 074308-5.

159. Ruff, O., Bretschneider, O., and Ebert, F. Z. 1934. Die Reaktionsprodukte der verschiedenen Kohlenstoffformen mit Fluor II (Kohlenstoff-monofluorid). *Anorg. Allgm. Chem.* 217:1–18.
160. Robinson, J. T., Burgess, J. S., Junkermeier, C. E., Badescu, S. C., Reinecke, T. L., Perkins, F. K., Zalalutdniov, M. K., Baldwin, J. W., Culbertson, J. C., Sheehan, P. E., and Snow, E. S. 2010. Properties of fluorinated graphene films. *Nano Letters* 10:3001-5.
161. van Noorden, R. 2011. The trials of new carbon. *Nature* 469:14–16.
162. Shao, Y., Wang, J., Wu, H., Liu, J., Aksay, I. A., and Lin, Y. 2010. Graphene based electrochemical sensors and biosensors: A review. *Electroanalysis* 22: 1027–1036.
163. Tang, Z., Wu, H., Cort, J. R., Buchko, G. W., Zhang, Y., Shao, Y., Aksay, I. A., Liu, J., and Lin, Y. 2010. Constraint of DNA on functionalized graphene improves its biostability and specificity. *Small* 6: 1205–1209.
164. Schedin, F., Geim, A. K., Morozov, S. V., Hill, E. W., Blake, P., Katsnelson, M. I., and Novoselov, K. S. 2007. Detection of individual gas molecules adsorbed on graphene. *Nature Materials* 6: 652–655.
165. Dan, Y., Lu, Y., Kybert, N. J., Luo, Z., and Charlie Johnson, A. T. 2009. Intrinsic response of graphene vapor sensors. *Nano Letters* 9: 1472–1475.
166. Ratinac, K. R., Yang, W., Ringer, S. P., and Braet, F. 2010. Toward ubiquitous environmental gas sensors—Capitalizing on the promise of graphene. *Environmental Science Technology* 44: 1167–1176.
167. Bae, S., Kim, H., Lee, Y., Xu, X., Park, J.-S., Zheng, Y., Balakrishnan, J., Lei, T., Kim, H. R., Song, Y. I., Kim, Y.-J., Kim, K. S., Özyilmaz, B., Ahn, J.-H., Hong, B. H., and Iijima, S. 2010. Roll-to-roll production of 30-inch graphene thin films for transparent electrodes. *Nanotechnology* 5: 574–578.
168. Liao, L., Bai, J., Cheng, R., Lin, Y.-C., Jiang, S., Qu, Y., Huang, Y., and Duan, X. 2010. Sub-100 nm channel length graphene transistors. *Nano Letters* 10: 3952–3956.

2 Graphene Synthesis

Santanu Das and Wonbong Choi

CONTENTS

2.1 INTRODUCTION

Since its invention, graphene, which consists of one to few graphitic layered sp^2 bonded 2D carbon allotropes, has become a unique material due to its extraordinary properties like electrical and thermal conductivity, high charge carrier density, carrier mobility, optical conductivity (Nair et al. 2008), and mechanical property (Geim and Kim 2008; Geim and Novoselov 2007; Choi et al. 2010; Lee et al. 2008). Hence it created unprecedented interest in research and industry to be used as next-generation electronic and photonic materials. The basic building blocks of all the carbon nanostructures are a single graphitic layer that is covalently functionalized sp^2 bonded carbon atoms in a hexagonal honeycomb lattice (as shown in Chapter 8, Figure 8.1), which forms 3D bulk graphite, when the layers of single honeycomb graphitic lattices are stacked and bound by a weak van der Waals force. When the single graphite layer forms a sphere, it is well known as 0-dimensional fullerene; when it is rolled up with respect to its axis, it forms a one-dimensional cylindrical structure called a *carbon nanotube*; and when it exhibits the planar 2D structure from one to a few layers stacked, it is called *graphene*. One graphitic layer is well known as mono-atomic or single-layer graphene and two and three graphitic layers are known as bilayer and trilayer graphene, respectively. More than 5-layer up to 10-layer graphene is generally called *few-layer graphene*, and ~20- to 30-layer graphene is referred to as *multilayer graphene, thick graphene*, or *nanocrystalline thin graphite*.

High-crystalline, pristine, mono-atomic graphene exhibits the unusual electronic property of semimetallic behavior due to the touching of π and π^* bands in a single point at the Fermi level (E_f) at the corners of the Brillouin zone (Novoselov, Geim, et al. 2005). Further, graphene became attractive to physicists for research because of its resemblance to the Dirac spectrum of massless fermions (Novoselov, Geim, et al. 2005) and its Landau-level quantization under a vertical magnetic field applied to the graphene basal plane (Castro Neto et al. 2009).

More interestingly, bilayer graphene exhibits a gapless state that occurs due to the parabolic bands touching at the K and K′ points at each point of the Brillouin zone, and at a particular approximation, a negligible band overlap (about 0.0016 eV) at higher energies makes bilayer graphene a gapless semiconductor (Castro Neto et al. 2009). Further, the physical properties of graphene strictly depend upon the number of stacked layers. For example, monolayer graphene exhibits ~97% optical transmittance (Li, Zhu, et al. 2009) with ~2.2 KΩ/sq sheet resistance. In turn, transmittance and sheet resistance decreases with increasing numbers of layers. More specifically, bilayer, trilayer, and 4-layer graphene possess ~95%, ~92%, and ~89% transmittance with corresponding sheet resistance of 1 KΩ/sq, ~700 Ω/sq, and ~400 Ω/sq, respectively (Li, Zhu, et al. 2009). On the other hand, graphene carrier density was found to be in the order of 10^{13} (Geim and Novoselov 2007) with charge carrier mobility ~15,000 $cm^2/V{\cdot}s$ (Geim and Novoselov 2007) and resistivity of ~10^{-6} ohm–cm, which are ideally fitted for field-effect transistors (FETs).

The exceptional electrical properties of graphene are attractive for applications in future electronics such as ballistic transistors, field emitters, components of integrated circuits, transparent conducting electrodes, and sensors. Graphene has a high electron (or hole) mobility as well as low Johnson noise,* allowing it to be utilized as the channel in an FET. The high electrical conductivity and high optical transparency promote graphene as a candidate for transparent conducting electrodes, which are required for applications in touch screens, liquid crystal displays, organic photovoltaic cells, and organic light-emitting diodes (OLEDs) (Choi et al. 2010; Geim and Novoselov 2007). Most of these interesting applications require growth of single-layer graphene on a suitable substrate and create controlled and practical bandgap, which are very difficult to control and achieve. Many reports are available on graphene synthesis and most of these are based on mechanical exfoliation from graphite, thermal graphitization of a SiC surface (Geim and Novoselov 2007; Berger et al. 2004), reduced graphite oxide (Park and Ruoff 2009), and recently, by chemical vapor deposition (Reina, Jia, et al. 2009; Li, Cai, An et al. 2009).

This chapter reviews graphene synthesis technologies, including exfoliation, chemical synthesis, unzipping nanotubes, thermal chemical vapor deposition, plasma enhanced chemical vapor deposition, and epitaxial growth on a SiC surface together with a discussion of their growth mechanism and feasibility.

---

* Electronic noise generated by the thermal agitation of the charge carriers inside an electrical conductor at equilibrium, which happens regardless of any applied voltage.

2.2 OVERVIEW OF GRAPHENE SYNTHESIS METHODS

To date, several techniques have been established for graphene synthesis. However, mechanical cleaving (exfoliation) (Novoselov et al. 2004), chemical exfoliation (Allen, Tung, and Kaner 2010; Viculis, Mack, and Kaner 2003), chemical synthesis (Park and Ruoff 2009), and thermal chemical vapor deposition (CVD) (Reina, Jia et al. 2009) synthesis are the most commonly used methods today. Some other techniques are also reported such as unzipping nanotube (Jiao et al. 2010; Kosynkin et al. 2009; Jiao et al. 2009) and microwave synthesis (Xin et al. 2010); however, those techniques need to be explored more extensively. An overview of graphene synthesis techniques is shown in the flow chart in Figure 2.1. In 1975, few-layer graphite was synthesized on a single crystal platinum surface via chemical decomposition methods, but was not designated as graphene due to a lack of characterization techniques or perhaps due to its limited possible applications (Lang 1975).

In 1999, mechanical cleaving of highly ordered pyrolytic graphite (HOPG) by atomic force microscopy (AFM) tips was first developed in order to fabricate graphene from a few layers down to a mono-atomic single layer (Lu et al. 1999). Nevertheless, mono-layer graphene was first produced and reported in the year 2004, where adhesive tape was used to repeatedly slice down the graphene layers on a substrate (Novoselov et al. 2004). This technique was found to be capable of producing different layers of graphene and is relatively easy to fabricate. Mechanical exfoliation using AFM cantilever was found equally capable of fabricating few-layer graphene, but the process was limited to producing graphene of a thickness ~10 nm, which is comparable to 30-layer graphene. Chemical exfoliation is a method where solution-dispersed graphite is exfoliated by inserting large alkali ions between the graphite layers. Similarly, chemical synthesis process consists of the synthesis of graphite oxide, dispersion in a solution, followed by reduction with hydrazine. Like carbon nanotube synthesis, catalytic thermal CVD has proved to be one of the best processes for large-scale graphene fabrication. Here, thermally dissociated carbon is deposited

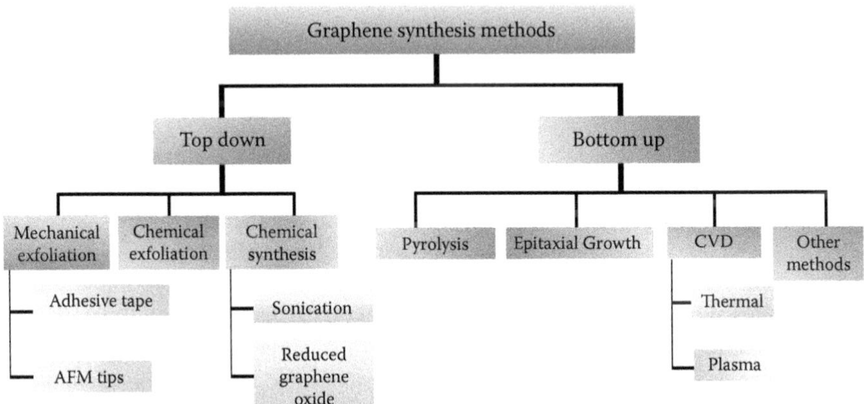

FIGURE 2.1 The schematic represents the different graphene synthesis methods.

onto a catalytically active transition metal surface and forms a honeycomb graphite lattice at elevated temperature under atmospheric or low pressures. When the thermal CVD process is carried out in a resistive heating furnace, it is known as *thermal CVD*, and when the process consists of plasma-assisted growth, it is called *plasma-enhanced CVD* or *PECVD*. As a whole, all the above techniques are standard in their respective fields. However, all synthesis methods have their own advantages as well as disadvantages depending upon the final application of graphene. For example, the mechanical exfoliation method is capable of fabricating different layers of graphene (from monolayer to few-layer), but the reliability of obtaining a similar structure using this technique is quite low. Moreover, large-area graphene fabrication using mechanical cleaving is a serious challenge at this moment. From monolayer to few-layer graphene can easily be obtained by the adhesive tape exfoliation method, but extensive research is a prerequisite for further device fabrication, which limits the feasibility of this process for industrialization. Furthermore, chemical synthesis processes (that involve the synthesis of graphite oxide and reducing it back to graphene in a solution dispersal condition) are low temperature processes that make it easier to fabricate graphene on various types of substrates at ambient temperature, particularly on polymeric substrates (those exhibit a low melting point). However, homogeneity and uniformity of large-area graphene synthesized by this method is not satisfactory. On the other hand, graphene synthesized from reduced graphene oxides (RGOs) often causes incomplete reduction of graphite oxide (which exhibits insulator characteristics) and results in the successive degradation of electrical properties depending on its degree of reduction. In contrast, thermal CVD methods are more advantageous for large-area device fabrication and favorable for future complementary metal-oxide semiconductor (CMOS) technology by replacing Si (Sutter 2009). Epitaxial graphene, thermal graphitization of a SiC surface, is another method, but the high process temperature and inability to transfer on any other substrates limit this method's versatility. In this context, the thermal CVD method is unique because a uniform layer of thermally chemically catalyzed carbon atoms can be deposited onto metal surfaces and can be transferred over a wide range of substrates. However, graphene layer controllability and low-temperature graphene synthesis are challenges for this technique. In the upcoming sections, a few of the graphene synthesis methods and their scientific and technological importance are described in detail.

2.2.1 Mechanical Exfoliation

Mechanical exfoliation is the first recognized method of graphene synthesis. This is a top-down technique in nanotechnology, by which a longitudinal or transverse stress is generated on the surface of the layered structure materials using simple scotch tape or AFM tip to slice a single layer or a few layers from the material onto a substrate. Graphite is formed when mono-atomic graphene layers are stacked together by weak van der Waals forces. The interlayer distance and interlayer bond energy is 3.34 Å and 2 eV/nm^2, respectively. For mechanical cleaving, ~300 nN/µm^2 external force is required to separate one mono-atomic layer from graphite (Zhang et al. 2005). In the year 1999, Ruoff et al. (Lu et al. 1999) first proposed the mechanical

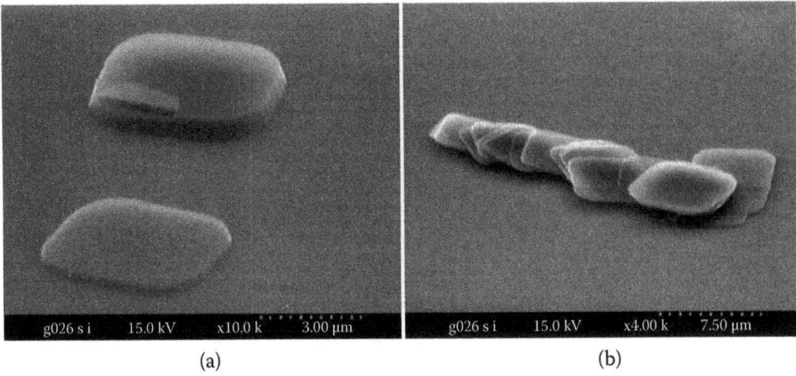

(a) (b)

FIGURE 2.2 (a) and (b): Scanning electron micrographs of mechanically exfoliated thin graphite layers from highly oriented pyrolytic graphite (HOPG) by AFM tip. (From X. Lu, K., M. F. Yu, H. Huang, and R. S. Ruoff. Tailoring graphite with the goal of achieving single sheets. *Nanotechnology* 10, no. 3 (1999):269–272. With permission.)

exfoliation technique of plasma-etched pillared HOPG using an AFM tip to fabricate graphene. As seen in Figures 2.2(a) and 2.2(b), the thin multilayered graphite was fabricated with the lowest thickness of ~200 nm, which consists of 500–600 layers of monolayer graphene. Later, the scientific field was significantly impacted by the creation of carbon-based nanomaterials when Novoselov and Geim first reported the isolation of single-layer up to few-layer graphene on a SiO_2/Si substrate and its electronics properties. They were awarded the Nobel Prize in Physics in 2010 for their novel synthesis approach and discovery of the extraordinary properties of thin flakes of simple graphite (see "The Rise and Rise of Graphene," 2010).

Novoselov et al. (2004) used adhesive tape to produce a single graphene layer by a mechanical cleaving technique from 1-mm-thick HOPG. First, they used dry etching by oxygen plasma to prepare graphite mesas of few millimeters thick graphite mesas on the top of the graphite platelets. The resulting graphite mesa surface was then compressed against a 1-mm-thick layer of wet photoresist over a glass substrate, followed by baking in order to firmly attach the HOPG mesas to the photoresist layer. Then, using scotch tape, they gradually peeled off the graphite flakes and released the flakes in acetone. Using a Si (n-doped Si with a SiO_2 top layer) wafer, the graphene (both single-layer and few-layer graphene) was transferred from the acetone solution to the Si substrate, followed by cleaning with water and propanol.

Finally, thin flakes of graphene (thickness less than 10 nm) were found to adhere on the surface of the wafer, and the adherence force between graphene and the substrate was claimed to be van der Waals and/or capillary forces. Figure 2.3 illustrates the optical micrograph of graphene flakes sliced down on a SiO_2/Si substrate using the scotch tape method. As shown in Figure 2.3(b), multiple graphene flakes with different layers can be produced simultaneously using mechanical exfoliation. Zhang et al. (2005) further tried to improve the graphene production method in large scale by cleaving the HOPG using a tipless AFM cantilever. The controlled exfoliation technique consisted of a cantilever with predetermined spring constant that

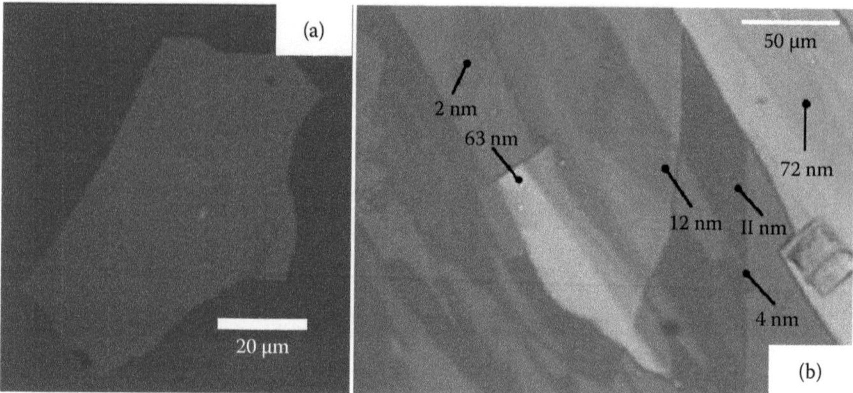

FIGURE 2.3 (a) Optical micrograph (under normal white light) of a few-layer graphene flake produced by the scotch tape method on a SiO_2/Si substrate (thickness ~3 nm); (b) graphitic films of various thicknesses measured by AFM. (From K.S. Novoselov, A. K. Geim, S. V. Morozov, et al. Electric field effect in atomically thin carbon films. *Science* 306, no. 5696 (2004):666–669. With permission.)

propagates the required shear stresses to peel off the graphite flakes. The thinnest flake produced by this technique was ~10 nm thick; however, the technique was unable to yield single- or bilayer graphene.

The graphene produced by these mechanical exfoliation techniques was used for fabrication of FET devices, which brought a research boom in the field of carbon nanoelectronics. Today, the number of publications concerning graphene has increased exponentially due to its scientific and technological consequences for future electronics applications. For that reason, the process was also extended for fabricating some of the other 2-D planar materials like boron nitride (BN), molybdenum disilicide (MoS_2), $NbSe_2$, and $Bi_2Sr_2CaCu_2O$ (Novoselov, Jiang, et al. 2005).

However, the mechanical exfoliation process needs to be improved further for large-scale, defect-free, high-purity graphene so that it can be used in feasible applications in nanoelectronics. In this regard, Liang et al. (Liang, Fu, and Chou 2007) proposed an interesting method for wafer-scale graphene fabrication by a cut-and-choose transfer printing method for integrated circuits, but uniform large-scale graphene fabrication with controlled layers is still a challenge.

2.2.2 Chemical Exfoliation

Like mechanical exfoliation, chemical exfoliation is one of the established methods for fabricating graphene. Chemical exfoliation is a process by which alkali metals are intercalated with the graphite structure to isolate few-layer graphene dispersed in solution. Alkali metals are the materials in the periodic table that can easily form graphite-intercalated structures with various stoichiometric ratios of graphite to alkali metals. One of the major advantages of alkali metals is their ionic radii, which are smaller than the graphite interlayer spacing; hence they fit easily in the interlayer spacing as shown in the schematic in Figure 2.4(a).

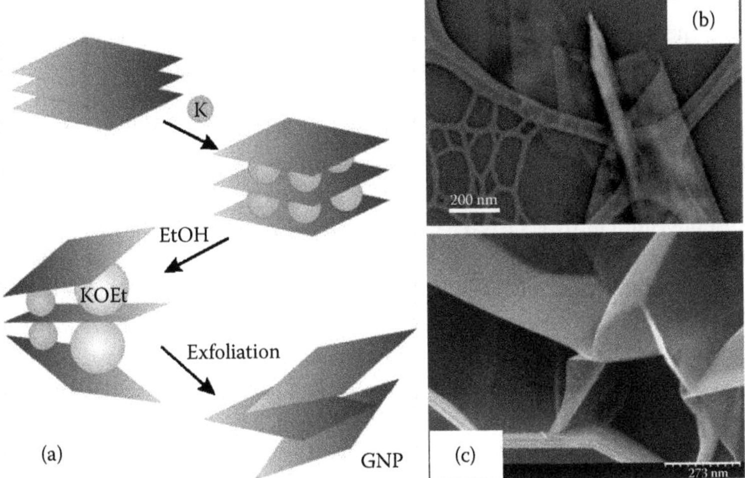

FIGURE 2.4 (a) Schematic illustrating the chemical exfoliation process, (b) transmission electron micrograph (TEM) of chemically exfoliated graphitic nanosheet, (c) SEM picture of thin graphite nanosheets after the exfoliation process, showing approximately a 10-nm thickness of ~30 layers of single graphite sheet. (From (a) Lisa M. Viculis, Julia J. Mack, Oren M. Mayer, H. Thomas Hahn, and Richard B. Kaner. Intercalation and exfoliation routes to graphite nanoplatelets. *Journal of Materials Chemistry* 15, no. 9 (2005):974–978; (b) Lisa M. Viculis, Julia J. Mack, and Richard B. Kaner. A chemical route to carbon nanoscrolls. *Science* 299, no. 5611 (2003):1361–1361; (c) Lisa M. Viculis, Julia J. Mack, Oren M. Mayer, H. Thomas Hahn, and Richard B. Kaner. Intercalation and exfoliation routes to graphite nanoplatelets. *Journal of Materials Chemistry* 15, no. 9 (2005):974–978. With permission.)

Kaner et al. first reported (Viculis, Mack, and Kaner 2003) chemically exfoliated few-layer graphite (later called "graphene") using potassium (K) as the intercalating compound forming alkali metal. Potassium (K) forms a KC_8 intercalated compound when reacting with graphite at 200°C under an inert helium atmosphere (less than 1 ppm H_2O and O_2). The intercalated compound KC_8 undergoes an exothermic reaction when it reacts with the aqueous solution of ethanol (CH_3CH_2OH) as per Equation (2.1).

$$KC_8 + CH_3CH_2OH\ 8C + KOCH_2CH_3 + 1/2H_2 \quad (2.1)$$

Hence, potassium ions dissolve into the solution forming potassium ethoxide, which is basic in nature, and the reaction leads to hydrogen generation, which helps to separate the graphite layers. Precaution must be taken with this type of reaction because alkali metals react vigorously with water and alcohol. For scalable production, the reaction chamber needs to be kept in an ice bath to dissipate the generated heat. Finally, the resultant few-layer exfoliated graphene was collected by a filtration process and purified by washing to bring it to pH 7. The formation of a few graphitic layers or few-layer graphene (FLG) is shown in Figure 2.4(b). Transmission electron microscopy (TEM) study showed that few-layer graphene produced by this method consisted of 40±15 layers of mono-atomic graphene. Later, the same researchers explored the exfoliation process using other alkali metals like Cs and NaK_2 alloy following the same process as reported by Viculis et al. (2005). Unlike Li and Na, the K ionization potential (4.34 eV) is less than graphite's electron affinity (4.6 eV); thus, K reacts directly with graphite to form intercalated compounds. Cs (3.894 eV) possesses a lower ionization potential than K (4.34 eV), and therefore reacts with graphite more violently than K, which creates a significant improvement in intercalation of graphite at a significantly low temperature and ambient pressure. Sodium-potassium alloy (Na-K_2) experiences eutectic melting at −12.62°C, and thus an exfoliation reaction is expected to occur at room temperature and ambient pressure. In particular, the graphene produced at room temperature using the Na-K_2 alloy graphite intercalated compounds exhibits a wide range of thicknesses from 2 nm to 150 nm. This process could produce large-scale exfoliation in a solution process at low ambient temperature conditions, which makes it distinct among other graphene fabrication processes. However, single-layer and bilayer graphene synthesis by the graphene intercalated route is yet to be explored and chemical contamination is one of the serious drawbacks of the process. A novel approach was proposed separately regarding the dispersion and exfoliation of pure graphite in organic solvents such as N-methyl-pyrrolidone. Hernandez et al. (2008) reported the exfoliation of pure graphite in N-methyl-pyrrolidone by a simple sonication process. The report showed high-quality, unoxidized monolayer graphene synthesis at yields of ~1%. Further improvement of the process could potentially improve yields by 7–12% of the starting graphite mass with sediment recycling (the details of the process are given in Hernandez et al. 2008). The morphology of graphite and graphene by the sonication process is shown in Figures 2.5(a) and 2.5(b), respectively. The proposed mechanism states that the exfoliation of a layered structure is possible upon the addition of mechanical energy, if the solute and solvent surface energy are the same. In this context, the energy required to exfoliate graphene should be equivalent to the

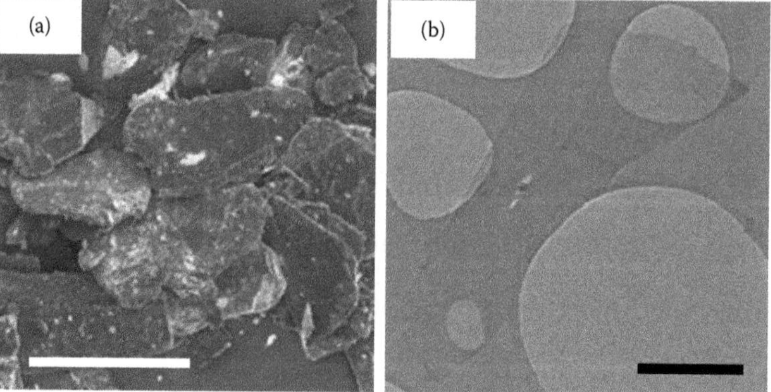

FIGURE 2.5 (a) SEM image of pristine graphite before sonication and (b) transmission electron microscopy of graphene flake prepared in N-methyl-pyrrolidone after the sonication process. (From Y. Hernandez, V. Nicolosi, M. Lotya, et al. High-yield production of graphene by liquid-phase exfoliation of graphite. *Nature Nanotechnology* 3, no. (2008):563–568. With permission.)

solvent–graphene interaction for the solvents whose surface energies are analogous to that of the suspended graphene. The process is versatile because it is a low-cost solution-phase method, is scalable, and would be capable of depositing graphene on a wide variety of substrates, which is not possible using other processes like cleavage or thermal deposition. Furthermore, the method can be extended to produce graphene-based composites and films, which are the key requirements for special applications, such as thin-film transistors, transparent conductive electrodes, and so on.

2.2.3 Some Other Novel Routes of Graphene Synthesis

In addition to the synthesis methods discussed previously, a few reports exist regarding some novel techniques for graphene fabrication. For example, mechanical exfoliation of graphene produced by high-velocity clusters impacting on a graphite surface (Sidorov et al. 2010). The graphene nanoribbon produced by this method was ~30 nm thick. In another report, Xin et al. (2010) indicated graphene exfoliated from microwave irradiation of graphite-intercalated compounds in a solution process followed by combining those exfoliated sheets with carbon nanotubes (CNTs). They claimed that the graphene CNT combined sheet resistance was 181 ohm/sq with 82.2% transmittance, which is equivalent to the commercially available indium tin oxide (ITO). Similarly, a green methodology was demonstrated using microwaves for graphene synthesis (Sridhar, Jeon, and Oh 2010). Plasma-assisted etching of graphite to form multilayered graphene and monolayer graphene was also demonstrated in another report (Hazra et al. 2011). This is another top-down approach that involves the gradual thinning process of graphite to graphene using plasma in an H_2 and N_2 atmosphere. In a different approach, de Parga et al. reported the epitaxial graphene formation on Ru(0001) under ultrahigh vacuum (UHV) conditions (~10–11 Torr) (de Parga et al. 2008). Furthermore, Zheng et al. (2010) reported the metal catalyzed formation of graphene using amorphous carbon at high temperature. However, all the processes discussed here are in rudimentary stages and need to be developed further to obtain low-cost, high-purity, reliable, and scalable graphene.

2.2.4 Chemical Synthesis: Graphene from Reduced Graphene Oxide

Chemical synthesis is a top-down indirect graphene synthesis method, and is the first method that demonstrated graphene synthesis by a chemical route. In the year 1962, Boehm et al. first demonstrated monolayer flakes of reduced graphene oxide, which was recently acknowledged by the graphene inventor Andre Geim. The method involves the synthesis of a graphite oxide (GO) by oxidation of graphite, dispersing the flakes by sonication, and reducing it back to graphene. There are three popular methods available for GO synthesis: the Brodie method (Brodie 1860), Staudenmaier method (Staudenmaier 1898), and Hummers and Offeman method (Hummers and Offeman 1958). All three methods involve oxidation of graphite using strong acids and oxidants. The degree of oxidation can be varied by the reaction conditions (e.g., temperature, pressure, etc.), stoichiometry, and the type of precursor graphite used as a starting material. Despite the wide range of research that has been carried out already to describe the chemical structure of GO, several models are suggested to

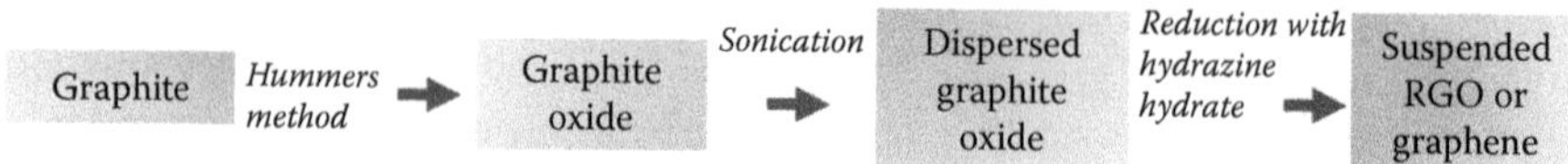

FIGURE 2.6 The process flow chart of graphene synthesis derived from graphite oxide.

explain this chemical structure. GO was first prepared by Brodie (1860), by mixing graphite with potassium chlorate and nitric acid. However, the process contains several steps that are time consuming, unsafe, and hazardous. In order to overcome those problems, Hummers (Hummers and Offeman 1958) developed a method for fabricating graphite oxide by mixing graphite with sodium nitrite, sulfuric acid, and potassium permanganate, well known as Hummers method.

When graphite turns into graphite oxide, the interlayer spacing is increased two or three times larger than in pristine graphite. For pristine graphite, the interlayer distance is 3.34 Å, which expanded up to 5.62 Å after 1 hour of oxidative reaction, and further interlayer expansion occurred to 7.0 ±0.35 Å upon prolonged oxidation for 24 hours. As reported by Boehm et al. (1962), the interlayer distances can be further increased by inserting polar liquids such as sodium hydroxide. As a result, interlayer distance is further expanded, which in fact separates a single layer from the GO bulk materials. Upon treatment with hydrazine hydrate, GO reduces back to graphene. The chemical reduction process is carried out using dimethylhydrazine or hydrazine in the presence of a polymer or surfactant to produce homogeneous colloidal suspensions of graphene. The process flow chart of the chemical synthesis of graphene is shown in a schematic in Figure 2.6.

The chemical synthesis method was brought into focus again in 2006 when Ruoff and his coworkers produced mono-atomic graphene by a chemical synthesis process (Stankovich, Dikin, et al. 2006; Stankovich, Piner, et al. 2006). They prepared GO by the Hummers method and chemically modified GO to produce a water-dispersible GO. GO is a stacked layer of squeezed sheets with AB stacking, which exhibits an oxygen containing functional group like hydroxyl and epixoide in their basal plane when it is highly oxidized (Jeong et al. 2008). The attached functional groups (carbonyl and carboxyl) are hydrophilic in nature, which facilitates the exfoliation of GO upon ultrasonication in an aqueous medium. Thus, the hydrophilic functional groups accelerate the intercalation of water molecules between the GO layers. In this process, functionalized GO is used as a precursor material for graphene production, which forms graphene upon reduction with dimethylhydrazine at 80°C for 24 h. Stankovich et al. (Stankovich, Piner, et al. 2006) showed that chemical functionalization of GO flakes by organic molecules leads to the homogeneous suspension of GO flakes in organic solvents. They reported that reaction of graphite oxide with isocyanate results in isocyanate-modified graphene oxide, which can be dispersed uniformly in polar aprotic solvents like dimethylformamide (DMF), N-methylpyrrolidone (NMP), dimethyl sulfoxide (DMSO), and hexamethylphosphoramide (HMPA). The proposed mechanism states that the reaction of isocyanate with hydroxyl and carboxyl groups generates the carbamate and amide functional groups, which become attached to the GO flakes (as shown in Figure 2.7).

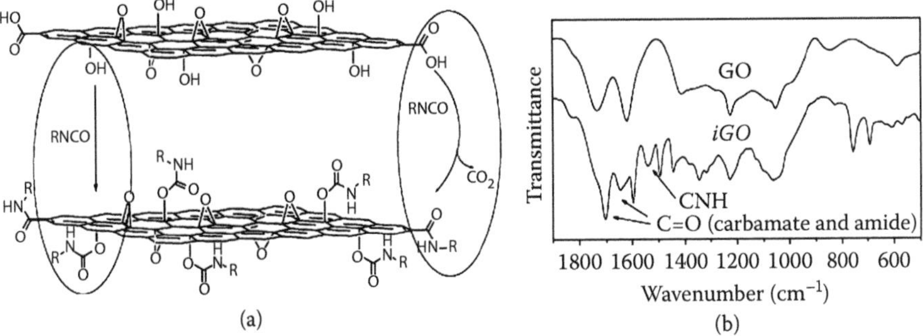

FIGURE 2.7 Mechanism proposed by Stankovich et al. on isocyanate-treated GO where organic isocyanates react with the hydroxyl (left oval) and carboxyl groups (right oval) of graphene oxide sheets to form carbamate and amide functionalities, respectively. (b) Representative FT-IR spectra of GO and phenyl isocyanate-functionalized GO. (From Stankovich, S., R. D. Piner, S. T. Nguyen, and R. S. Ruoff. Synthesis and exfoliation of isocyanate-treated graphene oxide nanoplatelets. *Carbon* 44, no. 15 (2006):3342–3347. With permission.)

Xu et al. (2008) further reported the colloidal suspensions of chemically modified graphene (CMG) decorated with small organic molecules or nanoparticles. They demonstrated graphene oxide sheets noncovalently functionalized with 1-pyrenebutyrate (PB^-). 1-pyrenebutyrate (PB^-), is an organic molecule with a strong adsorption affinity for the graphite basal plane via π stacking. PB^--functionalized graphene was prepared by dispersing GO in pyrenebutyric acid followed by reducing it with hydrazine monohydrate at 80°C for 24 h. The resultant product was a homogeneous black colloidal suspension, which is a PB^--functionalized graphene dispersed in water. However, dispersion of graphene needs stabilizers or surfactants that induce contamination during device fabrication. Moreover, removal of stabilizers or surfactants readily agglomerates the dispersed graphene; hence, obtaining a monolayer was difficult. Therefore, fabrication of stabilizer- or surfactant-free dispersed graphene via chemical synthesis was becoming important. Few reports have been found related to synthesis methods involving stabilizer- or surfactant-free colloidal suspensions of unagglomerated graphene sheets. Li et al. demonstrated the surfactant- and stabilizer-free aqueous suspension (0.5 mg ml^{-1}) of RGO sheets under basic conditions (pH 10) (Li, Muller, et al. 2008). They found that electrostatically stabilized dispersion is strongly dependent on pH. The highly negative surface charge (zeta potential) of as-prepared GO sheets that contain carboxylic acid and phenolic hydroxyl groups form a stable suspension when reduced by hydrazine in the presence of ammonia (pH ~10). As illustrated in Figure 2.8(a), at pH 10, the neutral carboxylic group converts into negatively charged carboxylate during the reduction reaction, which results in the retardation of further agglomeration of the suspended graphene. The report claimed that the reduced graphene exhibited a substantial amount of surface negative charge, which was confirmed by zeta potential measurement (as shown in Figure 2.8(b)). The thickness of the dispersed chemically converted graphene (CCG) on a SiO_2/Si wafer was reported as ~1 nm using tapping mode AFM as shown in

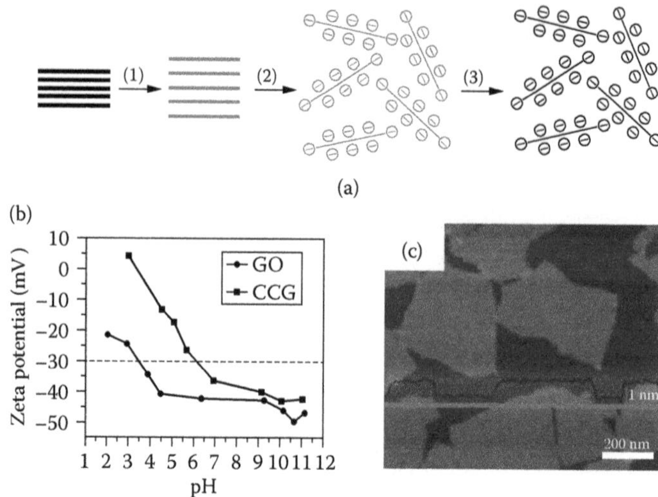

FIGURE 2.8 (a) Schematic showing the aqueous suspension of the graphene fabrication mechanism via chemical technique. The process steps consist of (1) graphite oxide production with greater interlayer distance, (2) sonication of GO in order to prepare a mechanically exfoliated colloidal suspension of GO in water, (3) conversion of GO to graphene using hydrazine reduction. (b) Representative data of the Zeta potential of GO and chemically converted graphene (CCG) as a function of pH. (c) Tapping mode atomic force micrograph of drop-casted CCG flakes on a silicon wafer. (From D. Li, M. B. Muller, S. Gilje, R. B. Kaner, and G. G. Wallace. Processable aqueous dispersions of graphene nanosheets. *Nature Nanotechnology* 3, no. 2 (2008):101–105. With permission.)

Figure 2.8(c). They concluded that the transformation of negatively charged stable GO colloids was due to the electrostatic repulsion, not due to just the hydrophilicity of GO as per the earlier report of Stankovich et al. (2007). Later, Tung et al. (2009) reported synthesis of large-scale (~20-mm × 40-mm) single sheets of graphene using graphite oxide paper, where they tried to remove oxygen functionalities from GO to restore the planar geometry of the single sheets of CCG. In this approach, reduction as well as dispersion of the GO film was done directly in hydrazine, which creates hydrazinium graphene (HG) through the formation of counter ions. HG composed of a negatively charged, reduced graphene sheet surrounded by $N_2H_4^+$ counter ions is shown in Figure 2.9. Finally, a single-layer stable graphene sheet of ~0.6 nm thickness was obtained by this process.

A few reports revealed that the reduction of GO occurs at significantly high temperature with faster heating rate during the chemical process (Schniepp et al. 2006; McAllister et al. 2007). One interesting chemical technique, the Langmuir-Blodgett assembly of GO single layers (GOSL), was successfully demonstrated by Cote et al. (Cote, Kim, and Huang 2008). The results are summarized as follows: (1) water supported mono layers of GOSL were suspended without any surfactant or stabilizing agent, (2) the single layers formed stable dispersion when bound at the 2-D air–water interface, (3) GOSL monolayers can be readily transferred to any substrate with tunable density from dilute, closepacked to overpacked monolayers of interlocking

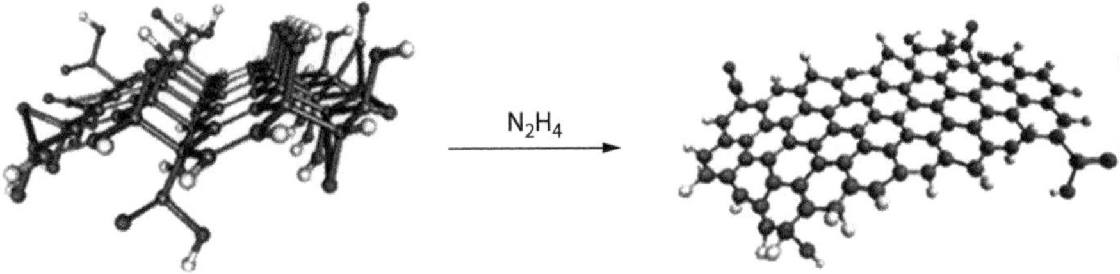

FIGURE 2.9 Schematic showing that the 3-D GO (carbon in grey, oxygen in dark grey, and hydrogen in white) restores its planar structure when reduced and dispersed with hydrazine. (From V. C. Tung, M. J. Allen, Y. Yang, and R. B. Kaner. High-throughput solution processing of large-scale graphene. *Nature Nanotechnology* 4, no. 1 (2009):25–29. With permission.)

sheets, and (4) the process is novel because one could obtain a GO sheet on solid substrates of ~1 nm thickness.

All of the previously mentioned processes comprise the chemical approach for synthesizing graphene. Direct graphene synthesis using electrochemical methods was reported by Liu et al. (Liu et al. 2008). The method is environment friendly and leads to the production of a colloidal suspension of imidazolium ion–functionalized graphene sheets by direct electrochemical treatment of graphite. The mechanism stated that the imidazolium ion covalently attached to the graphene nanosheets electrochemically through the breaking of the C–C π bond. As shown in Figure 2.10, 10–20 V potential was applied to originate graphene nanosheets from the graphite anode. Further, dispersion of the modified graphene nanosheet was done in N,N-dimethylformamide (DMF) and the measured thickness of the graphene nanosheets (GNSs) was found to be ~1.1 nm.

Several other reports are also found based on graphene functionalization with poly (m-phenylenevinylene-co-2,5-dioctoxy-p-phenylenevinylene) (PmPV) (Li, Wang, et al. 2008), 1,2-distearoyl-sn-glycero-3-phosphoethanolamine-N[methoxy

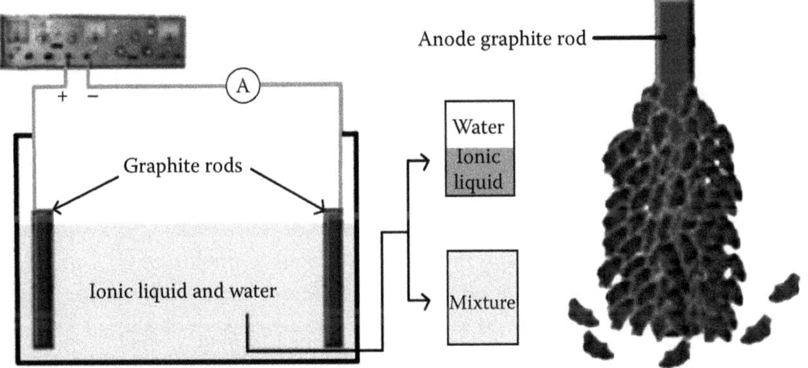

FIGURE 2.10 Schematic diagram of the electrochemical graphene synthesis process. (From N. Liu, F. Luo, H. X. Wu, Y. H. Liu, C. Zhang, and J. Chen. One-step ionic-liquid-assisted electrochemical synthesis of ionic-liquid-functionalized graphene sheets directly from graphite. *Advanced Functional Materials* 18, no. 10 (2008):1518–1525. With permission.)

(polyethyleneglycol)-5000] (DSPE-mPEG) (Li, Zhang, et al. 2008), poly(tert-butyl acrylate), and so on. In view of technological applications on the device level, several recent reports are also found based on the poly(N-vinyl pyrrolidone) graphene nanocomposite for humidity sensing (Zhang, Shen, et al. 2010), GO-polymer for organic solar cells, (Li, Tu, et al. 2010), dye-sensitized solar cells (Hasin, Alpuche-Aviles, and Wu 2010; Roy-Mayhew et al. 2010), organic memory devices (Li, Liu, et al. 2010), Li ion batteries (Chen and Wang 2010), and so on. Apart from the stable dispersion of graphene with different polymers and surfactants, some recent reports are also available based on the modification of graphene with inorganic nanoparticles like Au (Muszynski, Seger, and Kamat 2008), TiO_2 (Qian et al. 2009; Yang et al. 2010), Fe_3O_4 (Zhou et al. 2010), and CuO (Zhu et al. 2010).

Chemical synthesis of graphene has several advantages such as a low temperature process, and therefore could be readily processed on any substrate with much more flexibility. In situ, functionalized graphene with different functional groups can be easily synthesized via this route for chemical and biological applications. Further, the process is low in cost as graphite is abundant in nature (natural graphite supplies worldwide are estimated at 800M tons). In contrast, the chemical syntheses of graphene have several disadvantages such as small yield, defective graphene, and partially reduced GO, which readily deteriorates the properties of graphene. Moreover, the process involves too many tedious steps and uses hazardous explosive chemicals like hydrazine. During chemical reduction of GO, incomplete reduction dictates the possible deterioration of conductivity, charge carrier concentration, carrier mobility, and so on. Finally, graphene produced by chemical methods is not superior in grade compared to other methods, and therefore needs further development for applications in chemical and nano-biotechnology, nano-medicine, and so on.

2.2.5 Direct Chemical Synthesis: Pyrolysis of Sodium Ethoxide

All chemical synthesis processes described above are top-down approaches as the process involves the oxidation of bulk graphite, exfoliation of GO, and then reduction back to graphene. In contrast, a bottom-up approach of chemical synthesis of graphene, called the *solvothermal method*, is introduced (Choucair, Thordarson, and Stride 2009). In this method, laboratory-grade ethanol and sodium were used as starting materials to synthesize sodium ethoxide, followed by pyrolyzation, which yields a fused array of graphene sheets that can be easily dispersed using mild sonication. The solvothermal reaction involves a reaction of 1:1 molar ratio of sodium (2 g) and ethanol (5 ml) in a sealed reactor vessel at 220°C for 72 h, resulting in a yield of sodium ethoxide, which was used as a graphene precursor for further reaction. The resultant solid (sodium ethoxide) was rapidly pyrolyzed, vacuum filtered, and dried in a vacuum oven at 100°C for 24 h. The process yield was 0.1 g / 1 ml of ethanol, typically yielding ~0.5 g per reaction. Raman spectroscopy of the resultant sheet showed a broad D-band at 1353 cm^{-1}, a G-band at 1590 cm^{-1}, and the intensity ratio of I_G/I_D ~1.16, representative of defective graphene. Finally, a 4±1 Å thickness of graphene was obtained by this process, representing a mono-atomic graphene sheet. The advantage of this process is a low-cost and bottom-up process that can be further extended to the more controlled fabrication of high-purity, functionalized graphene. Moreover, it is a scalable, low-temperature process, which is an added advantage of the direct bottom-up chemical synthesis methods yielding high-purity graphene. However, the quality of graphene is still not satisfactory because it contains a large number of defects.

2.2.6 Unzipping of Nanotubes

A new graphene synthesis process was proposed that involves unzipping a carbon nanotube by using a chemical and plasma-etched method. The unzipping of carbon nanotubes (CNTs) yields a thin elongated strip of graphene that exhibits straight edges, called a graphene nanoribbon (GNR). Graphene, when narrowing along the width, deliberately transforms its electronic state from semimetal to semiconductor (Chen et al. 2007). Therefore, the electronic properties of thin strip graphene nanoribbons are presently under vigorous investigation (Jiao et al. 2010; Jiao et al. 2009). Depending upon whether the starting nanotube is multiwalled or single walled, the final product will be multilayered graphene or single-layer graphene, respectively.

Cano-Marquez et al. (2009) demonstrated a novel chemical route of longitudinal unwrapping of multiwalled carbon nanotubes (MWNTs) by intercalation of lithium (Li) and ammonia (NH_3) followed by exfoliation. They used CVD-grown MWNTs dispersed in dry tetrahydrofuran (THF) followed by adding liquid NH_3 (99.95%), while maintaining the dry ice bath temperature of −77°C. Li was added with the ratio of 10:1 (Li:C) and allowed the intercalation of MWNT to occur for a few hours. Subsequently, HCl was mixed in the solution containing intercalated MWNTs, which further facilitate complete exfoliation. They proposed a mechanism for intercalation that was initiated by electrostatic attraction between the negatively charged MWNTs and NH_3-solvated Li^+. The exothermic reaction occurs when the HCl reacts with Li ions and simultaneous neutralization of NH_3 causes further unwarping of the nanotubes. Some of the unexfoliated or partially exfoliated nanotubes were also obtained, which could be further exfoliated by thermal treatments. The process yields ~60% fully exfoliated MWNTs including a very small amount (0-5%) of partially exfoliated MWNTs. At the same time, Tour research group demonstrated the unzipping of nanotubes by a different chemical process. They reported the opening of the side walls of CNTs by a step-by-step solution-based oxidation process using H_2SO_4, $KMnO_4$, and H_2O_2 (Kosynkin et al. 2009). The successive increase in $KMnO_4$ (oxidizing agent) concentration (100% to 500%) in the solution resulted in a larger degree of opening the consecutive MWNT layers. However, the resultant product was an oxidized GNR, which needed a further reduction step using 1 vol% concentrated ammonium hydroxide (NH_4OH) and 1 vol% hydrazine monohydrate (N_2H_4, H_2O) in order to restore its electrical properties. Furthermore, the starting MWNT diameter was 40–80 nm, hence the thickness of GNR was increased to >100 nm after it was unwrapped, whereas the length of the GNRs was equivalent to the initial length of the MWNTs (~4 mm). The authors also demonstrated the unraveling of single-walled nanotubes (SWNTs) (average height of the SWNT was ~1.3 nm), which produced a narrow entangled GNR with a decreased average thickness of ~1 nm.

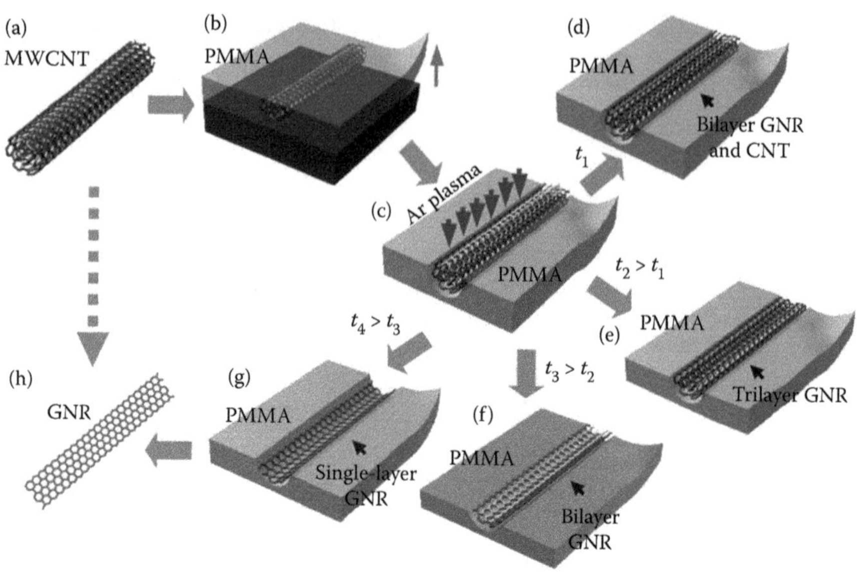

FIGURE 2.11 A process flow chart of graphene nanoribbon fabrication from a carbon nanotube by the plasma etching process. (From L. Jiao, Y., L. Zhang, X. R. Wang, G. Diankov, and H. J. Dai. Narrow graphene nanoribbons from carbon nanotubes. *Nature* 458, no. 7240 (2009):877–880. With permission.)

Jaio et al. (2009) reported a rather simplified and facile technique, which they called the *controlled unzipping technique*. The step-by-step fabrication process from nanotube to nanoribbon is shown in Figure 2.11. A pristine MWNT (dia. ~4–18 nm) suspension was deposited onto a Si substrate pretreated with 3-aminopropyltriethoxysilane. A polymethylmethacrylate (PMMA) solution was spin-coated with MWNTs on the substrate followed by baking at 170°C for 2 h. The PMMA-coated MWNT film was peeled off using 1M KOH solution at 80°C. After that, using Ar-plasma (10 watt, 40 mTorr), MWNT walls were etched away followed by removal of PMMA in acetone vapor. The average diameter of MWNTs was ~6–12 nm, which produced resultant GNRs having a width of 10–20 nm after plasma etching. The step heights of the resulting GNRs were 0.8 to 2.0 nm, which were representative single- to few-layer GNRs, respectively. They claimed that the process easily produces high-quality GNRs with different numbers of layers while maintaining high process yield <40% (single- and few-layer GNRs). In this context, the yield of single- to few-layer GNRs also depends on the diameter, number of concentric tubes of MWNTs, and the plasma etching time.

Simultaneously, several recent reports showed the future prospects of GNRs in the field of nanoelectronics, but further control over the fabrication process is required to achieve a high-purity, defect-free controlled synthesis process for scalable device fabrication.

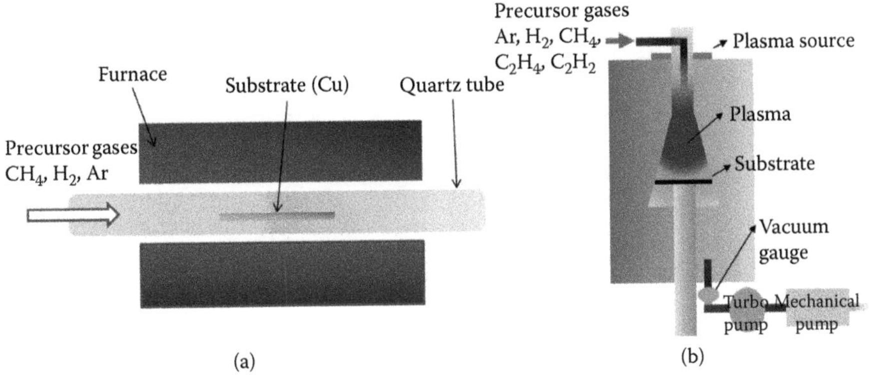

FIGURE 2.12 Schematic of (a) thermal CVD and (b) plasma-enhanced CVD (PECVD).

2.2.7 Thermal Chemical Vapor Deposition Process

Thermal chemical vapor deposition (CVD) is a chemical process by which a substrate is exposed to thermally decomposed precursors and the desired product deposited onto the substrate surface at high temperature. Because the high temperature is not desired in many cases, plasma-assisted decomposition and reaction may lower the process temperature. Figures 2.12(a) and 2.12(b) demonstrate the schematic of thermal and plasma-enhanced CVDs (PECVD), respectively.

There are numerous advantages to the thermal CVD process. The process yields high quality and high purity final products in large scale. Moreover, by controlling the CVD process parameters, control over the morphology, crystallinity, shape, and size of the desired product is possible. On the other hand, by applying a wide range of solid, liquid, and gaseous precursor materials, a large variety of nanomaterials and thin films are executable with this process. However, tailoring very high precision atomic-level properties using thermal and plasma CVD is still under investigation.

2.2.7.1 Thermal Chemical Vapor Deposition

Deposition of mono-layer graphitic materials on Pt by thermal CVD was first reported in 1975 by Lang et al. (1975). They found that the decomposition of ethylene onto platinum results in the formation of a graphitic overlayer and surface rearrangements of the substrate. Later, Eizenberg and Blakely (1979) reported graphite layer formation on Ni (111). The process involved the doping of single-crystal Ni (111) with carbon at an elevated temperature of 1200–1300 K for a significant period of time (~1 week), followed by quenching. The carbon phase condensation on Ni (111) was found with detailed thermodynamic analysis and the carbon phase segregation on Ni (111) is solely dependent upon the rate of quenching.

Since then (almost two decades), the field has not been explored further due to an inability to find possible applications of thin graphite film as a semiconductor, transparent conductor, and so on. In the early twenty-first century, discovery of graphene

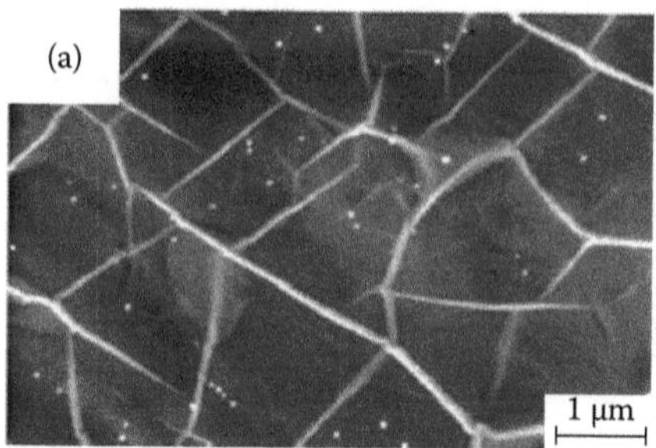

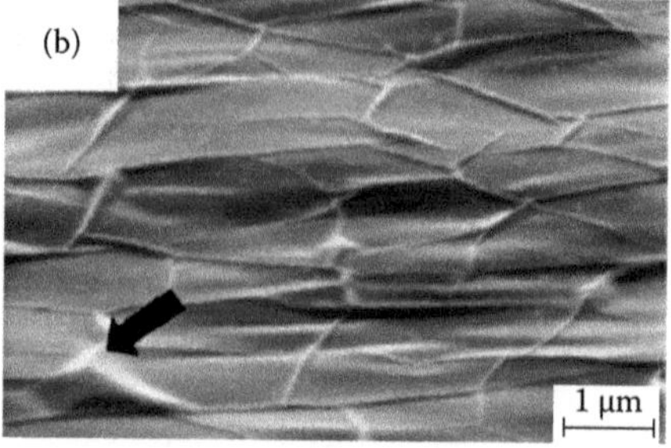

FIGURE 2.13 Scanning electron micrograph of graphene syntheses on Ni (111) by the DC discharge method. (From A. N. Obraztsov. Chemical vapour deposition: Making graphene on a large scale. *Nature Nanotechnology* 4, no. 4 (2009):212–213; and A. N. Obraztsov, E. A. Obraztsova, A. V. Tyurnina, and A. A. Zolotukhin. Chemical vapor deposition of thin graphite films of nanometer thickness. *Carbon* 45, no. 10 (2007):2017–2021. With permission.)

created a research boom because the unusual properties of thin monolayer or few-layer graphite films have been explored, attracting tremendous attention from the scientific and industrial communities. Furthermore, the physics and chemistry of graphene have been meticulously scrutinized to open a new area of graphene-based electronics (Novoselov et al. 2004; Katsnelson 2007; Geim and Kim 2008; Dreyer et al. 2010).

In 2006, the first attempt at graphene synthesis on Ni foil using CVD was done using camphor (terpinoid, a white transparent solid of chemical formula $C_{10}H_{16}O$) as the precursor material (Somani, Somani, and Umeno 2006). In this reference, graphene synthesis was carried out in a two-step process—camphor deposition on Ni foil at ~180°C followed by pyrolyzation at 700–850°C in Ar atmosphere. Using TEM, they found that the planar few-layer graphene consists of ~35 layers of stacked single graphene sheets with an interlayer distance of 0.34 nm. The study presents a new path toward large-scale graphene growth using thermal CVD. Nevertheless, large-scale mono or bilayer graphene growth using thermal CVD was still in demand until Obraztsov et al. (2007) reported the deposition of thin-layer graphite on Ni. Using DC discharge in a thermal CVD system, a thin layer (1–2 nm) of graphene was deposited at 40–80 mT pressure and 950°C. They used a gas mixture of H_2:CH_4 = 92:8 as a precursor under DC discharge of the current ~ 0.5 A/cm^2. As shown in Figure 2.13, the final graphene on Ni was ~1- to 2-nm-thick few-layer graphene covered with surface ridges, which they explained as ridge formation due to the thermal expansion coefficient mismatch between graphene and the Ni substrate. In particular, well-ordered few-layer graphene was found on the Ni surface; however, the same process was not able to yield well-ordered graphene on Si except amorphous carbon.

In 2008, Pei et al. (see Yu et al. 2008) demonstrated high-quality graphene formation on polycrystalline Ni during thermal CVD of methane. They fabricated few-layer graphene at 1000°C using CH_4: H_2: Ar =0.15:1:2 with a total gas flow rate of 315 sccm (standard cubic centimeters per minute) under ~1 atmospheric pressure. From high-resolution TEM (HRTEM) study (Figure 2.14(a)), they confirmed the formation of 3–4 layers of graphene on Ni and suggested that the graphene

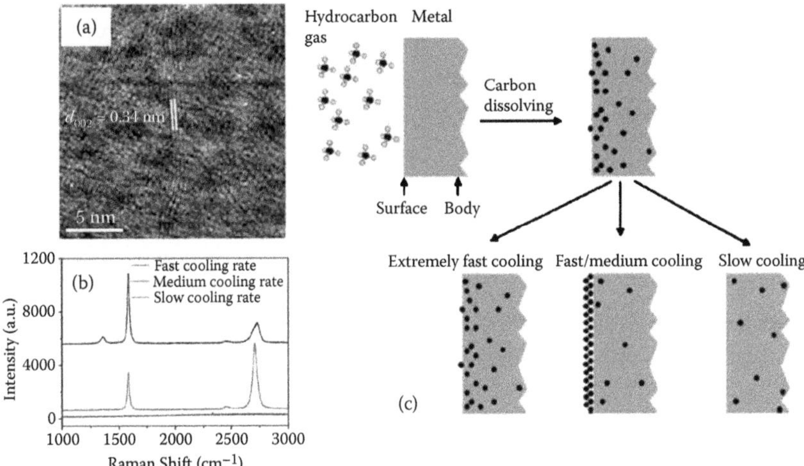

FIGURE 2.14 (a) HRTEM of graphene precipitated on Ni, (b) Raman spectra confirming the effect of cooling rate on graphene formation, and (c) schematic representing the mechanism of carbon segregation on Ni. (From Q. K. Yu, J. Lian, S. Siriponglert, H. Li, Y. P. Chen, and S. S. Pei. Graphene segregated on Ni surfaces and transferred to insulators. *Applied Physics Letters* 93, no. 11 (2008). With permission.)

formation is due to the segregation of carbon on Ni. They also reported that the cooling rates (fast ~20°C/s, medium ~10°C/ s, and slow ~0.1°C/s) significantly affect the formation of different numbers of graphene layers, which was further confirmed by Raman spectroscopy (as shown in Figure 2.14(b)). However, large-scale monolayer graphene production using thermal CVD was still in demand until an interesting report in 2009 (Kim et al. 2009) regarding large-scale patterned growth of graphene films as stretchable transparent electrodes. Kim et al. (2009) demonstrated graphene growth over e-beam evaporated Ni followed by thermal CVD of CH_4:H_2: Ar ~550:65:200 at 1000°C. Further, they transferred the graphene over a flexible, stretchable polydimethylsiloxane (PDMS) substrate by a wet chemical process. The transferred graphene over polymer substrate exhibits sheet resistance of ~280 ohm/sq with more than 80% transmittance at the visible light wavelength region. The film on SiO_2/Si substrate showed charge carrier mobility of ~3750 cm^{2}V^{-1}s^{-1} and carrier density of 5 x 10^{12} per cm^{-2}.

Subsequently, Reina et al. demonstrated the formation of single- to few-layer graphene on polycrystalline Ni and transfer of graphene to any substrate by the wet-etching method (Reina, Jia, et al. 2009). They used e-beam evaporated Ni on a SiO_2/Si substrate, annealed it at 900–1000°C under a flow of Ar and H_2 for 10–20 minutes, followed by graphene growth using diluted hydrocarbon gas flow at 900–1000°C and ambient pressure. They reported the formation of 1 to 10 layers of graphene over the Ni surface as shown in HRTEM in Figures 2.15(a), (b), and (c) and Raman spectroscopy in Figure 2.15(d). A wet chemical method was applied to transfer the CVD graphene onto any substrate. Wang et al. (2009) attempted a new approach of gram-scale graphene production in a thermal CVD at 1000°C by using a

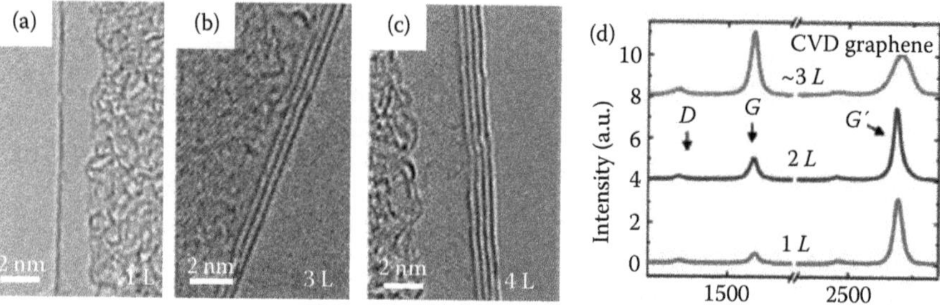

FIGURE 2.15 (a), (b) and (c) HRTEM images confirming the formation of thermal CVD-grown mono- to few-layer graphene on Ni and (d) representative Raman plot for successive layers of graphene. (From A. Reina, X. T. Jia, J. Ho, et al. Large area, few-layer graphene films on arbitrary substrates by chemical vapor deposition. *Nano Letters* 9, no. 1 (2009):30–35. With permission.)

Co-supported MgO catalyst in a ceramic boat and a precursor gas mixture of CH_4:Ar (1:4 volume ratio, total 375 ml/min flow rate). Finally, graphene was obtained after the catalyst particles were washed away with concentrated HCl; the process yield was 0.05 gm from 500 mg of catalyst powder. The authors claimed that the process is unique because it is a substrate-free, gram-scale yield of graphene and a low-cost process. However, the graphene produced by this method is 5-layered graphene sheets, which are crumpled and randomly aggregated.

Large-scale graphene synthesis took a unique direction when the Ruoff group discovered that catalytic graphene deposition on Cu occurs at elevated temperature, by decomposition of hydrocarbon gases (Li, Cai, An, et al. 2009). They showed single-layer uniform large-scale (1 cm^2) graphene growth on Cu foil by the thermal CVD technique. This method involves heating the quartz tube furnace to ~1000°C in a hydrogen atmosphere (~2 sccm flow) under an ambient pressure of 40 mTorr, annealing the Cu film at 1000°C, and injecting of 35 sccm of methane gas (CH_4) at ~500 mTorr ambient pressure. Further, they developed a graphene transfer method by solution etching of Cu and then transferring of the floated graphene onto any substrate. The process yields single-layer, two-layer, and three-layer graphene, which was further confirmed by HRTEM and Raman spectroscopy. The Ruoff group (Li, Cai, An, et al. 2009), also explained that the graphene deposition on a Cu surface is due to the surface catalyzed process associated with the limited solubility of carbon in copper. Therefore, the graphene precipitation mechanism on a Cu surface exclusively differs from the Ni surface; graphene deposition on Cu occurs due to the segregation process or surface adsorption process.

A breakthrough occurred in the field of large-scale graphene synthesis process when Cu foil (~15 cm × 5 cm) was rolled up and placed in a 2 in. quartz tube furnace in order to grow graphene and then transfer the graphene onto any flexible polymer substrate using the hot press lamination method. The Choi group at Florida International University (Verma et al. 2010) reported large-area graphene growth as large as a 15 cm × 5 cm rectangular Cu foil using the thermal CVD technique as shown in Figures 2.16(a) and 2.16(b). Verma and co-workers deposited graphene at 1000°C with a mixture of H_2: CH_4 (1:4) at ambient atmospheric pressure, and

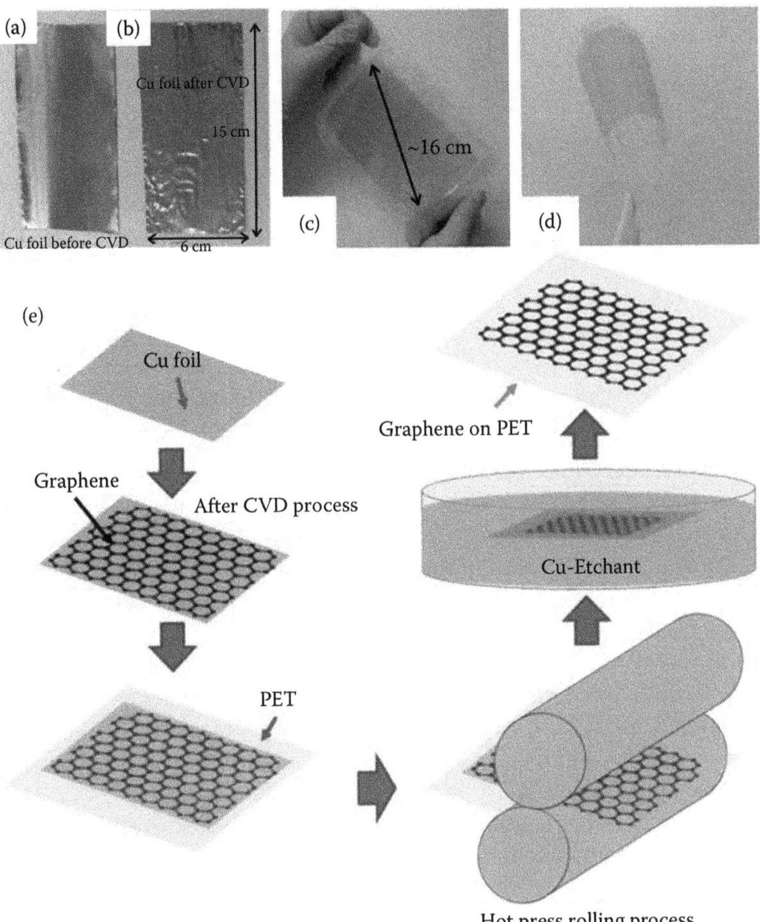

FIGURE 2.16 (a) and (b) Picture illustrates the large-scale before and after CVD of Cu foil, (c) and (d) demonstrate large-scale fabrication of graphene film on flexible PET, and (e) the hot press lamination process for graphene-PET film fabrication. (From V. P. Verma, S. Das, I. Lahiri, and W. Choi. Large-area graphene on polymer film for flexible and transparent anode in field emission device. *Applied Physics Letters* 96, no. 20 (2010). With permission.)

then graphene was transferred using a hot press lamination process (as shown in Figure 2.16(e)), which was proficient as well as industrially scalable. Their work demonstrated large-scale graphene on flexible film (as shown in Figures 2.16(c) and 2.16(d)) as transparent conducting anodes that can be used as a current collector in a flexible transparent field emission devices.

Bae et al. (2010) reported the roll-to-roll production of graphene on a flexible polymer with an area as large as 30 inches as shown in Figure 2.17(a) and used as a touch screen panel. They grew graphene using thermal CVD on Cu foil and the processing steps are as follows: (1) annealing of Cu in a thermal CVD in a H_2 environment (under ~90 mT pressure) at 1000°C, (2) processing of graphene growth at

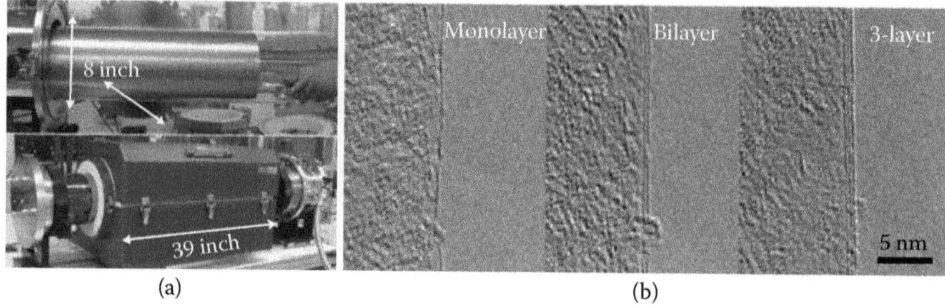

FIGURE 2.17 (a) Large-scale CVD of graphene on Cu foil; and (b) HRTEM images demonstrating the growth of single-layer, two-layer, and three-layer graphene on Cu. (From S. Bae, H. Kim, Y. Lee, et al. Roll-to-roll production of 30-inch graphene films for transparent electrodes. *Nature Nanotechnology* 5, no. 8 (2010):574–578. With permission.)

1000°C using the precursor gas mixture of CH_4 and H_2 (under the flow rates of 24 and 8 sccm, respectively) for 30 minutes at ~460 mT ambient pressure, and (3) cooling of the furnace at ~10°C/minutes under H_2 flow at 90 mT pressure. Formation of different layers of graphene was confirmed using HRTEM (as shown in Figure 2.17(b)) and Raman spectroscopy, though the predominant area was found to be covered with monolayer graphene.

Several recent reports were also found demonstrating graphene fabrication on Ni foil, polycrystalline nickel thin film, patterned Ni thin film (Reina, Thiele, et al. 2009), Cu foil, Cu thin film, patterned Cu thin film, and many other different transition metal substrates. Similarly, a few novel approaches are also recently reported on CVD graphene (De Arco et al. 2010; Gao et al. 2010; Sun et al. 2010). A CVD-grown graphene nanoribbon has also been reported (Campos-Delgado et al. 2009; Campos-Delgado et al. 2008). In principle, all these thermal CVD processes are related to the processes discussed previously except that the ratio and total gas flow differ. The rate and total gas flow in those processes correspond to the total volume of the furnace, ambient pressure of the furnace, and temperature.

Graphene growth directly onto any insulator substrate was first reported in 2010 by Ismach et al. (2010). In this process, they deposited Cu thin film on dielectric substrates by e-beam evaporation and then using thermal CVD, they grew graphene at 1000°C under 100–500 mT ambient pressure. Graphene precipitation on dielectric surfaces occurs due to the surface catalyzed process of Cu and the copper films de-wetted and evaporated from the surface, which leads to direct graphene deposition on dielectric substrates as shown in Figure 2.18. Further, the CVD process leads to the formation of well-crystallized graphene on the dielectric surfaces that contain low defects and exhibit thicknesses of 0.8–1 nm corresponding to 1- to 2-layer graphene. Later, Rümmeli et al. (2010) reported another technique to deposit graphene on insulator substrates using thermal CVD at a significantly low temperature. The low-temperature CVD process produced several-nanometer-thick to a few hundred-nanometer-thick graphene directly on magnesium oxide (MgO) nano crystal powder using cyclohexane, acetylene, and argon as a feedstock gas mixture at temperatures between 875°C and 325°C. The domain size of graphene

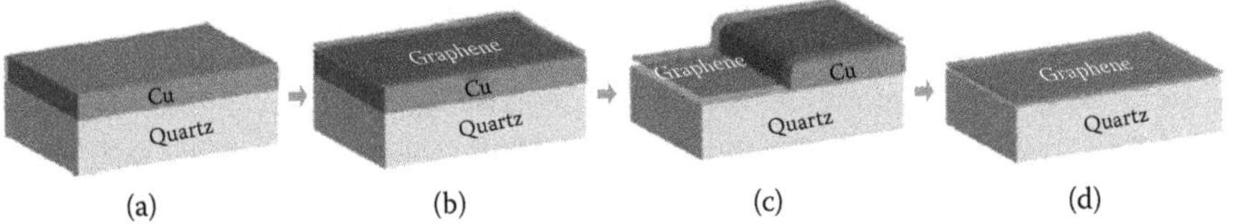

FIGURE 2.18 Schematic illustration of the CVD graphene formation mechanism on a direct dielectric substrate. (From A. Ismach, C. Druzgalski, S. Penwell, et al. Direct chemical vapor deposition of graphene on dielectric surfaces. *Nano Letters* 10, no. 5 (2010):1542–1548. With permission.)

was found to be from a few hundred nanometers to several micrometers yielding few-layer graphene.

Therefore, direct synthesis of graphene on dielectric substrates bypasses the post-synthesis graphene transfer process, hence avoiding the inclusion of defects and contamination in graphene. Furthermore, low-temperature graphene synthesis by CVD will be highly promising for easy device integration to reinstate present Si-based CMOS technology. Nonetheless, graphene deposition on dielectric surfaces needs to be explored further to achieve well-ordered, large-scale, defect-free graphene for electronic applications, and most importantly, to achieve band gap engineering capabilities.

2.2.7.2 Plasma-Enhanced Chemical Vapor Deposition

When a thermal CVD process involves chemical reactions of the reacting gases by generating plasma* inside a vacuum chamber, which leads to the deposition of thin film on the substrate surface, the process is known as plasma-enhanced chemical vapor deposition (PECVD). A schematic is shown in Figure 2.12b illustrating the PECVD chamber for graphene synthesis. The plasma can be generated inside a PECVD system using RF (AC frequency), microwave, and inductive coupling (electrical currents produced by electromagnetic induction). PECVD has a wide range of advantages over other conventional CVD methods. By this technique, a process can be carried out at relatively low temperature compared to the other thermal CVD processes; hence, it is more feasible for industrial-scale applications. Moreover, catalyst-free graphene growth (Shang et al. 2008) can be carried out by controlling the process parameters, which normally influence the properties of the final graphene product. However, the process is costly and only gas-phase precursor materials can be used in order to obtain the final products, which limits its applications for synthesis of a wide range of industrial products.

Synthesis of a thin graphitic layer using PECVD was first demonstrated by Obraztsov et al. (2003) by DC discharge CVD of a gas mixture containing CH_4 and H_2 (0–25% CH_4), at 10 to 150 Torr. In this report, a Si wafer and different metal sheets of Ni, W, and Mo were used as substrates for nanocrystalline graphite (NG) growth. Using optical emission spectra embedded in the PECVD system, they

* Plasma is an ionized state matter resembling a gas.

observed the presence of activated H, H_2, CH, and C species in the DC discharge plasma. They concluded that the presence of C_2 dimers in CH_4 plasma played a significant role in graphitic nanostructure formation. However, the graphene produced by this process was much thicker than single- to few-layer graphene. Similarly, Wang et al. (Wang, Zhu, Outlaw, Zhao, Manos, and Holloway 2004; Wang, Zhu, Outlaw, Zhao, Manos, Holloway, et al. 2004) attempted to deposit single- to few-layer graphene using PECVD on different substrates like Si, SiO_2, Al_2O_3, Mo, Zr, Ti, Hf, Nb, W, Ta, Cu, and 304 stainless steel. Using 900-watt RF power, 10 sccm total gas flow, and inside chamber pressure of ~12 Pa, they tried to deposit graphene by varying methane concentrations (5–100%) at different temperatures from 600–900°C. The typical deposition time was found to be from 5 minutes to 40 minutes. The free-standing graphene growth rates varied with the varying methane concentration and substrate temperature. It was also observed that the increase in % CH_4 concentration and substrate temperature resulted in an increase in the graphene growth rate. Finally, using HRTEM a corrugated few-layer graphene nanosheet with a thickness of ~1–2 nm was found with a dark folded ridge and fringes at the edge consisting of two graphene sheets.

Zhu et al. further reported the synthesis of a vertical free-standing graphene sheet (thickness ~1 nm) by an inductively coupled RF PECVD system on a variety of catalyst-free substrates. They carried out carbon nanotube (CNT) growth and vertical free-standing graphene growth using inductively coupled plasma and found that, depending upon the hydrocarbon and hydrogen concentration in the feed gas mixture, CNT and carbon nanosheet (CNS) growth occurred. According to them, increasing concentration of hydrocarbon and hydrogen gases in the precursor gas mixture causes graphene growth because of the larger accumulation of activated carbon species and the increase in vertical electric field intensifies the surface species diffusion.

In this context, several other reports were also found on graphene synthesis using microwave plasma (Yuan et al. 2009; Vitchev et al. 2010), PECVD graphene synthesis at atmospheric pressure (Jasek et al. 2010), petal-like graphene structures (Qi et al. 2009; Bhuvana et al. 2010), N-doped graphene-CNT hybrid structures (Lee et al. 2010), and so on. The PECVD process can only produce vertically oriented graphene, which has not yet been demonstrated using other graphene synthesis methods. The PECVD method produced high-purity and high-crystalline graphene; however, uniform large-area, single-layer graphene production using this method is still under vigorous investigation. Further studies on the PECVD process are needed to gain better control over graphene layer, morphology, growth rate, and height (only for vertically oriented graphene).

2.2.8 Epitaxial Growth of Graphene on SiC Surface

Epitaxial thermal growth on a single crystalline silicon carbide (SiC) surface is one of the most acclaimed methods of graphene synthesis and has been explored vigorously for the last 7 to 8 years. The term *epitaxy* can be defined as a method that allows deposition of a single crystalline film on a single crystalline substrate. The deposited film is referred to as *epitaxial film* or the *epitaxial layer* over the single crystalline substrate and the process is known as *epitaxial growth*. Epitaxial graphene growth is a process to fabricate high-crystalline graphene onto single-crystalline SiC substrates. When the film deposited on a substrate is of the same material it is known as a *homo-epitaxial layer*, and if the film and substrate are different materials it is called a *hetero-epitaxial layer*. For example, few-layer graphite or graphene formation on SiC is known as a hetero-epitaxial layer, which will be discussed in this section.

Investigations on the electronic properties of graphene have taken two directions. One direction is related to device fabrication based on exfoliated graphene on different substrates and the other is wafer-scale synthesis of epitaxial graphene, which is the most feasible and scalable approach to graphene-based electronics. Extensive conscientious research areas have been explored such as, (1) the electronic properties of graphene on SiC, (2) band gap formation, (3) graphene growth mechanism, and (4) graphene–SiC interfaces, focusing on pathways toward large-scale graphene-based electronics.

In 1975, Bommel et al. (Van Bommel, Crombeen, and Van Tooren 1975) first reported graphite formation on both the 6H–SiC(0001) and $(000\bar{1})$ surfaces. The work showed that heat treatment in the range of 1000–1500°C in an ultrahigh vacuum (~10^{-10} m bar) produced graphite on both of the SiC polar planes (0001) and $(000\bar{1})$. The crystal structure, crystal plane orientations, and crystal plane stacking information of SiC are described in detail elsewhere (Hass, de Heer, and Conrad 2008). In 2004, de Heer's group reported the fabrication of ultrathin graphite consisting of 1–3 mono-atomic graphene layers on the Si terminated (0001) face of single-crystal 6H-SiC and analyzed its electronic properties (Berger et al. 2004). Their detailed experimental procedure consisted of (1) surface preparation using oxidation or H_2 etching, (2) surface cleaning using electron bombardment at 1000°C at ~10^{-10} Torr pressure, and (3) heat treatment of the samples at 1250–1450°C for 1–20 minutes. In 2007, de Heer et al. reported the fabrication of an epitaxial graphene layer down to 1–2 layers on the (0001) face of a 6H–SiC wafer using the thermal decomposition method (de Heer et al. 2007). They also developed a quantitative characterization method using x-ray photoelectron spectroscopy (XPS) and angle-resolved photoemission spectroscopy (ARPES) to measure the number of epitaxial graphene layers formed on the SiC substrate. Similarly, a few recent studies have also demonstrated the exceptionally high-quality graphene production on the 4H–SiC C-face in an RF furnace under pressure ~10^{-4} to 10^{-3} Torr. After removal of surface oxides by preheat treatment at 1200°C, the epitaxial graphene formation occurred at 1420°C, which was significantly higher than the normal UHV graphitization on SiC. However, the growth rate on 4H–SiC was relatively higher than on the other SiC surface; therefore, it is difficult to produce very thin layers of graphene as well as to control the number of graphene layers by changing time and temperatures. In general, 4- to 5-layer graphene film is obtained at 1420°C within ~6 minutes by the thermal decomposition of SiC.

In this context, graphene growth on a Ni thin film-coated SiC substrate at a quite lower temperature was reported by Juang et al. (2009). The authors elucidated low-temperature graphene growth on a SiC substrate coated with Ni thin film. The process consisted of 200 nm Ni thin film deposition over single-crystalline 6H–SiC (0001) and 3C–SiC substrates followed by the growth process at ~750°C under ~10^{-7} Torr pressure. Graphene was grown over the continuous Ni thin film surface as well as on the patterned Ni thin film surface. The process of graphene formation at such a low temperature was reported as follows: (1) rapid heating causes the dissolution of Ni into the SiC at the Ni–SiC interface, (2) formation of a nickel silicide/carbon mixed phase then occurred, (3) followed by the diffusion of carbon atoms into the Ni thin film matrix (due to the low solubility of C in Ni), and (4) finally the carbon atoms segregated onto the surface of the Ni during the cooling process in the same manner as thermal CVD (explained in Section 2.3). The process was delineated as a versatile large-scale facile method because it is a comparatively lower-temperature process than others. Further, the graphene obtained by this process is easily transferrable to any substrate for other applications. The feasibility of graphene-based 2-D electronics is based on fabricating wafer-scale graphene with controlled thickness, width, and specified crystallographic orientations in order to achieve control over the electronic properties of graphene. The research on epitaxial graphene on SiC attracted huge attention both academically and industrially due to its scalability and production of high-quality graphene. The major advantage of this process is large-scale fabrication of graphene on an insulator or semiconductor surface, which can be used for future CMOS-based electronics (Sutter 2009). Moreover, epitaxial multilayered graphene over the SiC substrate behaves as an isolated graphene, which would be an added advantage in applications in graphene-based nanoelectronics. Nevertheless, the high growth temperature and very low process pressure are the major disadvantages of this process. Further, the final epitaxial graphene produced by this thermal decomposition method exhibits smaller grain size. However, a novel procedure for atmospheric pressure graphene synthesized on 6H–SiC (0001) with much larger domain size has been reported recently (Emtsev et al. 2009). Although high-quality superior-grade epitaxial graphene formation was reported, the graphene transfer from SiC to any other substrate is difficult, which seriously limits the versatility of the process in a wide range of electronic applications.

2.3 GRAPHENE GROWTH MECHANISM

At present, the CVD process is one of the major graphene fabrication processes, which involves carbon deposition on the transition metal surface by decomposition of hydrocarbon gases at an elevated temperature under low or ambient pressure. In the thermal CVD process, at high temperature, hydrocarbon gases are decomposed by reacting with hydrogen, which leads to the formation of carbon atoms; when deposited on the metal surface, these carbon atoms segregate and form single- or few-layer graphitic sheets called *graphene*. In the PECVD process, the decomposition reaction occurs in the presence of plasma and carbon deposition, and segregation takes place at a comparatively low substrate temperature. Therefore, PECVD is known as a lower-temperature process than conventional thermal CVD. Further, the rate of decomposition reaction is solely dependent upon the power of the plasma source and the rate of carbon ion deposition on the substrate. However, both processes have several common parameters like time, temperature, pressure, gas flow rates, and type of catalyst, which play important roles in graphene formation. Graphene segregation on Ni is also dependent upon the postdeposition cooling rate of the process, which tailors the morphology as well as the final properties of the graphene. Furthermore,

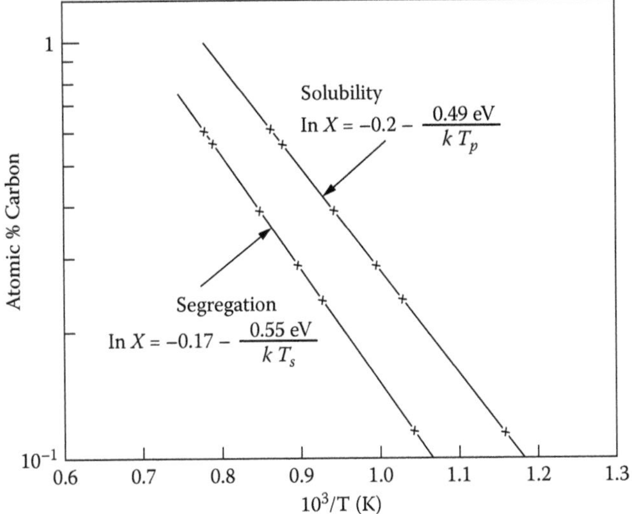

FIGURE 2.19 Solubility curve of C in Ni. (From M. Eizenberg and J. M. Blakely. Carbon monolayer phase condensation on Ni(111). *Surface Science* 82, no. 1 (1979):228–236.) Adapted from W. W. Dunn, R. B. McLellan, and W. A. Oates. *Transactions of AIME* 242(1968):2129.

the crystallinity (whether polycrystalline or single crystalline) of the metal also significantly affects graphene formation and grain size, and surface roughness also causes the formation of homogeneous graphene film. The detailed effects of the major parameters of graphene growth are described in the following text as per the current research reports.

Detailed thermodynamic calculations from the data (Eizenberg and Blakely 1979) related to C solubility in Ni is shown in Figure 2.19, where it was clearly stated that the lower % of surface atomic carbon condensed at a higher temperature rather than in low temperatures. The solubility curves were plotted from Equations (2.2) and (2.3) as follows:

$$\text{Solubility: } \ln x = -0.2 - \frac{(0.49\text{ eV})}{kT_P} \tag{2.2}$$

$$\text{Segregation: } \ln x = -0.17 - \frac{(0.55\text{ eV})}{kT_s} \tag{2.3}$$

Here,

k = Boltzmann constant

x = Atomic % carbon

T_S and T_P = the high and low temperatures, respectively (i.e., $T_S > T_P$)

As explained by Eizenberg and Blakely (1979), the slope of this curve and its intercept at $1/T = 0$ determines the values of partial atomic heat of segregation (ΔH_{seg})

and entropy of segregation (ΔS_{seg}), respectively. The ΔH_{seg} was found to be –.55 eV, which is ~10% lower than the energy/carbon atom in thick graphite. Therefore, monolayer condensation on Ni (111) is more favorable at high temperatures than in low temperatures. However, the entropy of segregation ΔS_{seg} does not show much difference in values; the researchers therefore concluded that monolayer and bulk graphite have the same degree of disorder.

Graphene formation on metal surfaces occurs due to the surface catalyzed process during chemical vapor deposition. To date, graphene formation has been reported on several transition metal surfaces such as Ni (Eizenberg and Blakely 1979), Co (Hamilton and Blakely 1980), Cu (Li, Cai, An, et al. 2009; Li, Cai, Colombo, et al. 2009), Ir (Coraux et al. 2009), Ru (Sutter et al. 2009), Rh (Castner, Sexton, and Somorjai 1978), Pd (Hamilton and Blakely 1980), and Pt (Lang 1975). Both single crystalline and polycrystalline metals were used as substrates for graphene fabrication. At high temperatures (and in the presence of plasma, in case of PECVD), hydrocarbon gases react with hydrogen, decompose, and form carbon. Obraztsov et al. (2003) reported that the DC discharge plasma then precursor gas mixture contains the dimers (C_2), which when deposited onto the substrate, forms surface-adsorbed graphitic layers. Graphitic layers readily nucleate and grow under the exposure of the transition metal surface to the hydrocarbon gas under wide range of ambient pressure conditions (atmospheric pressure-to low pressure to ultra-low pressure).

Ni is a well-known transition metal catalyst that segregates carbon atoms on its surface at high temperature. Therefore, Ni nanoparticles and thin films are widely used as catalysts for carbon nanotube growth using the CVD process. In this context, when Ni is used as a substrate for catalyzed decomposition of hydrocarbon under ambient pressure, it gives rise to ultrathin graphite film condensation over the Ni surface by the segregation mechanism. Eizenberg et al. explained that the graphite formation mechanism is due to carbon phase dissolution and segregation on Ni (111) plane. From low-energy electron diffraction (LEED) spots, they found that the graphitic unit cell has the same dimensions as the nickel unit cell, and thus carbon atoms accumulate on Ni (111) surface epitaxially. Another report described the mechanism of nucleation and growth of graphene layers on a Ni-supported catalyst by in situ TEM analysis along with density functional theory (DFT) calculations. Helveg et al. (2004) reported the mechanics of nucleation and growth of graphene layers on Ni (111) by the dynamic formation and restructuring of mono-atomic step edges at the nickel surface. They described that methane dissociation as well as carbon adsorption are facilitated at the step edges and preferentially at the Ni (111) step sites. They calculated the driving force for graphene formation on Ni (111) is associated with an energy gain of 0.7 eV per C atom. Further, they concluded that the Ni (111) step edges act as energetically preferable growth sites for graphene growth, which is primarily due to the higher binding energy of carbon atoms with those sites as compared to the other sites at the closely packed Ni facets. Furthermore, a recent report demonstrated their work on a comparison of mechanisms of graphene formation on single crystalline and polycrystalline Ni (Zhang, Gomez, et al. 2010). They proposed that formation of monolayer and bilayer graphene on the single crystal Ni surface is more preferable due to its atomically smooth surface and the absence of grain boundaries. However, polycrystalline Ni leads to the formation of a higher percentage of few-layer graphene (3 layers) because of the presence of grain boundaries, which can act as nucleation sites for multilayer growth. Further, micro Raman mapping showed that under the same CVD conditions, formation of monolayer or bilayer graphene on single-crystalline and polycrystalline surfaces is 91.4% and 72.8%, respectively. A recent discovery shows that graphene growth on Ni (111) is also associated with the complex carbide formation at the graphene Ni interface. Lahiri et al. (2011) proposed that the graphene formation mechanism on Ni (111) involves (1) the exchange of Ni and C atoms from the surface confined Ni_2C phase at the interface, and (2) the removal of Ni atoms via an external source of carbon from the Ni_2C phase.

Cu is another transition metal that acts as a catalyst to deposit graphene on its surface by the surface adsorption mechanism rather than by segregation or precipitation like Ni. Ruoff and co-workers first reported the precipitation of graphene on a Cu surface at high temperature by the surface catalyzed process associated with the limited solubility of carbon in copper (Li, Cai, An, et al. 2009; Li, Cai, Colombo, et al. 2009; Li, Magnuson, et al. 2010). Li, Cai, Colombo, et al. (2009) showed that by carbon isotope labeling, one can compare the graphene growth mechanism on Cu and Ni. Therefore, the graphene precipitation mechanism on a Cu surface differs from the Ni surface where deposition occurs due to the carbon segregation process or precipitation mechanism.

Li et al. (2009) proposed the mechanism shown in Figure 2.20, which illustrates the two-step process of graphene formation: (1) segregation and precipitation and (2) surface adsorption or surface-mediated growth. Figure 2.20 illustrates the step-by-step formation of graphene layers by the segregation process on a Ni substrate that consists of (1) the decomposition of CH_4 in the presence of hydrogen at

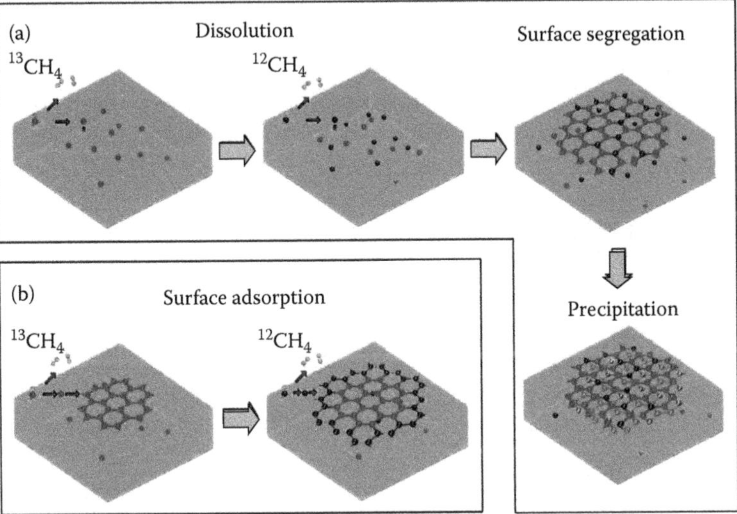

FIGURE 2.20 (a) Graphene formation mechanism by surface segregation and precipitation and (b) mechanism of surface adsorption as reported by Ruoff et al. (From X. S. Li, W. W. Cai, L. Colombo, and R. S. Ruoff. Evolution of graphene growth on Ni and Cu by carbon isotope labeling. *Nano Letters* 9, no. 12 (2009):4268–4272. With permission.)

elevated temperatures, (2) dissolution of carbon atoms in a metal matrix, (3) segregation of carbon atoms on a metal surface, and (4) precipitation during the cooling process. They also explained the mechanism of surface-mediated graphene growth on Cu as shown in Figure 2.20, which consists of the following steps: (1) carbon formation from methane decomposition, (2) surface nucleation and growth, (3) further spraying of nuclei throughout the entire surface, and (4) domain formation. As the surface is fully covered with graphene, the growth process terminates, which is described as the self-limiting process of graphene growth on a Cu surface. One very recent report explaining that there is much less influence of the Cu-crystal lattice on graphene growth, thus, single-crystalline graphene can readily be obtained on polycrystalline Cu. Yu et al. (2011) confirmed that there is almost no definite epitaxial relationship between the CVD grown graphene sheet with the underlying Cu substrate. Thus, they concluded their findings as follows: (1) occurrence of very weak graphene-Cu interaction (van der Walls), (2) multi-oriented graphene grains can be obtained on a Cu grain, and (3) graphene grain can be grown continuously across Cu grain boundaries.

According to the binary phase diagram of Ni–C and Cu–C, Ni has higher carbon solubility than Cu. Therefore, at an elevated temperature, Ni can readily dissolute more carbon in a solid solution than Cu. In attempting to obtain uniform monolayer graphene on Ni it is difficult to control the precipitation of some excess carbon on the Ni surface. The precipitation of extra carbon during the cooling process leads to the formation of thick graphene on Ni rather than mono-atomic layer formation. In order to control the larger carbon deposition, one can attempt applying a faster cooling rate (Yu et al. 2008) or thin Ni film (Reina et al. 2008) in order to avoid thick graphene formation. Yu et al. reported the rapid cooling rate effect on thin graphene formation on Ni.

When comparing graphene growth on Cu and Ni, we need to first compare the atomic and crystal structure of Cu and Ni. Both Cu and Ni exhibit the same face centered cubic (FCC) crystal structure, equal coordination numbers, and almost equivalent electronegativity, which is 1.9 and 1.91 for Cu and Ni, respectively. However the basic difference is the electronic structure of Cu and Ni. Cu possesses a completely filled 3d band, whereas Ni exhibits a partially filled 3d band. One recent report nicely correlates those properties with the calculated adsorption energy of both the Cu and Ni surfaces. Hu et al. (2010) reported the first principle calculations of the low-index Cu and Ni surfaces, namely (100), (110), and (111), by DFT. They studied and compared the adsorption energies of C atoms on those stable low-index sites of Cu and Ni as tabulated in Table 2.1. They concluded that (1) (100) sites for both Ni and Cu are the most stable adsorption sites, which can accommodate C atoms easily; (2) (111) exhibits the lowest diffusion barrier, which facilitates the easy movement of adsorbed C atoms; (3) the adsorption energy of carbon on Ni is ~2 eV higher than the Cu. According to their report, d-bands at the Fermi level play a significant role in C adsorption on Cu and Ni surfaces where partially filled d-bands in Ni hybridized with carbon atoms more strongly than the completely filled d-bands in Cu. Therefore, the binding energy of carbon on Ni is stronger than that of Cu.

Comparison between the kinetics of the graphene growth phenomenon at atmospheric pressure and low pressure and under different gas flow rates was demonstrated

TABLE 2.1
Carbon Adsorption Energy on Cu and Ni

Adsorption Site	E_{ads} (in eV) on Cu	E_{ads} (in eV) on Ni
100 (H)	−6.42	−8.48
110 (H)	−5.57	−7.74
111 hcp	−4.88	−7.09
111 fcp	−4.89	−7.14
111 bridge	−4.88	

Source: From Hu, T., Q. M. Zhang, J. C. Wells, X. G. Gong, and Z. Y. Zhang. A comparative first-principles study of the adsorption of a carbon atom on copper and nickel surfaces. *Physics Letters A* 374, no. 44 (2010):4563–4567. With permission.

recently. Bhaviripudi et al. (2010) reported the role of kinetic factors in variations in the uniformity of the resulting large-area graphene growth. The thermodynamics of the growth process showed independence of pressure and gas flow rates, whereas the kinetics of the process are solely dependent upon the ambient pressure and gas flow rates. Depending upon the gas flow rates, final graphene uniformity, thickness, and defect density is also changed. On the basis of their experimental results, they proposed one steady-state kinetic model for graphene growth on low carbon solubility metals like Cu. The steady-state kinetic model suggests the thickness variation in the boundary layer, which is readily brought into effect in the diffusivity of the decomposed active carbon species, hence controlling the rate of carbon deposition.

2.4 SUMMARY

In summary, to date, graphene synthesis methods have been well established using both top-down and bottom-up approaches. Each of the synthesis methods is highly acclaimed by researchers in their respective fields of expertise and corresponding to different applications. Mechanical exfoliation using scotch tape was the first method of fabricating graphene with different numbers of layers. This technique is simple and low cost, but the control of large-scale synthesis and reproducibility of the same structure is yet to be demonstrated. Similarly, chemical synthesis was manifested in low temperature, large-scale graphene synthesis methods, which are transfer-free processes and capable of fabricating graphene film on any substrate. Moreover, solution process synthesis methods are advantageous for easy fabrication of functionalized graphene. Nevertheless, the graphene produced using these techniques is defective, partially reduced graphene oxide, which seriously compromises the physical properties of graphene. A few other reports are available regarding the direct fabrication of graphene using the chemical process, but the properties of the graphene obtained via these methods are still not satisfactory. In contrast, thermal chemical vapor deposition has been proven to be a more industrially feasible and scalable process, which has myriad possibilities to obtain various morphologies of graphene over a wide range of substrates. However, the process is costly and incapable of producing graphene directly on polymer substrates, and involves a transfer process that could introduce defects and contaminants in graphene. Further investigation is also required in order to obtain valuable control over the number of graphene layers produced along with homogeneous electrical and optical properties. On the other hand, epitaxial growth on a SiC surface yields high-quality, high-purity graphene that exhibits good electrical properties. However, due to lack of reproducibility on different substrates, its application in a wide range of electronic and optoelectronic devices is seriously limited. Nevertheless, each of the graphene synthesis processes has advantages and disadvantages focusing on its field of application. Finally, this is a growing field of science, which needs more rigorous studies to obtain high quality graphene by controlling over the process parameters and more comprehensible scientific understanding.

2.5 FUTURE PERSPECTIVES

Graphene is a newly invented, attractive material that exhibits several unusual physical properties that can be applicable to future electronic and optoelectronic devices. From synthesis routes to its growth mechanism, or from its eccentric properties to possible applications, the subject is still under scholarly debate. However, the field is growing. Hence, it is worth emphasizing that although several process routes for graphene synthesis have been established, valuable control over the process parameters will be required for tailoring its sizable and well-defined bandgap along with reproducible properties. Similarly, detailed analysis of graphene growth mechanisms using thermal CVD will provide an ample opportunity for controlling its electronic properties. Furthermore, several novel methodologies for graphene synthesis have also been highlighted recently, which need more investigation to produce high-purity, good-quality and bandgap tailored graphene.

REFERENCES

Allen, M. J., V. C. Tung, and R. B. Kaner. 2010. Honeycomb carbon: A review of graphene. *Chemical Reviews* 110 (1):132–145.

Bae, S., H. Kim, Y. Lee, et al. 2010. Roll-to-roll production of 30-inch graphene films for transparent electrodes. *Nature Nanotechnology* 5 (8):574–578.

Berger, C., Z. M. Song, T. B. Li, et al. 2004. Ultrathin epitaxial graphite: 2D electron gas properties and a route toward graphene-based nanoelectronics. *Journal of Physical Chemistry B* 108 (52):19912–19916.

Bhaviripudi, S., X. T. Jia, M. S. Dresselhaus, and J. Kong. 2010. Role of kinetic factors in chemical vapor deposition synthesis of uniform large area graphene using copper catalyst. *Nano Letters* 10 (10):4128–4133.

Bhuvana, T., A. Kumar, A. Sood, et al. 2010. Contiguous petal-like carbon nanosheet outgrowths from graphite fibers by plasma CVD. *ACS Applied Materials & Interfaces* 2 (3):644–648.

Boehm, H. P., A. Clauss, G. O. Fischer, and U. Das Hofmann. 1962. Adsorptionsverhalten sehr dunner Kohlenstoff-Folien. *Anorg. Allg. Chem.* 316:119–127.

Brodie, B. C. 1860. Sur le poids atomique du graphite. *Annales des Chimie et des Physique* 59:466.

Campos-Delgado, J., Y. A. Kim, T. Hayashi, et al. 2009. Thermal stability studies of CVD-grown graphene nanoribbons: Defect annealing and loop formation. *Chemical Physics Letters* 469 (1–3):177–182.

Campos-Delgado, J., J. M. Romo-Herrera, X. T. Jia, et al. 2008. Bulk production of a new form of sp(2) carbon: Crystalline graphene nanoribbons. *Nano Letters* 8 (9):2773–2778.

Cano-Marquez, A. G., F. J. Rodriguez-Macias, J. Campos-Delgado, et al. 2009. Ex-MWNTs: Graphene sheets and ribbons produced by lithium intercalation and exfoliation of carbon nanotubes. *Nano Letters* 9 (4):1527–1533.

Castner, D. G., B. A. Sexton, and G. A. Somorjai. 1978. LEED and thermal desorption studies of small molecules (H_2, O_2, CO, CO_2, NO, C_2H_4, C_2H_2, AND C) chemisorbed on rhodium (111) and (100) surfaces. *Surface Science* 71 (3):519–540.

Castro Neto, A. H., F. Guinea, N. M. R. Peres, K. S. Novoselov, and A. K. Geim. 2009. The electronic properties of graphene. *Reviews of Modern Physics* 81 (1):109–162.

Chen, S. Q., and Y. Wang. 2010. Microwave-assisted synthesis of a Co_3O_4-graphene sheet-on-sheet nanocomposite as a superior anode material for Li-ion batteries. *Journal of Materials Chemistry* 20 (43):9735–9739.

Chen, Z. H., Y. M. Lin, M. J. Rooks, and P. Avouris. 2007. Graphene nano-ribbon electronics. *Physica E-Low-Dimensional Systems & Nanostructures* 40 (2):228–232.

Choi, Wonbong, Indranil Lahiri, Raghunandan Seelaboyina, and Yong Soo Kang. 2010. Synthesis of graphene and its applications: A review. *Critical Reviews in Solid State and Materials Sciences* 35 (1):52–71.

Choucair, M., P. Thordarson, and J. A. Stride. 2009. Gram-scale production of graphene based on solvothermal synthesis and sonication. *Nature Nanotechnology* 4 (1):30–33.

Coraux, J., A. T. N'Diaye, M. Engler, et al. 2009. Growth of graphene on Ir(111). *New Journal of Physics* 11:22.

Cote, Laura J., Franklin Kim, and Jiaxing Huang. 2008. Langmuir–Blodgett assembly of graphite oxide single layers. *Journal of the American Chemical Society* 131 (3):1043–1049.

De Arco, L. G., Y. Zhang, C. W. Schlenker, K. Ryu, M. E. Thompson, and C. W. Zhou. 2010. Continuous, highly flexible, and transparent graphene films by chemical vapor deposition for organic photovoltaics. *ACS Nano* 4 (5):2865–2873.

de Heer, W. A., C. Berger, X. S. Wu, et al. 2007. Epitaxial graphene. *Solid State Communications* 143 (1–2):92–100.

de Parga, A. L. V., F. Calleja, B. Borca, et al. 2008. Periodically rippled graphene: Growth and spatially resolved electronic structure. *Physical Review Letters* 100 (5).

Dreyer, D. R., S. Park, C. W. Bielawski, and R. S. Ruoff. 2010. The chemistry of graphene oxide. *Chemical Society Reviews* 39 (1):228–240.

Eizenberg, M., and J. M. Blakely. 1979. Carbon monolayer phase condensation on Ni(111). *Surface Science* 82 (1):228–236.

Emtsev, K. V., A. Bostwick, K. Horn, et al. 2009. Towards wafer-size graphene layers by atmospheric pressure graphitization of silicon carbide. *Nature Materials* 8 (3):203–207.

Gao, L. B., W. C. Ren, J. P. Zhao, L. P. Ma, Z. P. Chen, and H. M. Cheng. 2010. Efficient growth of high-quality graphene films on Cu foils by ambient pressure chemical vapor deposition. *Applied Physics Letters* 97 (18):3.

Geim, A. et al. 2010. Many pioneers in graphene discovery. *APS News* 19, 4.

Geim, A. K., and P. Kim. 2008. Carbon wonderland. *Scientific American* 298 (4):90–97.

Geim, A. K., and K. S. Novoselov. 2007. The rise of graphene. *Nature Materials* 6 (3):183–191.

Hamilton, J. C., and J. M. Blakely. 1980. Carbon segregation to single-crystal surfaces of PT, PD, and CO. *Surface Science* 91 (1):199–217.

Hanns-Peter, Boehm. 2010. Graphen—wie eine Laborkuriosität plötzlich äußerst interessant wurde. *Angewandte Chemie*: n/a.

Hanns-Peter, Boehm. 2010. Graphene—How a laboratory curiosity suddenly became extremely interesting. *Angewandte Chemie International Edition*: n/a.

Hasin, P., M. A. Alpuche-Aviles, and Y. Y. Wu. 2010. Electrocatalytic activity of graphene multi layers toward I-/I-3(-): Effect of preparation conditions and polyelectrolyte modification. *Journal of Physical Chemistry C* 114 (37):15857–15861.

Hass, J., W. A. de Heer, and E. H. Conrad. 2008. The growth and morphology of epitaxial multilayer graphene. *Journal of Physics-Condensed Matter* 20 (32):27.

Hazra, K. S., et al. 2011. Thinning of multilayer graphene to monolayer graphene in a plasma environment. *Nanotechnology* 22 (2):025704.

Helveg, S., C. Lopez-Cartes, J. Sehested, et al. 2004. Atomic-scale imaging of carbon nanofibre growth. *Nature* 427 (6973):426–429.

Hernandez, Y., V. Nicolosi, M. Lotya, et al. 2008. High-yield production of graphene by liquid-phase exfoliation of graphite. *Nature Nanotechnology* 3 (9):563–568.

Hu, T., Q. M. Zhang, J. C. Wells, X. G. Gong, and Z. Y. Zhang. 2010. A comparative first-principles study of the adsorption of a carbon atom on copper and nickel surfaces. *Physics Letters A* 374 (44):4563–4567.

Hummers, William S., and Richard E. Offeman. 1958. Preparation of graphitic oxide. *Journal of the American Chemical Society* 80 (6):1339.

Ismach, A., C. Druzgalski, S. Penwell, et al. 2010. Direct chemical vapor deposition of graphene on dielectric surfaces. *Nano Letters* 10 (5):1542–1548.

Jasek, O., P. Synek, L. Zajickova, M. Elias, and V. Kudrle. 2010. Synthesis of carbon nanostructures by plasma enhanced chemical vapour deposition at atmospheric pressure. *Journal of Electrical Engineering-Elektrotechnicky Casopis* 61 (5):311–313.

Jeong, H. K., Y. P. Lee, Rjwe Lahaye, et al. 2008. Evidence of graphitic AB stacking order of graphite oxides. *Journal of the American Chemical Society* 130 (4):1362–1366.

Jiao, L. Y., X. R. Wang, G. Diankov, H. L. Wang, and H. J. Dai. 2010. Facile synthesis of high-quality graphene nanoribbons. *Nature Nanotechnology* 5 (5):321–325.

Jiao, L. Y., L. Zhang, X. R. Wang, G. Diankov, and H. J. Dai. 2009. Narrow graphene nanoribbons from carbon nanotubes. *Nature* 458 (7240):877–880.

Juang, Z. Y., C. Y. Wu, C. W. Lo, et al. 2009. Synthesis of graphene on silicon carbide substrates at low temperature. *Carbon* 47 (8):2026–2031.

Katsnelson, Mikhail I. 2007. Graphene: Carbon in two dimensions. *Materials Today* 10 (1–2):20–27.

Kim, K. S., Y. Zhao, H. Jang, et al. 2009. Large-scale pattern growth of graphene films for stretchable transparent electrodes. *Nature* 457 (7230):706–710.

Kosynkin, D. V., A. L. Higginbotham, A. Sinitskii, et al. 2009. Longitudinal unzipping of carbon nanotubes to form graphene nanoribbons. *Nature* 458 (7240):872–876.

Lahiri, J., T. Miller, L. Adamska, I. I. Oleynik, and M. Batzill. 2011. Graphene growth on Ni(111) by transformation of a surface carbide. *Nano Letters* 11 (2):518–522.

Lang, B. 1975. A LEED study of the deposition of carbon on platinum crystal surfaces. *Surface Science* 53 (1):317–329.

Lee, C., X. D. Wei, J. W. Kysar, and J. Hone. 2008. Measurement of the elastic properties and intrinsic strength of monolayer graphene. *Science* 321 (5887):385–388.

Lee, D. H., J. A. Lee, W. J. Lee, D. S. Choi, and S. O. Kim. 2010. Facile fabrication and field emission of metal-particle-decorated vertical N-doped carbon nanotube/graphene hybrid films. *Journal of Physical Chemistry C* 114 (49):21184–21189.

Li, D., M. B. Muller, S. Gilje, R. B. Kaner, and G. G. Wallace. 2008. Processable aqueous dispersions of graphene nanosheets. *Nature Nanotechnology* 3 (2):101–105.

Li, G. L., G. Liu, M. Li, D. Wan, K. G. Neoh, and E. T. Kang. 2010. Organo- and water-dispersible graphene oxide-polymer nanosheets for organic electronic memory and gold nanocomposites. *Journal of Physical Chemistry C* 114 (29):12742–12748.

Li, S. S., K. H. Tu, C. C. Lin, C. W. Chen, and M. Chhowalla. 2010. Solution-processable graphene oxide as an efficient hole transport layer in polymer solar cells. *ACS Nano* 4 (6):3169–3174.

Li, X. L., X. R. Wang, L. Zhang, S. W. Lee, and H. J. Dai. 2008. Chemically derived, ultrasmooth graphene nanoribbon semiconductors. *Science* 319 (5867):1229–1232.

Li, X. L., G. Y. Zhang, X. D. Bai, et al. 2008. Highly conducting graphene sheets and Langmuir-Blodgett films. *Nature Nanotechnology* 3 (9):538–542.

Li, X. S., W. W. Cai, J. H. An, et al. 2009. Large-area synthesis of high-quality and uniform graphene films on copper foils. *Science* 324 (5932):1312–1314.

Li, X. S., W. W. Cai, L. Colombo, and R. S. Ruoff. 2009. Evolution of graphene growth on Ni and Cu by carbon isotope labeling. *Nano Letters* 9 (12):4268–4272.

Li, X. S., C. W. Magnuson, A. Venugopal, et al. 2010. Graphene films with large domain size by a two-step chemical vapor deposition process. *Nano Letters* 10 (11):4328–4334.

Li, X. S., Y. W. Zhu, W. W. Cai, et al. 2009. Transfer of large-area graphene films for high-performance transparent conductive electrodes. *Nano Letters* 9 (12):4359–4363.

Liang, X., Z. Fu, and S. Y. Chou. 2007. Graphene transistors fabricated via transfer-printing in device active-areas on large wafer. *Nano Letters* 7 (12):3840–3844.

Liu, N., F. Luo, H. X. Wu, Y. H. Liu, C. Zhang, and J. Chen. 2008. One-step ionic-liquid-assisted electrochemical synthesis of ionic-liquid-functionalized graphene sheets directly from graphite. *Advanced Functional Materials* 18 (10):1518–1525.

Lu, X. K., M. F. Yu, H. Huang, and R. S. Ruoff. 1999. Tailoring graphite with the goal of achieving single sheets. *Nanotechnology* 10 (3):269–272.

McAllister, M. J., J. L. Li, D. H. Adamson, et al. 2007. Single sheet functionalized graphene by oxidation and thermal expansion of graphite. *Chemistry of Materials* 19 (18):4396–4404.

Muszynski, R., B. Seger, and P. V. Kamat. 2008. Decorating graphene sheets with gold nanoparticles. *Journal of Physical Chemistry C* 112 (14):5263–5266.

Nair, R. R., P. Blake, A. N. Grigorenko, et al. 2008. Fine structure constant defines visual transparency of graphene. *Science* 320 (5881):1308.

Novoselov, K. S., A. K. Geim, S. V. Morozov, et al. 2004. Electric field effect in atomically thin carbon films. *Science* 306 (5696):666–669.

Novoselov, K. S., A. K. Geim, S. V. Morozov, et al. 2005. Two-dimensional gas of massless Dirac fermions in graphene. *Nature* 438 (7065):197–200.

Novoselov, K. S., D. Jiang, F. Schedin, et al. 2005. Two-dimensional atomic crystals. *Proceedings of the National Academy of Sciences USA* 102 (30):10451–10453.

Obraztsov, A. N. 2009. Chemical vapour deposition: Making graphene on a large scale. *Nature Nanotechnology* 4 (4):212–213.

Obraztsov, A. N., E. A. Obraztsova, A. V. Tyurnina, and A. A. Zolotukhin. 2007. Chemical vapor deposition of thin graphite films of nanometer thickness. *Carbon* 45 (10):2017–2021.

Obraztsov, A. N., A. A. Zolotukhin, A. O. Ustinov, A. P. Volkov, Y. Svirko, and K. Jefimovs. 2003. DC discharge plasma studies for nanostructured carbon CVD. *Diamond and Related Materials* 12 (3–7):917–920.

Park, S., and R. S. Ruoff. 2009. Chemical methods for the production of graphenes. *Nature Nanotechnology* 4 (4):217–224.

Qi, J. L., X. Wang, W. T. Zheng, et al. 2009. Effects of total CH_4/Ar gas pressure on the structures and field electron emission properties of carbon nanomaterials grown by plasma-enhanced chemical vapor deposition. *Applied Surface Science* 256 (5):1542–1547.

Qian, Jiangfeng, Ping Liu, Yang Xiao, et al. 2009. TiO_2-coated multilayered SnO_2 hollow microspheres for dye-sensitized solar cells. *Advanced Materials* 21 (36):3663–3667.

Reina, A., X. T. Jia, J. Ho, et al. 2009. Large area, few-layer graphene films on arbitrary substrates by chemical vapor deposition. *Nano Letters* 9 (1):30–35.

Reina, A., S. Thiele, X. T. Jia, et al. 2009. Growth of large-area single- and bi-layer graphene by controlled carbon precipitation on polycrystalline Ni surfaces. *Nano Research* 2 (6):509–516.

Reina, Alfonso, Xiaoting Jia, John Ho, et al. 2008. Large area, few-layer graphene films on arbitrary substrates by chemical vapor deposition. *Nano Letters* 9 (1):30–35.

Roy-Mayhew, Joseph D., David J. Bozym, Christian Punckt, and Ilhan A. Aksay. 2010. Functionalized graphene as a catalytic counter electrode in dye-sensitized solar cells. *ACS Nano* 4 (10):6203–6211.

Rummeli, M. H., A. Bachmatiuk, A. Scott, et al. 2010. Direct low-temperature nanographene CVD synthesis over a dielectric insulator. *ACS Nano* 4 (7):4206–4210.

Schniepp, H. C., J. L. Li, M. J. McAllister, et al. 2006. Functionalized single graphene sheets derived from splitting graphite oxide. *Journal of Physical Chemistry B* 110 (17):8535–8539.

Shang, N. G., P. Papakonstantinou, M. McMullan, et al. 2008. Catalyst-free efficient growth, orientation and biosensing properties of multilayer graphene nanoflake films with sharp edge planes. *Advanced Functional Materials* 18 (21):3506–3514.

Sidorov, A. N., T. Bansal, P. J. Ouseph, and G. Sumanasekera. 2010. Graphene nanoribbons exfoliated from graphite surface dislocation bands by electrostatic force. *Nanotechnology* 21 (19).

Somani, P. R., S. P. Somani, and M. Umeno. 2006. Planar nano-graphenes from camphor by CVD. *Chemical Physics Letters* 430 (1–3):56–59.

Sridhar, V., J. H. Jeon, and I. K. Oh. 2010. Synthesis of graphene nano-sheets using eco-friendly chemicals and microwave radiation. *Carbon* 48. (10):2953–2957.

Stankovich, S., D. A. Dikin, G. H. B. Dommett, et al. 2006. Graphene-based composite materials. *Nature* 442 (7100):282–286.

Stankovich, S., R. D. Piner, S. T. Nguyen, and R. S. Ruoff. 2006. Synthesis and exfoliation of isocyanate-treated graphene oxide nanoplatelets. *Carbon* 44 (15):3342–3347.

Stankovich, Sasha, Dmitriy A. Dikin, Richard D. Piner, et al. 2007. Synthesis of graphene-based nanosheets via chemical reduction of exfoliated graphite oxide. *Carbon* 45 (7):1558–1565.

Staudenmaier, L. 1898. Verfahren zur Darstellung der Graphitsaure. *Berichte der Deutschen Chemischen Gesellschaft* 31:1481–1499.

Sun, Z. Z., Z. Yan, J. Yao, E. Beitler, Y. Zhu, and J. M. Tour. 2010. Growth of graphene from solid carbon sources. *Nature* 468 (7323):549–552.

Sutter, P. 2009. Epitaxial graphene: How silicon leaves the scene. *Nature Materials* 8 (3):171–172.

Sutter, P., M. S. Hybertsen, J. T. Sadowski, and E. Sutter. 2009. Electronic structure of few-layer epitaxial graphene on Ru(0001). *Nano Letters* 9 (7):2654–2660.

The rise and rise of graphene. 2010. *Nature Nanotechnology* 5 (11):755.

Tung, V. C., M. J. Allen, Y. Yang, and R. B. Kaner. 2009. High-throughput solution processing of large-scale graphene. *Nature Nanotechnology* 4 (1):25–29.

Van Bommel, A. J., J. E. Crombeen, and A. Van Tooren. 1975. LEED and Auger electron observations of the SiC(0001) surface. *Surface Science* 48 (2):463–472.

Verma, V. P., S. Das, I. Lahiri, and W. Choi. 2010. Large-area graphene on polymer film for flexible and transparent anode in field emission device. *Applied Physics Letters* 96 (20).

Viculis, Lisa M., Julia J. Mack, and Richard B. Kaner. 2003. A chemical route to carbon nanoscrolls. *Science* 299 (5611):1361.

Viculis, Lisa M., Julia J. Mack, Oren M. Mayer, H. Thomas Hahn, and Richard B. Kaner. 2005. Intercalation and exfoliation routes to graphite nanoplatelets. *Journal of Materials Chemistry* 15 (9):974–978.

Vitchev, R., A. Malesevic, R. H. Petrov, et al. 2010. Initial stages of few-layer graphene growth by microwave plasma-enhanced chemical vapour deposition. *Nanotechnology* 21 (9):7.

Wang, J. J., M. Y. Zhu, R. A. Outlaw, X. Zhao, D. M. Manos, and B. C. Holloway. 2004. Synthesis of carbon nanosheets by inductively coupled radio-frequency plasma enhanced chemical vapor deposition. *Carbon* 42 (14):2867–2872.

Wang, J. J., M. Y. Zhu, R. A. Outlaw, et al. 2004. Free-standing subnanometer graphite sheets. *Applied Physics Letters* 85 (7):1265–1267.

Wang, X. B., H. J. You, F. M. Liu, et al. 2009. Large-scale synthesis of few-layered graphene using CVD. *Chemical Vapor Deposition* 15 (1–3):53–56.

Xin, G. Q., W. Hwang, N. Kim, S. M. Cho, and H. Chae. 2010. A graphene sheet exfoliated with microwave irradiation and interlinked by carbon nanotubes for high-performance transparent flexible electrodes. *Nanotechnology* 21 (40).

Xu, Y. X., H. Bai, G. W. Lu, C. Li, and G. Q. Shi. 2008. Flexible graphene films via the filtration of water-soluble noncovalent functionalized graphene sheets. *Journal of the American Chemical Society* 130 (18):5856+.

Yang, N. L., J. Zhai, D. Wang, Y. S. Chen, and L. Jiang. 2010. Two-dimensional graphene bridges enhanced photoinduced charge transport in dye-sensitized solar cells. *ACS Nano* 4 (2):887–894.

Yu, Q. K., L. A. Jauregui, W. Wu, et al. 2011. Control and characterization of individual grains and grain boundaries in graphene grown by chemical vapour deposition. *Nature Materials* 10 (6):443–449.

Yu, Q. K., J. Lian, S. Siriponglert, H. Li, Y. P. Chen, and S. S. Pei. 2008. Graphene segregated on Ni surfaces and transferred to insulators. *Applied Physics Letters* 93 (11).

Yuan, G. D., W. J. Zhang, Y. Yang, et al. 2009. Graphene sheets via microwave chemical vapor deposition. *Chemical Physics Letters* 467 (4–6):361–364.

Zhang, J. L., G. X. Shen, W. J. Wang, X. J. Zhou, and S. W. Guo. 2010. Individual nanocomposite sheets of chemically reduced graphene oxide and poly(N-vinyl pyrrolidone): Preparation and humidity sensing characteristics. *Journal of Materials Chemistry* 20 (48):10824–10828.

Zhang, Y., L. Gomez, F. N. Ishikawa, et al. 2010. Comparison of graphene growth on single-crystalline and polycrystalline Ni by chemical vapor deposition. *Journal of Physical Chemistry Letters* 1 (20):3101–3107.

Zhang, Y. B., J. P. Small, W. V. Pontius, and P. Kim. 2005. Fabrication and electric-field-dependent transport measurements of mesoscopic graphite devices. *Applied Physics Letters* 86 (7).

Zheng, M., K. Takei, B. Hsia, et al. 2010. Metal-catalyzed crystallization of amorphous carbon to graphene. *Applied Physics Letters* 96 (6).

Zhou, K. F., Y. H. Zhu, X. L. Yang, and C. Z. Li. 2010. One-pot preparation of graphene/Fe_3O_4 composites by a solvothermal reaction. *New Journal of Chemistry* 34 (12):2950–2955.

Zhu, J. W., G. Y. Zeng, F. D. Nie, et al. 2010. Decorating graphene oxide with CuO nanoparticles in a water-isopropanol system. *Nanoscale* 2 (6):988–994.

3 Quantum Transport in Graphene-Based Materials and Devices
From Pseudospin Effects to a New Switching Principle

Stephan Roche, Frank Ortmann, Alessandro Cresti, Blanca Biel, and David Jiménez

CONTENTS

3.1 INTRODUCTION

The understanding of transport properties in graphene has become a topic of great interest not only because of the underlying fascinating physics, but also because of the flourishing graphene-based technologies in fields such as flexible displays, high-frequency devices, composite materials, or photovoltaic applications. Graphene shows an exceptional ability to convey charge carriers (electrons and holes) and displays some of the most exotic quantum transport features of modern condensed matter physics (Geim and Novoselov 2007). The origin of these properties roots in the linear band dispersion of low-energy electrons (developed around two independent K and K′ valleys in the Brillouin zone), and in the presence of a pseudospin degree of freedom; both properties are shared with metallic carbon nanotubes, their one-dimensional counterparts (Charlier, Blase, and Roche 2007, Castro Neto et al. 2009). Low-energy electronic excitations behave as massless Dirac fermions, which yield unique quantum phenomena such as Klein tunneling (Katsnelson, Novoselov, and Geim 2006) and weak antilocalization (McCann et al. 2006). These fascinating properties of clean graphene-based materials can be further tuned and diversified in unprecedented ways by chemical modifications of the underlying π-conjugated network (Loh et al. 2010). The reported charge carrier mobilities in graphene layers can reach about 200.000 $cm^2\ V^{-1}s^{-1}$ at room temperature, which is several orders of magnitude larger than those of silicon. However, an undoped single-layer graphene acts as a zero-gap semiconductor, which is unsuitable for achieving competitive electrostatic gating efficiency and further developing all carbon-based nanoelectronics. Experimental measurements reported ratios between the current in the ON state and the current in the OFF state not higher than one order of magnitude. One possibility to increase the (zero) gap of two-dimensional graphene single layers is to reduce the lateral dimensions of the device, fabricating graphene nanoribbons with widths down to a few tens of nanometers, using state-of-the-art e-beam lithographic techniques and oxygen plasma etching, or ion-beam lithography. Graphene ribbons can also be engineering by chemically unzipping carbon nanotubes (Shimizu et al. 2011). These approaches allow some more or less efficient confinement gap engineering despite an interfering contribution of disorder effects (Cresti et al. 2008, Roche 2011). Besides, theoretical simulations suggest that the best accessible energy band gaps remain too small or very unstable in regards to edge reconstruction and defects (Dubois et al. 2010), to envision outperforming ultimate complementary metal-oxide semiconductor field-effect transistors (CMOS-FETs) (Lee 2010).

In this chapter, we first discuss the contribution of pseudospin effects in the magnetoconductance fingerprints of weakly disordered graphene, unveiling the underlying mechanism at the origin of large mobility and reduced multiple scattering effects in weakly damaged graphene. We then comment on the behaviors and limits of ultraclean graphene nanoribbon-based field effect transistors. A final section concerns the use of chemical doping to introduce controlled mobility gaps and electron-hole transport asymmetry, to considerably improve graphene transistor performance, whose fabrication remains within the reach of conventional technologies.

3.2 PSEUDOSPIN EFFECTS AND LOCALIZATION IN DISORDERED GRAPHENE

The peculiarity of quantum transport in graphene stems from both the linear energy dispersion relation versus momentum $E_k = \hbar v_F |k|$, and the additional pseudospin quantum number (which refers to the two A and B graphene sublattices), which enforces a description of electronic states as 4-component pseudospinors, whose behavior resembles that of massless relativistic particles. The full electron-hole symmetry of the electronic spectrum also mimics the charge conjugation symmetry of particles/antiparticles in high-energy physics and introduces novel transport properties. Indeed, this perfect electron-hole symmetry in an energy window about 1 eV around the charge neutrality point (or Dirac point) enables the possibility for a Klein tunneling mechanism that is a reflection-less quantum transmission of incoming fermions when crossing a potential barrier of arbitrary height and width (Katsnelson,

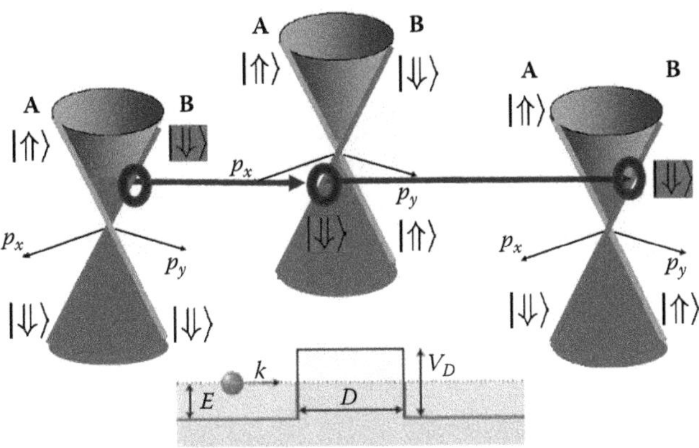

FIGURE 3.1 Schematic illustration of the Klein tunneling process through a potential barrier of height V_D and width D (bottom panel), inspired by M. I. Katsnelson, K. S. Novoselov, and A. K. Geim. Chiral tunnelling and the Klein paradox in graphene, *Nature Physics* 2 (2006): 620. Incoming electrons from the left-hand side tunnel through the barrier by occupying hole states with opposite momentum and equivalent pseudospin. Pseudospins are labeled by arrows, while A and B sites denote graphene sublattices. The wavefunctions' symmetry around one selected K-point are described by the additional pseudospin degree of freedom as

$$\left|\Psi_P^K\right\rangle = \frac{1}{\sqrt{2}}\begin{pmatrix}\psi_P^K(A)\\ \psi_P^K(B)\end{pmatrix}, \text{ defining } \left|\uparrow\right\rangle = \begin{pmatrix}1\\0\end{pmatrix} \text{ and } \left|\downarrow\right\rangle = \begin{pmatrix}0\\1\end{pmatrix}$$

Novoselov, and Geim 2006). This result (initially proposed for high-energy relativistic particles) remains robust for a single potential barrier and as long as charge transport is restricted to one of the two nonequivalent energy cones in momentum space. Figure 3.1 schematically illustrates such a reflection-less propagation. Assuming that incoming electrons are propagating from left to right with a well-defined momentum and pseudospin (the corresponding wavefunction spreads only on the B sublattice), then owing to the local energy upshift induced by the potential barrier, available hole states with opposite momentum and identical pseudospin serve as a wall pass for electrons, which cross the barrier without noticing it.

However, when multiple scattering comes into play, then quantum interferences contribute to the conductance, eventually yielding weak localization (WL) or weak antilocalization (WAL) effects (McCann et al., 2006). The effect of WAL is deeply connected with the additional pseudospin symmetry of graphene electronic states, and occurs provided the underlying disorder potential preserves some specific symmetries. Inspired by the pioneering work of Hikami, Larkin, and Nagaoka (1980) developed for conventional metals with strong spin-orbit coupling, a diagrammatic theory has been proposed to account for pseudospin-related phase interferences at the origin of a sign reversal of the quantum correction of the electronic conductance in graphene (McCann et al., 2006). This theory is, however, irreparably forced to introduce several phenomenological parameters (intravalley and intervalley elastic scattering times) to cut off the several contributions of Cooperons related to a given disorder symmetry class. No analytical derivation is possible for realistic and complex forms of disorder potential, as found in real materials, which can originate from many different sources (charges trapped in the oxide, ripples, adsorbed atoms and molecules, topological defects, vacancies, etc.).

Recent experiments (Tikhonenko et al. 2009) have reported on a very complex phase diagram of localization phenomena in graphene (versus temperature, magnetic field or charge energy, and screening effects), and despite the fitting possibilities using the previously mentioned theory, the deep connection between underlying disorders and the origin of crossovers between the different localization regimes remains totally elusive. Additionally, it is important to clarify the conditions for maintaining pseudospin-related properties, since this also pinpoints the possibility for an Anderson localization regime, where all states are exponentially localized, and temperature-dependent conductance described by a variable range hopping behavior (Moser et al. 2010, Leconte et al. 2010).

To explore quantum transport in disordered graphene under external weak magnetic fields, we have improved an efficient computational approach based on an order N real space numerical implementation of the Kubo-Greenwood conductivity σ_{dc} (for details, see Roche and Mayou 1997, Roche 1999, Roche and Saito 2011, Ortmann 2011). This method solves the time-dependent Schrödinger equation and computes diffusion coefficient

$$D(E_F,t)=\frac{\partial}{\partial t}\Delta X^2(E_F,t)$$

and Kubo conductivities

$$\sigma_{dc}=\frac{e^2}{2}\rho(E_F)\lim_{t\to\infty}\frac{\partial}{\partial t}\Delta X^2(E_F,t)$$

where $\Delta X(E_F,t)$ and $\rho(E_F)$, respectively, denote the time-dependent spreading of electronic wavepackets at the Fermi energy. The order N algorithm is implemented through an expansion of the spectral measure in a basis of Chebyshev polynomials and the use of Lanczos and recursion procedures.

We consider a disorder model with defect density n_i based on a screened long-range Coulomb potential mimicked by a Gaussian potential with strength taken at random within $[-W/2,W/2]$ and decay length ξ (see Figure 3.2(a) for illustration). By solving the time-dependent Schrödinger equation, one can follow the dynamics of quantum wavepackets and compute the conductance scaling properties. Figure 3.1(b) shows $D(E = 0,t)$ at three different defect densities and taking $W = 2\gamma_0$ in absence of an external magnetic field, with γ_0 the coupling energy between nearest neighbors. In all cases, $D(E = 0,t)$ are seen to saturate or decay after an initial increase of the wavepacket spreading. This corresponds to the transition from a ballistic to a diffusive-like regime, while the maximum value of $D(E,t)$ allows us to deduce the elastic mean free path (and corresponding semiclassical conductivity [Drude] and charge mobility).

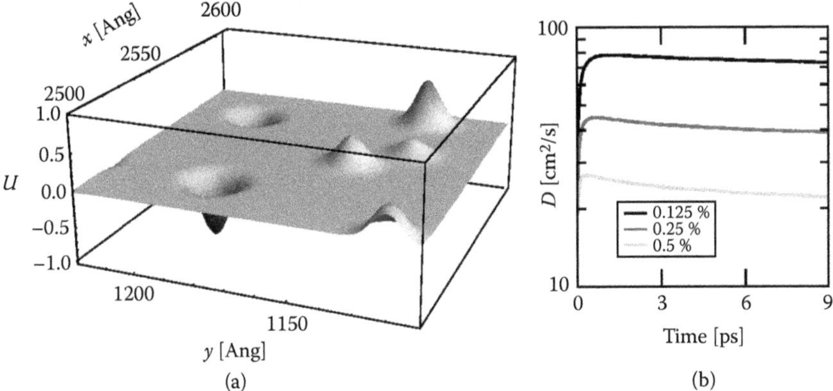

FIGURE 3.2 (a) Schematic of the disordered potential in graphene due to long-range Coulomb impurities. (b) Time-dependent diffusion coefficient at Dirac point for several impurity densities and $W = 2\,\gamma_0$.

By applying an external magnetic field, it is possible to unveil the origin of the crossover between WL and WAL. We first notice that the accuracy of our implementation of the magnetic field in the tight-binding model as evidenced by the Landau level structure obtained at $B = 10$ T (not shown here), is in full agreement with analytical results.

Figure 3.3 shows the main result of this study. On the top panels are given the time-dependent behaviors of the diffusion coefficients $D(t,B,W)$ as a function of external magnetic field strength and disorder strength W. It is seen that for $W = 2\gamma_0$, weak localization dominates with a corresponding positive magnetoconductance fingerprint. As W decreases from $W = 2\gamma_0$ to $W = 1.5\gamma_0$, magnetoconductance switches from an exclusive WL fingerprint, to a transition behavior, which pinpoints the increasing contribution of WAL effects. As W becomes close to γ_0, the electronic conduction becomes quasiballistic, evidencing an increasing contribution of Klein tunneling and strong suppression of backscattering effects. Note that the dashed lines shown on the bottom panels for the magnetoconductivity are fits using the phenomenology introduced in McCann et al. (2006), but keeping a single additional elastic scattering time (for details see Ortmann et al, 2011). This transition is also related to the relative contribution of intervalley versus intravalley scattering processes involved in conduction (Ortmann 2011).

The contribution of sample edges is analyzed by exploring the case of graphene nanoribbons (Cresti et al. 2008). We here focus on a 10 nm-wide armchair ribbon (84-aGNR) with impurity densities $n_i = 0.2\%$, $\xi = 0.426$ nm, and $W = 0.5\gamma_0$ (see Figure 3.4(a) for illustration). The calculation of the total transmission coefficient is performed using the Landauer-Büttiker formula and Green's function techniques. Figure 3.4(b) presents $T(E)$, averaged over 100 disorder configurations, from which one can deduce the elastic mean-free path (ℓe) (see Cresti et al. 2008). Figure 3.4(c) shows that ℓe exhibits important energy-dependent modulations, with systematic decay around each onset of new electronic subbands. At low energies, ℓe reaches a few microns for the considered low disorder potential value ($W = 0.5\,\gamma_0$).

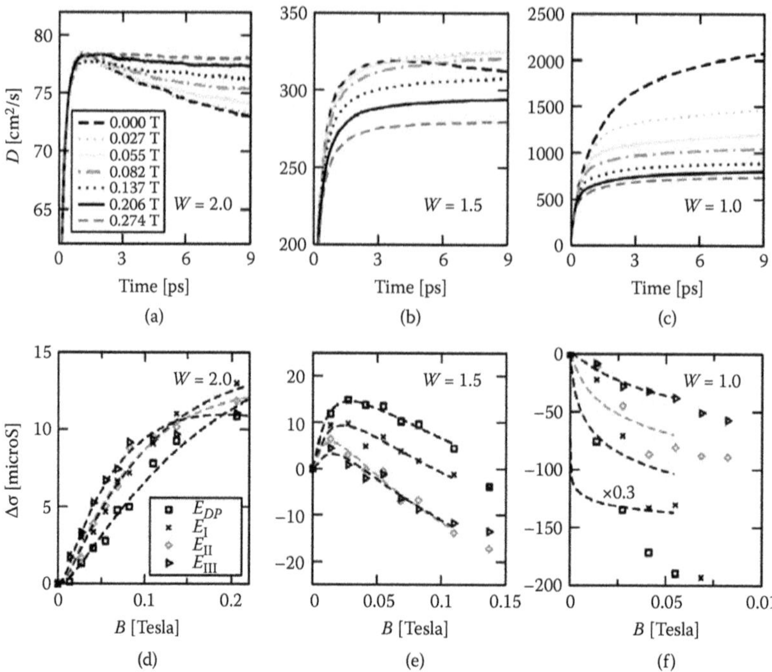

FIGURE 3.3 Top panels: diffusion coefficient at the Dirac point for various magnetic fields and W, $n_i = 0.125\%$, $\xi = 0.426$nm. Bottom panels: $\Delta\sigma$ for four different Fermi level positions $E_{DP} = 0$, $E_I = 0.049$ eV, $E_{II} = 0.097$ eV and $E_{III} = 0.146$ eV). Dashed lines are fits as explained in the text. For $W = 1.0$ in (f) data and fit at E_{DP} have been rescaled by 0.3 for clarity.

Accordingly, magnetotransport fingerprints need to be evaluated with care since for a given system length, the zero-field conduction regime can switch from quasiballistic to a strongly diffusive regime. The magnetotransport properties are then explored at low fields (up to 0.5 T) in a chosen diffusive regime, where the band structure does not change significantly with respect to the zero field case and where the main modulation of magnetofingerprints is driven by quantum interferences in a diffusive regime, prior to strong localization.

Figure 3.5 shows the evolution of the quantity $\Delta G(B) = G(B) - G(B = 0)$ (for $n_i = 0.2\%$) at two different selected energies. The small value of $W = 0.5$ considered here would prohibit valley mixing in the two-dimensional case. When $L/\ell e > 1$, a markedly positive magnetoconductance dominates ($\Delta G(B) > 0$) in agreement with the standard weak localization regime. The magnetoconductance variation increases for larger lengths, giving a stronger field-driven suppression of quantum interferences. The edges of the ribbon modify the band structure and reduce (if not suppress totally) the contribution of pseudospin-related phase interferences (Ortmann et al. 2011).

We have thus shown that these crossovers between WL to WAL can be rationalized by the atomistic potential profile of defects. By introducing a smooth disorder by means of a long-range disorder potential, changes from positive magnetoconductance to negative magnetoconductance have been found to be only driven by the strength

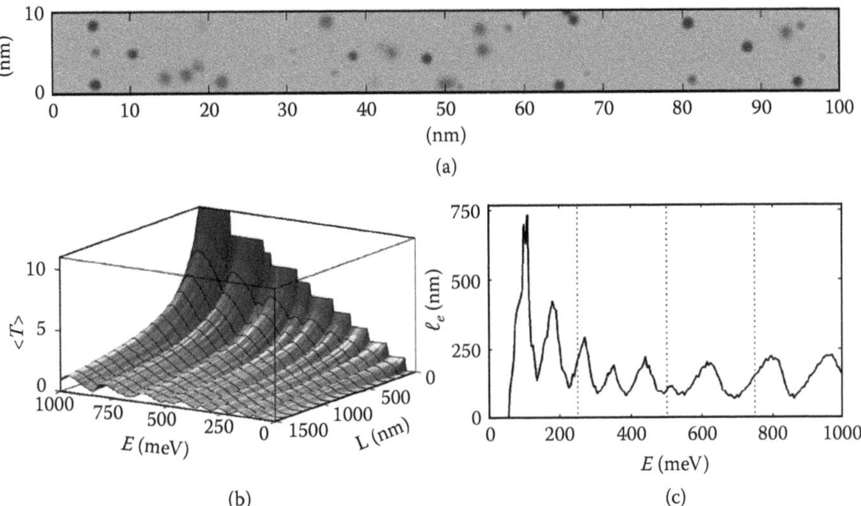

FIGURE 3.4 (a) Schematic of a short segment of a disordered nanoribbon with long-range scatters. Blue (red) spots correspond to positive (negative) values of the disorder potential. (b) 3D plot of the average total transmission coefficient for a 10 nm-wide armchair ribbon (84-aGNR) with $W = 1/2\gamma_0$, $n_i = 0.1\%$ for varying energy and length (up to 1.5 μm) of the disordered region (c) Energy-dependent mean free path for $n_i = 0.2\%$.

of the scattering potential (W). One reminds that the potential strength defined by W is directly related to the strength of Coulomb screening, which thus explains also why the magnetotransport fingerprints are significantly affected by a change of the Fermi level position.

These results rationalize the experimental phase diagram obtained recently (Tikhonenko et al. 2009).

3.3 GRAPHENE-BASED FIELD EFFECT TRANSISTORS: THE CLEAN CASE

This section presents a simple model to analyze (and design) the current-voltage (I-V) characteristics of graphene nanoribbon FETs (GNR-FETs) as a function of physical parameters, such as GNR width (W) or gate insulator thickness (t_{ins}), and electrical parameters, such as the Schottky barrier (SB) height (φ_{SB}) (Jiménez 2008). The model shares principles that are similar to the one formulated for carbon nanotube FETs (Jiménez et al. 2007). This approach prevents the computational burden required by self-consistent nonequilibrium Green's function-based methods (NEGF), by using a closed-form electrostatic potential from Laplace's equation. This simplification yields, however, fully accurate results compared with NEGF (Ouyang, Yoon, and Guo 2007) for the relevant limit dominated by the GNR quantum capacitance (Guo, Yoon, and Ouyang 2007) (C_{GNR}). Note that this aspect appears to be the relevant case for advanced applications because the ability of the gate to control the potential in the channel is maximized.

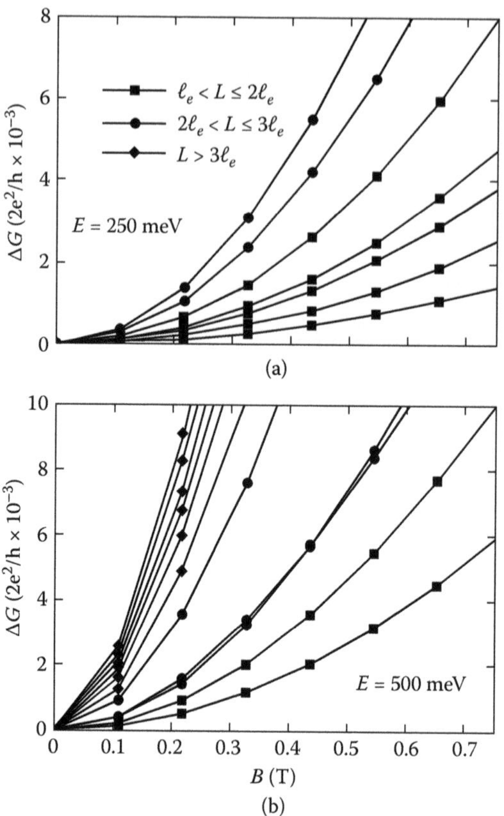

FIGURE 3.5 (a) Magnetoconductance for the disordered ribbon (W = 0.5) at two different selected energies for different ribbon lengths L. Different sets of curves are identified depending on the ratio L/ℓ*e*.

3.3.1 Graphene Nanoribbon Electrostatics

The first issue to consider for building up a model of the entire transistor is the GNR electrostatics. Let us assume a semiconducting GNR acting as the transistor channel contacted with metal electrodes serving as source/drain (S/D) reservoirs (Figures 3.6(a)–(b)). The resulting spatial band diagram along the transport direction has been sketched in Figure 3.6(c). For a long-channel transistor, the potential energy at the central region is exclusively controlled by the gate electrode and we assume that: (1) C_{GNR} dominates the total gate capacitance $C_G^{-1} = C_{ins}^{-1} + C_{GNR}^{-1} \approx C_{GNR}^{-1}$, where C_{ins} represents the geometrical capacitance; and (2) $C_{GNR} \approx 0$. The validity of the latter assumption depends on the quantum confinement strength. Downscaling W produces an increasing separation between adjacent peaks of the density of states versus energy. It is therefore more difficult to induce mobile charge (Q) into the GNR for reasonable values of gate voltage, and then $C_{GNR} = dQ/d\varphi_S \rightarrow 0$. In the quantum capacitance limit, the problem can be highly simplified because the electrostatics is governed by Laplace's equation, instead of the more involved Poisson's equation.

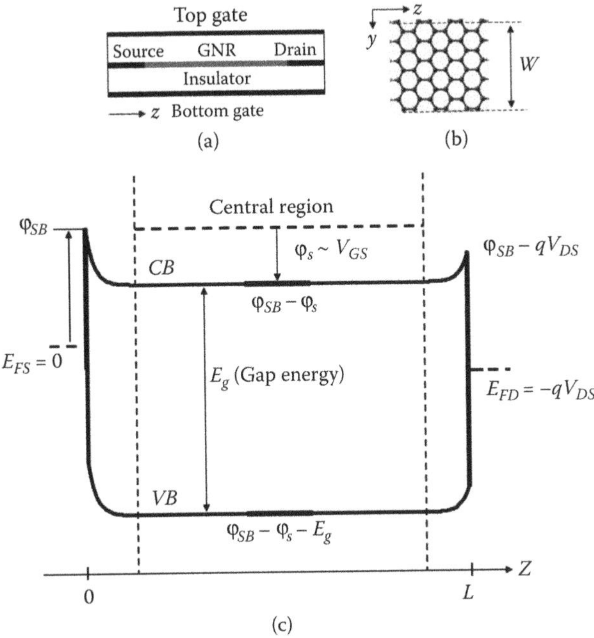

FIGURE 3.6 Geometry and band diagram of the GNR-FET: (a) cross section, (b) top view of the armchair GNRs forming the channel and (c) sketch of the spatial band diagram along transport direction.

This has two important consequences on the band diagram: (1) the central region shifts following the gate voltage in a 1:1 ratio or, equivalently, $\varphi_S = V_{GS}$; and (2) the band edge near the contact region has a simple analytical closed-form. For instance, the conduction band edge potential energy can be written as:

$$E_C(z) = \varphi_{SB} - \frac{2V_{GS}}{\pi}\arccos\left(e^{\frac{-z\pi}{2t_{\text{ins}}}}\right), \qquad 0 < z < \frac{L}{2}$$

$$E_C(z) = \left(\varphi_{SB} - V_{DS}\right) - \frac{2(V_{GS} - V_{DS})}{\pi}\arccos\left(e^{\frac{(z-L)\pi}{2t_{\text{ins}}}}\right), \qquad \frac{L}{2} \le z < L \tag{3.1}$$

where L is the channel length. This expression applies to a double-gate planar geometry in the long-channel limit with vanishing contact thickness (Morse and Feshbach 1953) (Figure 3.6[a]). The valence band can be written as $E_V(z) = E_C(z) - E_g$, where E_g is the energy gap. Analytical expressions of E_g for armchair-shaped edge GNRs with arbitrary chirality have been derived (Son, Cohen, and Louie 2006), presenting an inverse dependence with W.

An interesting question is why self-consistency is not needed for quantum capacitance-controlled devices. A simple model for self-consistency shows that the actual channel potential (U) is intermediate between the Laplace potential (U_L) and the potential needed to keep the channel neutral (U_N) (Datta 2005). Self-consistency means to determine U_N and U simultaneously. In the quantum capacitance limit, the channel potential is simply U_L and we can skip self-consistency. Next, we address the question about the design window where the quantum capacitance limit is relevant. The regime dominated by the quantum capacitance fulfills the condition $C_{GNR} < C_{ins}$, where

$$C_{ins} = N_G \varepsilon_r \varepsilon_0 \left(\frac{W}{t_{ins}} + \alpha \right) \tag{3.2}$$

N_G refers to the number of gates, ε_r is the insulator relative dielectric constant, and α is a fitting parameter ≈1 (Guo, Yoon, and Ouyang 2007). Hence, the quantum capacitance dominates the gate capacitance as long as:

$$N_G \varepsilon_r \left(\frac{W}{t_{ins}} + \alpha \right) > \frac{C_{GNR}}{\varepsilon_0} \tag{3.3}$$

Using, for instance, the quantum capacitance (C_{GNR}) value for a nanoribbon with $W = 5$ nm, of about 10 pF/cm (Guo, Yoon, and Ouyang 2007), the combination $N_G = 2$, $k = 16$, and $t_{ins} = 2$ nm just fulfills the above inequality, meaning that the quantum capacitance limit should be relevant, in general, for low-thickness and high-k insulators.

3.3.2 Graphene Nanoribbon Transport Model

The current along the channel can be calculated from Landauer's formula assuming a one-dimensional ballistic channel in between contacts that are further connected to external reservoirs, where dissipation takes place:

$$I = \frac{2q}{h} \sum_n \int_{-\infty}^{\infty} \mathrm{sgn}(E) T_n(E) \big(f(\mathrm{sgn}(E)(E - E_{FS})) - f(\mathrm{sgn}(E)(E - E_{FD})) \big) dE \tag{3.4}$$

where n is a natural number labeling subbands, $f(E)$ is the Fermi-Dirac distribution function, and T_n is the transmission probability of the nth-subband. This expression accounts for the spin degeneracy of the injected carriers. The current carried by each subband can be split into tunneling and thermionic components for carriers injected with energy within and above the barrier, respectively. Assuming phase-incoherent transport, transmission probabilities are computed through the *S/D* regions separately and then combined by using

$$T(E) = \frac{T_S T_D}{T_S + T_D + T_S T_D}$$

to obtain the total transmission to be included in Landauer's formula (Datta 1995). Tunneling transmission probability through a single SB is computed using the Wentzel-Kramers-Brillouin (WKB) approximation

$$T(E) = \exp\left(-2\int_{z_i}^{z_f} k(z)dz\right),$$

where the wavevector $k(z)$ is related to the energy by the GNR dispersion relation:

$$\pm\left(\left|E_{C,V}(z)\right| - |E|\right) + n\frac{E_g}{2} = \hbar v_F k(z), \tag{3.5}$$

where $v_F \sim 10^6$ m/s corresponds to the Fermi velocity of graphene, and the +/− sign applies to the calculation of tunneling and thermionic currents, respectively. The integration limits appearing in the transmission formula are the classical turning points. For computing tunneling transmission close to the source contact, note that as long as $|\varphi_S| < E_g$, the turning points satisfy $z_i = 0$ and $E_{C,V}(z_f) = E$ (for conduction and valence band, respectively). In the case of $|\varphi_S| > E_g$, the spatial band diagram curvature becomes high enough to trigger band-to-band tunneling (BTBT), and the turning points satisfy instead: $E_V(z_i) = E$ and $E_C(z_f) = E$ for electron BTBT; $E_C(z_i) = E$ and $E_V(z_f) = E$ for hole BTBT. Similar considerations must be made for tunneling through the drain contact barrier, but replacing φ_S by $\varphi_S - V_{DS}$. For energies $|E|$ above SB, the thermionic transmission probability can be computed using the WKB approach to yield (John 2006):

$$T(E) = \frac{16 k_C k_{\text{GNR}}^3}{\left(k'_{\text{GNR}}\right)^2 + 4\left(k_{\text{GNR}}^2 + k_C k_{\text{GNR}}\right)^2} \tag{3.6}$$

where k_C and k_{GNR} are the wave vectors in the contact and the GNR region close to the contact, respectively; the primed notation denotes a derivative respect to z. Assuming graphene metallic contacts $k_C = (|E| - E_F)/\hbar v_F$, with $E_F = E_{FS} = 0$ at the source contact and $E_F = E_{FD} = -qV_{DS}$ at the drain contact. The k_{GNR} wave vector at the S/GNR (D/GNR) interface can be easily obtained from Equations (3.1) and (3.5) at $z = 0$ ($z = L$). Using the approximation

$$E_C(z) = SB - \varphi_S\left(1 - e^{\frac{-\sqrt{2}z}{t_{ox}}}\right),$$

the derivative of $k(z)$ along the z-direction yields

$$\left|k'_{\text{GNR}}\right| = \left|\frac{dE_{C,V}(z)/dz}{\hbar v_F}\right| \approx \frac{\sqrt{2}\left|\varphi_S\right|}{\hbar v_F t_{\text{ins}}} \tag{3.7}$$

for the S/GNR interface. The same expression holds for the D/GNR interface, replacing φ_S by $\varphi_S - V_{DS}$.

3.3.3 Model Assessment

To assess the presented model, we have simulated the same nominal device as used in Ouyang, Yoon, and Guo (2007). It is formed by an armchair edge GNR channel with a ribbon index $N = 12$, presenting a width $W = \sqrt{3}d_{CC}(N-1)/2 \approx 1.35$ nm, where $d_{CC} = 0.142$ nm refers to the carbon–carbon bond distance. Room temperature and band gap of 0.83 eV were assumed for comparison purposes with the NEGF method (Ouyang, Yoon, and Guo 2007). This value was estimated using tight-binding methods, though a different band gap $E_g \approx 0.6$ eV results from a first-principles approach (Son, Cohen, and Louie 2006). A gate insulator thickness $t_{ins} = 2$ nm has been assumed. Note that the model, based on Laplace's equation, gives results that do not depend on the dielectric constant. The metallic S/D are directly attached to the GNR channel, and the SB height for both electrons and holes between S/D and channel are supposed to be half of the GNR band gap $\varphi_{SB} = E_g/2$. The flat band voltage is zero. A power supply of $V_{DS} = 0.5$ V has been assumed. The nominal device parameters have been varied to explore different scaling issues. The transfer characteristics exhibit two branches on the left and right from the minimum off-state current (Figure 3.7). This minimum occurs at $V_{GS} = V_{DS}/2$ for a half-gap SB height, which is the spatial band diagram symmetric for electrons and holes, and the respective currents are identical. This bias point is named the *ambipolar conduction point*. When V_{GS} is greater (smaller) than $V_{DS}/2$, the SB width for electrons (holes) is reduced, producing a dominant electron (hole) tunneling current. The effect of power

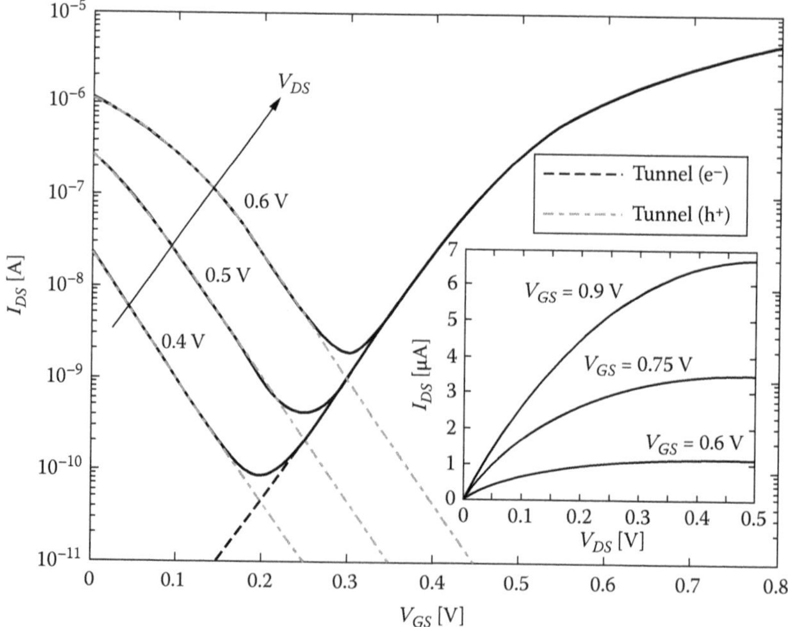

FIGURE 3.7 Transfer and output characteristics (inset) for the nominal GNR-FET. Decomposition of the total current in electron and hole tunneling contributions is shown.

supply up-scaling is to further reduce the SB width at the drain side, thus making it more transparent and allowing more turn-on current to flow.

The output characteristics of the SB GNR-FET are shown in the inset of Figure 3.7, with an overestimation of the current by a factor of 2 when compared to the NEGF-based model. The dominant current for the nominal device is due to electron tunneling and exhibits linear and saturation regimes. Increasing V_{GS} produces a larger saturation current and voltage due to further transparency of SB and the expansion of the energy window for carrier injection from the source into the channel. Moreover, downsizing W increases the gap and hence φ_{SB} in the simulation (assumed to be $E_g/2$) and further reduces the current due to less populated higher energy levels (Figure 3.8[a]). However, the resulting on-off current ratio, a figure-of-merit for digital circuits, is largely improved. Reducing SB height with respect to the half-gap case favors electron transport and results in a parallel shift of the ambipolar conduction point toward smaller gate voltages and asymmetries between the left and right branches of the transfer characteristic (Figure 3.8[b]). We also note that

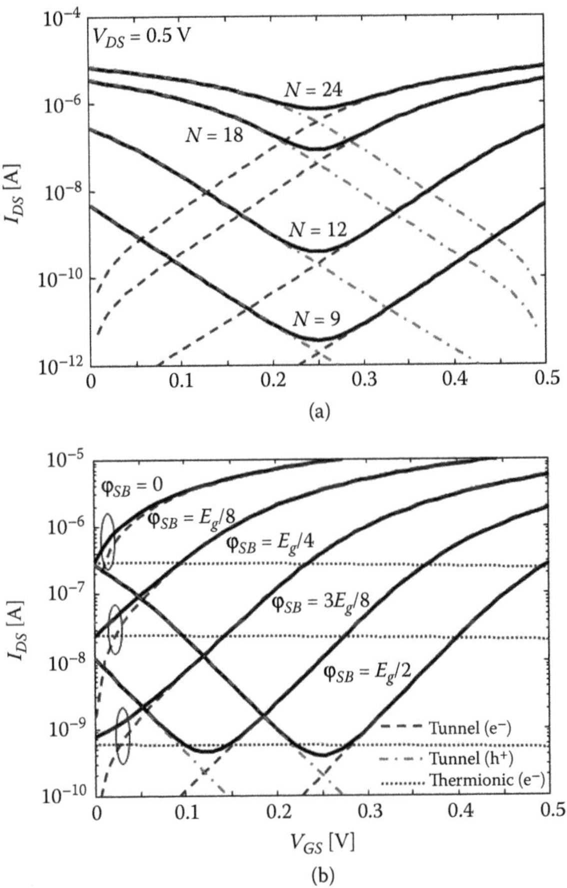

FIGURE 3.8 Influence of the GNR width (a) and SB height (b) in the transfer characteristics.

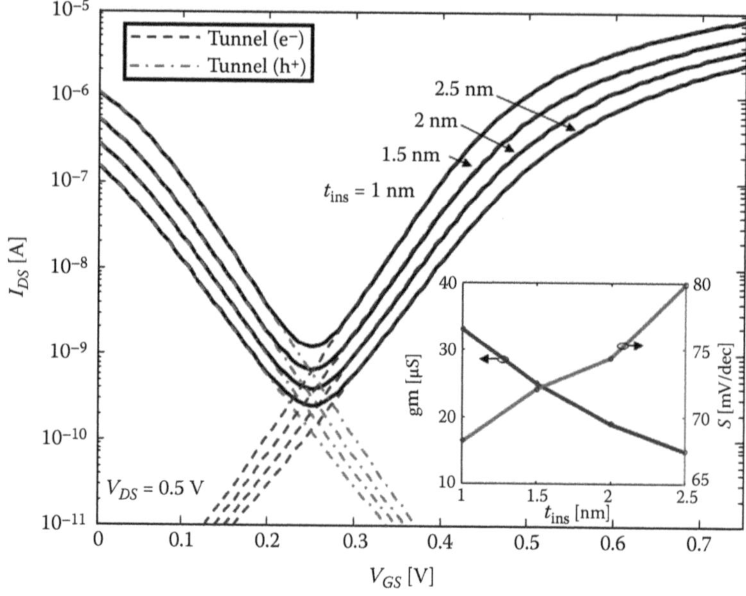

FIGURE 3.9 Impact of the gate insulator thickness scaling on the transfer characteristics. The inset shows the effect of scaling on the transconductance for V_{GS} = 0.75 V and subthreshold swing.

for low φ_{SB} and V_{GS}, the thermionic electron current exceeds the tunneling electron current, and this should be taken into account for computing the off-state current. It is worth pointing out that for the thin insulator considered here, the SB, whose thickness is roughly the insulator gate thickness, is nearly transparent, producing a small effect on the qualitative feature of the transfer characteristics (only a parallel shift). Hence, it does not seem feasible to further reduce the off-state current by engineering the SB height. The scaling of gate insulator thickness improves gate electrostatic control producing larger transconductances and smaller subthreshold swings, as shown in Figure 3.9. Also note that a thinner oxide produces a larger on-current and on-off current ratio. All results shown in Figures 3.7–3.9 are in close agreement with that obtained with the NEGF method, despite the fact that we assumed double-gate geometry for the simulations presented in Figures 3.7–3.9 instead of single-gate geometry (Ouyang, Yoon, and Guo 2007). This observation points out the limited influence of gate geometry for a quantum capacitance-controlled device.

In conclusion, this section has presented a simple model for the I-V characteristics of SB- graphene field-effect transistors that captures the main physical effects governing the operation of these devices. The results obtained by applying this model to prototype devices are in close agreement with a more rigorous treatment based on the NEGF approach, thus validating the approximations made. The presented model could assist at the design stage as well as for quantitative understanding of experiments involving GNR-FETs. We note, however, that the performances of clean graphene nanoribbon-based field-effect transistors have several drawbacks when compared to their metal-oxide-semiconductor field-effect transistor (MOSFET) counterparts. IN the following section we propose an alternative to improve graphene-based devices.

3.4 IMPROVING DEVICE PERFORMANCES: MOBILITY GAP ENGINEERING

We recently proposed a new device principle based on the chemical modification of graphene layers incorporating chemical substitutions (boron, nitrogen, phosphorous) (Biel et al. 2009a, Biel et al. 2009b, Roche et al. 2011). The new device principle is based on the engineering of electron-hole transport asymmetry and mobility gaps. The underlying concept extends to other types of controlled functionalization or intercalation with any other atomic element or molecular unit, provided their induced impurity levels present either a donor- or acceptor-type character with respect to carbon. The switching behavior efficiency and ON/OFF ratio of the chemically modified graphene-based material will be monitored by the induced mobility (or transport) gap, which has a different nature than the electronic band gap arising from the periodicity of the crystal and that might appear in cases of periodic disorder. Depending on the choice of chemical impurities, either hole or electron charge mobilities are markedly degraded, thus allowing enhanced current density modulations under electrostatic gating. This is very different from the conductance fluctuations due to natural defects occurring in small-width GNRs. Indeed, when the lateral size becomes too small, transport properties and device-to-device fluctuations become dominated by edge disorder, as we show here.

In Figure 3.10, we illustrate the fluctuation of the quantum conductance for several edge disorder profiles, albeit keeping the same density of removed edge carbon atoms. These fluctuations impact on the resulting elastic (and inelastic) mean free path, which will turn the graphene devices from a quasiballistic to a strongly diffusive regime, with considerable variability effects (Cresti and Roche 2009, Roche 2011). This is due to the increasing sensitivity of quantum transport in reduced dimension and the symmetry-breaking effects induced by certain types of short-range scatters. Additionally, there is convincing evidence that under large current flows, edge geometries of graphene ribbons significantly reconstruct. This phenomenon has been experimentally observed by recent Raman spectroscopy (Xu et al. 2011) and transport measurements (Shimizu et al. 2011), while detailed calculations of edge reconstruction and edge chemical passivation effects were previously theoretically discussed (Dubois et al. 2010). This complex edge profile reconstruction can be seen as a natural process taking place in situations of strong energy dissipation. Edges of graphene nanoribbons are, in contrast to the bulk graphene or carbon nanotubes, very sensitive to chemical dangling bonds and during current flow, high electron–phonon coupling can produce local geometrical (and topological) rearrangements. Conversely, chemical substitutions are stable and insensitive to energy dissipation and current flows. In that perspective, the induced mobility gaps discussed in the following text will be robust to device operation up to room temperature and high bias voltages.

The impact of boron or nitrogen substitutions (doping) and edge disorder have been investigated using first-principles methods based on the density functional

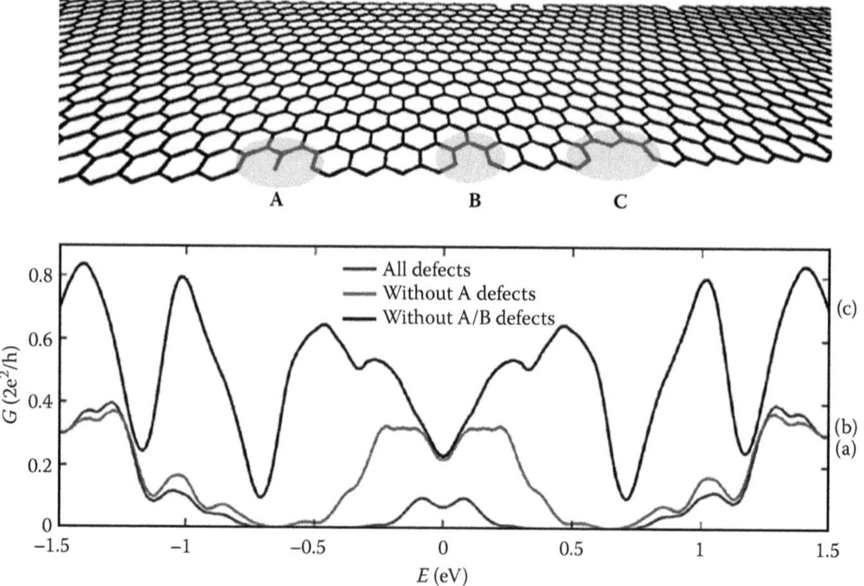

FIGURE 3.10 Top: schematic representation of the edge-disordered ribbon with A (dangling atoms), B (single missing hexagons), and C (double missing hexagons) defects, marked with surrounding solid circles. Bottom: conductance of a disordered 16-zGNR (with length L = 500 nm) with 7.5% of randomly removed edge carbon atoms. Case (a) includes A defects, B and C defects, whereas defects of type A are disallowed in case (b), and A and B defects are disallowed for case (c).

theory (DFT) method as implemented in the SIESTA code (Sánchez-Portal et al. 1997). A self-consistent calculation provides the profile of the scattering potential around the impurity or the edge defect, which generally produces quasibound states well localized in energy (Biel et al., 2009b). A tight-binding model is further elaborated by adjusting the onsite and hopping self-consistent Hamiltonian matrix elements (on a localized basis set) in order to reproduce the ab initio conductance fingerprints of a single impurity (Biel et al. 2009b, Cresti and Roche 2009). Therefore, this method combines the accuracy of ab initio calculations with the moderate computational cost of a tight-binding (TB) Hamiltonian.

We have simulated large-width graphene nanoribbons (above 10 nm in lateral sizes) and we have shown that it is possible to compensate for the loss of gain due to band-gap shrinking by triggering the mobility gaps through chemical doping. All calculations are performed within the Landauer-Büttiker formalism, and standard order (N) decimation procedures have been used to quickly calculate the conductance of disordered ribbons up to the micron scale by combining the scattering potentials of isolated impurities.

It has been shown that the scattering potential of a single impurity (dominated by the energy position of the quasibound state induced by the impurity) strongly depends on the dopant position with respect to the ribbon edges, and so the conductance profile of a single dopant will be different for a different location of the

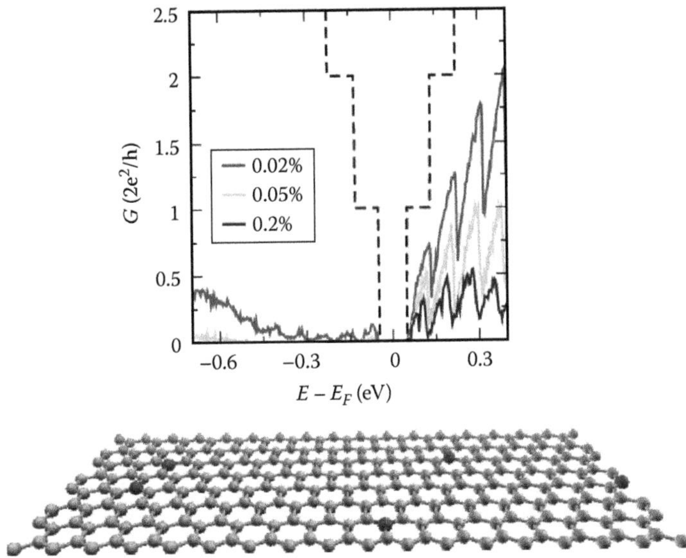

FIGURE 3.11 Top panel: average conductance as a function of energy for the semiconducting 81-aGNR and three selected doping rates (about 0.02%, 0.05% and 0.2%, from top to bottom). Bottom: Schematic plot of a randomly doped 34-aGNR.

impurity (Biel et al. 2009b). In the case of acceptor-like dopants, such as boron, the quasibound states are mainly localized in the valence band while the conduction band is not much affected at energies close to the Fermi level. The obtained mobility gaps in randomly boron-doped ribbons are thus unique consequences of a wide distribution of quasibound states over the entire valence band (for acceptor-type impurities) in the first conductance plateau, provided that dopants are randomly distributed not only along the ribbon length but also across the ribbon width.

Figure 3.11 shows the conductance of a 10-nm-wide armchair nanoribbon with a low boron doping. For a doping density of about 0.2%, the system presents a mobility gap of the order of 1 eV, caused by the degradation of the conductance for holes in the valence band and a good preservation of the electron transport in the conduction band. When lowering the doping level to 0.05%, the mobility gap reduces to about 0.5 eV and finally becomes less than 0.1 eV for lower density. The 0.2% case is obtained for a fixed nanoribbon width and length, so that adjustments need to be performed if up-scaling either lateral or longitudinal sizes, but the recipe for the gap creation is straightforward once the transport length scales (mean free paths, localization length) have been computed (Biel et al. 2009a).

3.5 CONCLUSION

Several electronic and transport features of clean and chemically modified graphene-based materials and devices (field-effect transistors) have been discussed, including pseudospin effects in weakly disordered graphene material, or the possibility to chemically tune transport properties of graphene ribbons by using chemical doping. The ultimate properties of clean graphene nanoribbon-based field-effect transistors have been discussed, while mobility gaps have been proposed to open a new perspective for efficient graphene-based nanoelectronics.

ACKNOWLEDGMENTS

B.B. acknowledges financial support from the Juan de la Cierva Program and the FIS2008-05805 Contract of the Spanish MICINN. A.C. acknowledges support from Fondation Nanosciences via the RTRA Dispograph project. F. O. acknowledges support by a Marie Curie fellowship within the 7th European Commission Framework Programme. This work was partly funded by the European Union under Contract No. 215752 "GRAND" and by the NANOSIM-GRAPHENE project n·ANR-09-NANO-016-01 funded by the French National Agency (ANR) in the frame of its 2009 programme in Nanosciences, Nanotechnologies and Nanosystems (P3N2009). Financial support of this work was provided by Ministerio de Ciencia e Innovación under project FR2009-0020 and TEC2009-09350. Support from the French Ministry of Foreign Affairs is also acknowledged within the PICASSO project.

REFERENCES

Biel, B., Triozon, F., Blase, X., and Roche, S. 2009a. Chemically induced mobility gaps in graphene nanoribbons. *Nano Lett.* 9, 2725–2728.

Biel, B., Blase, X., Triozon, F., and Roche, S. 2009b. Anomalous doping effects on charge transport in graphene nanoribbons. *Phys. Rev. Lett.* 102, 096803.

Castro Neto, A. H., Guinea, F., Peres, N. M. R, Novoselov, K. S., and Geim, A. K. 2009. The electronic properties of graphene. *Rev. Mod. Phys.* 81, 109.

Charlier, J. C., Blase, X., and S. Roche. 2007. Electronic and transport properties of carbon nanotubes. *Rev. Mod. Phys.* 79, 677.

Cresti, A., Nemec, N., Biel, B., Niebler, G., Triozon, F., Cuniberti, G., and Roche, S. 2008. Charge transport in disordered graphene-based low dimensional materials. *Nano Research* 1, 361.

Cresti, A., and Roche, S. 2009. Edge-disorder-dependent transport length scales in graphene nanoribbons: From Klein defects to the superlattice limit. *Phys. Rev. B* 79, 233404.

Datta, S. 1995. *Electronic transport in mesoscopic systems*, 63. Cambridge: Cambridge University Press.

Datta, S. 2005. *Quantum transport: Atom to transistor*, 172–173. Cambridge: Cambridge University Press.

Dubois, S. M. M., Lopez-Bezanilla, A., Cresti, A., Triozon, F., Biel, B., Charlier, J. C., and Roche, S. 2010. Quantum transport in graphene nanoribbons: Effects of edge reconstruction and chemical reactivity. *ACS Nano* 4, 1971–1976.

Geim, A. K., and Novoselov, K. S. 2007. The rise of graphene. *Nat. Mater.* 6, 183.

Guo, J., Yoon, Y., and Ouyang, Y. 2007. Gate electrostatics and quantum capacitance of graphene nanoribbons. *Nano Lett.* 7, 1935.

Hikami, A., Larkin, I., and Nagaoka, Y. 1980. Spin–orbit interaction and magnetoresistance in the two-dimensional random system. *Prog. Theor. Phys.* 63, 707.

Jiménez, D. 2008. A current–voltage model for Schottky-barrier graphene-based transistors. *Nanotechnology* 19, 345204.

Jiménez, D., Cartoixà, X., Miranda, E., Suñé, J., Chaves, F. A., and Roche S. 2007. A simple drain current model for Schottky-barrier carbon nanotube field effect transistors. *Nanotechnology* 18, 025201.

John, D. L., 2006. Simulation studies of carbon nanotube field-effect transistors, chap. 4, 23. PhD thesis, University of British Columbia.

Katsnelson, M. I., Novoselov, K. S., and Geim, A. K. 2006. Chiral tunnelling and the Klein paradox in graphene. *Nat. Phys.* 2, 620.

Leconte, N., Moser, J., Ordejon, P., Tao, H., Lherbier, A., Bachtold, A., Alzina, F., Sotomayor Torres, C. M., Charlier, J. C., and Roche, S., 2010. Damaging graphene with ozone treatment: A chemically tunable metal–insulator transition. *ACS Nano* 4 (7), 4033–4038.

Lemme, M. 2010. Graphene transistors. *Sol. State Phen.* 156–158, 499–509.

Loh, K. P., Bao, Q, Ang, P. K., and Yang, J. 2010. The chemistry of graphene. *J. Mat. Chem.,* 20, 2277–2289.

McCann, E., Kechedzhi, K., Falko, V., Suzuura, H., Ando, T., and Altshuler, B. L. 2006. Weak-localization magnetoresistance and valley symmetry in graphene. *Phys. Rev. Lett.* 97, 146805.

Morse, P. M., and Feshbach, H. 1953. *Methods of theoretical physics*, New York: McGraw-Hill.

Moser, J., Tao, H., Roche, S., Alzina, F., Sotomayor Torres, C. M., and Bachtold, A. 2010. Magnetotransport in disordered graphene exposed to ozone: From weak to strong localization. *Phys. Rev. B* 81, 205445.

Ortmann, F., Cresti, A., Montambaux, G., and Roche, S. 2011. Magnetoresistance in disordered graphene: The role of pseudospin and dimensionality effects unravelled. *Eur. Phys. Lett.* 94, 47006.

Ouyang, Y., Yoon, Y. and Guo, J. 2007. Scaling behaviors of graphene nanoribbon FETs: A 3D quantum simulation. *IEEE Trans Electron Devices* 54, 2223.

Roche, S. 1999. Quantum transport by means of O(N) real-space methods. *Phys. Rev. B* 59, 2284.

Roche, S. 2011. Graphene gets a better gap. *Nat. Nano.* 6, 8.

Roche, S., Biel, B., Cresti, A., and Triozon, F. 2011. Chemically enriched graphene-based switching devices: A novel principle driven by impurity-induced quasi-bound states and quantum coherence. *Physica E* (in press).

Roche, S., and Mayou, D. 1997. Conductivity of quasiperiodic systems: A numerical study. *Phys. Rev. Lett.* 79, 2518.

Roche, S., and Saito, R. 2011. Magnetoresistance of carbon nanotubes: From molecular to mesoscopic fingerprints. *Phys. Rev. Lett.* 87, 246803.

Sánchez-Portal, D., Ordejón, P., Artacho, E., and Soler, J. M. 1997. Density-functional method for very large systems with LCAO basis sets. *Int. J. Quant. Chem.* 65, 453.

Shimizu, T., Haruyama, J., Marcano, D. C., Kosinkin, D. V., Tour, J. M., Hirose, K., and Suenaga, K. 2011. Large intrinsic energy band gaps in annealed nanotube-derived graphene nanoribbons. *Nat. Nano.* 6, 45.

Son, Y. W., Cohen, M. L., and Louie, S. G. 2006. Energy gaps in graphene nanoribbons. *Phys. Rev. Lett.* 97, 216803.

Tikhonenko, F. V., Kozikov, A. A., Savchenko, A. K., and Gorbachev, R. V. 2009. Transition between electron localization and antilocalization in graphene. *Phys. Rev. Lett.* 103, 226801.

Xu, Y. N., Zhan, D., Liu, L., Suo, H., Ni, Z. H., Nguyen, T. T., Zhao, C., and Shen, Z. X. 2011. Thermal dynamics of graphene edges investigated by polarized raman spectroscopy. *ACS Nano* 5(1):147-52.

4 Electronic and Photonic Applications for Ultrahigh-Frequency Graphene-Based Devices

Taiichi Otsuji, Tetsuya Suemitsu, Akira Satou, Maki Suemitsu, Eiichi Sano, Maxim Ryzhii, and Victor Ryzhii

CONTENTS

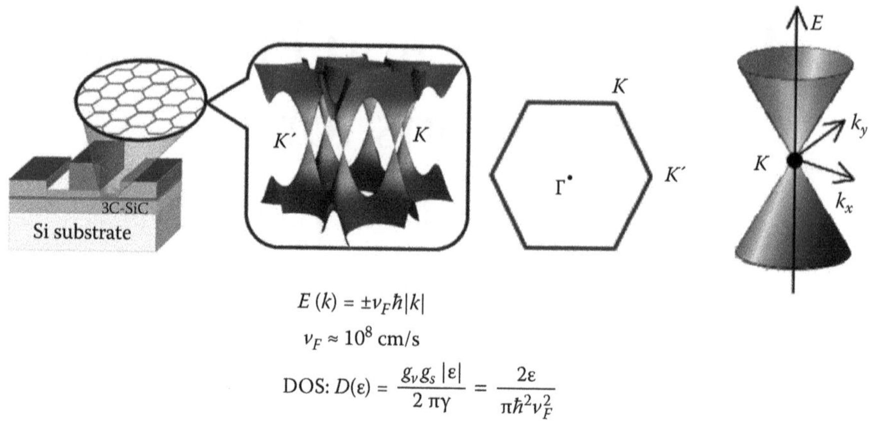

FIGURE 4.1 Energy band structures and dispersion relations for graphene.

4.1 INTRODUCTION

Graphene, a monolayer of sp^2-bonded carbon atoms in a honeycomb crystal lattice, has attracted considerable attention due to its unique carrier transport and optical properties [1–4]. Figure 4.1 depicts the energy band structures and dispersion relations for graphene. The conduction band and the valence band of graphene take a symmetrical corn shape around the K and K′ points and contact each other at the K and K′ points. Electrons and holes in graphene hold a linear dispersion relation with zero band gap, resulting in peculiar features like massless relativistic Fermions with backscattering-free ultrafast transport [2–8] as well as the negative-dynamic conductivity in the terahertz (THz) spectral range under optical pumping [9–11].

The history of the graphene synthesis began in the late 1990s with the thermal decomposition of epitaxially grown SiC on a SiC substrate [12] and even earlier in the late 1960s with chemical vapor deposition using a metallic catalyst [13]. The first success in the synthesis of monolayer graphene was made by A. Geim and K. Novoselov in 2004 using mechanical exfoliation from bulk graphite [1]. The unusual properties of exfoliated monolayer graphene, which had been theoretically studied for more than 60 years [5–8,14–17], were experimentally observed and verified in 2005 by A. Geim et al. [2] and P. Kim et al. [3], almost at the same time. These groundbreaking achievements prompted research and development of graphene-based electronic, optoelectronic, and photonic devices.

The electronic properties of graphene, the significant mobilities of massless electrons/holes (due to its linearly dispersive band structure), and real two-dimensional electron/hole systems (due to the thin monolayer structure) are superior advantages beyond any other semiconductor materials [1–8]. Thanks to the linear dispersion relation, the density of states in graphene is proportional to the energy, which creates extremely high saturation density of electrons and holes. Sheet electron/hole density on the order of 10^{13} cm^{-2} is easily obtainable, which is more than one order of magnitude higher than those of conventional semiconductor materials. Furthermore, the saturation velocities of electrons and holes are quite high because no valley exists around the K and K′ points and optical phonon energy is high enough that the optical phonon scattering becomes weaker than scattering in conventional semiconductor materials. When graphene is introduced in field-effect transistors (FETs) as the channel material, it will exceed the limits on conventional planar transistor performance, so that it could become a booster technology for making short-channel-free ultimately fast transistors.

However, the gapless energy spectrum of graphene is an obstacle for creating transistor digital circuits based on graphene-channel FETs (G-FETs) due to the nature of ambipolar behavior and relatively strong interband tunneling in the FET off state [18–20], and suffers from a poor on/off ratio of its switching current. Therefore, graphene-based structures like graphene nanoribbons [5, 21–23], graphene nanomeshes [24], and graphene bilayers [25–27] that can open the energy band gap should be introduced to fabricate G-FETs with a sufficiently large on/off ratio. The band gap opening sacrifices the electron transport properties; its energy dispersion becomes parabolic, yielding a nonzero effective mass.

Let us look at the photonic properties of graphene. When we consider the nonequilibrium carrier relaxation–recombination dynamics of optically pumped graphene, a very fast energy relaxation of photoexcited electrons/holes via the optical phonon emission and a relatively slow recombination will lead to the population inversion in the wide THz range under sufficiently high pumping intensity. This will make it possible to obtain negative dynamic conductivity or gain in the THz range [9,10]. Such an active mechanism can be used for creating graphene-based coherent sources of THz radiation.

In this chapter, the recent advances in theoretical and experimental studies on applications of graphene materials to electronic and photonic devices that have been done by the chapter authors' group will be reviewed.

4.2 GRAPHENE-BASED ELECTRONIC DEVICES

4.2.1 Band Gap Engineering for Graphene

Monolayer graphene shows a so-called ambipolar characteristic, where electrons and holes coexist symmetrically against the Fermi level (also called the Dirac point). Thus, when monolayer graphene is introduced as a channel material of the FET, the channel current cannot be turned off. The formation of the band gap is necessary to cope with this problem. Patterning of monolayer graphene into a nanoribbon can open the band gap by the spatial confinement of electrons. Chiral stacking of monolayer graphene into multiple layers is another way to open the band gap by deforming the orbital of pi electrons forced by the interfacial potential difference between graphene and graphene or between graphene and the substrate.

Due to the honeycomb lattice structure, graphene has two types of carrier transport: one is along with the armchair edge and the other is along the zigzag edge, as shown in Figure 4.2(a). Graphene nanoribbons exhibit an energy spectrum with a gap

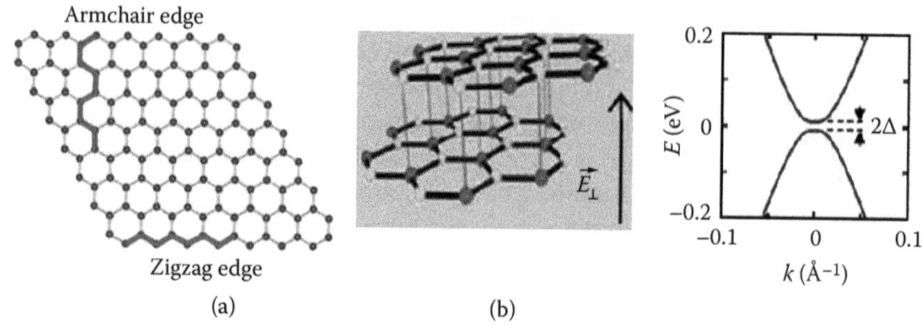

FIGURE 4.2 (a) Carrier transport for monolayer graphene and (b) band gap opening for chirally stacked bilayer graphene. (From J. B. Oostinga et al. 2008. *Nature Materials* 7: 151. With permission.)

between the valence and conduction bands for armchair transport depending on the nanoribbon width d_r [19]:

$$\varepsilon^{\mp}_{p,n} = \pm v\sqrt{p^2 + (\pi\hbar/d_r)^2 n^2}\,, \tag{4.1}$$

where $v \approx 10^8$ cm/s is the characteristic velocity of the electron (upper sign) and hole (lower sign) spectra, p is the momentum along the nanoribbon, $\hbar$ is the reduced Planck constant, and $n = 1, 2, 3, \ldots$ is the subband index. The quantization corresponding to Equation (1.1) of the electron and hole energy spectra in nanoribbons due to the electron and hole confinement in one of the lateral directions results in the appearance of the band gap between the valence and conduction bands and a specific density of states (DOS) as a function of the energy.

On the other hand, as shown in Figure 4.2(b), chirally A-B stacked bilayer graphene exhibits an energy gap E_g between the valence and conduction bands [28]:

$$E_g = \frac{edV_g}{W} \tag{4.2}$$

where $d \approx 0.36$ nm is the effective spacing between the graphene layers in the graphene bilayer (GBL) which accounts for the screening of the electric field between these layers, W is the distance between the gate and the graphene layer, and V_g is the gate-source voltage.

The authors numerically simulated the energy band structures and corresponding electron effective mass for monolayer and bilayer armchair graphene nanoribbons epitaxially grown on SiC substrates using a nearest-neighbor tight-binding approximation with two distinct models (interlayer coupling [ILC] [29] and substrate-induced asymmetry [SIS] models [30]). The results are plotted in Figure 4.3 [31]. In the case of monolayer graphene, the strong band-gap hopping is clearly seen, even for wider ribbon width conditions with a narrower band gap <100 meV for the ILC model according to the number of carbon atoms n in the ribbon width direction [31]. This indicates that the monolayer nanoribbon graphene cannot practically be engineered if the ILC model is valid since controlling the number of atoms is almost

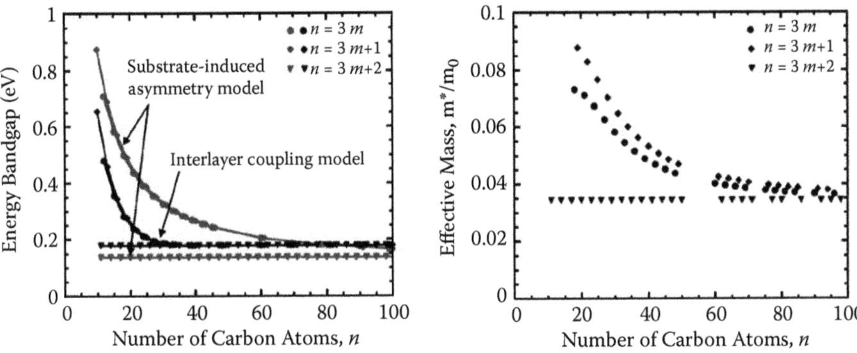

FIGURE 4.3 The dependence of (a) the energy band gap and (b) the effective mass of chirally stacked bilayer graphene on ribbon width. The horizontal axis takes the number of carbon atoms along with the ribbon width direction. (After E. Sano and T. Otsuji. 2009. *Japanese Journal of Applied Physics* 48: 041202. With permission.)

impossible with today's fabrication technology. On the other hand, in the case of bilayer graphene, the band-gap hopping decreases with increased ribbon width. In a wider width range $n > 80$ (~ >20 nm in width), the band-gap hopping disappears with a saturated band gap of ~180 meV for the SIS model or shrinks to ~30 meV with a saturated band gap of ~135 or ~180 meV. As a consequence, it is suggested that FET channels should be constructed with wide films of bilayer armchair graphene on SiC substrates.

4.2.2 Analytical Modeling for G-FETs

Let us consider a G-FET with a highly conducting substrate serving as the backgate, and with the topgate, as well as with two contacts (source and drain) to the channel as shown in Figure 4.4. Energy distribution of electrons and holes in graphene is similar to those in normal semiconductors and is given by the Fermi-Dirac statistics. The spatial distribution of electron and hole densities in graphene is also given by the Poisson equation, as it is in normal semiconductors. Assuming that the gate width L_w (taking to the y axis) is sufficiently wide compared to the gate length L_g (taking

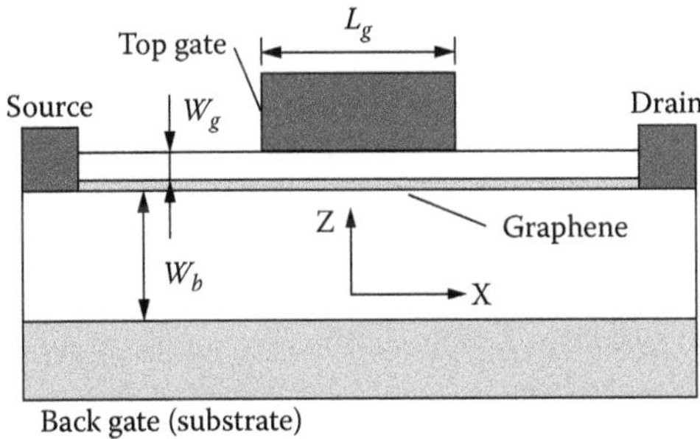

FIGURE 4.4 Cross-sectional view of a model for graphene-channel FETs.

to the x axis), the transport of electrons and holes is formulated by the semiclassical Boltzmann kinetic equation [28]:

$$\frac{\partial f_{\mathbf{p}}}{\partial t} + v_x \frac{\partial f_{\mathbf{p}}}{\partial x} = \int d^2\mathbf{q}w(q)(f_{\mathbf{p+q}} - f_{\mathbf{p}})\delta(\varepsilon_{\mathbf{p+q}} - \varepsilon_{\mathbf{p}}) \tag{4.3}$$

where $f_{\mathbf{p}}(x,t)$ is the electron/hole distribution function, v_x is the electron/hole characteristic velocity, $w(q)$ is the probability of the electron/hole scattering on disorder and acoustic phonons with the variation of the electron/hole momentum by quantity q, and $\varepsilon_{\mathbf{p}}$ is the electron/hole energy with momentum $\mathbf{p}$. The density of the electron (thermionic) current, $J = J(x,t)$, in the gated section of the channel (per unit length in the y-direction) can be calculated using the following formula [28]:

$$J = \frac{4e}{(2\pi\hbar)^2}\int d^2\mathbf{p}v_x f_{\mathbf{p}}. \tag{4.4}$$

When the gate bias V_g is below the threshold voltage (which is called the Dirac voltage V_{th}) and the drain bias V_d (>0) is applied, the G-FET channel constitutes a lateral n-p-n structure. When the band gap is not sufficiently wide, the tunneling current J_T should be considered [18]:

$$J_T = G_T V_\varphi,\ G_T = \frac{2e^2}{\pi\hbar}\frac{L_w}{4\pi}\sqrt{\frac{F}{\hbar v_x}},\ V_\varphi = V_d - \frac{\pi}{4\sqrt{3}}\frac{\sqrt{W_b W_g}}{e\hbar v}E_g{}^2 \tag{4.5}$$

where

$$F = e\left|\frac{d\varphi}{dx}\right|$$

is the slope of the potential energy curve at the tunneling points, W_b and W_g are the distance between the channel and the backgate and the topgate, respectively, and E_g is the band-gap energy.

According to the previously mentioned formulations, device models of the DC and AC characteristics for graphene nanoribbon FETs (GNR-FETs) and graphene bilayer FETs (GBL-FETs) have been derived by V. Ryzhii and his collaborators [28,32–34]. Figures 4.5 and 4.6 show typical DC current-voltage characteristics and maximum transconductance simulated for a GBL-FET having a relatively thin gate stack W_t, W_b = 5 or 10 nm [34]. As is seen in Figure 4.5, thanks to the backgate-biased high background carrier concentration in the channel, an excellent current density over 3 A/mm as well as a high *on/off* current ratio of more than two orders of magnitude can be obtained even for relaxed dimensions L_g = 100 nm and $W_t = W_b$ = 10 nm. Correspondingly, a maximum transconductance of >1.5 S/mm is expected, as seen in Figure 4.6. In the case of an ultrathin topgate stack of W_g ~1 nm, an extremely

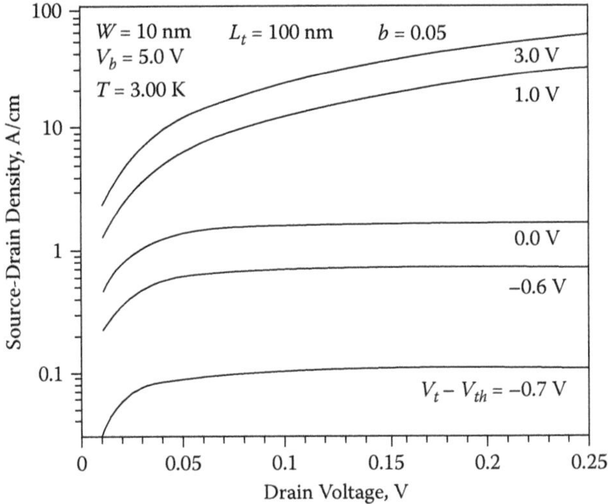

FIGURE 4.5 Simulated DC current voltage characteristics in GBL-FETs. W_b: the backgate insulator thickness, W_t: the topgate insulator thickness, L_t: the topgate length, V_b: the backgate bias, V_t: the topgate bias, and V_{th}: the threshold level. (After V. Ryzhii et al. 2011. *Journal of Applied Physics* 109: 064508. With permission.)

high drain current density $I_d \approx$ 10 A/mm and the maximum transconductance $G_m \approx$ 10 S/mm are expected. Compared to the performance projection for GNR-FETs, these values for GBL-FETs are superior. This is due to the higher area density of the channels for GBL-FETs than for GNR-FETs.

4.2.3 State-of-the-Art G-FETs

IBM has led the development of high-frequency G-FETs. They demonstrated a highest current-gain-cutoff frequency f_T of 100 GHz at the time of publication by using a 240-nm-long epitaxial-graphene channel [35]. The graphene channel layer is grown by thermal decomposition of a SiC epitaxial layer grown on a bulk SiC wafer. A gate insulator of 10-nm-thick HfO_2 is formed by the atomic layer deposition technique. A drain current saturation characteristic that originates from the band-gap opening is not clearly confirmed, although the graphene channel is specified to be monolayer or bilayer. The field-effect mobility is characterized as ~1500 cm^2/(Vs).

Recently UCLA has demonstrated a record 300-GHz f_T performance by a G-FET featuring a dedicated Co_2/Si composite nanowire gate wrapped with a 5-nm-thick Al_2O_3 insulator [36]. Graphene is mechanically exfoliated from bulk graphite and transferred to a SiO_2/Si substrate to form the channel. After aligning the Co_2/Si/Al_2O_3 core-shell wire onto the graphene channel, 10-nm-thick Pt is deposited over the active area. Then the source and drain electrodes are formed in a self-aligned manner, which is the key to minimizing the parasitic access resistance that severely deteriorates the FET performance. The equivalent gate length is characterized to be 140 nm. Thanks to the inert nanowired gate

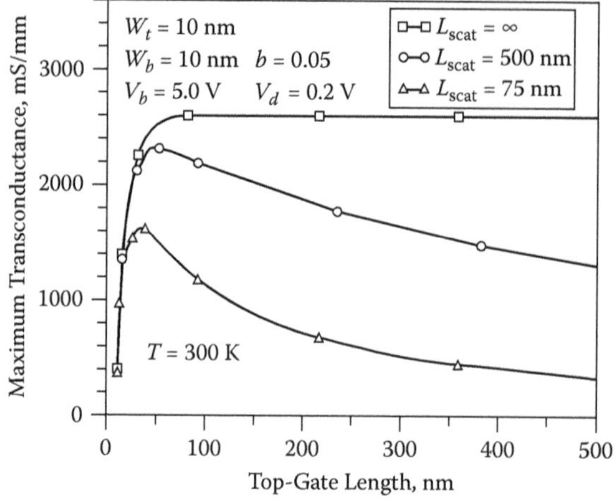

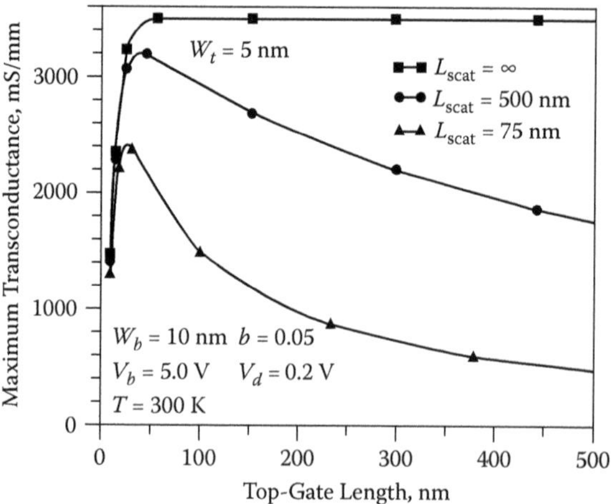

FIGURE 4.6 Simulated maximum transconductance vs. topgate length for GBL-FETs. W_b: the backgate insulator thickness, W_t: the topgate insulator thickness, L_{scat}: the characteristic scattering length, V_b: the backgate bias, V_d: the drain bias, and V_{th}: the threshold level. (After V. Ryzhii et al. 2011. *Journal of Applied Physics* 109: 064508. With permission.)

stacking, an excellent high field-effect mobility of 20,000 cm^2/(Vs) is also demonstrated. Although such an acrobatic gate stack technique is premature out of the standard planar process technology, the result has extended high-frequency performance of G-FETs, approaching the level of InP-based high electron mobility transistors (HEMTs).

Figure 4.7 plots f_T versus L_g for various types of FETs. The original figure in Schwierz [37] is modified with additional plots for the previously mentioned G-FETs. It is clearly seen that the real performance of G-FETs is just now on the same level as that for InP-based HEMTs and needs more study to demonstrate the superior performance expected from the original nature of graphene.

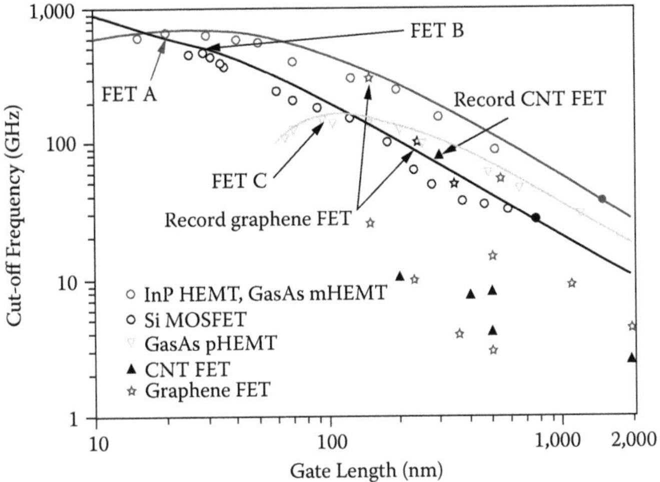

FIGURE 4.7 Technology trend in cutoff frequency vs. gate length for various FETs. Data for graphene channel FETs were added to the original figure in F. Schwierz. 2010. *Nature Nanotechnology* 5: 487–496. (Used with permission.)

4.2.4 Graphene-on-Silicon FETs

4.2.4.1 Epitaxial Graphene-on-Silicon Growth Technology

Exfoliation from highly oriented pyrolytic graphite and surface decomposition of epitaxial SiC are well known as graphene formation technologies [38]. However, when it is introduced to post-complementary metal-oxide semiconductor very-large-scale integration (CMOS VLSIs), low-temperature and reproducible growth technology starting with a Si substrate is mandatory. Recently, chemical vapor deposition of graphene with a metallic catalyst has made it possible to synthesize graphene in a wide area onto a Si substrate [39–41]. However, metallic catalysts like Fe, Ni, and Cu suffer from contamination in the FET fabrication process.

To cope with those issues, heteroepitaxial graphene-on-silicon (GOS) technology has been developed [42]. Epitaxial graphene is formed on the surface of 3C-SiC(110) grown on a B-doped p-type Si(110) substrate. 3C-SiC was grown by gas-source molecular beam epitaxy (GSMBE) using monomethyl silane (MMS) [43] at a pressure of 3.3×10^{-3} Pa. The SiC growth process consists of two stages. First, a buffer layer is formed (for 5 min) and a subsequent SiC growth (for 120 min). For this growth condition, the thickness of the SiC layer is typically 80 nm. After that, the sample is annealed in vacuum at 1200°C for 30 minutes to form a graphene film at the SiC surface. For the fabricated epitaxial graphene sample, we confirmed that few-layer graphene (FLG) was formed at the SiC surface by evaluating (1) the sample surface with a Normarski microscope and Raman-scattering microscopy (Renishaw, Ar 514 nm) [44] and (2) the sample cross section with a transmission electron microscope (TEM). The Raman spectra and the TEM image are shown in Figure 4.8. The G band in the Raman spectra reflects the direct lattice structure of graphene.

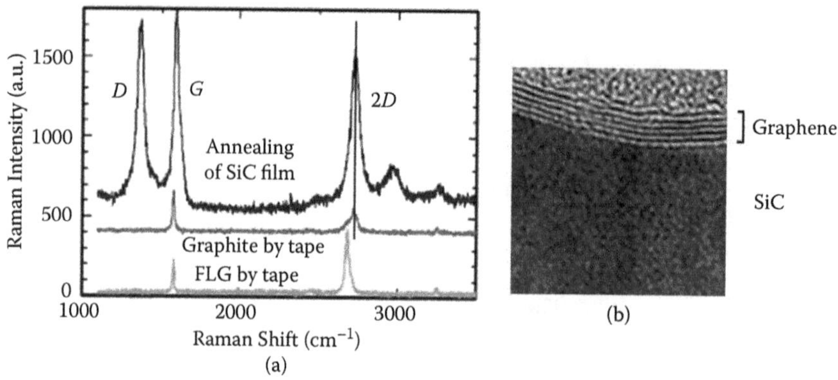

FIGURE 4.8 (a) Raman spectra for a GOS sample (top) in comparison with graphite (middle) and mechanically exfoliated few-layer graphene (bottom). (b) TEM image for a GOS sample.

The D band originates from the defects in graphene. The ratio of the spectral peak intensity, G/D, corresponds to the grain size, which is identified as ~20 nm. The G′ band provides important information about the wave functions of the pi electrons in graphene, which originate from the double resonance of optical phonons at the K and K′ points. When G′ consists of a single peak, it proves the existence of monolayer graphene. When G′ consists of multiple peaks at slightly different wavenumbers, it proves the existence of A-B chirally stacked multilayers of graphite. As shown in Figure 4.8, the measured G′ peak of the fabricated graphene on 3C–SiC(110)/Si(110) shows the monopeak shape. This is quite similar to the case for graphene that is epitaxially grown on the C-face of 6H–SiC [38]. This is believed to be the so-called non-Bernal stacking of multiple layers of monolayer graphene. From these discussions, the fabricated GOS sample is identified as non-Bernal stacked multilayer graphene with gapless and linear dispersive band structures.

Further study of the GOS growth technology, in terms of the chirality and the crystal orientation, has been made by M. Suemitsu and co-workers. It has been found that graphene can be grown on various crystal orientations: 3C–SiC(111)/Si(111), 3C–SiC(110)/Si(110), 3C–SiC(100)/ Si(110), and 3C–SiC(111)/Si(110) and that the graphene on 3C–SiC(111)/Si(111) only exhibits the A-B chiral (Bernal) stacking, whereas the graphene on the other crystal orientations exhibits non-Bernal stacking [45]. This is quite an important finding from a technological viewpoint. This is because most of the electron devices like FETs need a band gap while most of the photonic devices like lasers do not need any band gap and the GOS technology can selectively control the stacking formation, Bernal or non-Bernal.

4.2.4.2 Backgate GOS-FETs

By using the GOS material, a graphene-channel backgate FET was first fabricated [46–48]. A schematic of the device cross section is shown in Figure 4.9. The device process starts with the formation of ohmic contacts. Ti/Au is evaporated and lifted off. After the channel pattern is defined by standard optical lithography, the sample is exposed to oxygen plasma to remove the graphene layer outside the channel

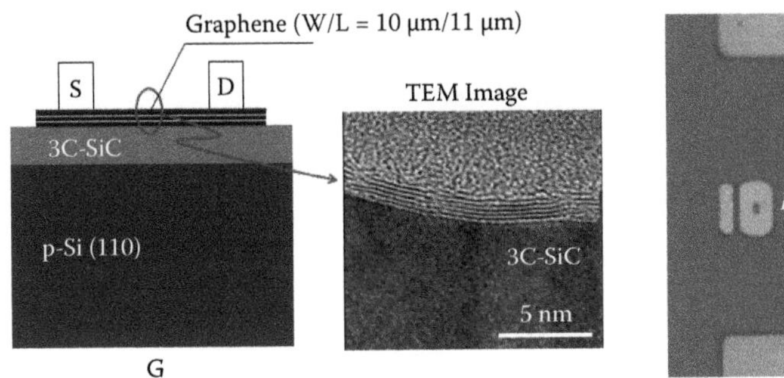

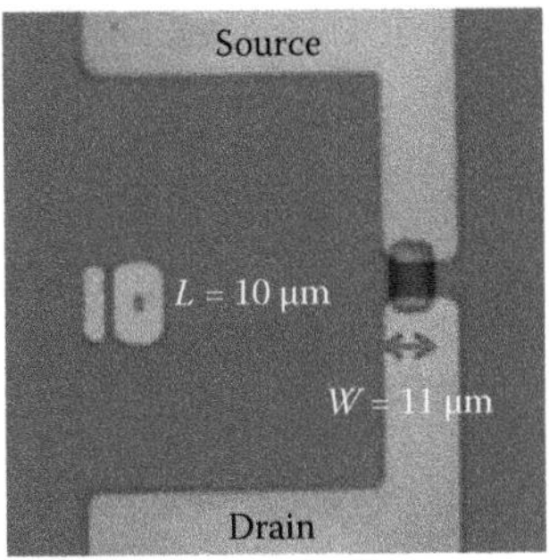

FIGURE 4.9 Device cross section and top view of a fabricated GOS-FET.

region. The typical dimension of the graphene channel is 11-μm long (source-drain separation) and 10-μm wide.

The sample exhibited a severe backgate leakage current, which is presumably due to postformation defects (e.g., void) in the 3C–SiC layer during the graphitization (annealing at 1200°C) process. This annealing temperature is much higher than the growth temperature of the 3C–SiC layer [42]. Recently the gate leakage current has been reduced by growing a thicker SiC epilayer than ever. It is also conceivable that silicon-on-insulator (SOI) wafers can be used as a starting substrate to form graphene. The drain current I_D of the measured GOS-FET includes not only the intrinsic channel current I_{DS} but also the backgate leakage current I_{BG}. To characterize the intrinsic FET performance, I_{BG} is de-embedded from the total drain current I_D. The detailed procedure and the equivalent circuit model for de-embedding I_{BG} are described in detail in Kang et al. [48].

Figure 4.10(a) shows the typical ambipolar behaviors in I_{DS}-V_{BG} characteristics near the Dirac voltage [47] confirming normal operation of the conduction

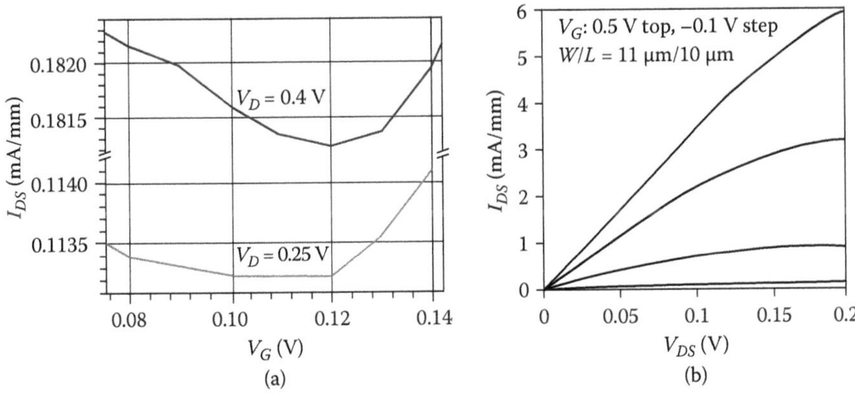

FIGURE 4.10 Measured (a) ambipolar behavior (from R. Olac-bow et al. 2010. *Japanese Journal of Applied Physics* 49: 06GG01), and (b) drain current-voltage characteristics (from H.-C. Kang et al. 2010. *Japanese Journal of Applied Physics* 49: 04DF17). (Figures used with permission.)

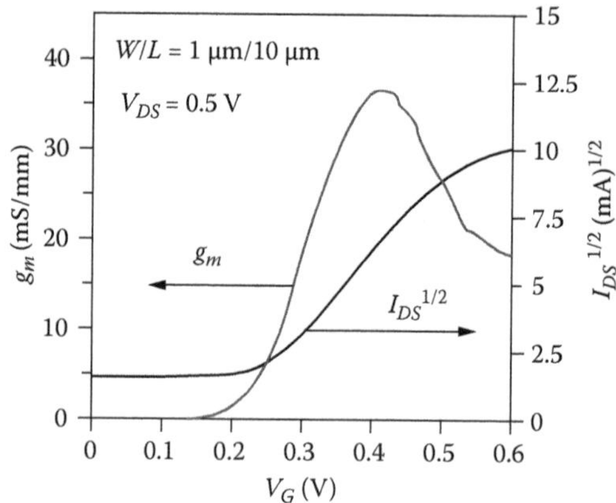

FIGURE 4.11 Extracted transfer characteristic of the intrinsic part of the GOS-FET. (After H.-C. Kang et al. 2010. *Solid-State Electronics* 54: 1010. With permission.)

through the graphene channel. The results exhibit considerable asymmetric properties, with larger/smaller I_{DS} in the electron/hole mode in a V_{BG} that is larger/smaller than the Dirac voltage V_{Dirac} (the V_{BG} point at the minimal I_{DS}). The smaller I_{DS} in the hole mode is caused by backgate leakage; hole carriers easily tunnel through the graphene/3C–SiC/*p*-Si heterointerface due to the band offset of the *p*-type Si substrate and the leaky thin 3C–SiC [47].

Figure 4.10(b) shows the I_{DS}-V_{DS} characteristics in the electron mode of the fabricated GOS-FET for different V_{GS} conditions. In spite of poor GOS quality, with grain size of 20~50 nm, fairly good channel current (on the order of mA/mm) is obtained. Further increases in V_{BG} do not change I_{DS}. In general, G-FETs with monolayer graphene or non-Bernal stacked multilayer graphene don't show the current saturation in I_{DS} due to the gapless band structure of those types of graphene. On the contrary, the channel current I_{DS} shown in Figure 4.10(b) seems to be saturated as V_{DS} increases. There are three possible factors causing the drain current saturation. One is the overestimation of de-embedding I_{BG} in a high V_{DS} region, which may occur when the source/drain contact/access resistances cannot be negligible. The second factor is optical phonon scattering at the graphene–substrate interface, which becomes severe enough to saturate the electron drift velocity under high field conditions at high V_{DS}. The third factor is a band-gap opening due to an unintentional process-dependent factor. Further investigation is needed for quantitative discussion.

Figure 4.11 shows the transfer characteristics of the FET when V_{DS} = 0.5 V. It is seen that the GOS-FET has a maximum transconductance G_m of 37 mS/mm. Assuming a simple linear scaling row with respect to L_g, shrinking L_g down to 100 nm can create increase in G_m to beyond 4 S/mm, which is more than twice as large as that for the state-of-the-art InP-based HEMTs with the same L_g value.

The field-effect mobility of the GOS-FET is next extracted from the measured data. When we consider an *n*-channel metal-oxide-semiconductor field-effect transistor (MOSFET) of gate length L and width W, the drain-source current I_{DS} is calculated as a combination of drift and diffusion currents:

$$I_{DS} = \frac{W\mu_{eff} e n_s V_{DS}}{L} - W\mu_{eff}\frac{kT}{e}\frac{d(en_s)}{dx} \tag{4.6}$$

where n_s is the mobile sheet carrier density (1/cm^2) and μ_{eff} is the effective mobility. μ_{eff} is usually measured at low drain voltages of typically 50 ~ 100 mV because the channel charge is more uniform from the source to drain, allowing the second term to be dropped:

$$\mu_{eff} = \frac{LI_{DS}}{Wen_s V_{DS}} \approx \frac{1}{en_s}\left(\frac{W}{L}\cdot\left.\frac{dV_{DS}}{dI_{DS}}\right|_{V_{GS}=const.}\right)^{-1} \equiv \frac{1}{\rho_{sh} e n_s} \tag{4.7}$$

where ρ_{sh} is the sheet resistivity of the FET channel. From the I_{DS}-V_{DS} data in Figure 4.10(b), ρ_{sh} is calculated to be 2.84 ~ 215 kΩ/□ depending on V_G values from 0.1 to 0.5 V. The sheet carrier density n_s is approximated as an ideal parallel-plate MOS capacitor:

$$n_s = \frac{C_{\mathrm{SiC}}\left|V_{GS} - V_T\right|}{e} \tag{4.8}$$

where C_{SiC} is the gate capacitance per unit area of the GOS-FET and V_T is the threshold voltage. C_{SiC} is calculated to be ~107 nF/cm^2, assuming a relative dielectric constant of 3C–SiC of 9.72 and a SiC thickness of 80 nm. Then n_s is obtained with values of 0.67 to 4.46 × 10^{11} cm^2 depending on V_G values from 0.1 to 0.5 V. As a consequence, μ_{eff} is identified as varying from 430 to 6200 cm^2/(Vs) with increasing V_G from 0.1 to 0.5 V as shown in Figure 4.12. It should be noted that in spite of poor GOS quality with grain size ~20 nm, fairly large mobility (1200 ~ 6000 cm^2/Vs) is obtained. This undoubtedly gives us brighter expectations of promising FET performance with further improvement of GOS quality.

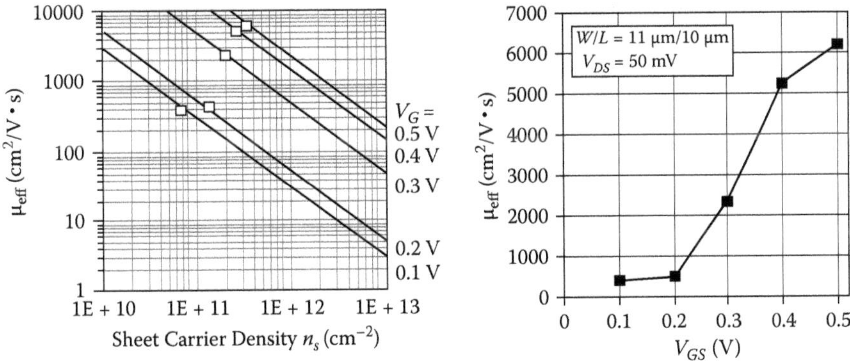

FIGURE 4.12 Extracted field-effect mobility μ_{eff} of the GOS-FET. (a) μ_{eff} vs. sheet carrier density, (b) μ_{eff} vs. gate bias.

4.2.4.3 Topgate GOS-FETs

Topgate GOS-FETs are fabricated using GOS materials grown in 3C–SiC(110)/Si(110) and 3C–SiC(111)/Si(111) [49]. The schematics of the cross section and the photo image of the top view for fabricated samples are shown in Figures 4.13(a) and (b), respectively. The device fabrication starts with Ti/Au lift off for the ohmic electrodes. The graphene channel is then defined and oxygen plasma ashing is carried out to etch the graphene out of the active device area. As for the gate stack, SiN is deposited by plasma-enhanced chemical vapor deposition (PECVD). The gate metallization is done by e-beam evaporation of Ti/Au and lift-off process. Typical gate length is 10 μm for definition by photolithography and 0.5 μm for definition by electron-beam lithography, respectively. The SiN thickness was set at 200 nm for 10-μm topgate GOS-FETs and at 50 nm for 0.5-μm topgate GOS-FETs. All other parts pattern definition processes are done by a conventional optical lithography with a mask aligner. TEM images of the cross section of the 10-μm topgate GOS-FETs on Si(110) and Si(111) substrate are shown in Figures 4.13(c). From the TEM images, the number of graphene layers excluding the interfacial layer (0th layer) is estimated to be about two for Si(110) and three for Si(111). A detailed structural characterization of the graphene sample by Raman spectroscopy and TEM shows that all graphene samples formed on two types of Si substrates have a grain size of ~20 nm [14].

Figures 4.14(a) and (b) show the output characteristics of the 10-μm topgate GOS-FET on Si(110) and Si(111) substrates, respectively. The topgate voltage (V_G)

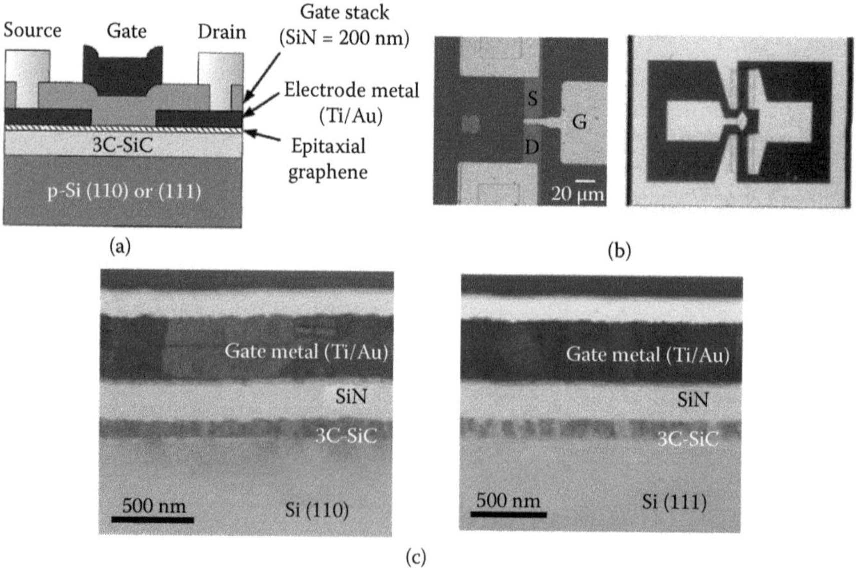

FIGURE 4.13 Topgate GOSFET. (a) schematic of the cross section. (From R. Olac-bow et al. 2010. *Japanese Journal of Applied Physics* 49: 06GG01. With permission.) (b) Photo image of the top view of a 10-μm topgate FET (left) and a 0.5-μm topgate FET. (c) TEM image for a 10-μm gate FET sample fabricated on Si(110) substrate (left) and Si(111) substrate (right). (From H.-C. Kang et al. 2010. *Solid-State Electronics* 54: 1071. With permission.)

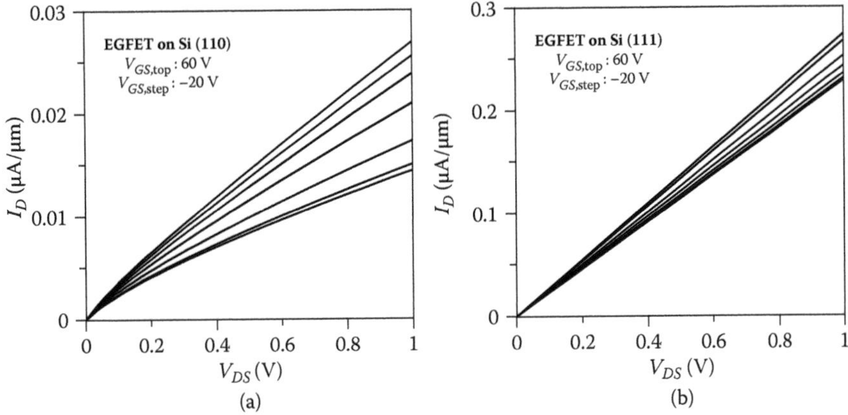

FIGURE 4.14 Drain current-voltage characteristics measured from 10-μm topgate GOS-FETs (a) on Si(110) substrate and (b) on Si(111) substrate. (After H.-C. Kang et al. 2010. *Solid-State Electronics* 54: 1071. With permission.)

in both devices was swept from 60 V to −60 V with a −20 V step. In both devices, a further decrease in V_G does not change I_{DS}. The results show an *n*-type transistor operation by the V_G modulation in all GOS-FETs on the two types of Si substrates. Note that the drain current of the GOS-FET on Si(111) is one order of magnitude larger than that on Si(111). This result is consistent with the measured sheet resistance.

The Dirac voltage (equivalent to the threshold voltage for unipolar FETs) in both devices is negatively shifted up to −40 V. This is possibly due to the positive fixed charge within the SiN gate insulator and/or substrate-induced *n*-type doping in graphene [50]. On the other hand, as shown in Figure 4.15, the Dirac voltage of the 0.5-μm topgate GOS-FET with a 50-nm-thick SiN gate insulator stays near the neutral point. We believe the thinner SiN can dramatically reduce the undesirable fixed charge. Compared to the backgate GOS-FETs shown in Figure 4.11, estimated transconductance G_m is far smaller by almost two orders of magnitude than that for

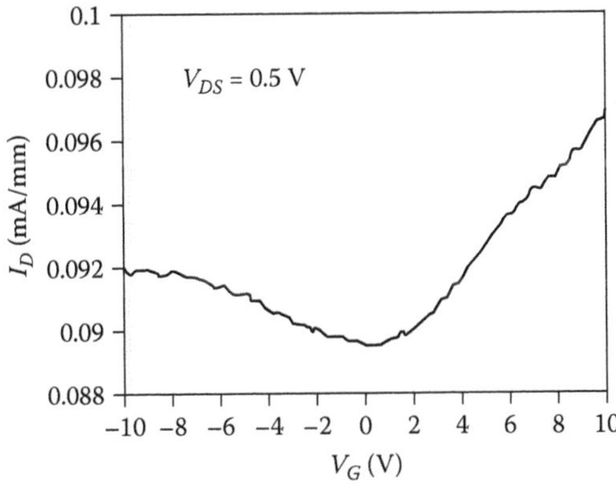

FIGURE 4.15 Ambipolar characteristics measured from a 0.5-μm topgate GOS-FET.

backgate GOS-FETs. This is considered to be attributable to imperfection of the SiN deposition; residual oxygen in the PECVD chamber may damage the graphene and/or interface defects or traps may increase the scattering in carrier transport.

4.2.5 Issues for G-FETs Device/Process Technology

4.2.5.1 Gate Stack/Insulator

Since graphene is so easily oxidized, direct deposition of oxide materials like SiO_2 or HfO_2 onto graphene may cause Fermi-level pinning, preventing gate modulation of the channel conductance. Recently an inert, nonreactive gate insulation technique using atomic-layer deposition (ALD) has been developed in which a thin noncovalent functionalization catalytic polymer layer (NCFL) is introduced as the noninterfacing film between graphene and a thin, high-*k* oxide insulator of Al_2O_3 or HfO_2 [51,52]. Recent results by the IBM group demonstrated successful channel current modulation and resultant transconductance G_m ~150 mS/mm for a 0.24-μm gate G-FET with an effective oxide thickness (EOT) of 1.6 nm of the HfO_2 [35]. The obtained G_m vs. EOT performance is still inferior to the present Si-CMOS technology. The major drawback is the existence of the low-*k* and nm-thick NCFL, which hinders the effectiveness of the high-*k* dielectrics.

On the other hand, overturning the previously mentioned common sense, a group of HRL Labs recently demonstrated a superior G_m of 600 mS/mm for a 3-μm gate GFET with 35-nm EOT SiO_2 insulator deposited by e-beam evaporation [53]. The physical/chemical mechanism of the nonreaction should be clarified.

The authors recently succeeded in fabricating carbonaceous G-FETs in which the diamond-like carbon (DLC) is introduced as the gate insulator directly stacked onto the graphene channel. Photoemission-assisted plasma-enhanced CVD (PA-PECVD) is the key for high-quality DLC deposition [54]. A 10-μm gate G-FET with 100-nm DLC (EOT 76 nm) produced the maximal G_m of 37 mS/mm. With simple linear geometric scaling, one can expect an extremely high G_m value of > 10 S/mm for G-FETs having a 100-nm gate and a 50-nm-thick DLC. Further thinning the DLC insulator for shorter channel G-FETs will be the future direction.

4.2.5.2 Source/Drain Ohmic Contact

The formation of the source/drain ohmic contact is one of the most fundamental and difficult key issues in the G-FET fabrication process. Compared to its excellent in-plane conductivity, graphene has extremely low out-of-plane conductivity. Hence, current conduction is concentrated on the edges at the metal–graphene interface, resulting in high resistivity. Normal semiconductors utilize the heavy doping technique to make the metal-semiconductor Schottky contacts ohmic, but this technique cannot be used for graphene because no doping techniques that are chemically, mechanically, and electrically stable have yet been developed. Unintentional doping between the metal–graphene contacts originating from the difference of the work functions of these materials is currently a method of making an ohmic contact with low resistance. If there is no formation of the interfacial states at the metal–graphene contact, electrons (holes) are unintentionally doped from the metal to graphene when the work function of the metal is larger (smaller) than that of graphene. Recent theoretical study on group X metals like Pd, Ni, and Pt suggests that those metals will form interfacial states when they contact graphene, so that the Fermi level may be pinned and no doping effect is obtained [55,56]. Further investigation with experimental proof will be future research subjects.

4.2.5.3 On/Off Ratio

Nagashio et al. reported that G-FETs with gapless monolayer graphene can obtain an even higher *on/off* current ratio than those with bilayer graphene [57]; the massive carriers in bilayer graphene considerably reduce the *on* current, which can well exceed the reduced *off* current due to the band gap opening. However, as is discussed in Figure 4.5 one can expect a higher *on/off* current ratio for G-FETs with bilayer graphene by increasing the background carrier concentration with a high backgate bias. Recently the IBM group demonstrated a high *on/off* current ratio of more than 100 from a G-FET with bilayer graphene [58], supporting the previous discussion.

4.2.5.4 Carrier Doping and Unipolar Operation

The ambipolar characteristic of graphene can theoretically be suppressed to originate the unipolarity by opening the band gap and doping the acceptor/donor impurities. Presently available methods for carrier doping, however, are intercalation of alkaline metals like potassium [59] and chemical doping with a vapor phase of NH_3 and NO_2 gases [60] or soaking in an organic solution of polyethylene imines [61]. There is no technology available for providing thermally, chemically, and mechanically stable carrier doping like thermal diffusion of substitutional impurities and/or ion implantation. This is one of the critical issues. In light of the immature level of doping technology, formation of electric potential barriers for one side of carriers, electrons, or holes at the source/drain electrodes by using heterostructure band engineering will be effective for obtaining unipolar operation [62].

4.3 GRAPHENE-BASED PHOTONIC DEVICES

4.3.1 Carrier Relaxation and Recombination Dynamics in Optically Pumped Graphene

The optical conductance of graphene-based systems in the terahertz (THz) to far-infrared spectral range has been a topic of intense interest due to the ongoing search for viable terahertz detectors and emitters. Research for applications of graphene in THz science and technology has been carried out and suggests that graphene can be used in building innovative devices for THz optoelectronics. The gapless and linear energy spectra of electrons and holes in graphene can lead to nontrivial features such as negative dynamic conductivity in the THz spectral range [9], which may lead to the development of a new type of THz laser [63,64].

To realize such THz graphene-based devices, understanding the nonequilibrium carrier relaxation/recombination dynamics is critical. Figure 4.16 presents the carrier relaxation/recombination processes and the nonequilibrium energy distributions of photoelectrons/photoholes in optically pumped graphene at specific times from

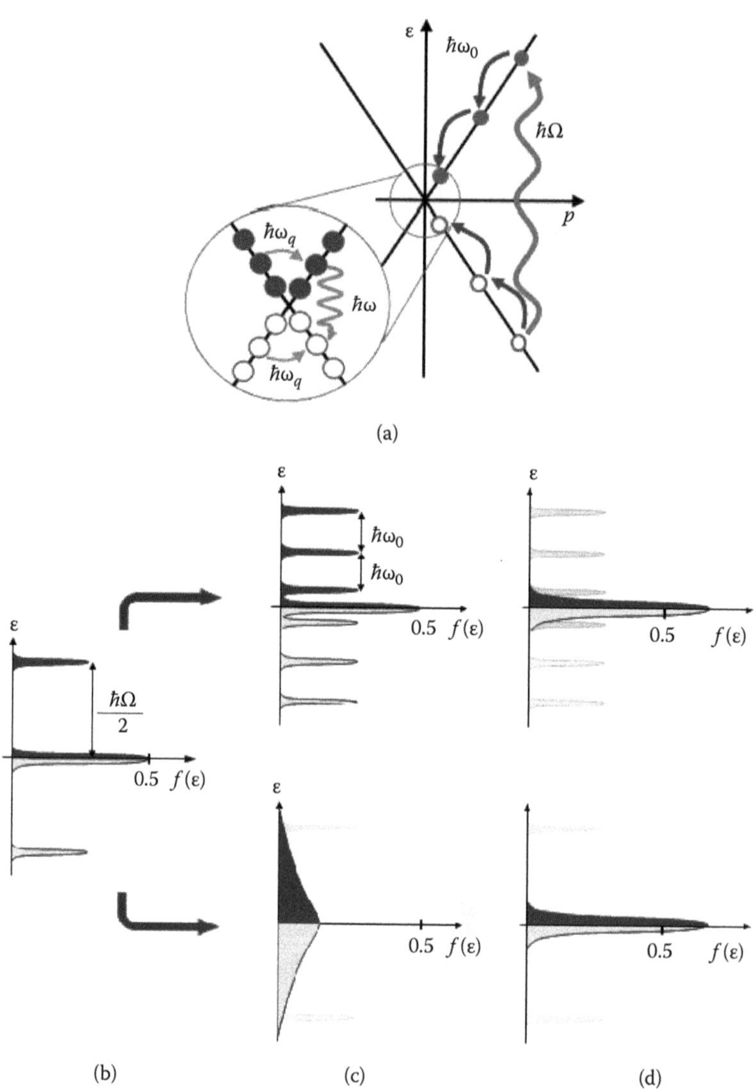

FIGURE 4.16 Schematic view of graphene band structure (a) and energy distributions of photogenerated electrons and holes (b)–(d). Arrows denote transitions corresponding to optical excitation by photons with energy $\hbar\Omega$, cascade emission of optical phonons with energy $\hbar\omega$, and radiative recombination with emission of photons with energy $\hbar\omega_0$. (b) After ~20 fs from optical pumping, (c) after 200~300 fs from optical pumping, upper: phonon-cascade-emission dominant case, lower: cc-scattering-dominant case for high electronic temperature, (d) after a few ps from optical pumping, upper: phonon-cascade-emission dominant case, lower: cc-scattering-dominant case for high electronic temperature.

~10 fs to picoseconds after pumping. It is known that photoexcited carriers are first cooled and thermalized mainly by intraband relaxation processes on femtosecond to subpicosecond time scales, and then by interband recombination processes. Recently, time-resolved measurements of fast nonequilibrium carrier relaxation dynamics have been carried out for multilayers and monolayers of graphene that were epitaxially grown on SiC [65–69] and exfoliated from highly oriented pyrolytic graphite (HOPG) [69,70]. Several methods for observing the relaxation processes have been reported. Dawlaty et al. [65] and Sun et al. [66] used an optical-pump/optical-probe technique and George et al. [67] used an optical-pump/THz-probe technique to evaluate the dynamics starting with the main contribution of carrier-carrier (cc) scattering in the first 150 fs, followed by observation of carrier-phonon (cp) scattering on the picosecond time scale. Ultrafast scattering of photoexcited carriers by optical phonons has been theoretically predicted by Ando [72], Suzuura and Ando [73], and Rana et al. [74]. Kampfrath et al. [70] observed strongly coupled optical phonons in the ultrafast carrier dynamics for a duration of 500 fs by optical-pump/THz-probe spectroscopy. Wang et al. [69] also observed ultrafast carrier relaxation via emissions from hot-optical phonons for a duration of ~500 fs by using an optical-pump/optical-probe technique. The measured optical phonon lifetimes found in these studies were ~7 ps [70], 2–2.5 ps [69], and ~1 ps [67], respectively, some of which agreed fairly well with theoretical calculations by Bonini et al. [75]. A recent study by Breusing et al. [71] more precisely revealed ultrafast carrier dynamics with a time resolution of 10 fs for exfoliated graphene and graphite.

4.3.2 Population Inversion and Negative Conductivity in Optically Pumped Graphene

4.3.2.1 Low Electronic Temperature Case

First we consider the case of cold electronic temperature conditions (such as a cryogenic temperature environment with weak optical pumping). It has been shown that the intraband carrier equilibration in optically excited graphene (with pumping photon energy $\hbar\Omega$)establishes quasi-equilibrium distributions of electrons and holes at around the level $\pm\hbar\Omega/2$within 20–30 fs after the excitation (see Figure 4.16[b]) first. It is followed by cooling these electrons and holes mainly by emission of a cascade (N times) of optical phonons ($\hbar\omega_0$) within 200~300 fs to occupy the states $\varepsilon_N \approx \pm\hbar(\Omega/2 - N\omega_0)$, $\varepsilon_N < \hbar\omega_0$ (see Figure 4.16[c] upper). Then, further equilibration occurs via electron-hole recombination as well as intraband Fermization due to cc scattering and cp scattering (as shown with energy $\hbar\omega_q$ in Figure 4.16[a]) on a few picoseconds time scale (see Figure 4.16[d] upper), while the interband cc scattering and cp scattering are slowed by the density of states effects and Pauli blocking.

Let us consider the THz optical conductivity due to such relaxation/recombination processes that are responsible for the carriers staying at $\hbar\omega/2$. The electron and hole distribution functions at the Dirac point are $f_e(0) = f_h(0) = 1/2$. This implies that at even weak photoexcitation, the values of the distribution functions at low energies ε can be $f_e(\varepsilon) = f_h(\varepsilon) > 1/2$, which corresponds to the population inversion [9]. Such a population inversion might lead to the interband transitions related to negative AC conductivity at THz frequencies. However, the intraband processes determined by Drude conductivity provide the positive contribution to the AC conductivity. One might expect that under sufficiently strong optical excitation resulting in photogeneration of electron-hole pairs, total AC conductivity becomes "negative."

The real part of the net AC conductivity Re σ_ω is proportional to the absorption of photons with frequency ω and comprises the contributions of both interband and intraband transitions [9],

$$\mathrm{Re}\,\sigma_\omega = \mathrm{Re}\,\sigma_\omega^{inter} + \mathrm{Re}\,\sigma_\omega^{intra}.$$

Let us assume a relatively weak optical excitation $\varepsilon_F < k_B T$, where ε_F is the non-equilibrium quasi-Fermi energy and $k_B T$ the thermal energy. In this case, for the THz frequencies $\omega < 2\,k_B T/\hbar$, Re σ_ω^{inter} can be presented as

$$\mathrm{Re}\,\sigma_\omega^{inter} = \frac{e^2}{4\hbar}\left[1 - f_e\left(\frac{\hbar\omega}{2}\right) - f_h\left(\frac{\hbar\omega}{2}\right)\right] \approx \frac{e^2}{8\hbar}\left(\frac{\hbar\omega}{2k_B T} - \eta_F\right),$$

where e is the elementary charge and $\eta_F \equiv \varepsilon_F / k_B T$ is the normalized quasi-Fermi energy [9].

On the other hand, Re σ_ω^{intra} can be presented as

$$\mathrm{Re}\,\sigma_\omega^{intra} \approx \frac{(\ln 2 + \eta_F / 2)e^2}{\pi\hbar}\,\frac{T\tau}{\hbar(1+\omega^2\tau^2)},$$

where τ is the momentum relaxation time of electrons/holes [9]. Since η_F corresponds to the ratio of the excess electron/hole concentration δn (equivalent to photoelectron/photohole concentration) to the thermally equilibrium electron/hole concentration, η_0, η_F is given by

$$\eta_F = \frac{\delta n}{n_0} = \frac{\tau_R \alpha_\Omega I_\Omega}{n_0} \approx \frac{12e^2}{\hbar c}\left(\frac{\hbar v_F}{k_B T}\right)^2 \frac{\tau_R I_\Omega}{\hbar\Omega},$$

where τ_R is the electron/hole recombination time, α_Ω the interband absorption coefficient, I_Ω the pumping intensity, c the speed of light, and v_F the Fermi velocity [9].

As a consequence, Re σ_ω becomes

$$\begin{aligned}\mathrm{Re}\,\sigma_\omega &\approx \frac{e^2}{8\hbar}\left[\frac{\hbar\omega}{2k_B T} + \frac{8\ln 2}{\pi}\,\frac{k_B T\tau}{\hbar(1+\omega^2\tau^2)} - \frac{12e^2}{\hbar c}\left(\frac{\hbar v_F}{k_B T}\right)^2 \frac{\tau_R I_\Omega}{\hbar\Omega}\right] \\ &= \frac{e^2\bar{g}}{8\hbar}\left[1 + \frac{3}{2}\left(\frac{\omega - \bar{\omega}}{\bar{\omega}}\right)^2 - \frac{I_\Omega}{\bar{I}_\Omega}\right],\end{aligned}$$

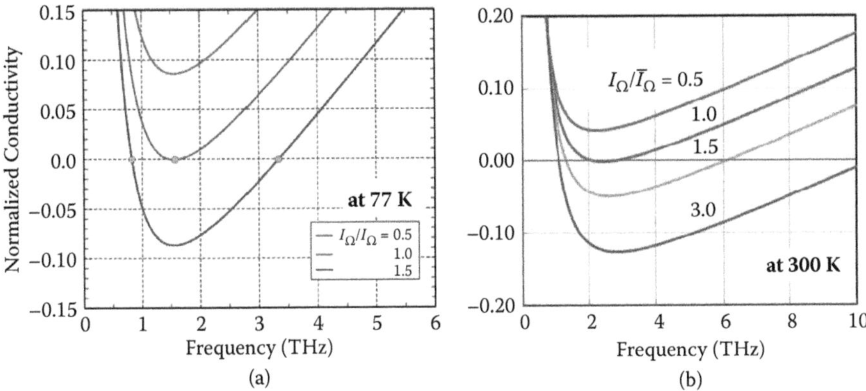

FIGURE 4.17 Calculated AC conductivity for various pumping intensities (a) at 77K (from V. Ryzhii, M. Ryzhii, and T. Otsuji. 2007. *J. Appl. Phys.* 101: 083114) and (b) at 300K (from T. Otsuji, H. Karasawa, T. Watanabe, T. Suemitsu, M. Suemitsu, E. Sano, W. Knap, and V. Ryzhii. 2010. *C. R. Phys.* 11: 421–432). The vertical scale is normalized to the characteristic conductivity. (Graphs used with permission.)

where

$$\overline{\omega} \approx \left(\frac{k_B T \tau}{\hbar}\right)^{2/3} \frac{1.92}{\tau}, \quad \overline{I_\Omega} \approx 11\left(\frac{\hbar}{k_B T \tau}\right)^{1/3} \left(\frac{k_B T}{\hbar v_F}\right)^2 \frac{\hbar\Omega}{\tau_R}.$$

When the pumping intensity exceeds the threshold, $I_\Omega > \overline{I_\Omega}$, Re σ_ω becomes negative in a certain range around $\overline{\omega}$. When $T = 300$ K, $\tau = 10^{-12}$ s, $\tau_R = 10^{-9} \sim 10^{-11}$ s, $\overline{I_\Omega} \approx 600 \sim 600$ W/cm^2. Assuming the device size of 100 μm × 100 μm, we find that the required pumping intensity, which provides the negative dynamic conductivity, $\overline{I_\Omega} \approx 6 \sim 600$ mW.

Figure 4.17 plots the calculated Re σ_ω for various pumping intensities when $T =$ 77K [9] and 300K [76] in case $\tau = 10^{-12}$ s, and $\tau_R = 10^{-9}$ s. The vertical scale is normalized to the characteristic conductivity $e^2/2\hbar$. It is clearly seen that the frequency range where the conductivity becomes negative widens with pumping intensity.

4.3.2.2 High Electronic Temperature Case

When the photogenerated electrons and holes are heated in a room temperature environment and/or strong pumping, collective excitations due to the carrier-carrier (cc) scattering, for example, intraband plasmons, should have a dominant role in performing an ultrafast carrier redistribution along the energy as shown in Figures 4.16(b), (c) upper, and (d) upper. Then optical phonons (ops) are emitted by carriers on the high-energy tail of the electron and hole distributions. This energy relaxation process accumulates the nonequilibrium carriers around the Dirac points as shown in Figure 4.16(d) upper. Due to a fast intraband relaxation (ps or less) and relatively slow interband recombination (>>1 ps) of photoelectrons/holes, one can obtain the population inversion under a sufficiently high pumping intensity [9]. Due to the gapless symmetrical band structure of graphene, photon emissions over a wide THz frequency range are expected if the pumping infrared (IR) photon energy is properly chosen.

We consider an intrinsic graphene under optical pulse excitation in the case where the cc scattering is dominant and carriers always take quasi-equilibrium [77,78]. We take into account both the intra- and interband ops [73,74]. The carrier distribution (equivalent electron and hole distributions) is governed by the following equations for the total energy and concentration of carriers:

$$\frac{d\Sigma}{dt} = \frac{1}{\pi^2}\sum_{i=\Gamma,K}\int d\mathbf{k}\left[(1-f_{\hbar\omega_i - v_w\hbar k})(1-f_{v_w\hbar k})/\tau_{iO,\text{inter}}^{(+)} - f_{v_w\hbar k}f_{\hbar\omega_i - v_w\hbar k}/\tau_{iO,\text{inter}}^{(-)}\right],$$

$$\frac{dE}{dt} = \frac{1}{\pi^2}\sum_{i=\Gamma,K}\int d\mathbf{k}v_w\hbar k\left[(1-f_{\hbar\omega_i - v_w\hbar k})(1-f_{v_w\hbar k})/\tau_{iO,\text{inter}}^{(+)} - f_{v_w\hbar k}f_{\hbar\omega_i - v_w\hbar k}/\tau_{iO,\text{inter}}^{(-)}\right]$$
$$+\frac{1}{\pi^2}\sum_{i=\Gamma,K}\int d\mathbf{k}h\omega_i\left[f_{v_w\hbar k}(1-f_{v_w\hbar k+h\omega_i})/\tau_{iO,\text{intra}}^{(+)} - f_{v_w\hbar k}(1-f_{v_w\hbar k-h\omega_i})/\tau_{iO,\text{intra}}^{(-)}\right],$$

where Σ and E are the carrier concentration and energy density, f_ε is the quasi-Fermi distribution, $\tau_{iO,\text{inter}}^{(\pm)}$ and $\tau_{iO,\text{intra}}^{(\pm)}$ are the inverses of the scattering rates for inter- and intraband ops ($i = \Gamma$ for ops near the Γ point with ω_Γ = 196 meV, i = K for ops near the zone boundary with ω_Γ = 161 meV, + for absorption, and – for emission). Time-dependent quasi-Fermi energy ε_F and the carrier temperature T_c are determined by these equations. Figure 4.18 shows the typical results for fs pulsed laser pumping with photon energy 0.8 eV [78]. It is clearly seen that ε_F rapidly increases when the carrier is cooled and it becomes positive when the pumping intensity exceeds a certain threshold level. This result proves the occurrence of the population inversion. After that, the recombination process follows more slowly (~10 ps).

4.3.3 Observation of Amplified Stimulated THz Emission from Optically Pumped Graphene

We observed the carrier relaxation and recombination dynamics in optically pumped graphene [11,76,79] using THz time-domain spectroscopy based on an optical pump/THz-and-optical-probe technique [67]. An exfoliated monolayer graphene/SiO_2/Si sample or heteroepitaxial graphene/3C–SiC/Si sample is placed on the stage and a 0.12-mm-thick (100)-oriented cadmium telluride (CdTe) crystal is placed on the sample, acting as a THz probe pulse emitter as well as an electrooptic sensor. A single 80-fs, 1550-nm fiber laser beam having 4 mW average power and 20 MHz repetition is split into two: one for optical pumping and generating the THz probe beam in the CdTe crystal, and one for optical probing. The pumping laser, which is linearly polarized, is simultaneously focused at normal incidence from the back surface on the graphene sample to induce population inversion and the CdTe to induce

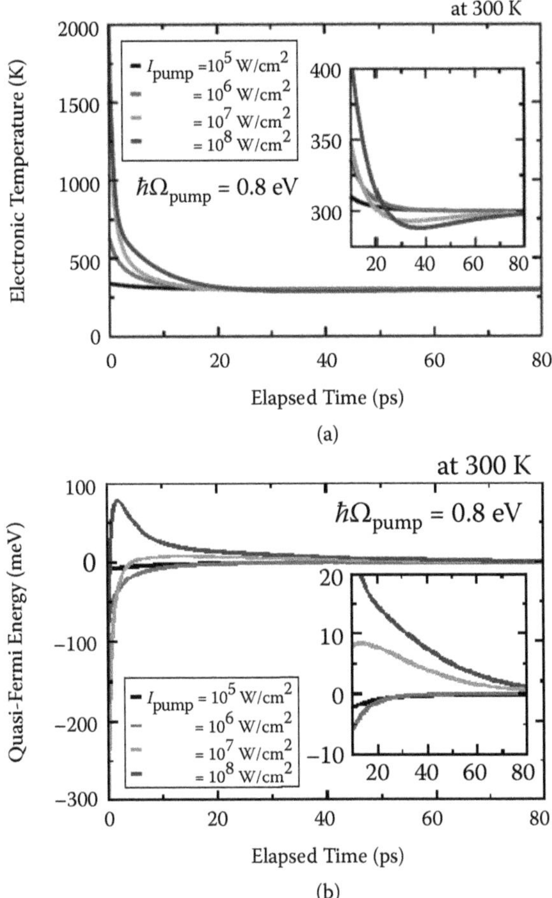

FIGURE 4.18 Temporal evolution of (a) the electronic temperature and (b) quasi-Fermi energy after impulsive pumping. (After A. Satou, T. Otsuji, and V. Ryzhii. 2010. Theoretical study of population inversion in graphene under pulse excitation. *Technical Digest of the International Symposium on Graphene Devices* (ISGD, Sendai, Japan, October): 80–81. With permission.)

optical rectification and emission of a THz pulse (the primary pulse is marked as "1" in Figure 4.19). This THz beam, reflecting back in part at the CdTe top surface, stimulates the THz emission in graphene, which is electrooptically detected as a THz photon echo signal (the secondary pulse is marked as "2" in Figure 4.19).

First the experiment was carried out for the exfoliated monolayer graphene sample. Figures 4.20(a) and 4.20(b) plot the measured results [79]. Figure 4.20(a) shows a typical temporal response under the maximal pumping intensity of 3×10^7 W/cm^2. The black/gray curve is the response when the pumping beam is focused onto the sample with or without graphene. The second pulse (the THz photon echo signal) that is obtained with graphene is more intense compared with that obtained without

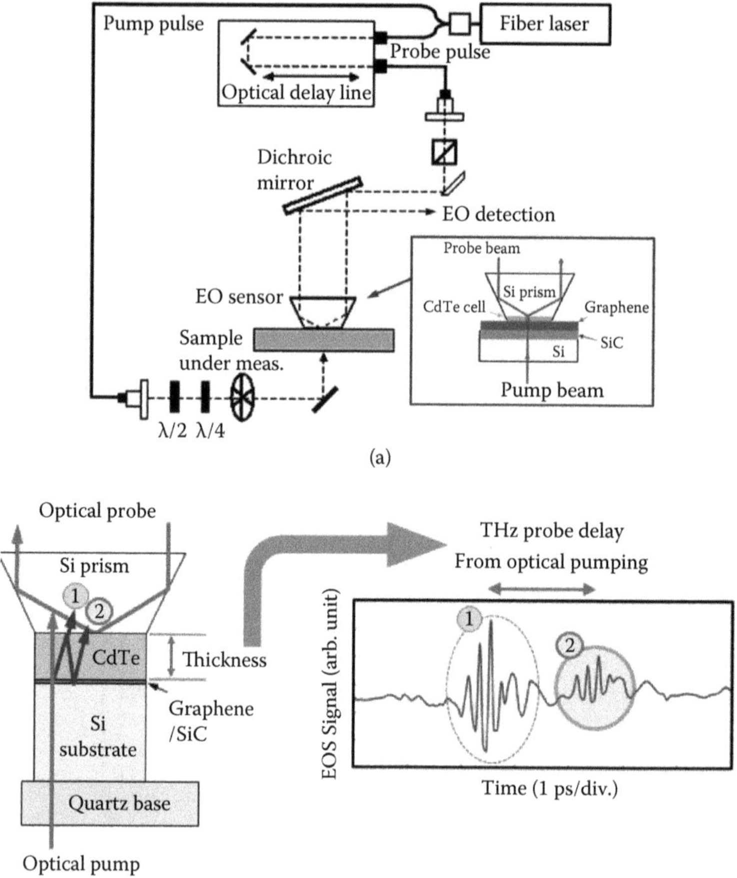

FIGURE 4.19 Experimental setup (a) and the pump-and-probe geometry (b) of coherent emission from graphene by an optical-pump/THz-probe technique. Time-resolved electric field intensity is electrooptically sampled by the probe beam throughout the CdTe sensor crystal in total reflection geometry. The CdTe also works as a THz probe beam source. The secondary pulsation is the photon echo signal representing stimulated emission of THz radiation from graphene.

graphene. This indicates that graphene acts as an amplifying medium. Figure 4.20(b) shows the emission spectra from graphene after normalization to the one without graphene. The inset in Figure 4.20(b) shows the measured gain as a function of the pumping power. The emission is drastically reduced when the optical pump power is decreased. One can also notice that below 1×10^5 W/cm^2, the emission completely disappears and only attenuation can be seen. The measured gain pump-power dependency is presented in Figure 4.20(b). A threshold-like behavior can be seen, which confirms the occurrence of negative conductivity and THz light amplification by stimulated emission of radiation.

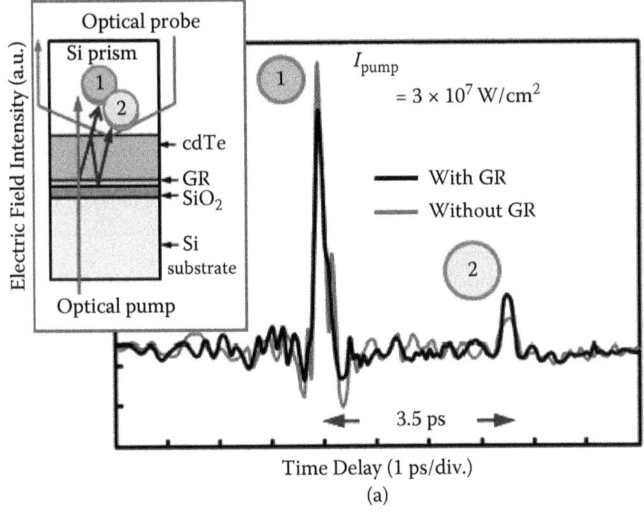

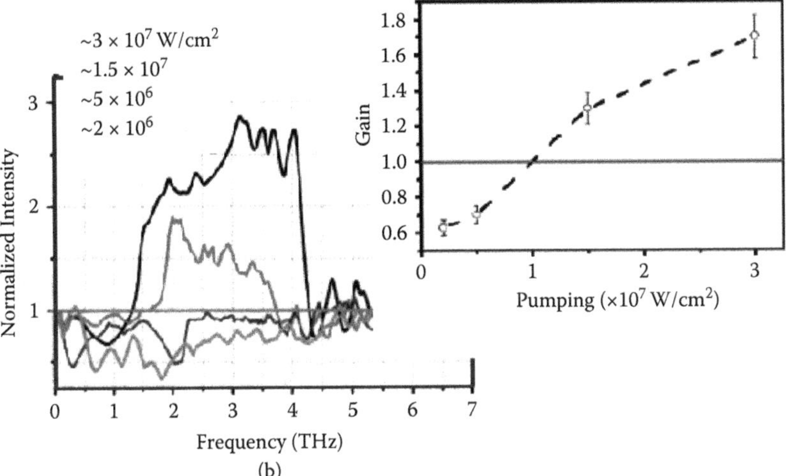

FIGURE 4.20 Measured results for an exfoliated monolayer graphene sample. (a) Temporal profile. The secondary pulse is the THz photon echo transmitted and reflected through graphene. (b) Normalized Fourier spectra and gain profile. (After S. Boubanga Tombet, S. Chan, A. Satou, T. Watanabe, V. Ryzhii, and T. Otsuji. 2010. Amplified stimulated terahertz emission at room temperature from optically pumped graphene. Paper presented at EOS Annual Meeting, TOM2_4011_10, Paris, France, October 27, 2010. With permission.)

Next, a similar measurement was carried out for the heteroepitaxial graphene-on-silicon (GOS) sample [11,76]. The optical pumping intensity condition is set at the maximum level. Typical measured results are shown in Figure 4.21. The emission from the CdTe without graphene exhibits a temporal response similar to optical rectification with a single peak at around 1 THz and an upper weak-side lobe

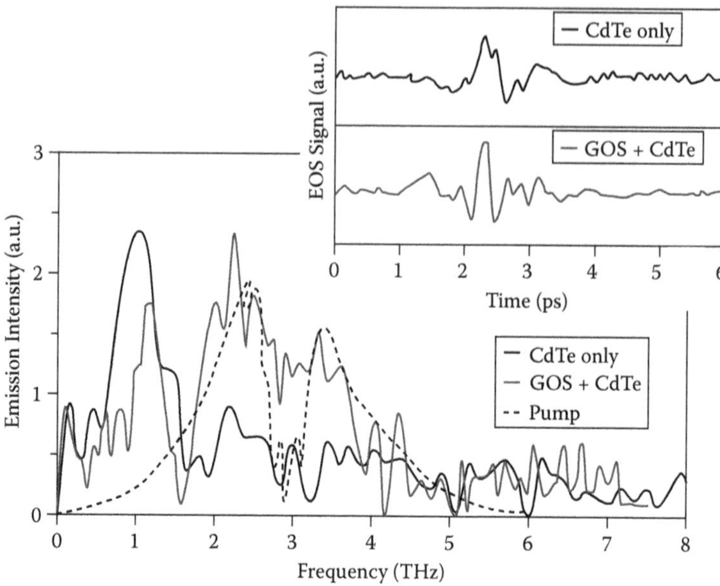

FIGURE 4.21 Measured results for a heteroepitaxial graphene-on-silicon sample (inset: temporal responses; main plot: Fourier spectrum). Dashed line is the photoemission spectrum predicted from the pumping laser spectrum. (After T. Otsuji, H. Karasawa, T. Watanabe, T. Suemitsu, M. Suemitsu, E. Sano, W. Knap, and V. Ryzhii. 2010. Emission of terahertz radiation from two-dimensional electron systems in semiconductor nano-heterostructures. *C. R. Physique* 11: 421–432. With permission.)

extending to around 7 THz (solid black lines in Figure 4.21). On the other hand, the results with graphene agree well with the pumping photon spectrum and include an additional peak around 1 THz from the original CdTe spectrum (gray lines in Figure 4.21). The obtained temporal profile is not identical to the ones for the exfoliated monolayer graphene but are instead noisy or shaggy. This is because the GOS sample under measurement includes the following factors that are different from the exfoliated monolayer graphene sample: (1) a thin 3C–SiC epilayer that absorbs or rectifies the THz radiation, which may cause additional artifacts on the temporal response, (2) multiple layered non-Bernal stacked gapless graphene with small grain size ~20 nm. However, their Fourier spectra show tendencies similar to those for the exfoliated graphene sample. It is thought that the THz emissions from graphene are stimulated by the coherent THz probe radiation that originates from the CdTe excited by the pump laser beam. The THz emissions are amplified by photoelectron–hole recombination in the range of negative dynamic conductivity.

From these results and discussion, we have successfully observed coherent amplified stimulated THz emissions arising from the carrier relaxation–recombination dynamics of graphene. The results provide evidence of the occurrence of negative dynamic conductivity, which can potentially be applied to a new type of THz lasers.

4.4 SUMMARY AND MORE

This chapter covers the authors' recent advances in theoretical and experimental studies on applications of graphene materials to electronic and photonic devices. Due to its unique carrier transport and optical properties, including massless and gapless energy spectra, graphene will exceed many technological limits on conventional electronic and photonic devices. One of the most promising uses of graphene in electronic devices is as the channel material in FETs. Graphene-channel FETs will exceed the speed limits of any type of conventional semiconductor FET in the very near future, although several critical issues concerning process technology including carrier doping, gate stacking, and so on must be solved. The optical properties of graphene can provide many advantages in optoelectronic applications. One typical example is ultrasensitive, ultrafast photodetector and phototransistor operation of graphene in junction and graphene-channel FETs. Detailed discussions [80–82] and experimental studies [83] can be seen in the work of Ryzhii et al. On the other hand, when we consider the ultrafast carrier relaxation and relatively slow recombination dynamics in optically pumped graphene, one dramatic feature of negative dynamic conductivity in the terahertz range can be derived. We succeeded in experimental observation of amplified stimulated emission of terahertz radiation from fs laser-pumped graphene. Although the observed stimulated terahertz emission is a nonequilibrium ultrafast phenomenon due to the lack of laser cavity structure [62,63], the results encourage us to forge ahead to create a new type of real THz lasers that can operate at room temperature. In addition to optical pumping, current injection–type lasing is also feasible. A dual-gate-type graphene-channel FET structure has been proposed for this purpose [84]. In the optical frequency range, a new wave of graphene-based device development is emerging, that is, graphene saturable absorbers [85,86]. Graphene naturally exhibits a broadband optical response with a constant absorbance of 2.3%, which is determined only by universal physical constants [87]. However, if intense radiation with a specific photon energy saturates the absorbance due to the Pauli blocking, graphene becomes transparent to the photon radiation, resulting in a saturable absorber. Hence, it can work for femtosecond pulse compressors. These are typical examples of some of the broader possible applications of graphene in electronic and photonic devices.

REFERENCES

1. K. S. Novoselov et al. 2004. Electric field effect on atomically thin carbon films. *Science* 306: 666.
2. K. S. Novoselov et al. 2005. Two-dimensional gas of massless Dirac fermions in graphene *Nature* 438: 197.
3. P. Kim et al. 2005. Experimental observation of the quantum Hall effect and Berry's phase in graphene *Nature* 438: 201–204.
4. A. K. Geim and K. S. Novoselov. 2007. The rise of graphene. *Nat. Mater*. 6: 183.
5. M. Fujita et al. 1996. Peculiar localized state at zigzag graphite edge. *J. Phys. Soc. Jpn*. 65: 1920.

6. T. Ando et al. 1998. Berry's phase and absence of back scattering in carbon nanotubes. *J. Phys. Soc. Jpn.* 67: 2857.
7. R. Saito et al. 1998. Raman intensity of single-wall carbon nanodes. *Phys. Rev. B* 57: 4145.
8. K. Wakabayashi et al. 1998. Spin wave mode of edge-localized magnetic states in nanographite zigzag ribbons. *J. Phys. Soc. Jpn.* 67: 2089.
9. V. Ryzhii, M. Ryzhii, and T. Otsuji. 2007. Negative dynamic conductivity of graphene with optical pumping, *J. Appl. Phys.* 101: 083114.
10. V. Ryzhii et al. 2008. Population inversion of photoexcited electrons and holes in graphene and its negative terahertz conductivity. *Phys. Stat. Sol. (c)* 5: 261.
11. H. Karasawa et al. 2011. Observation of amplified stimulated terahertz emission from optically pumped heteroepitaxial graphene-on-silicon materials. *J. Infrared Milli. Terahz. Waves* 32: 655–665.
12. I. Forbeaux et al. 1998. Heteroepitaxial graphite on 6H-SiC(0001): Interface formation through conduction-band electronic structure. *Phys. Rev. B* 58: 16396.
13. J. W. May. 1969. Platinum surface LEED rings. *Surf. Sci.* 17: 267–270.
14. P. R. Wallace. 1947. The band structure of graphite. *Phys. Rev.* 71: 622–634.
15. H. P. Boehm, A. Clauss, G. O. Fischer, and U. Hofmann. 1962. Das Adsorptionsverhalten sehr dünner Kohlenstoffolien. *Zeitschrift für anorganische und allgemeine Chemie* 316 (3–4): 119–127.
16. G. W. Semenoff. 1984. Condensed-matter simulation of a three-dimensional anomaly. *Phys. Rev. Lett.* 53: 5449.
17. D. P. DiVincenzo and E. J. Mele. 1984. Self-consistent effective mass theory for intralayer screening in graphite intercalation compounds. *Phys. Rev. B* 295: 1685.
18. V. Ryzhii, M. Ryzhii, and T. Otsuji. 2008. Tunneling current–voltage characteristics of graphene field-effect transistor. *Appl. Phys. Express* 1: 013001.
19. V. Ryzhii, M. Ryzhii, A. Satou, and T. Otsuji. 2008. Current-Voltage Characteristics of a Graphene Nanoribbon Field-Effect Transistor. *J. Appl. Phys.* 103: 094510.
20. V. Ryzhii, M. Ryzhii, T. Otsuji. 2008. Thermionic and tunneling transport mechanisms in graphene field-effect transistors. *Phys. Stat. Sol. (a)* 205:1527–1533.
21. K. Nakada et al. 1996. Edge state in graphene ribbons: Nanometer size effect and edge shape dependence. *Phys. Rev. B* 54: 17954.
22. B. Obradovic et al. 2006. Analysis of graphene nanoribbons as a channel material for field-effect transistors. *Appl. Phys. Lett.* 88: 142102.
23. G. Liang et al. 2007. Performance projections for ballistic graphene nanoribbon field-effect transistors. *IEEE Trans. Electron Devices* 54: 677–682.
24. J. Bai et al. 2010. Graphene nanomesh. *Nat. Nanotech.* 5: 190–194.
25. T. Ohta et al. 2006. Controlling the electronic structure of bilayer graphene. *Science* 313: 951–954.
26. E. McCann. 2006. Asymmetry gap in the electronic band structure of bilayer graphene. *Phys. Rev. B* 74: 161403(R).
27. J. B. Oostinga et al. 2088. Gate-induced insulating state in bilayer graphene devices. *Nat. Mater.* 7: 151–157.
28. V. Ryzhii, M. Ryzhii, A. Satou, T. Otsuji, and N. Kirova. 2009. Device model for graphene bilayer field-effect transistor. *J. Appl. Phys.* 105:104510.
29. S. Y. Zhou, G.-H. Gweon, A. V. Fedorov, P. N. First, W. A. De Heer, D.-H. Lee, F. Guinea, A. H. Castro Neto, and A. Lanzara. 2007. Substrate-induced bandgap opening in epitaxial graphene. *Nat. Mater.* 6: 770–775.
30. A. Bostwick, T. Ohta, J. L. McChesney, K. V. Emtsev, T. Seyller, K. Horn, and E. Rotenberg. 2008. Symmetry breaking in few layer graphene films. *New J. Phys.* 9: 385.
31. E. Sano and T. Otsuji. 2009. Theoretical evaluation of channel structure in graphene field-effect transistors. *Jpn. J. Appl. Phys.* 48: 041202.

32. M. Ryzhii, A. Satou, V. Ryzhii, and T. Otsuji. 2008. High-frequency properties of a graphene nanoribbon field-effect transistor. *J. Appl. Phys.* 104: 114505.
33. V. Ryzhii, M. Ryzhii, A. Satou, T. Otsuji, and N. Kirova. 2009. Device model for graphene bilayer field-effect transistor. *J. Appl. Phys.* 105: 104510.
34. V. Ryzhii, M. Ryzhii, A. Satou, T. Otsuji, and V. Mitin. 2011. Analytical device model for graphene bilayer field-effect transistors using weak nonlocality approximation. *J. Appl. Phys.* 109: 064508.
35. Y.-M. Lin et al. 2010. 100-GHz transistors from wafer-scale epitaxial graphene. *Science* 327: 662.
36. L. Liao, Y.-C. Lin, M. Bao, R. Cheng, J. Bai, Y. Liu, Y. Qu, K. L. Wang, Y. Huang, and X. Duan. 2010. High-speed graphene transistors with a self-aligned nanowire gate. *Nature* 467: 305–308.
37. F. Schwierz, 2010. Graphene transistors. *Nature Nanotech.* 5: 487–496.
38. W. A. de Heer et al. 2007. Epitaxial graphene. *Solid State Comm.* 143: 92–100.
39. Y. Gamo et al. 1997. Atomic structure of monolayer graphite formed on Ni(111). *Surf. Sci.* 374: 61–64.
40. D. Kondo et al. 2010. Low-temperature synthesis of graphene and fabrication of top-gated field effect transistors without using transfer processes. *Appl. Phys. Express* 3: 025102.
41. S. Bae et al. 2010. Roll-to-roll production of 30-inch graphene films for transparent electrodes. *Nature Nanotech.* 5: 574–578.
42. M. Suemitsu et al. 2009. Graphene formation on a 3C-SiC(111) thin film grown on Si(110) substrate. *e-J. Surface Sci. and Nanotech.* 7: 311–313.
43. H. Nakazawa and M. Suemitsu. 2003. Formation of quasi-single-domain 3C-SiC on nominally on-axis Si(001)...substrate using organosilane buffer layer. *J. Appl. Phys.* 93: 5282–5286.
44. Y. Miyamoto et al. 2009. Raman-scattering spectroscopy of epitaxial graphene formed on SiC film on Si substrate. *e-J. Surface Sci. and Nanotech.* 7: 107–109.
45. M. Suemitsu and H. Fukidome. 2010. Epitaxial graphene on silicon substrates. *J. Phys. D: Appl. Phys.* 43: 374012.
46. H.-C. Kang et al. 2010. Extraction of drain current and effective mobility in epitaxial graphene channel field-effect transistors on SiC layer grown on silicon substrates. *Jpn. J. Appl. Phys.* 49: 04DF17.
47. R. Olac-bow et al. 2010. Ambipolar behavior in epitaxial graphene-based field-effect transistors on Si substrate. *Jpn. J. Appl. Phys.* 49: 06GG01.
48. H.-C. Kang et al. 2010. Epitaxial graphene field-effect transistors on silicon substrates. *Solid State Electron.* 54: 1010–1014.
49. H.-C. Kang et al. 2010. Epitaxial graphene top-gate FETs on silicon substrates. *Solid State Electron.* 54: 1071–1075.
50. S. Takabayashi et al. private communication.
51. D. B. Farmer and R. G. Gordon. 2006. Atomic layer deposition on suspended single-walled carbon nanotubes via gas-phase noncovalent functionalization. *Nano Lett.* 6: 699–703.
52. D. B. Farmer et al. 2009. Utilization of a buffered dielectric to achieve high field-effect carrier mobility in graphene transistors. *Nano Lett.* 9: 4474–4478.
53. J. S. Moon et al. 2010. Top-gated epitaxial graphene FETs on Si-face SiC wafers with a peak transconductance of 600 mS/mm. *IEEE Electron Device Lett.* 31: 260–262.
54. H. Sumi, S. Ogawa, M. Sato, A. Saikubo, E. Ikegami, M. Nihei, Y. Takakuwa. 2010. Effect of the carrier gas on the crystallographic quality of network nanographite grown on Si substrates by photoemission-assisted plasma-enhanced chemical vapor deposition. *Jpn. J. Appl. Phys.* 49: 076201.

55. G. Giovannetti et al. 2008. Doping graphene with metal contacts. *Phys. Rev. Lett.* 101: 026803.
56. T. Mueller et al. 2009. Role of contacts in graphene transistors: a scanning photocurrent study. *Phys. Rev. B* 79: 245430.
57. K. Nagashio et al. 2010. Systematic investigation of the intrinsic channel properties and contact resistance of monolayer and multilayer graphene field-effect transistor. *Jpn. J. Appl. Phys.* 49: 051304.
58. F. Xia et al. 2010. Graphene field-effect transistors with high on/off current ratio and large transport band gap at room temperature. *Nano Lett.* 10: 715–718.
59. J.-H. Chen et al. 2008. Charged-impurity scattering in graphene. *Nature Phys.* 4: 377–381.
60. F. Schedin et al. 2007. Detection of individual gas molecules adsorbed on graphene. *Nature Mat.* 6: 652–655.
61. D. B. Farmer et al. 2009. Chemical doping and electron-hole conduction asymmetry in graphene devices. *Nano Lett.* 9: 388–392.
62. E. Sano and T. Otsuji. 2009. Source and drain structures for suppressing ambipolar characteristics of graphene field-effect transistors. *Appl. Phys. Express* 2: 061601.
63. A. A. Dubinov, V. Y. Aleshkin, M. Ryzhii, T. Otsuji, and V. Ryzhii. 2009. Terahertz laser with optically pumped graphene layers and Fabri–Perot resonator. *Appl. Phys. Express* 2: 092301.
64. V. Ryzhii, M. Ryzhii, A. Satou, T. Otsuji, A. A. Dubinov, and V. Y. Aleshkin. 2009. Feasibility of terahertz lasing in optically pumped epitaxial multiple graphene layer structures. *J. Appl. Phys.* 106: 084507.
65. J. M. Dawlaty, S. Shivaraman, M. Chandrashekhar, F. Rana, and M. G. Spencer. 2008. Measurement of ultrafast carrier dynamics in epitaxial graphene. *Appl. Phys. Lett.* 92: 042116.
66. D. Sun, Z.-K. Wu, C. Divin, X. Li, C. Berger, W. A. de Heer, P. N. First, and T. B. Norris. 2008. Ultrafast relaxation of excited Dirac fermions in epitaxial graphene using optical differential transmission spectroscopy. *Phys. Rev. Lett.* 101: 157402.
67. P. A. George, J. Strait, J. Dawlaty, S. Shivaraman, M. Chandrashekhar, F. Rana, and M. G. Spencer. 2008. Ultrafast optical-pump terahertz-probe spectroscopy of the carrier relaxation and recombination dynamics in epitaxial graphene. *Nano Lett.* 8: 4248.
68. H. Choi, F. Borondics, D. A. Siegel, S. Y. Zhou, M. C. Martin, A. Lanzara, and R. A. Kaindl. 2009. Broadband electromagnetic response and ultrafast dynamics of few-layer epitaxial graphene. *Appl. Phys. Lett.* 94: 172102.
69. H. Wang, J. H. Strait, P. A. George, S. Shivaraman, V. B. Shields, M. Chandrashekhar, J. Hwang, F. Rana, M. G. Spencer, C. S. Ruiz-Vargas, and J. Park. 2009. Ultrafast relaxation dynamics of hot optical phonons in graphene. *Arxiv* 0909: 4912.
70. T. Kampfrath, L. Perfetti, F. Schapper, C. Frischkorn, and M. Wolf. 2005. Strongly coupled optical phonons in the ultrafast dynamics of the electronic energy and current relaxation in graphite. *Phys. Rev. Lett.* 95: 187403.
71. M. Breusing, C. Ropers, and T. Elsaesser. 2009. Ultrafast carrier dynamics in graphite. *Phys. Rev. Lett.* 102: 086809.
72. T. Ando 2006. Anomaly of optical phonon in monolayer graphene. *J. Phys. Soc. Jpn.* 75: 124701.
73. H. Suzuura and T. Ando. 2008. Zone-boundary phonon in graphene and nanotube. *J. Phys. Soc. Jpn.* 77: 044703.
74. F. Rana, P. A. George, J. H. Strait, J. Dawlaty, S. Shivaraman, M. Chandrashekhar, and M. G. Spencer. 2009. Carrier recombination and generation rates for intravalley and intervalley phonon scattering in graphene. *Phys. Rev. B* 79: 115447.
75. N. Bonini, M. Lazzeri, N. Marzari, and F. Mauri. 2007. Phonon anharmonicities in graphite and graphene. *Phys. Rev. Lett.* 99: 176802.

76. T. Otsuji, H. Karasawa, T. Watanabe, T. Suemitsu, M. Suemitsu, E. Sano, W. Knap, and V. Ryzhii. 2010. Emission of terahertz radiation from two-dimensional electron systems in semiconductor nano-heterostructures. *C. R. Phys.* 11: 421–432.
77. A. Satou, T. Otsuji, and V. Ryzhii. 2010. Study of hot carriers in optically pumped graphene. *Ext. Abstract Int. Conf. Solide State Devices and Materials* (JSAP, Tokyo, Japan, September): 882–883.
78. A. Satou, T. Otsuji, and V. Ryzhii. 2010. Theoretical study of population inversion in graphene under pulse excitation. *Tech. Dig. Int. Symp. Graphene Devices* (ISGD, Sendai, Japan, October): 80–81.
79. S. Boubanga Tombet, S. Chan, A. Satou, T. Watanabe, V. Ryzhii, and T. Otsuji. 2010. Amplified stimulated terahertz emission at room temperature from optically pumped graphene. Paper presented at EOS Annual Meeting, TOM2_4011_10, Paris, France, October 27, 2010.
80. V. Ryzhii, V. Mitin, M. Ryzhii, N. Ryzbova, and T. Otsuji. 2008. Device model for graphene nanoribbon phototransistor. *Appl. Phys. Express* 1(6): 063002.
81. V. Ryzhii, M. Ryzhii, N. Ryabova, V. Mitin, and T. Otsuji. 2009. Graphene nanoribbon phototransistor: Proposal and analysis. *Jpn. J. Appl. Phys.* 48(4): Part 2, 04C144-1-5.
82. V. Ryzhii, M. Ryzhii, V. Mitin, and T. Otsuji. 2010. Terahertz and infrared photodetection using p-i-n multiple-graphene-layer structures. *J. Appl. Phys.* 107(5): 054512-1-7.
83. F. Xia, T. Mueller, Y.-M. Lin, A. Valdes-Garcia, and P. Avouris. 2009. Ultrafast graphene photodetector. *Nature Nanotech.* 4: 839–843.
84. M. Ryzhii and V. Ryzhii. 2007. Injection and population inversion in electrically induced p–n junction in graphene with split gates. *Jpn. J. Appl. Phys.* 46(8): L151–L153.
85. G. Xing et al. 2010. The physics of ultrafast saturable absorption in graphene. *Opt. Express* 18: 4564–4573.
86. Z. Sun et al. 2010. Graphene modelocked ultrafast lasers. *ACS Nano* 4: 803–810.
87. R. R. Nair, P. Blake, A. N. Grigorenko, K. S. Novoselov, T. J. Booth, T. Stauber, N. M. R. Peres, and A. K. Geim. 2008. Fine structure constant defines visual transparency of graphene. *Science* 320: 1308.

5 Graphene Thin Films for Unusual Format Electronics

Chao Yan, Houk Jang, Youngbin Lee, and Jong-Hyun Ahn

CONTENTS

5.1 INTRODUCTION

Graphene, the thinnest elastic material, has attracted lots of attention due to its outstanding electrical, mechanical, optical, and thermal properties [1–7]. Its superb carrier mobility (up to 200,000 $cm^2/V{\cdot}s$ at room temperature) [2] and low resistivity (up to 30 $\Omega/\square$) [5] suggest the potential to outperform established inorganic materials for certain applications in high-speed transistors and transparent conductive films, respectively. Many experts believe that graphene with a 2D film format, in contrast to 1D format carbon nanotubes, offers fabrication methods that are compatible with a batch microfabrication process, which is essential to realizing practical devices or systems. In addition, graphene has a distinctive mechanical property with fracture strains of ~25% and Young's modulus of ~1 TPa [4], which is much better than that of other known electronic materials. As a result, graphene is particularly suitable for unusual format electronic systems such as flexible, conformal, and stretchable electronic devices with demanding high mechanical requirements. In particular, graphene has a molecular structure basically similar to that of organic electronic materials, and the strong interaction between graphene and organic materials could result in excellent interface contact. This suggests that graphene is a good candidate as a transparent electrode for flexible organic devices such as organic photovoltaics and organic light-emitting diodes [8–15].

This chapter provides an introduction to graphene films for electronic applications, focusing on growth and transfer techniques that can be used to synthesize and fabricate them on unusual—flexible, stretchable—substrates. The content is organized into six main sections. Section 5.1 introduces graphene and graphene applications that have been explored for high-performance electronics. Section 5.2 summarizes the production of high-quality graphene thin films using the chemical vapor deposition method and high-throughput transfer printing approach for large-area electronics. Section 5.3 describes representative results of high-performance radio frequency transistors and flexible electronic systems on plastic substrates. Section 5.4 presents the integration of graphene conductive films into layouts for electronic devices including touch-screen panels to organic solar cells and light-emitting diodes. Section 5.5 describes graphene-based gas barrier films for applications not only for electronic devices, but for food preservation or antioxidation coating on reactive metal surfaces. The last section summarizes the overall content and provides some perspectives on the trends for future work.

5.2 LARGE-AREA PRODUCTION OF GRAPHENE THIN FILMS

5.2.1 Large-Area Graphene Synthesis

The most attractive technique for growing large-area graphene is chemical vapor deposition (CVD) on Ni or Cu substrate. In the CVD method, the hydrocarbon gas precursor is injected into a chamber at high temperature, around 1000°C. At a high temperature, hydrocarbon atoms are adsorbed on the catalyst layer and leave carbon atoms. In the cooling process, the carbon atoms get the energy to form a 2-dimensional atomic structure [5,6,16–18]. Figure 5.1(a) shows a scanning electron microscope (SEM) image of few-layer graphene indicating various numbers of graphene layers grown on a Ni catalyst layer. The number of graphene layers is estimated by transmission electron microscope (TEM) in Figure 5.1(b). The number of graphene layers depends on the solubility of carbon atoms in Ni catalyst grain [6,16]. The Ni grain, which has many carbon atoms, results in thick graphene film. To avoid graphite crystal, the amount of carbon source absorbed into the Ni should be reduced by controlling the thickness of Ni and the reaction time in high temperature. The distribution of graphene layers is shown in Figures 5.1(c) and 5.1(d). After transferring the graphene film onto a SiO_2 (300 nm)/Si substrate, the optical and confocal scanning Raman microscopic images are observed. The brightest area in Figure 5.1(d) corresponds to monolayer, and the darkest area represents thickness of more than ten layers of graphene. The Raman spectroscopy data (Figure 5.1[e]) shows the characterization of number of graphene and their quality. All spectroscopic data show less intense D-band peaks, which indicate low defect density. The relative peak ratio in G/2D shows the number of graphene layers at the measuring point marked in

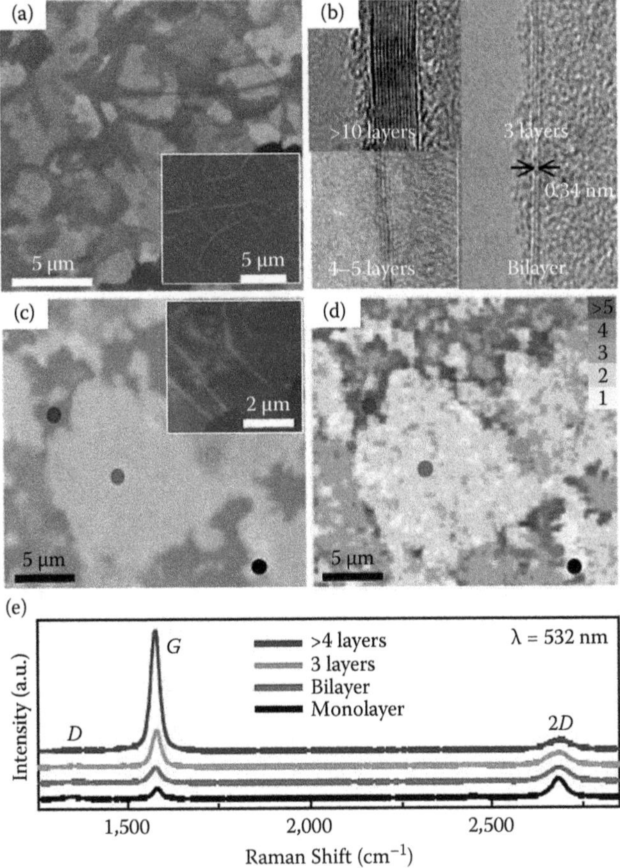

FIGURE 5.1 Graphene film synthesized on a nickel catalyst layer using the CVD method. (a) SEM image of graphene film grown on nickel layer. (b) Thickness and interlayer distance of graphene film estimated by HRTEM. (c) Optical microscope image of transferred graphene film on SiO_2 300-nm layer. (d) Confocal scanning Raman image corresponding to (c). (e) Raman spectroscopy of each point indicating different number of layers. (From Kim, K. S., Zhao, Y., Jang, H., Lee, S. Y., Kim, J. M., Kim, K. S., Ahn, J.-H., Kim, P., Choi, J.-Y., and Hong, B. H. 2009. Large-scale pattern growth of graphene films for stretchable transparent electrodes. *Nature* 457 (7230): 706–710. Copyright 2009 Nature Publishing Group. With permission.)

Figure 5.1(c). Two methods for transferring multilayered graphene grown on a Ni layer have been introduced in the literature. The first method involves the application of a polydimethylsiloxane (PDMS) or polymethylmethacrylate (PMMA) supporting layer to graphene film while the catalyst layer is being etched. After stamping the graphene onto useful substrates, the supporting layer on graphene film is removed [18,19]. The second method is to transfer graphene film without any supporting layer during the etching and cleaning process. Through this method we can obtain a clean surface, even though the graphene film can be easily broken during the process.

For high-quality graphene film production, technology that is able to produce uniform thickness distribution should be developed. It is nearly impossible to directly fabricate field-effect transistors using graphene grown on Ni, even though it can be synthesized on a large scale. For those kinds of applications, the uniform large-area graphene in monolayer thickness should be developed. Copper has less solubility of carbon atoms compared with the previous catalyst, Ni [6,16,17]. The monolayer graphene film could be synthesized by using a Cu catalyst layer with a similar CVD growing procedure, mentioned previously. Because the evaporation of Cu occurs at a relatively low temperature, thick Cu foil is used for CVD growth at 1000°C. The vacuum process results in low accumulation of amorphous carbon on the Cu surface. Figures 5.2(a) and (c) show SEM and optical microscope images of graphene film

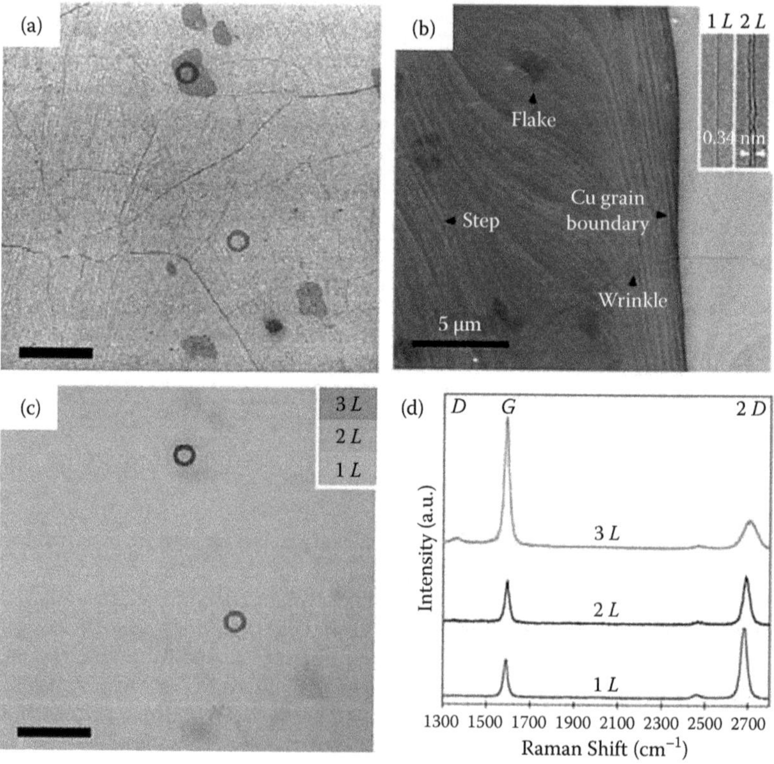

FIGURE 5.2 (*See color insert.*) Graphene film synthesized on a copper catalyst layer using the CVD method. (a) SEM image of graphene film grown on copper foil. (b) Morphological image of graphene film flowing copper surface conditions (inset denotes the monolayer and bilayer graphene estimated by TEM). (c) Optical microscope image of the graphene film transferred on SiO_2 layer showing different color indicating different number of layers. (d) Raman spectroscopy for each number of graphene layers on SiO_2 300-nm layer. (From Li, X., Cai, W., An, J., Kim, S., Nah, J., Yang, D., Piner, R., Velamakanni, A., Jung, I., Tutuc, E., Banerjee, S. K., Colombo, L., and Ruoff, R. S. 2009. Large-area synthesis of high-quality and uniform graphene films on copper foils. *Science* 324 (5932): 1312–1314. Copyright 2009 American Association for the Advancement of Science. With permission.)

synthesized on Cu film. These images show a uniform distribution of monolayer graphene. The dark parts pointed by circles in the upside of Figure 5.2(a) and (c) represent bilayered graphene and the arrows in the underside of Figure 5.2(a) and (c) indicate trilayered graphene film. The surface morphology of graphene film strongly depends on the topography of the Cu surface. The Raman spectra from bottom to top in Figure 5.2(d) are from the marked places in Figure 5.2(a) and (c) the middle circle, upside circle and underside arrow, respectively. The uniformity of the graphene films are evaluated via color contrast under optical microscope and Raman spectra. An analysis of the intensity of the optical image over the whole sample and corresponding Raman spectra shows the monolayer graphene distributed more than 95% of the Cu surface [17,20,21].

5.2.2 Large-Area Graphene Transfer Methods

Large-area, high-quality graphene production has been achieved by epitaxial growth on a silicon carbide (SiC) substrate and CVD on a metal surface of Cu and Ni. However, most applications require graphene to be located on an insulator. This means that graphene must be transferred to another appropriate substrate or processed in some other way if graphene is synthesized on a metal. The important thing is to transfer the graphene without significant deformation during the process. It is a challenge to transfer large-area graphene film to a target substrate without further degradation. Much effort has been focused on this field.

The schematic illustration of the wafer-scale graphene transfer process is presented in Figure 5.3 [18]. The graphene grown on a Ni catalyst layer in wafer requires effective removal of the catalyst layer. Here, the detaching of graphene film and catalyst layer from the SiO_2/Si mother substrate is introduced by using the wettability difference between the catalyst layer and SiO_2. Then the catalyst layer can be removed instantaneously because a few-nm-scale catalyst layer is exposed to etchant. The graphene attached to a polymer support is then state ready to transfer onto a useful substrate. After the graphene film is transferred onto target substrates, they can be patterned by photolithography with short oxygen plasma etching. Prepatterned graphene film can be transferred through this method as well.

Another transfer printing method has been successfully developed using PMMA to aid the transfer of graphene grown on Cu foil to a target substrate [16,17,19]. After graphene is grown on Cu foil, PMMA is spin coated on the top and baked for a short period of time. The Cu foil is then etched away and the remaining PMMA–graphene film is washed in deionized water to remove the etchant residue. At this stage, PMMA–graphene thin film is ready to transfer to an arbitrary substrate before removing the PMMA using acetone. To minimize the density of cracks caused during the transfer process, an improved transfer process for the preparation of large-area graphene films is explored [19], as shown in Figure 5.4(a).

The procedures for transferring graphene from a SiC growth wafer to another substrate [22,23], as depicted in Figure 5.4(b), is similar to those reported for the transfer of random networks and the alignment of single-walled carbon nanotubes [24]. In the first, the graphene–SiC sample is deposited with a layer of Au (or Pd) and polyimide. The baked polyimide thin layer serves as a strong support for the

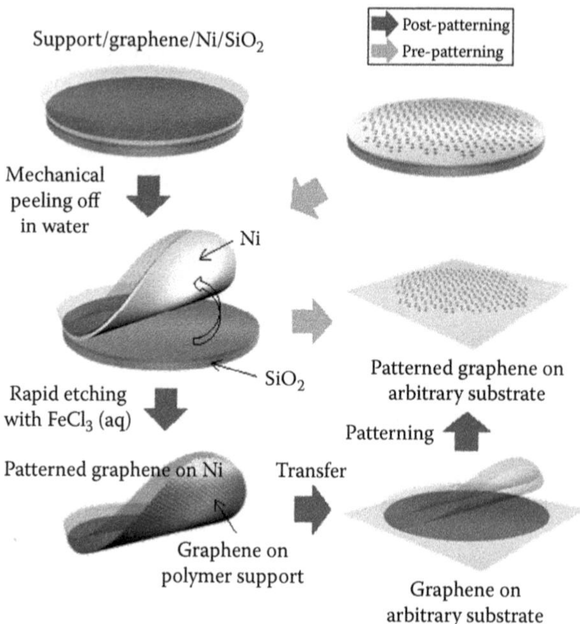

FIGURE 5.3 Schematic illustration of large-area graphene transfer process. Transfer process of both neat graphene film and patterned graphene film onto target substrate in a wafer-scale through dry transfer printing with a polymer support layer.

mechanical peeling process. After transfer to the target substrate, the polyimide and Au are removed by oxygen plasma reactive ion and wet chemical etching.

A large-scale (30-inch) roll-to-roll transfer method has been developed recently [5]. Figure 5.5 shows a schematic of the roll-based production of graphene films. This roll-to-roll method can enable the continuous production of graphene films in large scale. The first step is the adhesion of thermal release tape on graphene film grown on Cu foil using the roll process. Next, Cu foil is etched in a bath filled with copper etchant and rinsed in deionized water to remove residual etchant. In the final step, the graphene film on the thermal release tape is roll-transferred onto a desired substrate by exposure to the release temperature of thermal release tape. Wet chemical doping, which considerably enhances the electrical properties of graphene films grown on roll-type Cu substrates by chemical vapor deposition, can be carried out using a setup similar to etching. The repetition of this process results in randomly stacked graphene. The electrical resistance of these graphene films will be discussed later.

Figure 5.6 shows photographs of graphene film on the wafer and various target substrates after transfer. Wafer-scale graphene synthesized on Ni or Cu layers is shown in Figure 5.6(a). Transparency of graphene film is demonstrated in Figure 5.6(b), and graphene film transferred onto stretchable and flexible substrates is displayed in Figures 5.6(c) and 5.6(d).

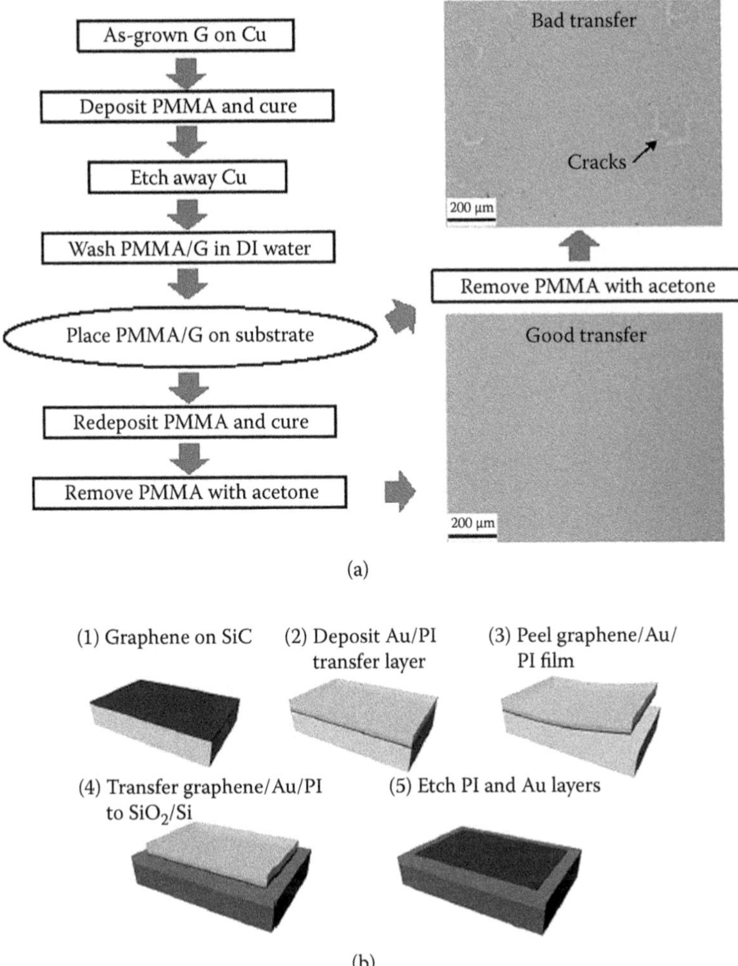

FIGURE 5.4 (a) Processes for transfer of graphene films. The two insets on the right are the optical micrographs of graphene transferred on SiO_2/Si wafers (285-nm-thick SiO_2 layer) with "bad" (top) and "good" (bottom) transfer, respectively. (From Li, X., Zhu, Y., Cai, W., Borysiak, M., Han, B., Chen, D., Piner, R. D., Colombo, L., and Ruoff, R. S. 2009. Transfer of large-area graphene films for high-performance transparent conductive electrodes. *Nano Letters* 9 (12): 4359–4363. Copyright 2009 American Chemical Society. With permission.) (b) Schematic illustration of the steps for transferring graphene grown on a SiC wafer to another substrate (SiO_2/Si in this case). (From Unarunotai, S., Murata, Y., Chialvo, C. E., Kim, H.-S., MacLaren, S., Mason, N., Petrov, I., and Rogers, J. A. 2009. Transfer of graphene layers grown on SiC wafers to other substrates and their integration into field effect transistors. *Applied Physics Letters* 95 (20): 202101-3. Copyright 2009 American Institute of Physics. With permission.)

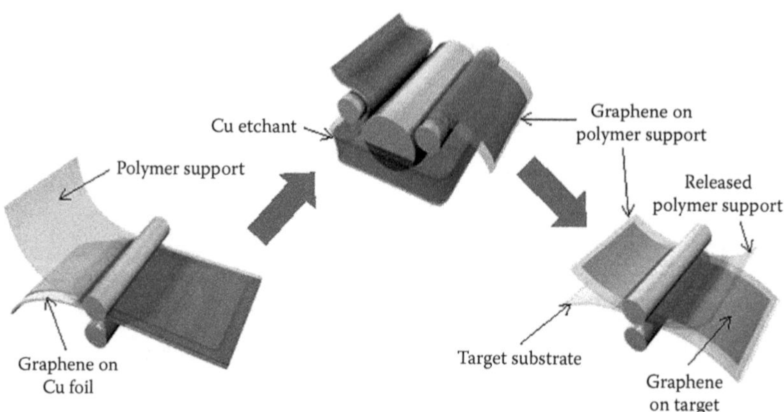

FIGURE 5.5 Schematic illustration of large-area graphene transfer process. Roll-to-roll transfer of graphene grown on copper foil with thermal release tape.

5.3 FIELD-EFFECT TRANSISTORS

The first graphene-based field-effect transistor (FET) was fabricated in 2004, which led to an explosion of interest in the electronic properties of graphene [1]. The charge carriers in graphene FETs can change from electrons to holes with the application of gate voltage. Graphene exhibits high intrinsic carrier mobility [2,25], recorded exceeding 200,000 cm^2/V·s, and high saturation velocity of 5.5 × 10^7 cm/s [26]. This makes graphene a great material for high-speed electronic devices, particularly those offering excellent radio-frequency characteristics with very high cutoff frequency (f_T). The outstanding mechanical properties of graphene highlight the fabrication of flexible and stretchable electronic devices.

5.3.1 RF Transistor

Graphene shares many advantages such as the high intrinsic mobility in room temperature over carbon nanotubes [27]. Many theoretical studies have been performed for these carbon-based materials and suggest the possibility of response on a picosecond time scale, which corresponds to a terahertz frequency regime [28]. Transistors with operating speeds in this THz range requiring high current gain are of considerable interest in terms of imaging and sensors. Although graphene FETs show different DC characteristics from those of silicon, including on/off ratio, because of a different principle, it follows that AC characteristics of graphene are quite similar. In addition, the high mobility of graphene and the 2D nature of the material, unlike carbon nanotubes, make it the best material for high-frequency operation [29].

Cutoff frequency, corner frequency, or break frequency, which indicate the boundary in a system's frequency response at which energy flowing through the system begins to be reduced rather than passing through, is one of the most accepted procedures for determining the frequency response in individual transistors, which is explained by the following equation [29];

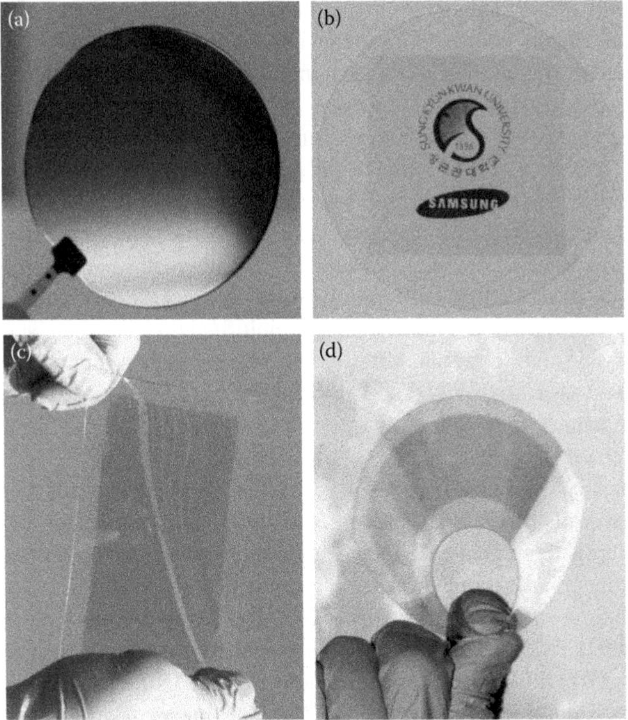

FIGURE 5.6 (a) Wafer-scale graphene film grown on Ni/SiO_2 layers. (b) Graphene film transferred onto transparent substrate. (c) Graphene films on stretchable PDMS substrate. (d) Graphene film formed on flexible substrate. ([a] and [b] from Lee, Y., Bae, S., Jang, H., Jang, S., Zhu, S.-E., Sim, S. H., Song, Y. I., Hong, B. H., and Ahn, J.-H. 2010. Wafer-scale synthesis and transfer of graphene films. *Nano Letters* 10 (2): 490–493. Copyright 2010 American Chemical Society. With permission.)

$$f_T = \frac{g_m}{2\pi C}$$

In this equation, f_T, g_m, and C are cutoff frequency, maximum gain, and capacitance, respectively. From the equation, a transistor showing high current gain value and low capacitance shows a high cutoff frequency value, which means the possibility of operation at a higher frequency. The current gain value means the ability of a circuit to increase the current from the input to the output. The value of g_m can be expressed by the following equation [30];

$$g_m = \mu \times \frac{W}{L} \times C_{ox} \times V_D = \mu \times \varepsilon \times \frac{t_{ox} \times W}{L} \times V_D$$

where μ, W/L, C_{ox}, V_D, ε, and t_{ox} are mobility, width/length ratio, oxide capacitance, drain voltage, dielectric constant, and thickness, respectively.

Three kinds of parameters compose the equation: the property of the material, geometry, and drain voltage. Graphene has significant advantages in the material and geometry properties while carbon nanotubes have challenges with respect to proper geometry for high-frequency operating FETs, namely large channel width. Using these two equations, we can find that to operate transistors at high frequencies, a high mobility value, high gate capacitance, low total capacitance, and a large W/L value are required [31].

High-k material and a thin dielectric layer can be used to form high gate capacitance while accurate lithography is necessary for removing parasitic capacitance to reduce total capacitance through removal of the overlapped region between the gate and source/drain (S/D). In addition, double or multifinger patterns can be used to create a large W/L ratio value. The major difference between graphene and a carbon nanotube is the formation of a large W/L ratio value. Because graphene is a 2D material, it can have a large channel width while the carbon nanotube, a 1D material, is too narrow to form a wide channel.

Among these processes, the problem is to form a thin dielectric layer using high-k material because of pinholes that generate *leakage current*. Researchers have attempted to fabricate top-gate graphene FETs and suggest a method that involves changing the material or functionalization with chemicals or a self-assembled monolayer.

These solutions offer breakthroughs to high-frequency devices with operating frequencies of GHz range. Figure 5.7 shows a graphene transistor operating with a frequency of 26 GHz [29]. Single-layer graphene is made using mechanical exfoliation confirmed by Raman spectroscopy. Source and drain electrodes consisting of 1 nm Ti and 50 nm Pd are defined through e-beam lithography. A functionalized layer consisting of 50 cycles of NO_2-TMA is deposited, followed by atomic layer deposition of a 12-nm Al_2O_3 layer as a gate insulator. 10 nm and 50 nm Pd and Au are deposited as gate electrodes, which are 40 μm in width and 500 nm in length (Figure 5.7[a]).

Figure 5.7(c) shows conductance as a function of gate voltage before and after (inset) depositing the top-gate dielectric. The Dirac (the point which is defined as the gate voltage at minimum conductance) is moved dramatically. Mobility, which is 400 $cm^2/V{\cdot}s$ before deposition, is also significantly reduced. This degradation of mobility is due to charged impurity scattering associated with the NO_2 layer and phonon scattering at the interface.

Graphene radio frequency (RF) transistors have been successfully fabricated in wafer size by epitaxial growth as well [32]. Graphene is epitaxially formed on the SiC semi-insulating wafer by thermal annealing, and shows a Hall effect mobility over 1000 $cm^2/V{\cdot}s$. Polyhydroxystyrene is used as an interfacial polymer layer before deposition of a 10-nm HfO_2 insulating layer. This deposition of thin high-k material maintains the Hall effect carrier mobility, over 900 $cm^2/V{\cdot}s$, resulting in a cutoff frequency of 100 GHz, which is a larger value compared to silicon-based devices with similar structure. Figure 5.8 shows a graphene transistor array operating with a frequency of 100 GHz. The device array is fabricated using graphene epitaxially grown on a SiC insulating wafer. In addition, this operation frequency of 100 GHz is the value before optimization in terms of geometry and mobility, indicating a possibility of operation of graphene in the THz range. Therefore, there are plenty of possibilities for graphene in THz devices.

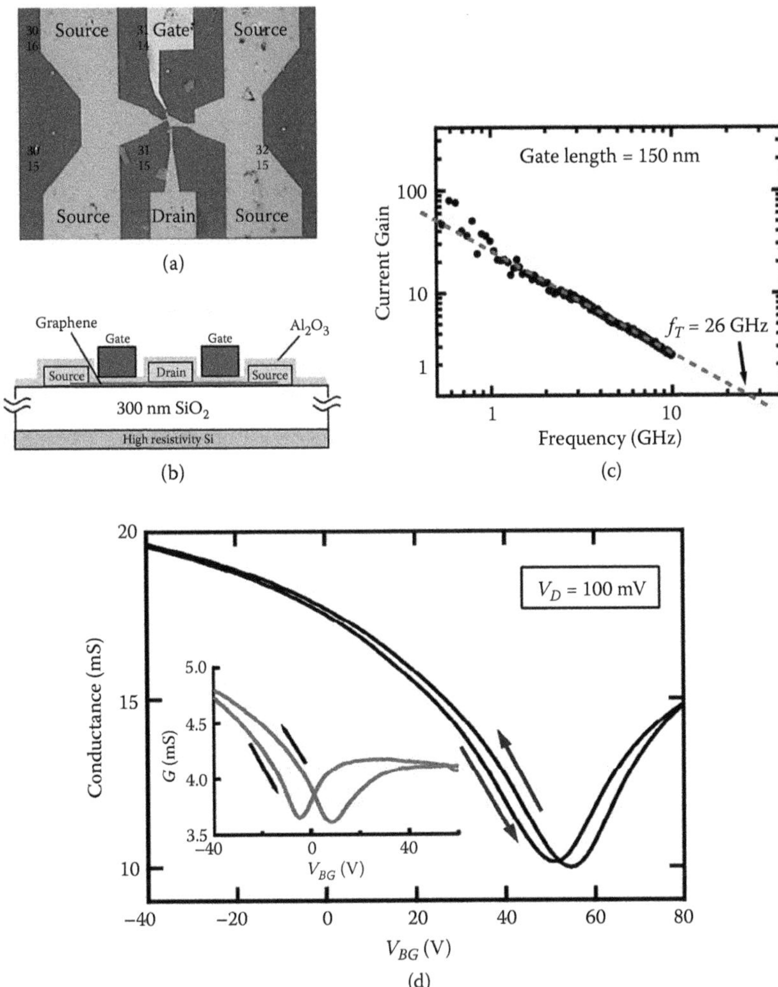

FIGURE 5.7 (a) Optical image of the device layout with ground-signal-ground accesses for the drain and the gate. (b) Schematic cross section of the graphene transistor. Note that the device consists of two parallel channels controlled by a single gate in order to increase the drive current and device transconductance. (c) Measured current gain h_{21} as a function of frequency of a GFET with L_G = 150 nm, showing a cut-off frequency at 26 GHz. The dashed line corresponds to the ideal 1/f dependence for h_{21}. (d) Measured conductance as a function of back-gate voltage, V_{BG}, of the graphene transistor before depositing the top-gate dielectric. The inset shows the same device after the deposition of Al_2O_3 by ALD. The two arrows represent the sweeping direction of the gate voltage. (From Lin, Y.-M., Jenkins, K. A., Valdes-Garcia, A., Small, J. P., Farmer, D. B., and Avouris, P. 2008. Operation of graphene transistors at gigahertz frequencies. *Nano Letters* 9 (1): 422–426. Copyright 2008 American Chemical Society. With permission.)

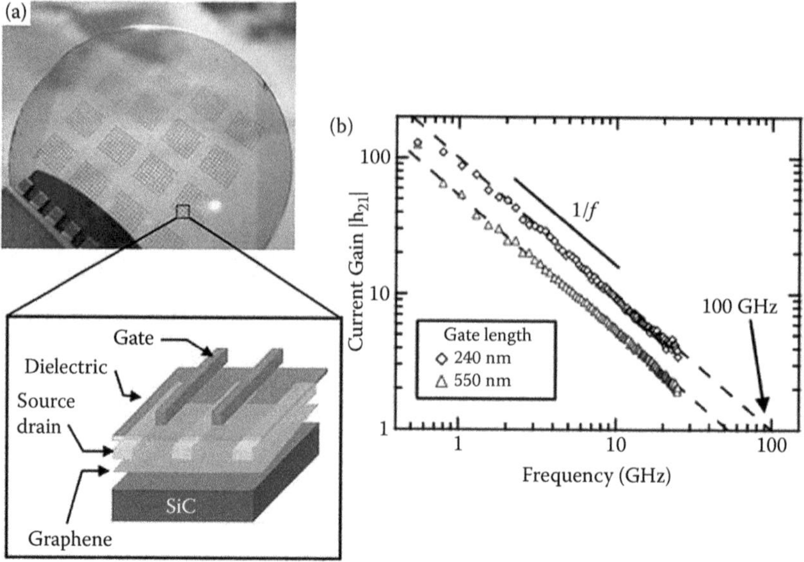

FIGURE 5.8 (a) Image of devices fabricated on a 2-inch graphene wafer and schematic cross-sectional view of a top-gated graphene FET. (b) Measured small-signal current gain $|h_{21}|$ as a function of frequency f for a 240-nm-gate (◇) and a 550-nm-gate (△) graphene FET at $V_D = 2.5$ V. Cutoff frequencies, f_T, were 53 and 100 GHz for the 550-nm and 240-nm devices, respectively. (From Lin, Y.-M., Dimitrakopoulos, C., Jenkins, K. A., Farmer, D. B., Chiu, H.-Y., Grill, A., and Avouris, P. 2010. 100-GHz transistors from wafer-scale epitaxial graphene. *Science* 327 (5966): 662. Copyright 2010 American Association for the Advancement of Science. With permission.)

5.3.2 Flexible Graphene-Based Transistor

Graphene is a very attractive material for a large range of applications, including displays, sensors, and solar cells because of its outstanding mechanical, optical, and electrical properties [33–36]. For these reasons, much research has been done on graphene transistors with respect to substrates, electrodes, and dielectrics [17,37–39]. Since the problem of synthesizing large-area, high-quality graphene has been solved, graphene films can now be considered for many kinds of applications such as FETs, transparent electrodes, and flexible and stretchable electrodes [6,16,17].

Figure 5.9 shows the first graphene transistor arrays in a 3-inch wafer size with mobility of around 2000 cm^2/V·s [18]. Generally, there are two ways to synthesize high-quality graphene in large scale for FET applications [18,32,40]. The first method uses graphene grown directly on a SiC wafer. The second method involves transferring graphene films synthesized on a metal catalyst to a useful substrate, such as a flexible substrate. The latter approach is attractive in terms of the possibility of device fabrication over large areas, flexible substrates. Although several studies have been reported about flexible graphene FETs [41], there are still remarkable obstacles in fabricating large-scale, flexible graphene FETs.

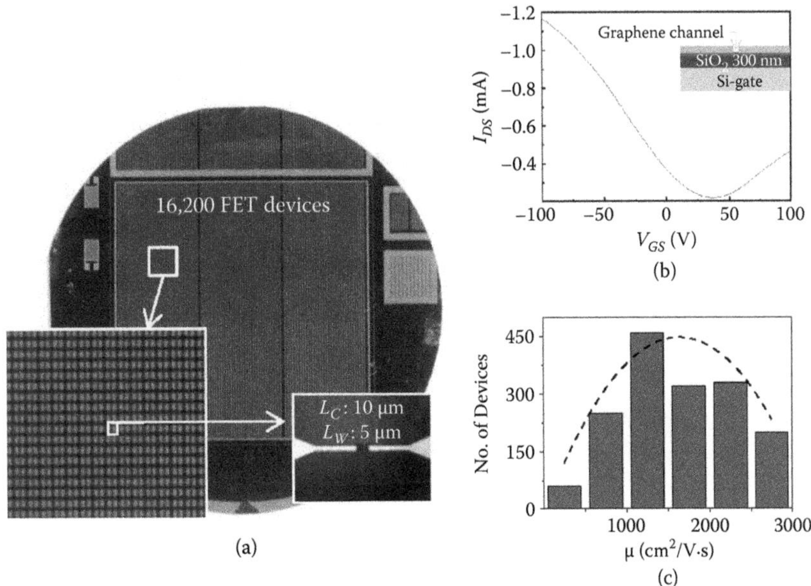

FIGURE 5.9 Electrical properties of graphene FETs and strain sensors. (a) Images of graphene FET arrays (~16,200 devices) fabricated on a 3-inch SiO_2/Si wafer. Source and drain electrodes are formed by 100-nm-thick Au. Inset denotes an image of the representative transistor. (b) A transfer curve of the transistor whose channel length and width are both 5 μm. Inset denotes a schematic cross section of a device. (c) Distribution of the hole and electron mobility of graphene FETs. (From Lee, Y., Bae, S., Jang, H., Jang, S., Zhu, S.-E., Sim, S. H., Song, Y. I., Hong, B. H., and Ahn, J.-H. 2010. Wafer-scale synthesis and transfer of graphene films. *Nano Letters* 10 (2): 490–493. Copyright 2010 American Chemical Society. With permission.)

To fabricate graphene FETs, a high-capacitance gate dielectric that forms a good interface with graphene films is required. Although several high-k materials including HfO_2, Al_2O_3, and ZrO_2 have been applied to graphene FETs, they are not available for flexible devices based on plastic substrates due to their high processing temperature [29,32,42]. Therefore, a high-capacitance gate dielectric that forms a good interface with graphene film through a low-temperature process is demanded. This problem can be solved by a high-capacitance, solution processable ion-gel gate dielectric [43–45].

Figure 5.10 shows the process for fabricating a flexible graphene FET on a polyethylene terephthalate (PET) substrate with an ion-gel gate dielectric [46]. First of all, they formed source and drain electrodes with Cr/Au through conventional photolithography. Graphene film synthesized on Cu foil was transferred and isolated to make the channel region. Ion-liquid with gelating triblock copolymer was drop casted to form an ion-gel dielectric layer followed by Au deposition using a shadow mask to form the gate electrode.

Figure 5.11 illustrates the performance of the flexible graphene FET. Figure 5.11(a) is an image of the bent device array. The transfer and current-voltage characteristics

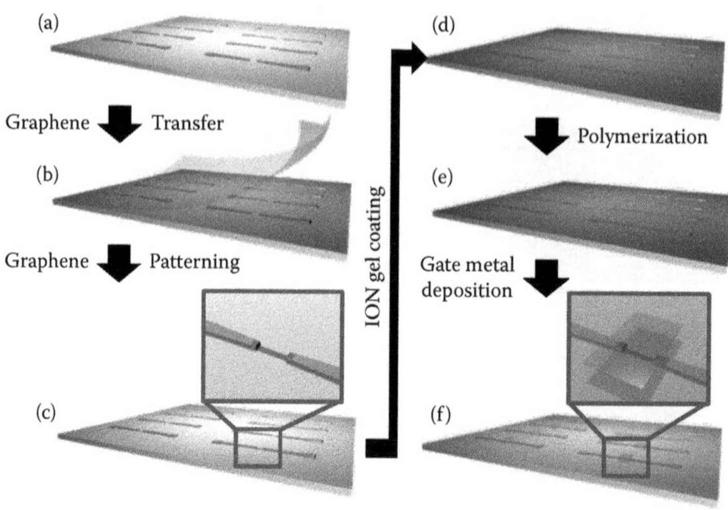

FIGURE 5.10 Schematic diagram of the steps used to fabricate the ion-gel gated graphene transistor array on a plastic substrate. (From Kim, B. J., Jang, H., Lee, S.-K., Hong, B. H., Ahn, J.-H., and Cho, J. H. 2010. High-performance flexible graphene field effect transistors with ion gel gate dielectrics. *Nano Letters* 10 (9): 3464–3466. Copyright 2010 American Chemical Society. With permission.)

of the device show low-voltage operation and high on-current (Figure 5.11[b]). This is due to the large capacitance of the ion-gel, which consists of a room-temperature ionic liquid and a gelating triblock copolymer with a large capacitance value, as high as 5.17 μF/cm^2. In addition, unlike other oxide-based dielectric materials, the robust flexibility of ion-gel provides higher flexibility, which is presented in Figure 5.11(c), and stable operation of the device with a small bending radius up to 5 mm. Figure 5.11(d) shows the statistic distribution of mobility around 200 cm^2/V·s and 100 cm^2/V·s for hole and electron, respectively. The results offer opportunities for graphene in devices requiring unusual form factors, including mechanical flexibility and stretchability.

Future electronic devices require good mechanical properties as well as optical transmittance. When a metal film is applied as an electrode, it is impossible to build a transparent, flexible device system. Recently, a transparent thin film transistor (TFT) has been developed [47] that uses graphene as a conducting electrode and single-walled carbon nanotubes (SWNTs) as the semiconducting channel. These SWNTs and graphene films are printed on flexible plastic substrates using a printing method. The resulting devices exhibit excellent optical transmittance and electrical properties. A schematic illustration of a transparent TFT device and its morphology are presented in Figures 5.12(a) and 5.12(b), respectively. The photograph of the devices (Figure 5.12[c]) demonstrates excellent optical transmittance and mechanical bendability. The resulting TFT devices exhibit mobility of ~2 cm^2/V·s and an on/off ratio of ~10^2, and the performance is independent on each channel length (Figure 5.12[d]).

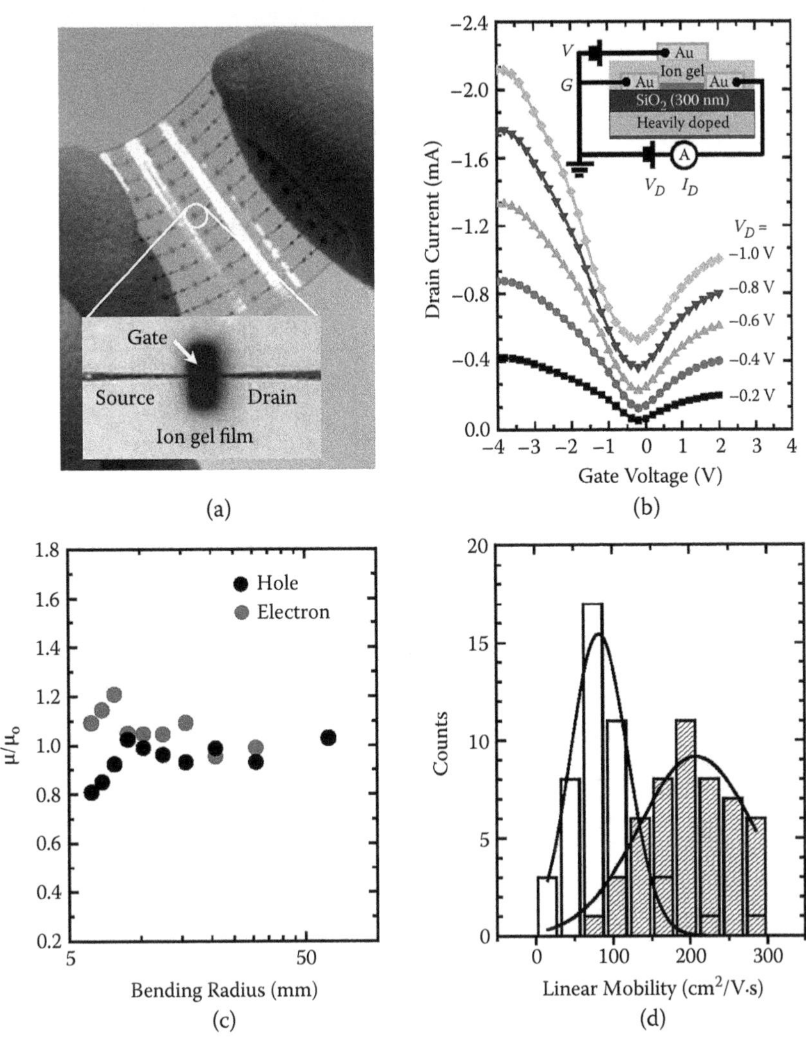

FIGURE 5.11 Electrical and mechanical properties of ion-gel gated graphene FETs fabricated on a flexible plastic substrate. (a) Optical images of an array of devices on a plastic substrate. (b) Transfer characteristics of graphene FETs on plastic substrate. In the output curve, the gate voltage was varied between +2 and −4 V in steps of −1 V. The insert is the schematic illustration of flexible TFTs. (c) Normalized effective mobility (μ/μ_0) as a function of the bending radius. (d) Distribution of the hole and electron mobility of graphene FETs. (From Kim, B. J., Jang, H., Lee, S.-K., Hong, B. H., Ahn, J.-H., and Cho, J. H. 2010. High-performance flexible graphene field effect transistors with ion gel gate dielectrics. *Nano Letters* 10 (9): 3464–3466. Copyright 2010 American Chemical Society. With permission.)

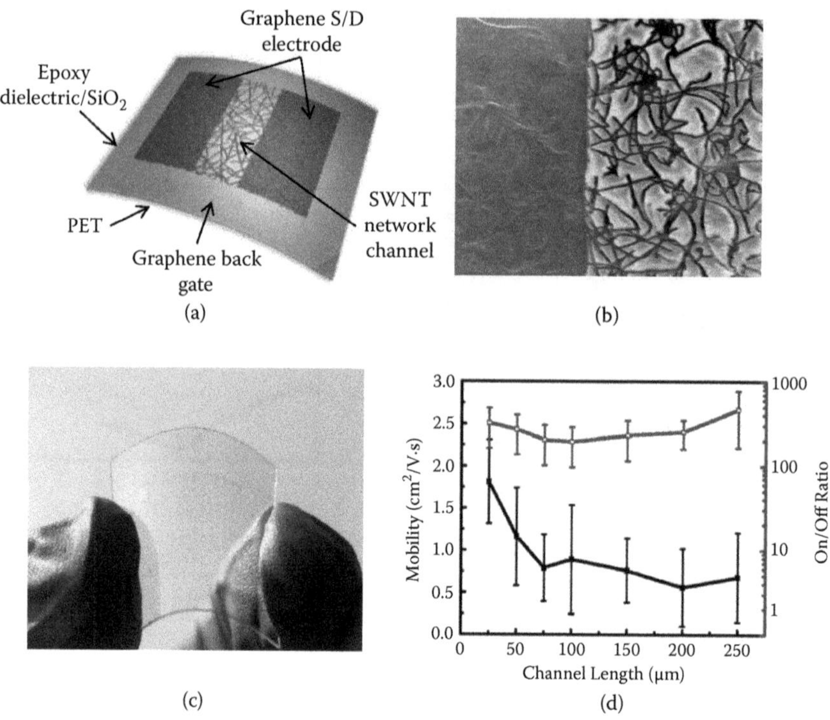

FIGURE 5.12 (a) Schematic structure of the flexible TTFTs on the plastic substrate. (b) SEM image of the interface between a source/drain graphene electrode (left) and SWNT network channel layer (right). (c) Optical images of a completed array of the TTFTs on the PET substrate. (d) Effective device mobility and on/off ratios as a function of the channel length. (From Sukjae, J., et al. 2010. Flexible, transparent single-walled carbon nanotube transistors with graphene electrodes. *Nanotechnology* 21 (42): 425201. Copyright 2010 UK Institute of Physics. With permission.)

5.4 TRANSPARENT ELECTRODES

In addition to being conductive and transparent, the next generation of optoelectronic devices requires that transparent conductive electrodes are lightweight, flexible, cheap, and compatible with large-scale fabricating methods. Graphene has a great potential for the electrode application of organic electronic devices where low sheet resistance and high transparency are essential. The conventional transparent electrode for optoelectronic devices is indium tin oxide (ITO) or fluorine-doped tin oxide, which suffer from the scarcity of indium reserves and the brittle nature of metal oxide [12,48,49]. Carbon nanotubes and nanowires have been used as transparent electrodes for optoelectronic devices as alternative materials [50–52]. However, the significant roughness of such films imposes serious limitations on their application. For example, the morphology of the transparent electrode of an organic photovoltaic (OPV) is very important in order to reduce the possibility of leakage current and short circuits and improve performance. The roughness of graphene grown by

CVD is less than 1 nm, which is comparable to that of commercial ITO and much smoother than that of carbon nanotubes.

The optical and electrical characterization of large-scale graphene film is shown in Figure 5.13. The randomly stacked graphene layers show that the optical transparency proportionally decreased according to the number of layers [5,6,19,52,53]. The transmittance of 1- to 4-layer graphene film is distributed from 97.4% to 90.1% at the wavelength of 550 nm with the decrease around 2.5% per layer. To increase the conductance of graphene film, doping using acid such as HNO_3 is performed after the transfer process. This doping process may induce little decrease in transmittance, as shown in the graph in Figure 5.13(c). Figure 5.13(b) shows the tendency

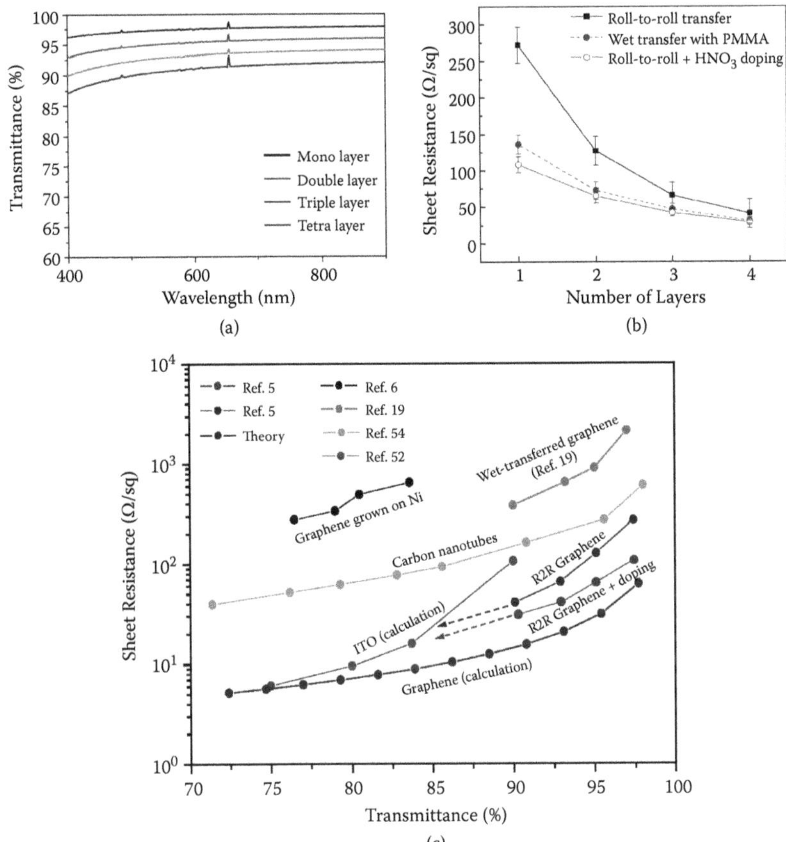

FIGURE 5.13 (a) Optical transmittance of randomly stacked graphene sheet on Quartz crystal. (b) Sheet resistances of graphene sheets using roll-to-roll transfer and that combined with acid doping and wet transfer with PMMA supporting layer. (c) Comparison of sheet resistances from other references and theory. (From Bae, S., Kim, H., Lee, Y., Xu, X., Park, J.-S., Zheng, Y., Balakrishnan, J., Lei, T., Ri Kim, H., Song, Y. I., Kim, Y.-J., Kim, K. S., Ozyilmaz, B., Ahn, J.-H., Hong, B. H., and Iijima, S. 2010. Roll-to-roll production of 30-inch graphene films for transparent electrodes. *Nature Nanotechnology* 5 (8): 574–578. Copyright 2010 Nature Publishing Group. With permission.)

of resistance with respect to an increase in the number of stacked graphene layers. Because the defect produced during the transfer process is corrected when another layer is stacked on it, the resistance is remarkably decreased. The resistances of graphene films that are transferred using different methods are grouped with similar values around 50 Ω/□ when graphene is stacked in 4 layers, which is independent of the transfer method. Figure 5.13(c) summarizes the relationship between sheet resistance and transparency of ITO, carbon nanotubes, and graphene prepared by different techniques. The resistance of 4-layer graphene is lower than that of ITO and carbon nanotubes at transmittance of 90%, whether it is doped or not, as shown in Figure 5.13(c). The transparent electrode made by CVD has a higher transmittance in the visible and infrared (IR) region and is more stable under bending, compared with ITO, which has a sheet resistance of 5–60 Ω/□ and ~85% transmittance in the range of 400–900 nm. To replace ITO electrodes, it is generally agreed that graphene should at least present a sheet resistance of less than 100 Ω/□, coupled with a transmittance of more than 90% in the visible range [52,54,55]. With this result we can argue that graphene film is a proper material to replace ITO in the near future. Moreover, the calculated resistance value of graphene is much lower than roll-to-roll doped graphene film. If the quality of graphene and the transfer process are improved continuously, highly conductive transparent electrodes for applications that require them will be achieved.

5.4.1 Touch-Screen Panels

A representative application of transparent electrodes is the touch screen, which has been adapted to many kinds of electrical devices, such as cell phones and monitors. The resistive type touch screen is operated by short transparent electrodes between the top and bottom that are activated when mechanical force is applied to the system [56]. This type of touch screen requires a resistance value up to 550 Ω/□, which can be fabricated with monolayer graphene. The area of graphene electrodes is defined by photolithography or shadow masking with oxygen plasma etching in a few seconds. After screen printing (using silver paste) to make the x-axis and y-axis on both the top and bottom graphene/PET films, the UV or thermally curable spacer dots are coated on the bottom graphene electrode to prevent shorting problems between layers. Before attaching the top and bottom films, contact via a hole is formed to extract a signal. Finally, the graphene-based touch screen is finished after attaching the top and bottom films, as shown in Figure 5.14 [5].

Figure 5.15 shows a resistive type touch screen that uses graphene electrodes. Figures 5.15(a) and 5.15(b) represent the screen printing process onto the graphene electrode and the final touch screen product with good flexibility. This graphene-based touch screen shows reliable operation even after many bending tests, due to the outstanding mechanical properties of graphene electrodes, as shown in Figure 5.15(c). A comparison of resistance change according to bending strain between graphene film [5] and ITO [57] is shown in Figure 5.15(d). Resistance of ITO is significantly increased under a strain value of 2~3%. Graphene exhibits little change in resistance up to 6% strain. This limitation value is not a property of

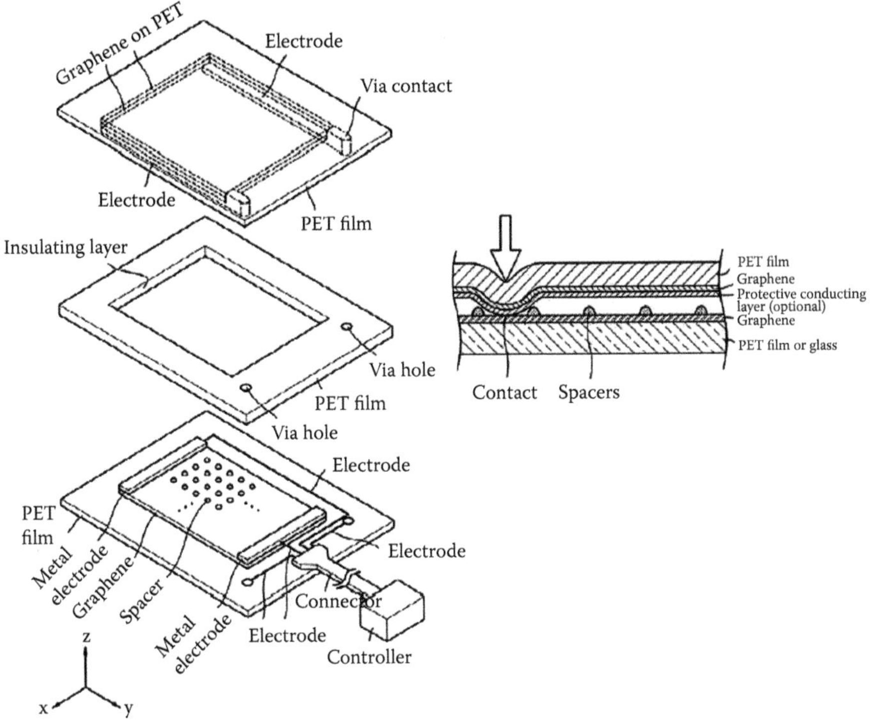

FIGURE 5.14 The structure and working principle of a graphene film-based touch screen. (From Bae, S., Kim, H., Lee, Y., Xu, X., Park, J.-S., Zheng, Y., Balakrishnan, J., Lei, T., Ri Kim, H., Song, Y. I., Kim, Y.-J., Kim, K. S., Ozyilmaz, B., Ahn, J.-H., Hong, B. H., and Iijima, S. 2010. Roll-to-roll production of 30-inch graphene films for transparent electrodes. *Nature Nanotechnology* 5 (8): 574–578. Copyright 2010 Nature Publishing Group. With permission.)

graphene, but is instead a result of printed silver electrodes. This demonstration indicates that the flexible graphene electrode is close to being used in real applications.

5.4.2 Electrodes for Organic Devices

As indicated, graphene is an attractive material for the conductive transparent electrodes of flexible optoelectronic devices, such as OPVs [8–13], organic light-emitting diodes (OLEDs) [14,15], and bulk heterojunction polymer memory [58,59], because of its unique optical transmittance and good electric conductivity. Researchers have developed various methods to prepare graphene or reduced graphene oxide suspensions for fabrication of transparent conductive electrodes [6,16,17,60–69].

Recently, graphene films, including reduced graphene oxide (GO) and CVD graphene, have been explored for application as transparent conductive electrodes in solar cells. In dye-sensitized solar cells, exfoliated graphene sheets with a thickness of 10 nm and 30 nm were used as electrodes and the resulting device had a power conversion efficiency of 0.26% [8].

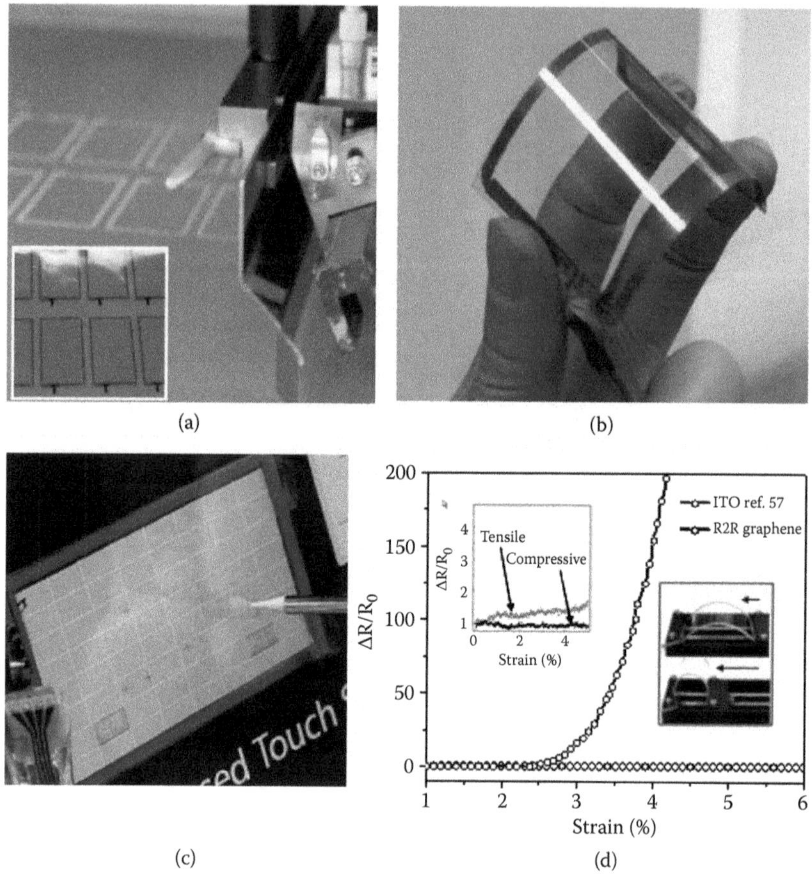

FIGURE 5.15 (a) Screen printing process of silver paste onto graphene/PET. (b) Image of a graphene-based flexible touch screen. (c) Draw test of a graphene-based touch screen using software connected to a computer. (d) Relative resistance change with respect to tensile strain. (From Bae, S., Kim, H., Lee, Y., Xu, X., Park, J.-S., Zheng, Y., Balakrishnan, J., Lei, T., Ri Kim, H., Song, Y. I., Kim, Y.-J., Kim, K. S., Ozyilmaz, B., Ahn, J.-H., Hong, B. H., and Iijima, S. 2010. Roll-to-roll production of 30-inch graphene films for transparent electrodes. *Nature Nanotechnology* 5 (8): 574–578. Copyright 2010 Nature Publishing Group. With permission.)

An illustration of polymeric P3HT:PCBM-based bulk heterojunction organic solar cells is presented in Figure 5.16(a), in which the architecture of the device is substrate/ graphene (or GO)/ electron blocking layer (normally, PEDOT:PSS)/ active layer/ hole blocking layer (e.g., LiF, Ca, and Alq3, etc.)/ metal. Conventionally, graphene or GO works as an anode and light can pass through on this side. The devices using chemically made graphene or graphene oxide as an electrode exhibit rather moderate performance [9,70,71]. A power conversion efficiency (PCE) of around 0.1% is obtained by using doped reduced graphene oxide films as electrodes [9], which had a sheet resistance of 40 kΩ/□ and a transparency of 64%. The poor performance is

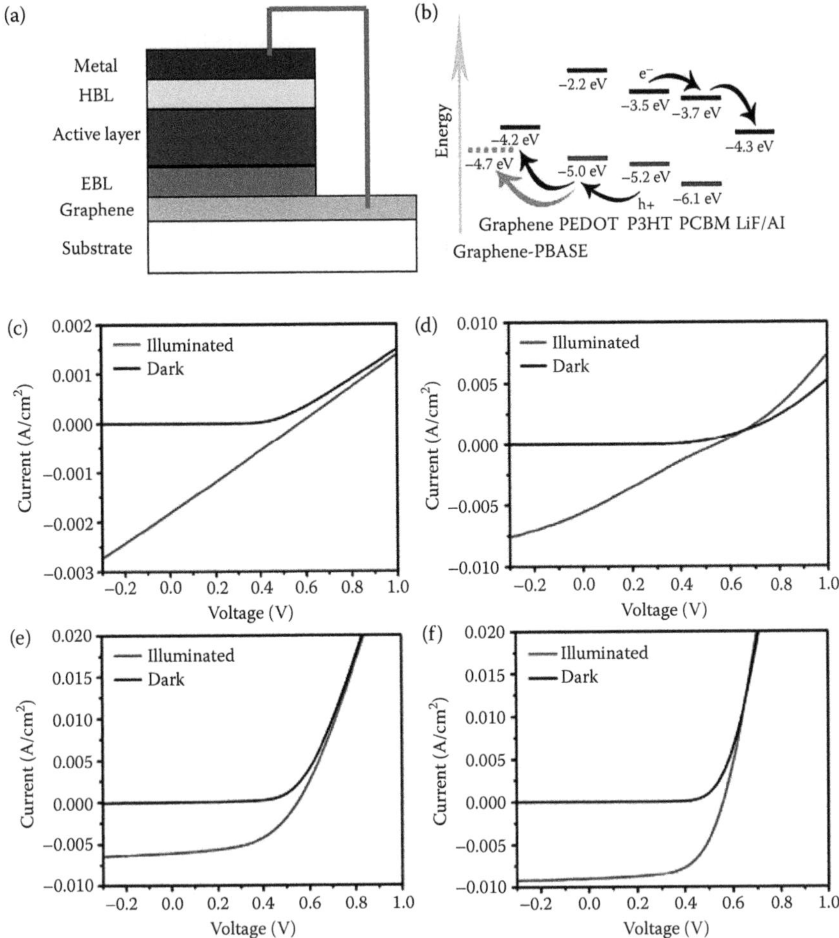

FIGURE 5.16 (a) Schematic structure of OPVs with graphene electrodes. (b) Energy diagram of the fabricated device with the structure: graphene/PEDOT:PSS/P3HT:PCBM/LiF/Al. (c)–(f) Current-voltage characteristics of photovoltaic devices based on graphene films in dark and under illumination, where (c) is from pristine graphene film, (d) graphene film treated by UV light, (e) graphene film modified by PBASE, (f) ITO anode for comparison. (From Wang, Y., Chen, X., Zhong, Y., Zhu, F., and Loh, K. P. 2009. Large area, continuous, few-layered graphene as anodes in organic photovoltaic devices. *Applied Physics Letters* 95 (6): 063302-3. Copyright 2009 American Institute of Physics. With permission.)

caused by the huge contact resistance of small graphene flakes and the insulating property of graphene that is reduced from graphene oxide. The structural defects and lateral disorder of such exfoliated graphene negatively affects the carrier mobility of the film. A nanocomposite comprised of chemically converted graphene and carbon nanotubes had a sheet resistance of 240 Ω/□ at 86% transmittance. The OPV devices constructed with this kind of electrode demonstrate 0.85% PCE [72].

CVD-grown graphene with extremely high intrinsic mobility can result in improved performance. The average sheet resistance varied from 1350 Ω/□ to 210 Ω/□ with a transparency from 91% to 72% in the visible range for 6- to 30-nm-thick graphene grown on a Ni substrate by CVD [11]. A Ni graphene is transferred to the substrate and used as an anode. A PEDOT:PSS thin film of 40 nm working as a hole injecting layer of the device is spin coated and annealed at 140°C for 10 minutes. The third layer is the active layer of bulk heterojunction composites of P3HT:PCBM. Finally, LiF and Al are deposited as a hole blocking layer and cathode, respectively. The current density-voltage characteristic of OPV devices is presented in Figure 5.16. When a pristine graphene is applied as an anode, the device has an open-circuit voltage (Voc) of 0.32V, a short-circuit current density (Jsc) of 2.39mA/cm^2, a fill factor (FF) of 27%, and overall PCE of 0.21%. The poor performance is attributed to the hydrophobic property of graphene film, which prevents the spread of PEDOT:PSS on the graphene surface from being uniform. After being treated by UV/ozone, the graphene electrode changes the wettability and improves PCE to 0.74%. A self-assembled pyrenebutanoic acid succinimidyl ester (PBASE) is performed to modify the surface property of graphene. The resulting device shows excellent characteristics of Voc = 0.55V, Jsc = 6.05A, FF = 51.3%, and overall PCE = 1.71%, which is about 55.2% of the PCE of the ITO electrode device (3.1%). It should be noted that the self-assembled PBASE film not only improves the hydrophilic property, but also effectively tunes the work function of graphene (Figure 5.16). When a CVD multilayer graphene is used as an anode and a hole blocking TiOx layer is inserted, the PCE of OPVs is enhanced to ~2.6% [13], which is considerably high efficiency compared with other OPVs adopting graphene as an electrode.

For bilayer small-molecule (CuPc/C60) OPV devices, moderate PCE of less than 0.4% is obtained when solution processed graphene oxide films are applied as electrodes [10], in which the thickness of graphene films is 4–7 nm and the corresponding sheet resistance and transmittance are 100–500k Ω/□ and 85–95%, respectively. The PCE value in this work is just half of ITO electrode devices. The devices with CVD-grown graphene as electrodes show much better performance [73]. For pristine CVD-grown graphene electrode OPV devices, the overall performance is only slightly inferior to their counterparts with ITO electrodes, which is supposed to be limited by the high sheet resistance and hydrophobic property of graphene. Much effort has been devoted to exploring ways to decrease the sheet resistance and improve the surface wetting property between graphene and the hole-injecting layer of PEDOT:PSS [74–76]. The sheet resistance is reduced by a factor of 3 after chemical doping by nitric acid, resulting in transparent graphene films with Rs = 90 Ω/□ at a transmittance of 80% at 550 nm. $AuCl_3$ was found to be a wonderful dopant in both cases. $AuCl_3$ doping on graphene can significantly reduce the sheet resistance of graphene electrodes and change the graphene surface from hydrophobic to hydrophilic, enabling a uniform hole blocking layer to be achieved. As a result, the PCE of 1.63% is observed [73] for $AuCl_3$-doped graphene electrode OPV devices.

An important aspect of modern optoelectronic thin film devices is flexibility, which is also a critical advantage of transparent graphene electrodes compared to its ITO counterpart. A chemically reduced graphene oxide film is transferred onto a PET substrate and the resulting flexible transparent film is used as an electrode for flexible polymeric bulk heterojunction solar cells [77]. The active layer of P3HT:PCBM is spin coated on the surface of graphene films. The device performance is more sensitive to sheet resistance than transmittance of graphene electrodes. The sheet resistance of graphene films can be controlled by thickness. The obtained graphene films have a sheet resistance of 720 Ω/□ with a corresponding transmittance of 40% and 16k Ω/□ with transmittance of 88% when the thickness is 28 nm and 4 nm, respectively. The lower sheet resistance of graphene film enhances the current density of devices and thus the overall power conversion efficiency. The best PCE of 0.78% is achieved for flexible OPVs using a chemically reduced graphene film as an electrode. Figure 5.17 depicts the performance of flexible OPVs changing with the bending test. The devices present excellent stability even after bending a thousand times. As mentioned, the sheet resistance of graphene electrodes can be significantly decreased when using CVD-grown graphene instead of reduced graphene. The CVD-grown graphene shows a sheet resistance of 230 Ω/□ at 72% transparency. The solar cell devices fabricated with such CVD-grown graphene electrodes exhibit PCEs of 1.18% for small molecule (CuPc/C60) OPVs, which is comparable to that of ITO electrode devices of 1.27% [12]. Moreover, the stability of graphene electrode devices is proved by applying bending conditions up to 138°, which completely surpasses that of ITO devices that just survive bending conditions at 36° (Figure 5.17).

Solution-processed [14] and CVD-grown graphene [15] thin films are also applied in organic light-emitting diodes. The device structure is substrate/graphene (or ITO)/ PEDOT:PSS / N, N′-1-naphthyl-N, N′-diphenyl-1,1′-biphenyl-4, 4′diamine (NPD)/ tris(8-hydroxyquinoline) aluminum (Alq3)/ LiF/ Al, as shown in the inset of Figure 5.18(a). OLED devices with graphene electrodes result in current drive and light emission intensity comparable to those with ITO electrodes in the low current density range (<10mA/cm^2). In addition, the turn-on voltage for graphene-electrode and ITO-electrode OLEDs is 4.8V and 3.8, respectively. Figure 5.18(b) depicts the external quantum efficiency (EQE) and luminous power efficiency (LPE) of both graphene- and ITO-electrode OLEDs. The remarkable results that have been achieved indicate that the EQE and LPE for graphene-based OLEDs nearly matches ITO-based devices despite the fact that graphene electrodes have a sheet resistance nearly 2 orders of magnitude higher than the ITO electrode.

5.5 GRAPHENE-BASED GAS BARRIER FILM

One of most important concepts in the future electronics market is flexible electronics. To realize this aim, many researchers have attempted to solve the substrate problem. There are three representative candidates for flexible electronics: thin glass, stainless steel, and plastic substrates. Thin glass substrates have advantages such as high transparency and low surface roughness. Stainless steel also has benefits such as less deformation and high elastic modulus. Although both substrates have benefits, plastic substrates are likely to be adopted in the roll-to-roll process, which is a low-cost production process. However, it is hard to realize flexible electronics on plastic substrates at this moment because of high gas permeation behavior. A high gas permeation rate results in degradation of device performance in a short time, so the reliability of devices will be poor. Table 5.1 shows the required water vapor transmission

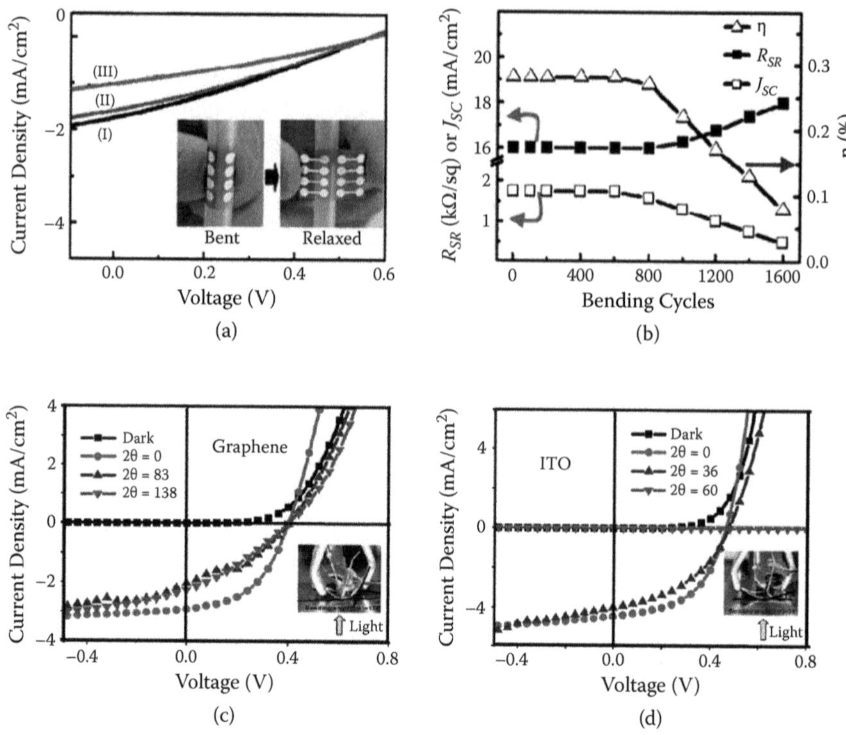

FIGURE 5.17 (a) J-V curves of device after applying (i) 400, (ii) 800, and (iii) 1200 cycles of bending. The short-circuit current density (Jsc), overall power conversion efficiency (η), and the sheet resistance (RS) of the device are plotted as a function of bending cycles in (b). Inset in (a): photograph of the bending-relaxing experiments of the OPV device. Current density vs. voltage characteristics of CVD graphene. (c) or ITO (d) photovoltaic cells under 100 mW/cm^2 AM1.5G spectral illumination for different bending angles. Insets in (c) and (d) show the experimental setup employed in the experiments. ([a] and [b] reprinted with permission from Li, S.-S., Tu, K.-H., Lin, C.-C., Chen, C.-W., and Chhowalla, M. 2010. Solution-processable graphene oxide as an efficient hole transport layer in polymer solar cells. *ACS Nano* 4 (6): 3169–3174. Copyright 2010 American Chemical Society; [c] and [d] reprinted with permission from Gomez De Arco, L., Zhang, Y., Schlenker, C. W., Ryu, K., Thompson, M. E., and Zhou, C. 2010. Continuous, highly flexible, and transparent graphene films by chemical vapor deposition for organic photovoltaics. *ACS Nano* 4 (5): 2865–2873. Copyright 2010 American Chemical Society.)

rate (WVTR) value for different applications [78–80]. Electric devices using organic materials require a high level of impermeability, as much as 10^{-6} g/m^2/day level. PET film represents a 1 ~ 10 g/m^2/day value, which is a much higher value than that shown in the table. Many researchers have been focused on reducing this effect. One representative method is coating multiple layers of polymer and oxide materials in a vacuum. Despite the fact that this structure can reduce the WVTR value to around ~10^{-6} g/m^2/day level, there are some disadvantages, such as deformation under strain and high process cost as a result of UHV process pressure [80].

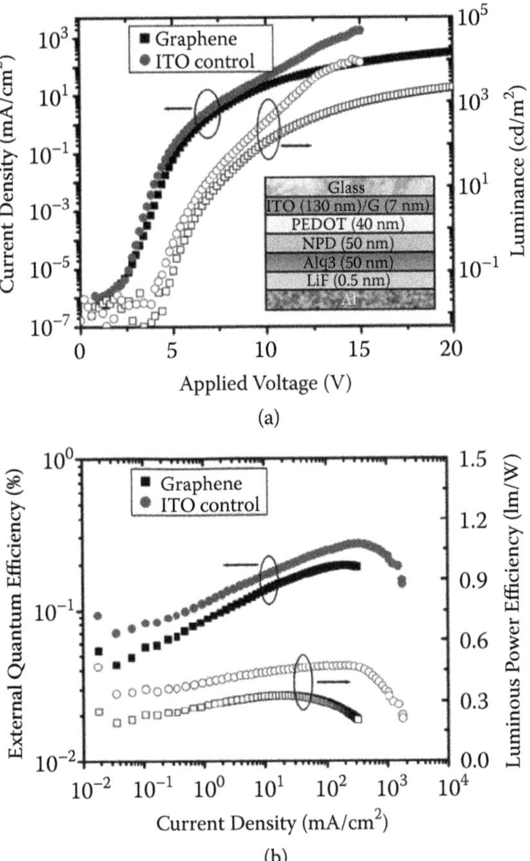

FIGURE 5.18 (a) Current density and luminance level of both OLED based on solution-processed graphene and ITO. (b) Comparison of external quantum efficiency and luminous power efficiency for solution-processed graphene electrode OLED and ITO electrodes. (From Wu, J., Agrawal, M., Becerril, H. A., Bao, Z., Liu, Z., Chen, Y., and Peumans, P. 2009. Organic light-emitting diodes on solution-processed graphene transparent electrodes. *ACS Nano* 4 (1): 43–48. Copyright 2009 American Chemical Society. With permission.)

Graphene has a 2-dimensional densely packed atomic structure with short atomic spacing that can prohibit the diffusion of foreign atoms through this lattice [34]. Moreover, there is a report introducing gas impermeability for smaller gas molecules such as He. Figure 5.19 shows a graphene bubble that occurred by pressure difference between two parts separated by a graphene membrane. This bubble can stand in −93 kPa pressure with a bubble height around 175 nm [81]. This suggests that large-scale graphene film has potential as a gas barrier layer of plastic substrates. Through the large-scale synthesis and transfer method introduced above, large-scale randomly stacked graphene can be formed onto any kind of useful plastic substrate, and the optical transparency of barrier film is proportional to the thickness of the graphene film.

TABLE 5.1
Required Water Vapor Transmission Rate Value for Different Applications

Application	WVTR (g/m²/day)
OLED	10^{-6}
Solar cell	10^{-4}
LCD	
RFID tag	10^{-2}
EL display	

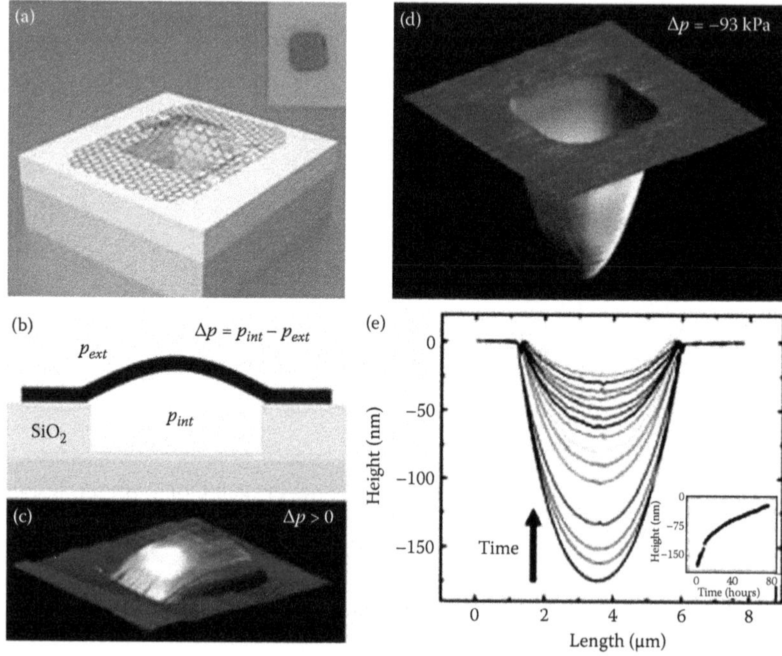

FIGURE 5.19 (a) Scheme of a microchamber created by graphene membrane. The inset image denotes the graphene membrane suspended on 440-nm SiO_2 with an area of 4.75 × 4.74 μm². (b) Cross-sectional view of microchamber. (c) AFM image of multilayer graphene drumhead with $\Delta p > 0$. (d) The AFM image of graphene membrane of (a) when the pressure difference is $\Delta p = -93$kPa. The minimum z position of this membrane is 175 nm. (e) Change in z shape of graphene membrane with respect to the time until 71.3 h in ambient condition. Inset graph denotes the center position of the graphene membrane according to time. (From Bunch, J. S., Verbridge, S. S., Alden, J. S., van der Zande, A. M., Parpia, J. M., Craighead, H. G., and McEuen, P. L. 2008. Impermeable atomic membranes from graphene sheets. *Nano Letters* 8 (8): 2458–2462. Copyright 2008 American Chemical Society. With permission.)

Above all, large-scale graphene films grown on Cu foil can act as an antioxidation layer for the Cu foil. This possibility can be easily measured through the oxidation level with certain conditions for two different copper foils where one has a graphene layer and the other has no graphene layer. These samples are being exposed to air in two months with ambient pressure and room temperature. The Cu foil with graphene on the surface experiences lower oxidation, while the Cu foil without graphene experiences remarkable surface change. The result indicates the gas impermeability of large-area graphene film that is synthesized using the CVD method [82].

A graphene barrier layer with dimensions larger than 10 × 10cm^2 is used to test WVTR. To demonstrate the possibility of a barrier layer using graphene, the WVTR is measured via tritiated water (^3H_2O), which can be detected by a radioactive analysis system, as shown in Figure 5.20(a). The tritiated water that has penetrated the graphene film is detected by a beta-ray detector that is as sensitive as 10^{-6} g/m^2/day. Figure 5.20(b) shows the WVTR result for 6-layer graphene film in 500 minutes. The initial WVTR value is around 10^{-4} g/m^2/day, which is comparable to that of inorganic barrier coating materials such as Al_2O_3 and SiO_2 [83].

The microcracks induced during the transfer process and particles result in the increase of the WVTR value of the graphene film barrier layer. After enhancing the technology of the transfer process, the quality of synthesis, and clean environment, the gas impermeability of graphene films may be improved. Graphene film has fascinating benefits such as high transmittance, which is essential for optoelectronic systems, and printable process for roll-to-roll process compared with other barrier materials. Considering these properties, graphene film barriers can be used for food packaging and flexible electronic systems.

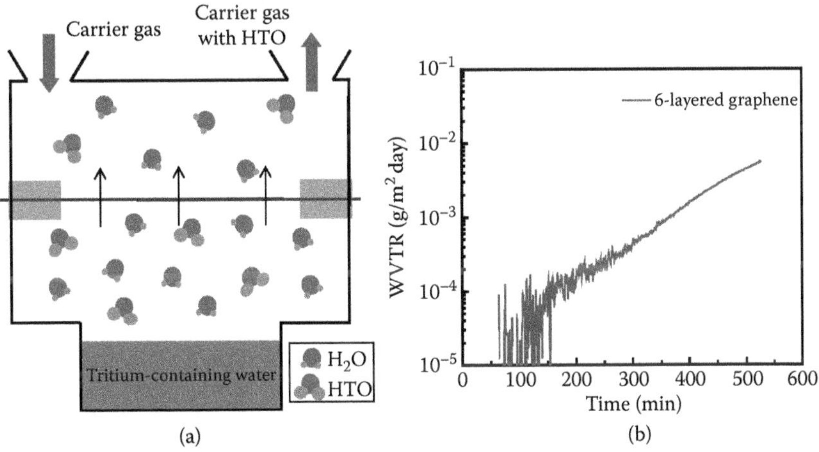

FIGURE 5.20 (a) Structure and principle of water vapor transmission rate (WVTR) measurement system. (b) WVTR for 6-layer graphene-covered PET film within 500 minutes.

5.6 CONCLUDING REMARKS

In this chapter we reviewed some recent work involving graphene films for high-performance transistors on rigid and even flexible substrates. The chemical vapor deposition approaches for producing high-quality, large-area graphene films were discussed. These materials and approaches could enable high-performance electronics on plastic substrates, as demonstrated in several different application examples. Successful commercial implementation of such techniques represents a significant engineering challenge, but could create interesting opportunities for developing next-generation electronic applications such as flexible OLED displays, touch screens, and high-speed RF transistors.

REFERENCES

1. Novoselov, K. S., Geim, A. K., Morozov, S. V., Jiang, D., Zhang, Y., Dubonos, S. V., Grigorieva, I. V., and Firsov, A. A. 2004. Electric field effect in atomically thin carbon films. *Science* 306 (5696): 666–669.
2. Bolotin, K. I., Sikes, K. J., Jiang, Z., Klima, M., Fudenberg, G., Hone, J., Kim, P., and Stormer, H. L. 2008. Ultrahigh electron mobility in suspended graphene. *Solid State Communications* 146 (9–10): 351–355.
3. Bunch, J. S., van der Zande, A. M., Verbridge, S. S., Frank, I. W., Tanenbaum, D. M., Parpia, J. M., Craighead, H. G., and McEuen, P. L. 2007. Electromechanical resonators from graphene sheets. *Science* 315 (5811): 490–493.
4. Lee, C., Wei, X., Kysar, J. W., and Hone, J. 2008. Measurement of the elastic properties and intrinsic strength of monolayer graphene. *Science* 321 (5887): 385–388.
5. Bae, S., Kim, H., Lee, Y., Xu, X., Park, J.-S., Zheng, Y., Balakrishnan, J., Lei, T., Ri Kim, H., Song, Y. I., Kim, Y.-J., Kim, K. S., Ozyilmaz, B., Ahn, J.-H., Hong, B. H., and Iijima, S. 2010. Roll-to-roll production of 30-inch graphene films for transparent electrodes. *Nature Nanotechnology* 5 (8): 574–578.
6. Kim, K. S., Zhao, Y., Jang, H., Lee, S. Y., Kim, J. M., Kim, K. S., Ahn, J.-H., Kim, P., Choi, J.-Y., and Hong, B. H. 2009. Large-scale pattern growth of graphene films for stretchable transparent electrodes. *Nature* 457 (7230): 706–710.
7. Balandin, A. A., Ghosh, S., Bao, W., Calizo, I., Teweldebrhan, D., Miao, F., and Lau, C. N. 2008. Superior thermal conductivity of single-layer graphene. *Nano Letters* 8 (3): 902–907.
8. Wang, X., Zhi, L., and Mullen, K. 2007. Transparent, conductive graphene electrodes for dye-sensitized solar cells. *Nano Letters* 8 (1): 323–327.
9. Eda, G., Lin, Y.-Y., Miller, S., Chen, C.-W., Su, W.-F., and Chhowalla, M. 2008. Transparent and conducting electrodes for organic electronics from reduced graphene oxide. *Applied Physics Letters* 92 (23): 233305-3.
10. Wu, J., Becerril, H. A., Bao, Z., Liu, Z., Chen, Y., and Peumans, P. 2008. Organic solar cells with solution-processed graphene transparent electrodes. *Applied Physics Letters* 92 (26): 263302-3.
11. Wang, Y., Chen, X., Zhong, Y., Zhu, F., and Loh, K. P. 2009. Large area, continuous, few-layered graphene as anodes in organic photovoltaic devices. *Applied Physics Letters* 95 (6): 063302-3.
12. Gomez De Arco, L., Zhang, Y., Schlenker, C. W., Ryu, K., Thompson, M. E., and Zhou, C. 2010. Continuous, highly flexible, and transparent graphene films by chemical vapor deposition for organic photovoltaics. *ACS Nano* 4 (5): 2865–2873.

13. Choe, M., Lee, B. H., Jo, G., Park, J., Park, W., Lee, S., Hong, W.-K., Seong, M.-J., Kahng, Y. H., Lee, K., and Lee, T. 2010. Efficient bulk-heterojunction photovoltaic cells with transparent multi-layer graphene electrodes. *Organic Electronics* 11 (11): 1864–1869.
14. Wu, J., Agrawal, M., Becerril, H. A., Bao, Z., Liu, Z., Chen, Y., and Peumans, P. 2009. Organic light-emitting diodes on solution-processed graphene transparent electrodes. *ACS Nano* 4 (1): 43–48.
15. Sun, T., Wang, Z. L., Shi, Z. J., Ran, G. Z., Xu, W. J., Wang, Z. Y., Li, Y. Z., Dai, L., and Qin, G. G. 2009. Multilayered graphene used as anode of organic light emitting devices. *Applied Physics Letters* 96 (13): 133301-3.
16. Reina, A., Jia, X., Ho, J., Nezich, D., Son, H., Bulovic, V., Dresselhaus, M. S., and Kong, J. 2008. Large-area, few-layer graphene films on arbitrary substrates by chemical vapor deposition. *Nano Letters*, 9 (1): 30–35.
17. Li, X., Cai, W., An, J., Kim, S., Nah, J., Yang, D., Piner, R., Velamakanni, A., Jung, I., Tutuc, E., Banerjee, S. K., Colombo, L., and Ruoff, R. S. 2009. Large-area synthesis of high-quality and uniform graphene films on copper foils. *Science* 324 (5932): 1312–1314.
18. Lee, Y., Bae, S., Jang, H., Jang, S., Zhu, S.-E., Sim, S. H., Song, Y. I., Hong, B. H., and Ahn, J.-H. 2010. Wafer-scale synthesis and transfer of graphene films. *Nano Letters* 10 (2): 490–493.
19. Li, X., Zhu, Y., Cai, W., Borysiak, M., Han, B., Chen, D., Piner, R. D., Colombo, L., and Ruoff, R. S. 2009. Transfer of large-area graphene films for high-performance transparent conductive electrodes. *Nano Letters* 9 (12): 4359–4363.
20. Ferrari, A. C., Meyer, J. C., Scardaci, V., Casiraghi, C., Lazzeri, M., Mauri, F., Piscanec, S., Jiang, D., Novoselov, K. S., Roth, S., and Geim, A. K. 2006. Raman spectrum of graphene and graphene layers. *Physical Review Letters* 97 (18): 187401.
21. Hass, J., Varchon, F., Millán-Otoya, J. E., Sprinkle, M., Sharma, N., de Heer, W. A., Berger, C., First, P. N., Magaud, L., and Conrad, E. H. 2008. Why multilayer graphene on 4H–SiC(000-1) behaves like a single sheet of graphene. *Physical Review Letters* 100 (12): 125504.
22. Unarunotai, S., Koepke, J. C., Tsai, C.-L., Du, F., Chialvo, C. E., Murata, Y., Haasch, R., Petrov, I., Mason, N., Shim, M., Lyding, J., and Rogers, J. A. 2010. Layer-by-layer transfer of multiple, large area sheets of graphene grown in multilayer stacks on a single SiC wafer. *ACS Nano* 4 (10): 5591–5598.
23. Unarunotai, S., Murata, Y., Chialvo, C. E., Kim, H.-s., MacLaren, S., Mason, N., Petrov, I., and Rogers, J. A. 2009. Transfer of graphene layers grown on SiC wafers to other substrates and their integration into field effect transistors. *Applied Physics Letters* 95 (20): 202101-3.
24. Kang, S. J., Kocabas, C., Kim, H.-S., Cao, Q., Meitl, M. A., Khang, D.-Y., and Rogers, J. A. 2007. Printed multilayer superstructures of aligned single-walled carbon nanotubes for electronic applications. *Nano Letters* 7 (11): 3343–3348.
25. Bolotin, K. I., Sikes, K. J., Hone, J., Stormer, H. L., and Kim, P. 2008. Temperature-dependent transport in suspended graphene. *Physical Review Letters* 101 (9): 096802.
26. Meric, I., Han, M. Y., Young, A. F., Ozyilmaz, B., Kim, P., and Shepard, K. L. 2008. Current saturation in zero-bandgap, top-gated graphene field-effect transistors. *Nature Nanotechnology* 3 (11): 654–659.
27. Avouris, P., Chen, Z., and Perebeinos, V. 2007. Carbon-based electronics. *Nature Nanotechnology* 2 (10): 605–615.
28. Zhong, Z., Gabor, N. M., Sharping, J. E., Gaeta, A. L., and McEuen, P. L. 2008. Terahertz time-domain measurement of ballistic electron resonance in a single-walled carbon nanotube. *Nature Nanotechnology* 3 (4): 201–205.

29. Lin, Y.-M., Jenkins, K. A., Valdes-Garcia, A., Small, J. P., Farmer, D. B., and Avouris, P. 2008. Operation of graphene transistors at gigahertz frequencies. *Nano Letters* 9 (1): 422–426.
30. Streetman, B. G., and Banerjee, S. K. 2006. *Solid state electronic devices*, 287. Upper Saddle River, NJ: Pearson Education International.
31. Kocabas, C., Dunham, S., Cao, Q., Cimino, K., Ho, X., Kim, H.-S., Dawson, D., Payne, J., Stuenkel, M., Zhang, H., Banks, T., Feng, M., Rotkin, S. V., and Rogers, J. A. 2009. High-frequency performance of submicrometer transistors that use aligned arrays of single-walled carbon nanotubes. *Nano Letters* 9 (5), 1937–1943.
32. Lin, Y.-M., Dimitrakopoulos, C., Jenkins, K. A., Farmer, D. B., Chiu, H.-Y., Grill, A., and Avouris, P. 2010. 100-GHz transistors from wafer-scale epitaxial graphene. *Science* 327 (5966): 662.
33. Geim, A. K. 2009. Graphene: Status and prospects. *Science* 324 (5934): 1530–1534.
34. Novoselov, K. S., Geim, A. K., Morozov, S. V., Jiang, D., Katsnelson, M. I., Grigorieva, I. V., Dubonos, S. V., and Firsov, A. A. 2005. Two-dimensional gas of massless Dirac fermions in graphene. *Nature* 438 (7065): 197–200.
35. Zhang, Y., Tan, Y.-W., Stormer, H. L., and Kim, P. 2005. Experimental observation of the quantum Hall effect and Berry's phase in graphene. *Nature* 438 (7065): 201–204.
36. Rogers, J. A. 2008. Electronic materials: Making graphene for macroelectronics. *Nature Nanotechnology* 3 (5): 254–255.
37. Lafkioti, M., Krauss, B., Lohmann, T., Zschieschang, U., Klauk, H., Klitzing, K. V., and Smet, J. H. 2010. Graphene on a hydrophobic substrate: Doping reduction and hysteresis suppression under ambient conditions. *Nano Letters* 10 (4): 1149–1153.
38. Venugopal, A., Colombo, L., and Vogel, E. M. 2010. Contact resistance in few and multilayer graphene devices. *Applied Physics Letters* 96 (1): 013512-3.
39. Zhu, W., Neumayer, D., Perebeinos, V., and Avouris, P. 2010. Silicon nitride gate dielectrics and band gap engineering in graphene layers. *Nano Letters* 10 (9): 3572–3576.
40. Kedzierski, J., Pei-Lan, H., Reina, A., Jing, K., Healey, P., Wyatt, P., and Keast, C. 2009. Graphene-on-insulator transistors made using C on Ni chemical-vapor deposition. *Electron Device Letters, IEEE* 30 (7): 745–747.
41. Chen, J. H., Ishigami, M., Jang, C., Hines, D. R., Fuhrer, M. S., and Williams, E. D. 2007. Printed graphene circuits. *Advanced Materials* 19 (21): 3623–3627.
42. Liao, L., Bai, J., Lin, Y.-C., Qu, Y., Huang, Y., and Duan, X. 2010. High-performance top-gated graphene-nanoribbon transistors using zirconium oxide nanowires as high-dielectric-constant gate dielectrics. *Advanced Materials* 22 (17): 1941–1945.
43. Cho, J. H., Lee, J., He, Y., Kim, B. S., Lodge, T. P., and Frisbie, C. D. 2008. High-capacitance ion gel gate dielectrics with faster polarization response times for organic thin film transistors. *Advanced Materials* 20 (4): 686–690.
44. Cho, J. H., Lee, J., Xia, Y., Kim, B., He, Y., Renn, M. J., Lodge, T. P., and Frisbie, C. D. 2008. Printable ion-gel gate dielectrics for low-voltage polymer thin-film transistors on plastic. *Nature Materials* 7 (11): 900–906.
45. Lee, J., Kaake, L. G., Cho, J. H., Zhu, X. Y., Lodge, T. P., and Frisbie, C. D. 2009. Ion gel-gated polymer thin-film transistors: Operating mechanism and characterization of gate dielectric capacitance, switching speed, and stability. *The Journal of Physical Chemistry C* 113 (20): 8972–8981.
46. Kim, B. J., Jang, H., Lee, S.-K., Hong, B. H., Ahn, J.-H., and Cho, J. H. 2010. High-performance flexible graphene field effect transistors with ion gel gate dielectrics. *Nano Letters* 10 (9): 3464–3466.
47. Sukjae, J., et al. 2010. Flexible, transparent single-walled carbon nanotube transistors with graphene electrodes. *Nanotechnology* 21 (42): 425201.
48. Andersson, A., Johansson, N., Bröms, P., Yu, N., Lupo, D., and Salaneck, W. R. 1998. Fluorine tin oxide as an alternative to indium tin oxide in polymer LEDs. *Advanced Materials* 10 (11): 859–863.

49. Boehme, M., and Charton, C. 2005. Properties of ITO on PET film in dependence on the coating conditions and thermal processing. *Surface and Coatings Technology* 200 (1–4): 932–935.
50. Wu, Z., Chen, Z., Du, X., Logan, J. M., Sippel, J., Nikolou, M., Kamaras, K., Reynolds, J. R., Tanner, D. B., Hebard, A. F., and Rinzler, A. G. 2004. Transparent, conductive carbon nanotube films. *Science* 305 (5688): 1273–1276.
51. Rowell, M. W., Topinka, M. A., McGehee, M. D., Prall, H.-J., Dennler, G., Sariciftci, N. S., Hu, L., and Gruner, G. 2006. Organic solar cells with carbon nanotube network electrodes. *Applied Physics Letters* 88 (23): 233506-3.
52. Lee, J.-Y., Connor, S. T., Cui, Y., and Peumans, P. 2008. Solution-processed metal nanowire mesh transparent electrodes. *Nano Letters* 8 (2): 689–692.
53. Nair, R. R., Blake, P., Grigorenko, A. N., Novoselov, K. S., Booth, T. J., Stauber, T., Peres, N. M. R., and Geim, A. K. 2008. Fine structure constant defines visual transparency of graphene. *Science* 320 (5881): 1308.
54. Geng, H.-Z., Kim, K. K., So, K. P., Lee, Y. S., Chang, Y., and Lee, Y. H. 2007. Effect of acid treatment on carbon nanotube-based flexible transparent conducting films. *Journal of the American Chemical Society* 129 (25): 7758–7759.
55. Kim, H., Horwitz, J. S., Kushto, G., Pique, A., Kafafi, Z. H., Gilmore, C. M., and Chrisey, D. B. 2000. Effect of film thickness on the properties of indium tin oxide thin films. *Journal of Applied Physics* 88 (10): 6021–6025.
56. Hecht, D. S., Thomas, D., Hu, L., Ladous, C., Lam, T., Park, Y., Irvin, G., and Drzaic, P. 2009. Carbon-nanotube film on plastic as transparent electrode for resistive touch screens. *Journal of the Society for Information Display* 17 (11): 941–946.
57. Cairns, D. R., Witte Ii, R. P., Sparacin, D. K., Sachsman, S. M., Paine, D. C., Crawford, G. P., and Newton, R. R. 2000. Strain-dependent electrical resistance of tin-doped indium oxide on polymer substrates. *Applied Physics Letters* 76 (11): 1425–1427.
58. Liu, J., Yin, Z., Cao, X., Zhao, F., Lin, A., Xie, L., Fan, Q., Boey, F., Zhang, H., and Huang, W. 2010. Bulk heterojunction polymer memory devices with reduced graphene oxide as electrodes. *ACS Nano* 4 (7): 3987–3992.
59. Liang, J., Chen, Y., Xu, Y., Liu, Z., Zhang, L., Zhao, X., Zhang, X., Tian, J., Huang, Y., Ma, Y., and Li, F. 2010. Toward all-carbon electronics: Fabrication of graphene-based flexible electronic circuits and memory cards using maskless laser direct writing. *ACS Applied Materials & Interfaces* 2 (11): 3310–3317.
60. Green, A. A., and Hersam, M. C. 2009. Solution phase production of graphene with controlled thickness via density differentiation. *Nano Letters* 9 (12): 4031–4036.
61. De, S., King, P. J., Lotya, M., O'Neill, A., Doherty, E. M., Hernandez, Y., Duesberg, G. S., and Coleman, J. N. 2010. Flexible, transparent, conducting films of randomly stacked graphene from surfactant-stabilized, oxide-free graphene dispersions. *Small* 6 (3): 458–464.
62. Park, S., and Ruoff, R. S. 2009. Chemical methods for the production of graphenes. *Nature Nanotechnology* 4 (4): 217–224.
63. Bourlinos, A. B., Georgakilas, V., Zboril, R., Steriotis, T. A., and Stubos, A. K. 2009. Liquid-phase exfoliation of graphite towards solubilized graphenes. *Small* 5 (16): 1841–1845.
64. Hernandez, Y., Lotya, M., Rickard, D., Bergin, S. D., and Coleman, J. N. 2009. Measurement of multicomponent solubility parameters for graphene facilitates solvent discovery. *Langmuir* 26 (5): 3208–3213.
65. Hernandez, Y., Nicolosi, V., Lotya, M., Blighe, F. M., Sun, Z., De, S., McGovern, I. T., Holland, B., Byrne, M., Gun'Ko, Y. K., Boland, J. J., Niraj, P., Duesberg, G., Krishnamurthy, S., Goodhue, R., Hutchison, J., Scardaci, V., Ferrari, A. C., and Coleman, J. N. 2008. High-yield production of graphene by liquid-phase exfoliation of graphite. *Nature Nanotechnology* 3 (9): 563–568.

66. Hamilton, C. E., Lomeda, J. R., Sun, Z., Tour, J. M., and Barron, A. R. 2009. High-yield organic dispersions of unfunctionalized graphene. *Nano Letters* 9 (10): 3460–3462.
67. Lotya, M., Hernandez, Y., King, P. J., Smith, R. J., Nicolosi, V., Karlsson, L. S., Blighe, F. M., De, S., Wang, Z., McGovern, I. T., Duesberg, G. S., and Coleman, J. N. 2009. Liquid phase production of graphene by exfoliation of graphite in surfactant/water solutions. *Journal of the American Chemical Society* 131 (10): 3611–3620.
68. Coraux, J., N'Diaye, A. T., Busse, C., and Michely, T. 2008. Structural coherency of graphene on Ir(111). *Nano Letters* 8 (2): 565–570.
69. De Arco, L. G., Yi, Z., Kumar, A., and Chongwu, Z. 2009. Synthesis, transfer, and devices of single- and few-layer graphene by chemical vapor deposition. *Nanotechnology, IEEE Transactions on Nanotechnology* 8 (2): 135–138.
70. Valentini, L., Cardinali, M., Bittolo Bon, S., Bagnis, D., Verdejo, R., Lopez-Manchado, M. A., and Kenny, J. M. 2010. Use of butylamine modified graphene sheets in polymer solar cells. *Journal of Materials Chemistry* 20 (5): 995–1000.
71. Kalita, G., Matsushima, M., Uchida, H., Wakita, K., and Umeno, M. 2010. Graphene constructed carbon thin films as transparent electrodes for solar cell applications. *Journal of Materials Chemistry* 20 (43): 9713–9717.
72. Tung, V. C., Chen, L.-M., Allen, M. J., Wassei, J. K., Nelson, K., Kaner, R. B., and Yang, Y. 2009. Low-temperature solution processing of graphene–carbon nanotube hybrid materials for high-performance transparent conductors. *Nano Letters* 9 (5): 1949–1955.
73. Hyesung, P. et al. 2010. Doped graphene electrodes for organic solar cells. *Nanotechnology* 21 (50): 505204.
74. Ki Kang, K., et al. 2010. Enhancing the conductivity of transparent graphene films via doping. *Nanotechnology* 21 (28), 285205.
75. Kasry, A., Kuroda, M. A., Martyna, G. J., Tulevski, G. S., and Bol, A. A. 2010. Chemical doping of large-area stacked graphene films for use as transparent, conducting electrodes. *ACS Nano* 4 (7): 3839–3844.
76. Wehling, T. O., Novoselov, K. S., Morozov, S. V., Vdovin, E. E., Katsnelson, M. I., Geim, A. K., and Lichtenstein, A. I. 2007. Molecular doping of graphene. *Nano Letters* 8 (1): 173–177.
77. Li, S.-S., Tu, K.-H., Lin, C.-C., Chen, C.-W., and Chhowalla, M. 2010. Solution-processable graphene oxide as an efficient hole transport layer in polymer solar cells. *ACS Nano* 4 (6): 3169–3174.
78. Mandlik, P., Gartside, J., Han, L., Cheng, I. C., Wagner, S., Silvernail, J. A., Ma, R.-Q., Hack, M., and Brown, J. J. 2008. A single-layer permeation barrier for organic light-emitting displays. *Applied Physics Letters* 92 (10): 103309-3.
79. Lewis, J. S., and Weaver, M. S. 2004. Thin-film permeation-barrier technology for flexible organic light-emitting devices. *IEEE Journal of Selected Topics in Quantum Electronics* 10 (1): 45–57.
80. Sprengard, R., Bonrad, K., Daeubler, T. K., Frank, T., Hagemann, V., Koehler, I., Pommerehne, J., Ottermann, C. R., Voges, F., and Vingerling, B. 2004. In *OLED devices for signage applications: A review of recent advances and remaining challenges*, ed. Kafafi, Z. H., and Lane, P. A., 173–183. Denver, CO: SPIE.
81. Bunch, J. S., Verbridge, S. S., Alden, J. S., van der Zande, A. M., Parpia, J. M., Craighead, H. G., and McEuen, P. L. 2008. Impermeable atomic membranes from graphene sheets. *Nano Letters* 8 (8): 2458–2462.
82. Poulston, S., Parlett, P. M., Stone, P., and Bowker, M. 1996. Surface oxidation and reduction of CuO and Cu_2O studied using XPS and XAES. *Surface and Interface Analysis* 24 (12): 811–820.
83. Lee, Y. B., and Ahn, J.-H. Graphene-based gas impermeable barrier film (in preparation).

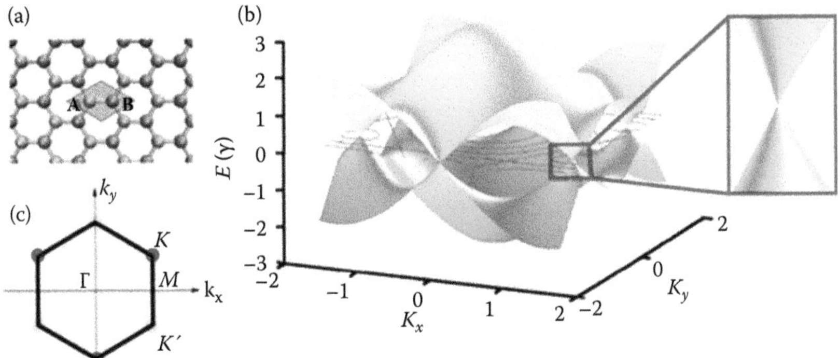

COLOR FIGURE 1.1 (a) Honeycomb lattice of graphene. The shadowed area delineates the unit cell of graphene with its two nonequivalent atoms labeled by A and B. (b) Band energy dispersion obtained via tight binding approximation. The inset highlights the conical-shape dispersion around the charge neutrality point. (c) First Brillouin zone.

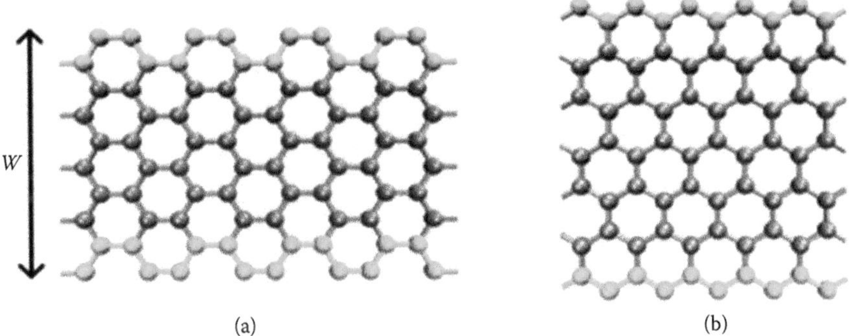

COLOR FIGURE 1.2 Atomic structure of an (a) armchair- and a (b) zigzag-edge graphene nanoribbon. Green color atoms delineate the respective edge-shape and W denotes the width of the ribbon.

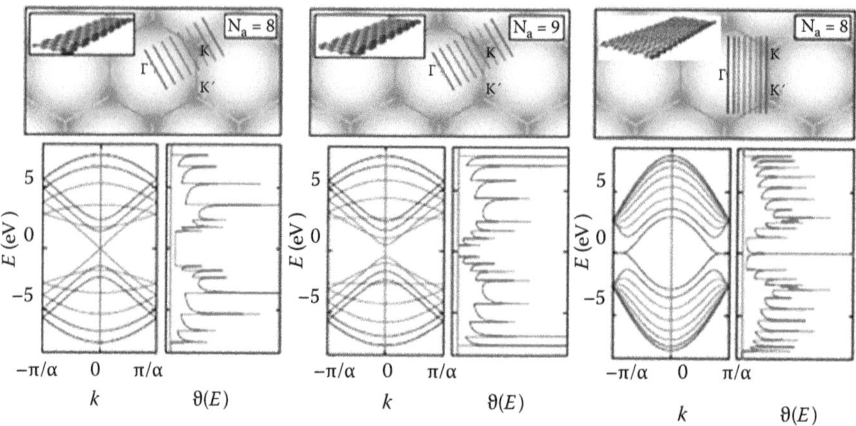

COLOR FIGURE 1.3 (Top panels) Zone-folding diagram for three different graphene nanoribbons: left, AGNR(8); middle, AGNR(9); and right, ZGNR(8). Their respective energy band structures and density of states curves are displayed on the lower panels. (Adapted from N. Nemec, Quantum transport in carbon-based nanostructures. Dr. rer. nat. (equiv. PhD) thesis, University of Regensburg, September 2007.)

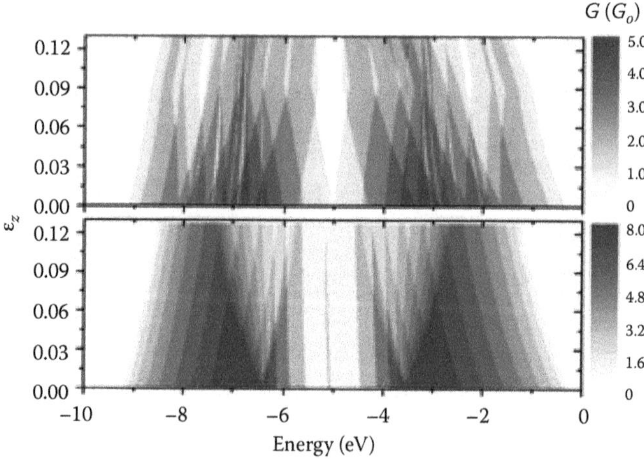

COLOR FIGURE 1.5 Contour plots of conductance as a function of Fermi energy and mechanical strain for (top panel) an AGNR(11) and (lower panel) a ZGNR(10). (Adapted from M. Poetschke, C.G. Rocha, L.E.F. Foa Torres, S. Roche, and G. Cuniberti. Modeling graphene-based nanoelectromechanical devices. *Physical Review B* 81 (2010): 193404-1-4.)

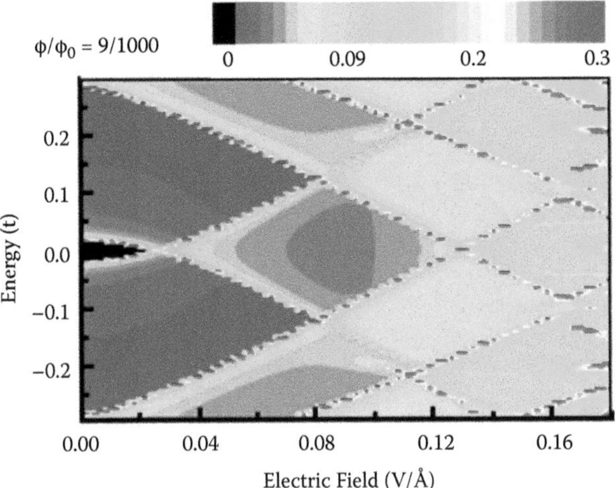

COLOR FIGURE 1.6 Local density of states contour plot for a 24-AGNR as a function of the electric field intensity for a fixed magnetic flux of $\phi/\phi_0 = 9/1000$. Black corresponds to null density of states while the highest LDOS value is highlighted by red color. (Adapted from C. Ritter, S.S. Makler, and A. Latgé. Energy-gap modulations of graphene nanoribbons under external fields: A theoretical study. *Physical Review B* 77: 195443-1-5; *Physical Review B* 82 (2008): 089903-1-2.)

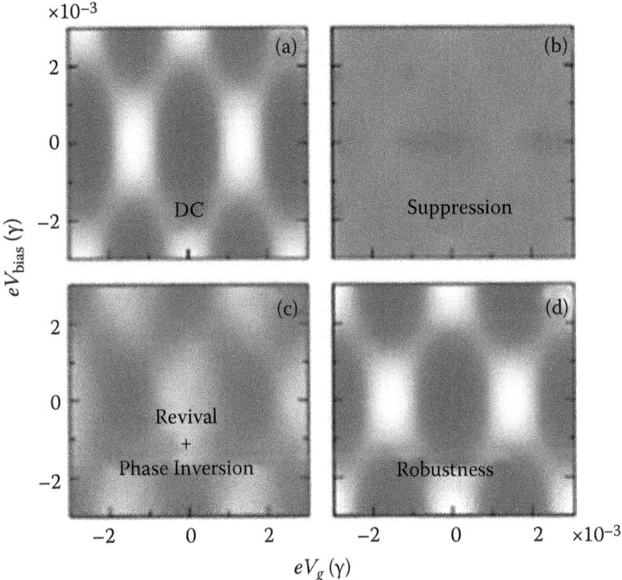

COLOR FIGURE 1.7 Fabry-Pérot conductance interference patterns for an AGNR as a function of bias and gate voltages calculated for different driving frequencies and amplitudes associated with a time-dependent gate potential that follows a harmonic time dependency. White and dark blue colors correspond to maximum and minimum conductance values, respectively. (Adapted from C.G. Rocha, L.E.F. Foa Torres, and G. Cuniberti. Ac transport in graphene-based Fabry-Perot devices. *Physical Review B* 81 (2010): 115435-1-8.)

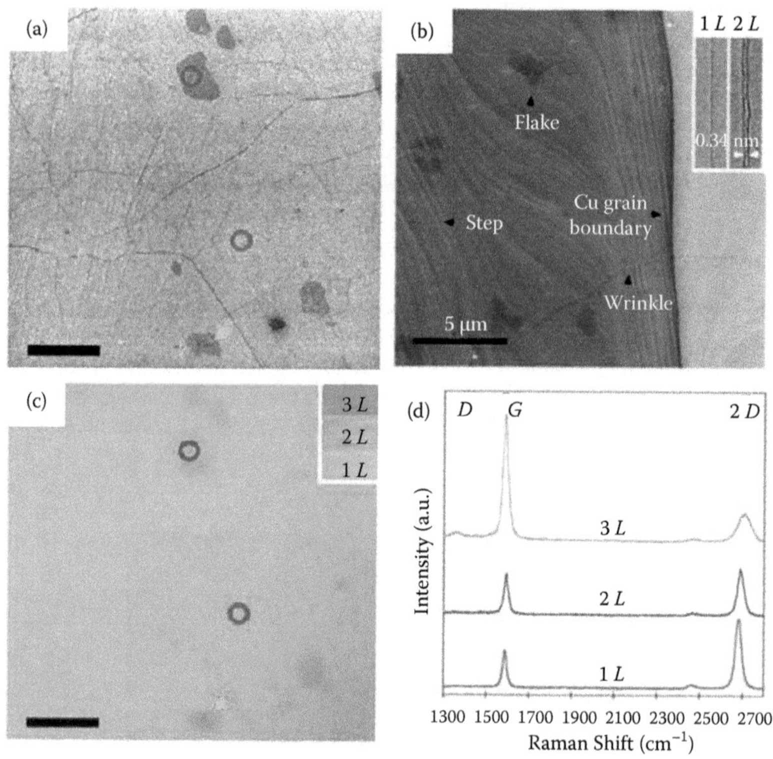

COLOR FIGURE 5.2 Graphene film synthesized on a copper catalyst layer using the CVD method. (a) SEM image of graphene film grown on copper foil. (b) Morphological image of graphene film flowing copper surface conditions (inset denotes the monolayer and bilayer graphene estimated by TEM). (c) Optical microscope image the graphene film transferred on SiO_2 layer showing different color indicating different number of layers. (d) Raman spectroscopy for each number of graphene layers on SiO_2 300-nm layer. (From Li, X., Cai, W., An, J., Kim, S., Nah, J., Yang, D., Piner, R., Velamakanni, A., Jung, I., Tutuc, E., Banerjee, S. K., Colombo, L., and Ruoff, R. S. 2009. Large-area synthesis of high-quality and uniform graphene films on copper foils. *Science* 324 (5932): 1312–1314. Copyright 2009 American Association for the Advancement of Science. With permission.)

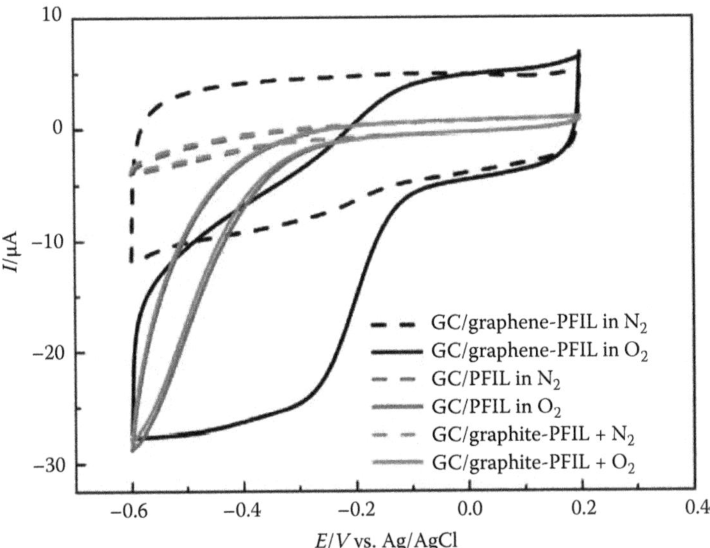

COLOR FIGURE 8.2 (a)Cyclic voltammograms at PFIL (blue), graphite-PFIL (magenta), and graphene-PFIL (black) modified electrodes in 0.05 M PBS solution saturated with O_2 (solid) and degassed with pure N_2 (dashed). (From C.S. Shan, H.F. Yang, J.F. Song, D.X. Han, A. Ivaska, and L. Niu. 2009. Direct electrochemistry of glucose oxidase and biosensing for glucose based on graphene. *Anal. Chem.* 81: 2378–2382.)

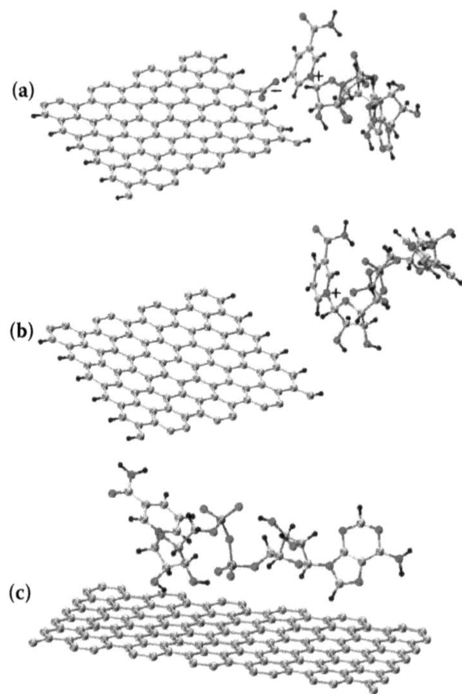

COLOR FIGURE 8.5 Model for the adsorption of NAD^+ on a (a) graphene basal plane, (b) graphene edge fully terminated by hydrogen atoms, and (c) graphene edge terminated by hydrogen atoms and containing one $-COO^-$ group *via* Car–Parrinello molecular dynamics. Gray, C; blue, N; red, O; yellow, P; black, H. (From M. Pumera, R. Scipioni, H. Iwai, T. Ohno, Y. Miyahara, and M. Boero. 2009. A mechanism of adsorption of β-Nicotinamide adenine dinucleotide on graphene sheets: Experiment and theory. *Chem. Eur. J.* 15, 10851.)

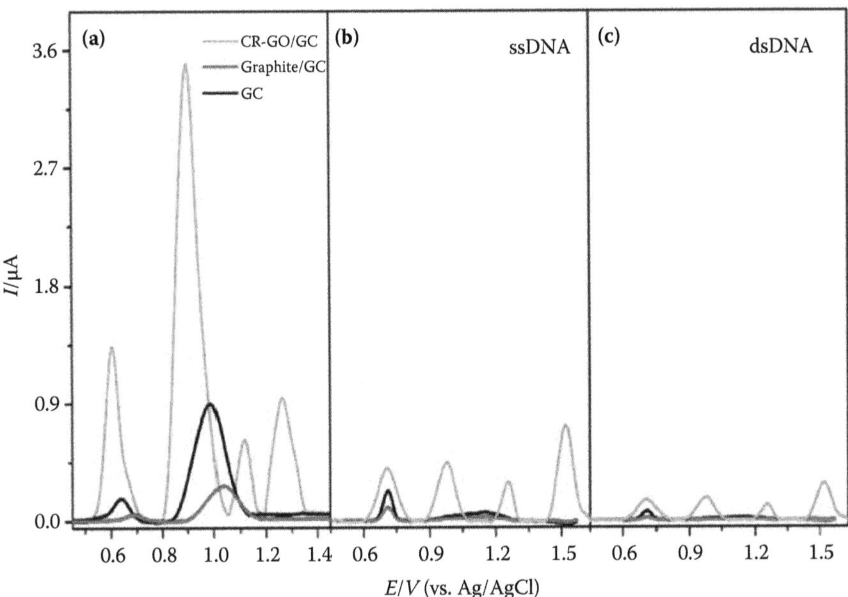

COLOR FIGURE 8.7 Differential pulse voltammograms (DPV) for (a) a mixture of DNA free base (G, A, T, and C), (b) ssDNA, and (c) dsDNA in 0.1 M pH 7.0 PBS at graphene/GC (green), graphite/GC (red), and bare GC electrodes (black). Concentrations G, A, T, C, ssDNA, or dsDNA: 10 mg mL–1. (From Y. Liu, D. Yu, C. Zeng, Z-C. Miao, and L. Dai. 2010. Biocompatible graphene oxide-based glucose biosensors. *Langmuir* 26, 6158–6160.)

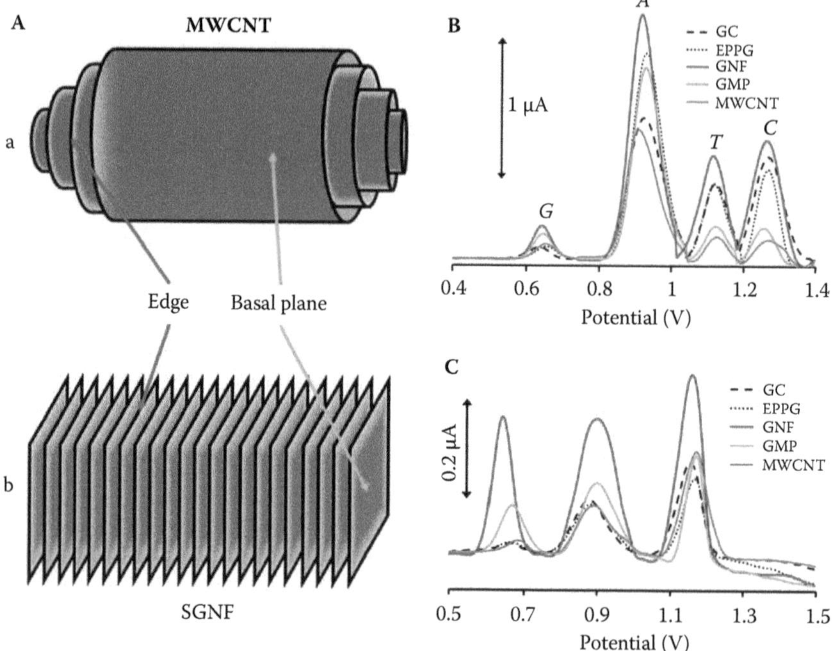

COLOR FIGURE 8.8 Stacked graphene nanofibers (SGNFs) outperform carbon nanotubes and graphite in the detection of free DNA bases and A (H1N1) DNA strand. (A) Graphene-sheet orientation in multiwalled carbon nanotubes (MWCNTs) (a) and SGNFs (b). The highly electroactive edge portion of the sheets is represented in yellow. (B) differential pulse voltammograms (DPVs) for a mixture of guanine, adenine, thymine, and cytosine at SGNF (red), graphene platelet (GNP) (green), and MWCNT (blue) electrodes. For comparison, glass-carbon electrode (GCE) (black dashed) and edge-plane pyrolytic graphite (EPPG) (black dotted) electrode signals are also shown. (C) DPVs for ssDNA A (H1N1) at SGNF (red), GNP (green), and MWCNT (blue) electrodes. For comparison, GCE (black dashed) and EPPG (black dotted) electrode signals are also shown. (From A. Ambrosi and M. Pumera. 2010. Stacked graphene nanofibers of electrochemical oxidation of DNA bases. *Phys. Chem. Chem. Phys.* 12, 8943–8947. With permission.)

6 Nanosized Graphene
Chemical Synthesis and Applications in Materials Science

Chongjun Jiao and Jishan Wu

CONTENTS

6.1 INTRODUCTION

Graphene can be regarded as a flat monolayer of carbon atoms tightly packed into a two-dimensional honeycomb lattice, and a basic building block for graphitic materials of all other dimensionalities. Despite its short history, graphene has attracted considerable attention and significant progress has been made in both academic research and application studies, because of its exceptional electronic and mechanical properties. Among numerous widely investigated graphene-based materials, nanosized graphenes represent an intriguing class of compounds that are somewhat different from the infinite graphene sheet. Generally speaking, *nanosized graphene* is the name given to a family of polycyclic aromatic hydrocarbons (PAHs) with average diameters between 1 and 10 nm, which can be regarded as two-dimensional graphene segments consisting of all-sp^2 carbons. Although certain types of nanosized graphenes can be found in the residue of domestic and natural combustion of coal, wood, and other organic materials, development of new methods toward functionalized nanosized graphenes, especially for aromatics with high molecular weights, is crucial. The history of synthesis and characterization of nanosized graphenes can be dated back to the first half of the twentieth century, when Scholl et al. (1910) and Clar (1964) made pioneering contributions to this area. Harsh conditions (such as high temperature and strong oxidants and bases) were inevitably used at that time, and some earlier conclusions have gradually proven to be incorrect, which is a result of the limited characterization method and technique.

Currently, nanosized graphene is one of the most widely studied families of organic compounds. A variety of synthetic strategies have been applied to prepare compounds of this family. Many inspiring molecules of this type have been discovered and more are waiting to be explored. Compared with other graphene-based materials (e.g., infinite graphene sheet and graphene nanoribbon), these types of molecules possess advantages such as physical processability, structural perfection, and interesting electronic and self-assembling properties, all of which make them promising candidates for practical applications including organic electronic devices, organic dyes, bioimaging, and so on. Therefore, a comprehensive review on the synthesis and application of nanosized graphene is necessary.

The basic building block for all kinds of nanosized graphenes is the benzene ring containing six delocalized π-electrons (also called the *aromatic sextet benzenoid ring*). As shown in Figure 6.1, fusion of benzene rings in different modes leads to a variety of nanosized graphenes with different shapes and properties. For example,

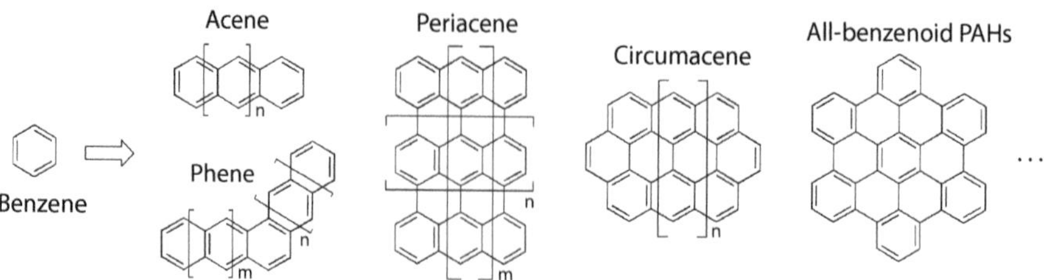

FIGURE 6.1 Nanosized graphenes by fusion of benzene rings in different modes.

linear fusion of benzene rings results in a series of molecules called *acene*, while angular annellation gives a type of molecule called *phene*. Further fusion of benzene rings into two-dimensional (2D) structures affords large PAHs with different edge structures and physical properties, such as the *peri*-fused acenes (*periacenes*), circularly fused acenes (*circumacenes*), the all-benzenoid PAHs such as hexa-*peri*-hexabenzocoronenes, and other higher-order nanosized graphene molecules. This review first summarizes the modern synthesis of these nanosized graphenes with different sizes, shapes, and edge structures and their basic physical properties. It is important to understand the fundamental structure–property relationship of the nanosized graphenes, which has been discussed and followed by an introduction to their applications in materials science.

6.2 CHEMICAL SYNTHESIS OF NANOSIZED GRAPHENES

6.2.1 Linear Acene

Acenes are a class of PAHs with linearly fused benzene rings. According to Clar's aromatic sextet rule, only one aromatic sextet benzenoid ring (labeled as a six-membered ring with a circle inside) can be drawn for this type of molecule (Figure 6.2). As a result, the energy gaps of acenes as well as their stabilities successively decease upon increasing the number of fused benzene rings. Acene has gained recognition as one of the most outstanding classes of semiconductors for electronic devices such as organic field-effect transistors (OFETs) and organic light-emitting diodes (OLEDs). Benzene and the first two members of the acene family (i.e., naphthalene, anthracene) can be obtained from coal-tar and petroleum distillates while the higher homologues can only be prepared by multistep synthesis. Naphthacene (also known as tetracene) can be readily prepared by reduction from the corresponding naphthacenequinone (Fieser 1931). Herein, we will mainly discuss the challenging synthetic chemistry of pentacene and other higher-order acenes, which were successfully achieved in recent years.

6.2.1.1 Pentacene and Its Derivatives

Pentacene (**1**, Scheme 6.1), the largest sufficiently stable acene for device studies, stands out as a benchmark among organic semiconductors, exhibiting charge carrier mobility larger than 3 $cm^2\ V^{-1}\ s^{-1}$. The general method to obtain pentacene is base-mediated fourfold Aldol condensations of *o*-phthalaldehyde **2** with cyclohexane-1,4-dione **3**, followed by reduction of the as-formed 6,13-pentacenedione **4** with Meerwein-Pondorff reagent (Scheme 6.1) (Bailey et al. 1953). The as-formed **4** also can undergo nucleophilic addition reaction with aryl- or alkynyl lithium regents or Grignard reagents, depending on the nature of the substrate, followed by reductive

FIGURE 6.2 Structure of acene.

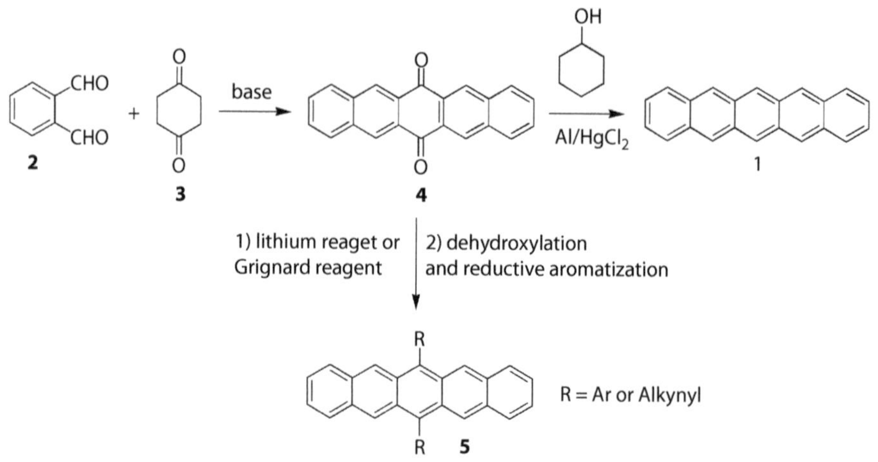

SCHEME 6.1 General synthesis of pentacene and its derivatives.

aromatization to afford 6,13-substituted pentacenes **5** (Miao et al. 2006). This *meso*-functionalization approach using bulky groups significantly improves the solubility and stability of pentacene derivatives, which has been envisioned as a plausible strategy to achieve stable and soluble higher acenes with tunable electronic properties (Bendikov et al. 2004; Anthony 2006 and 2008).

6.2.1.2 Hexacene, Heptacene, and Their Derivatives

Acenes larger than pentacene can easily degrade under ambient conditions. Therefore, their resistance to photo-oxidation should be improved by chemical functionalization. The synthesis of hexacene, the next higher homologue of pentacene, was firstly reported in 1939. Clar (1939) and Marschalk (1939) independently found that treatment of the same precursor dihydrohexacene **7** with copper or Pd/C via dehydrogenation formed hexacene **6** (Scheme 6.2), which is extremely unstable, however. Since then, the synthesis of stable hexacene turned out to be a bottleneck until a recent report from Anthony's group (Payne et al. 2005). A two-step reaction sequence

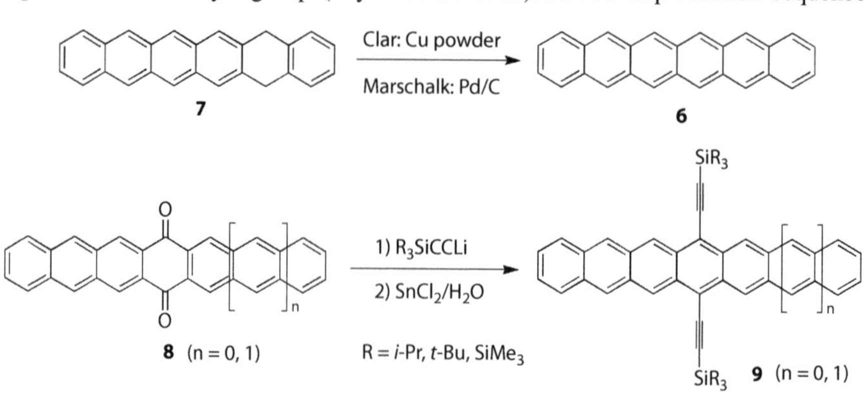

SCHEME 6.2 Synthesis of hexacene, heptacene, and their derivatives.

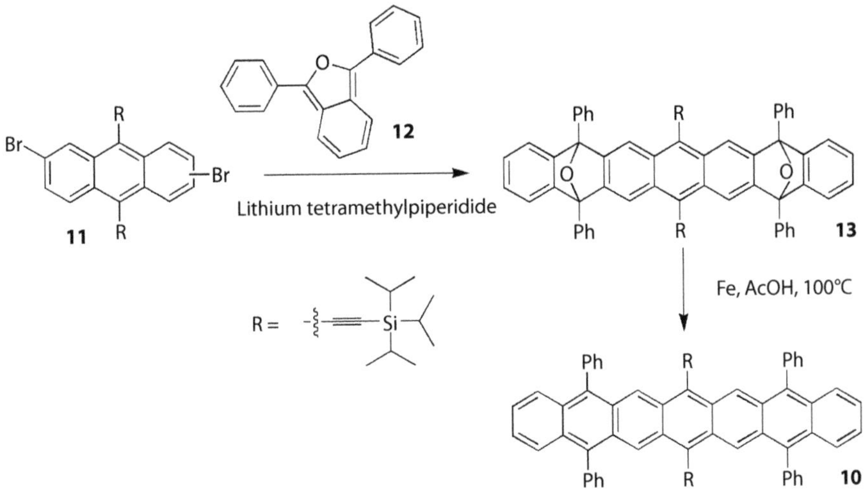

SCHEME 6.3 Synthesis of phenyl- and triisopropylsilylethynyl-substituted heptacene derivative reported by Wudl (Chan et al. 2008).

containing silylethynylation and subsequent reductive aromatization was used to prepare hexacene derivatives **9** ($n = 0$) from the corresponding hexacenequinone **8** ($n = 0$, Scheme 6.2). The introduction of bulky acetylene is required to stabilize the hexacene and improve its solubility. In particular, this synthetic methodology was further extended to prepare heptacene derivatives (**9**, $n = 1$) with reasonable stability.

Wudl and coworkers applied a different synthetic method to synthesize a silylethynyl and phenyl-substituted heptacene derivative **10** through twofold Diels-Alder cycloaddition between *bisanthracyne* and diphenylisobenzofuran **12**, followed by reduction of the cycloaddition product **13** with iron powder in acetic acid (Scheme 6.3) (Chun et al. 2008). The bisanthracyne intermediate was in situ generated by treatment of **11** with lithium tetramethylpiperidide. Due to the introduction of four phenyl groups and two bulky triisopropylethynyl groups, the obtained heptacene **10** is more stable than **9** ($n = 1$) and its presence was still detectable by UV/Vis/NIR spectroscopy in degassed toluene after 41 hours of exposure to air.

Additional progress was achieved by Miller's group and an aryl and arylthio-substituted heptacene **14** was prepared from the corresponding tetraone **15** by a similar synthetic method (Scheme 6.4) (Kaur et al. 2009). This design is based on the fact that arylthio groups are proved to be good substituents to enhance the photo-oxidative resistance of pentacene (Kaur et al. 2008). The combination of *p*-(*tert*-butyl)phenylthio-substituents at positions 7 and 16 (i.e., arylthio substituents attached to the most reactive ring) and *o*-dimethylphenyl substituents at positions 5, 9, 14, and 18 (i.e., steric resistance on neighboring rings) make heptacene derivative **14** an especially persistent species that endures for weeks as a solid, 1–2 days in solution if shielded from light, and several hours in solution when directly exposed to both light and air.

Very recently, Chi's group successfully synthesized a heptacene derivative **18** substituted with four electron-deficient trifluoromethylphenyl and two triisopropylsilylethynyl (TIPSE) groups, which was described to be the most stable heptacene

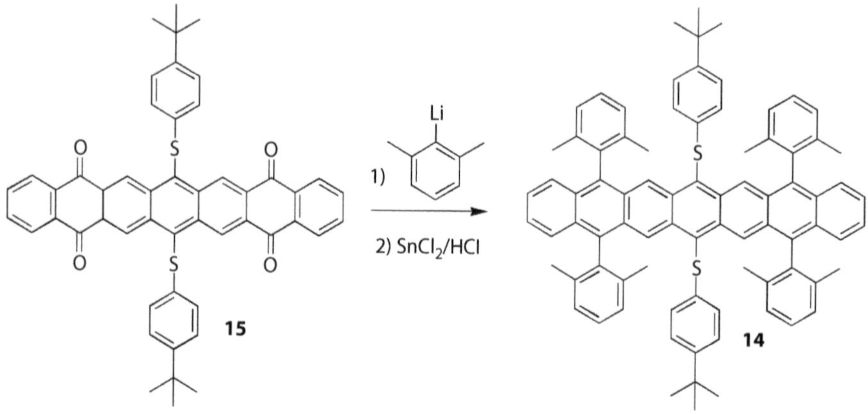

SCHEME 6.4 Synthesis of aryl- and arylthio-substituted heptacene reported by Kaur et al. (2009).

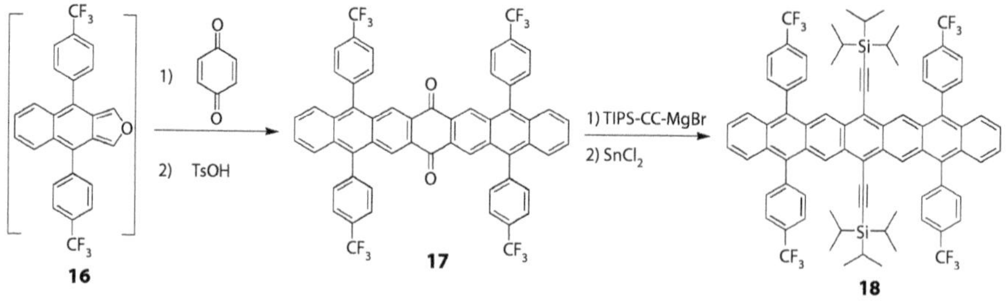

SCHEME 6.5 Synthesis of trifluoromethylphenyl- and triisopropylsilylethynyl-substituted heptacene reported by Qu et al.

to date (Qu et al. 2010). Synthesis of trifluoromethylphenyl- and triisopropylsilylethynyl-substituted heptacene is mainly based on twofold Diels-Alder cycloaddition between the in situ-generated isonaphthofuran **16** and cyclohexa-2,5-diene-1,4-dione, followed by acid treatment with *p*-toluenesulfonic acid (TsOH) to give the heptacenequinone **17** (Scheme 6.5). Nucleophilic addition of **17** with triisopropylsilylethynyl magnesium bromide and subsequent reduction/aromatization afforded the desired heptacene derivative **18**.

6.2.1.3 Octacene and Nonacene

The preparation of unsubstituted octacene and nonacene was reported by Bettinger's group by using a cryogenic matrix-isolation technique (Tönshoff and Bettinger 2010). The approach relies upon a protecting-group strategy based on the photochemically induced bisdecarbonylation of bridged diketones (Scheme 6.6). The removal of diketone bridges in tetraketone precursors **19** by UV irradiation under matrix-isolation conditions can afford UV/Vis/NIR-detectable unsubstituted octacene and nonacene **20**.

A fully characterized nonacene derivative was recently reported by Miller's group (Kaur et al. 2010). As shown in Scheme 6.7, the key step involved the Diels-Alder reaction of arylthio-substituted 1,4-anthracene quinone **21** with bis-*o*-quinodimethane

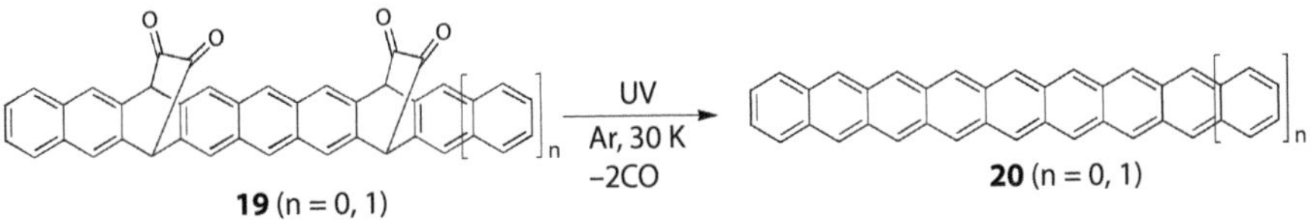

19 (n = 0, 1) UV Ar, 30 K −2CO 20 (n = 0, 1)

SCHEME 6.6 Synthesis of parent octacene and nonacene by a cryogenic matrix-isolation technique.

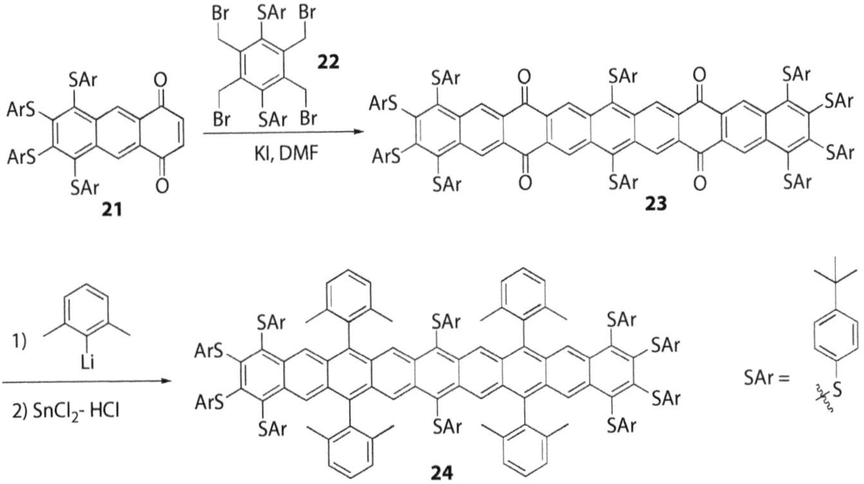

21 22 KI, DMF 23 1) Li 2) $SnCl_2$- HCl 24 SAr =

SCHEME 6.7 Synthesis of soluble and stable arylthio- and aryl-substituted nonacene derivatives reported by Kaur et al.

precursor **22** to produce a nonacene skeleton in the form of diquinone (**23**). Nucleophilic addition of **23** with aryl lithium reagent followed by reduction/aromatization gave the substituted nonacene **24** in good yield. Despite the narrow optical energy gap (1.12 eV), which is the smallest experimentally measured highest occupied molecular orbital–lowest unoccupied molecular orbital (HOMO–LUMO) gap for any acene, the nonacene derivative **24** is stable as a solid in the dark for at least 6 weeks because of the closed-shell electronic configuration resulting from arylthio-substituent effects and it was fully characterized by a suite of solution-phase techniques, including ^{1}H NMR, ^{13}C NMR, UV-Vis-NIR, and fluorescence spectroscopy.

6.2.2 Phene and Starphene

Phene refers to a type of polycyclic aromatic hydrocarbons consisting of benzenoid rings fused in an angular arrangement, which can also be regarded as linearly annelated benzologues of phenanthrene (Figure 6.3). Two aromatic sextet benzenoid rings can be drawn for each phene molecule, and as a result, the phene shows higher stability in comparison to the corresponding acene, according to Clar's aromatic sextet rule. The absorption spectra of phene and acene are also very different from each other due to their different annellation modes. The acenes give spectra in which

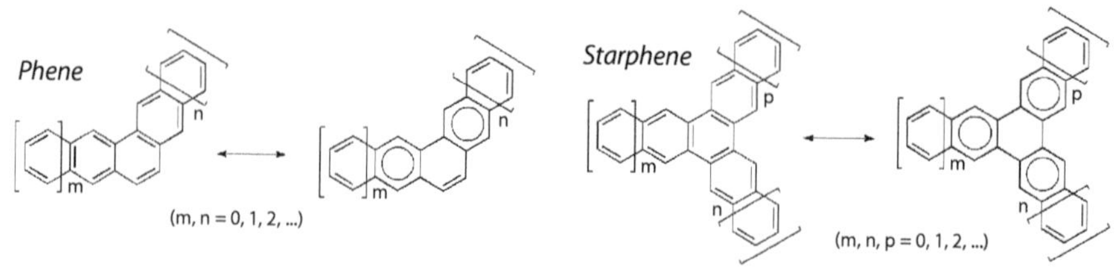

FIGURE 6.3 Structures of phene and starphene.

all three groups of bands, α-, β- and *p*-bands, show regular shifts toward the red with the fusion of each successive ring. In the spectra of the phenes, only the α- and β-bands shift in the same way toward the red, while the *p*-bands show a small shift toward shorter wavelengths (Clar 1964). If three branches are annellated to a central ring, radiating linearly from it, benzologues of triphenylene are formed, which are called *starphenes* (starlike phenes). In this case, only two branches are in aromatic conjugation during the time of light absorption and the two longest branches determine the long wave part of the spectrum (Clar and Mullen 1968). The synthesis of phenes (up to nonaphene) and starphenes (up to decastarphene) have been well developed by Clar and co-workers (1964); however, the synthesis and material applications of their functional derivatives have rarely been studied.

Very recently, Wu's group reported a series of electron-deficient triphenylene carboximides **26** and trinaphthylene carboximides **28** by using a different approach (Scheme 6.8). The synthesis relies on Diels-Alder cycloaddition reaction of the in

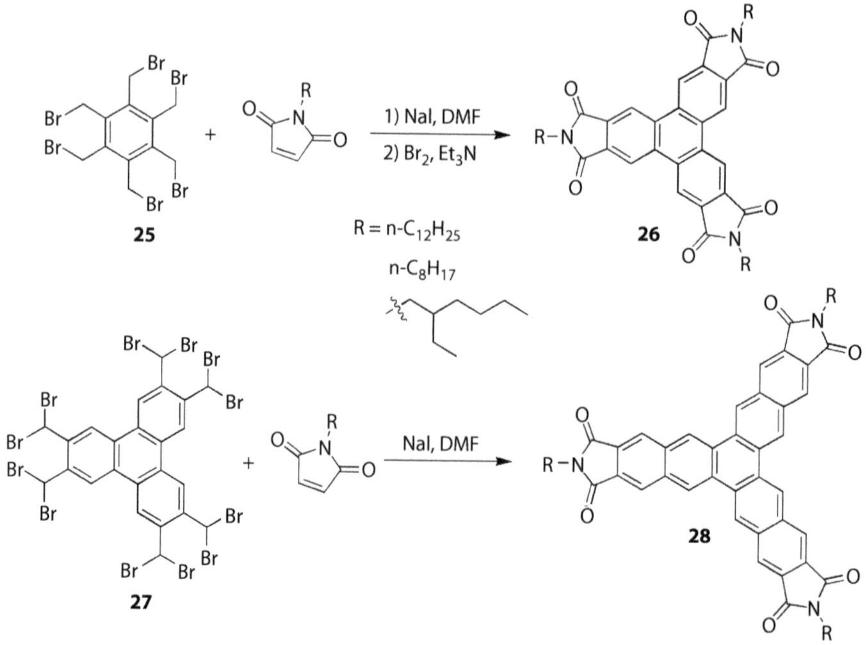

SCHEME 6.8 Synthesis of electron-deficient triphenylene and trinaphthylene carboximides.

situ-generated reactive radialenes (from **25** and **27**) with maleimides followed by an aromatization process (Yin et al. 2009). These imide-substituted triphenylenes **26** and trinaphthylenes **28** display high electron affinity and long-range ordered columnar stacking, making them promising *n*-type semiconductors in electronic devices such as field-effect transistors.

6.2.3 Peri-condensed Nanosized Graphenes with Zigzag Edges

Two main types of edges exist in nanographene: armchair and zigzag. The band gap of a nanosized graphene molecule is dependent on not only the molecular size, but also the edge structure. The zigzag-edged nanosized graphenes with a smaller amount of aromatic sextet benzenoid component would show a low band gap. Recent theoretical studies have proven that diradical structures do exist in some nanosized graphenes and the unpaired electrons are predominantly located at the zigzag edges (Jiang 2007a, Jiang et al. 2008). In this section, the focus is on the recent synthesis and properties of *peri*-condensed nanosized graphenes with zigzag edges, including rylene, bisanthene, teranthene, bistetracene, bispentacene, and circumacenes.

6.2.3.1 Rylene and Its Derivatives

In pursuit of stable dyes with high extinction coefficients and long-wavelength absorption/emission, rylenes have received a great deal of attention. Rylenes are large polycyclic aromatic hydrocarbons in which two or more naphthalene units are *peri*-fused together by single bonds. Only one aromatic sextet benzenoid ring can be drawn for each naphthalene unit and two zigzag edges exist at the terminal naphthalene units (Figure 6.4). Perylene **29a**, the first member of the perylene series ($n = 0$), has been intensively studied due to its outstanding chemical, thermal, and photochemical inertness, its nontoxicity, and its low cost. Extension of the conjugation length along the long molecular axis of perylene by incorporation of additional naphthalene units to form its higher homologue has proven to be an effective method to obtain long-wavelength absorbing rylene dyes. For example, terrylene **29b** ($n = 1$) (Avlasevich et al. 2006) and quaterrylene **29c** (Bohnen et al. 1990) absorb in the visible region with absorption maxima at 560 nm and 662 nm, respectively. In search of mild conditions to synthesize higher rylenes, a variety of intramolecular cyclodehydrogenation methods have been developed. These include oxidative cyclodehydrogenation using $FeCl_3$, or a combination of $CuCl_2$-$AlCl_3$ as Lewis acid as well as oxidant, and reductive cyclodehydrogenation via anion radical mechanism promoted by base. A typical example involving both processes is the synthesis of a quaterrylene derivative

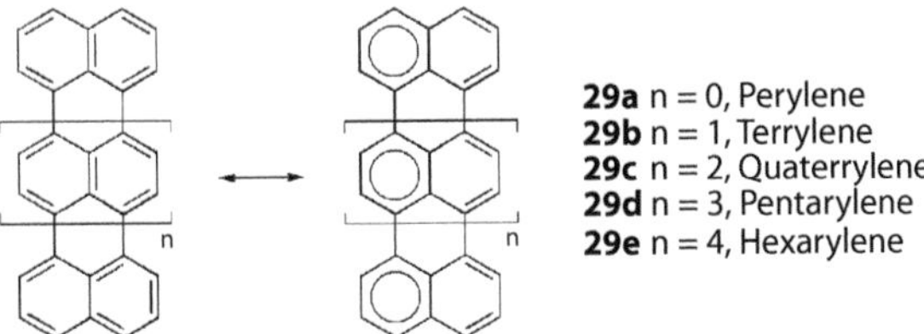

FIGURE 6.4 A general structure of rylene molecules.

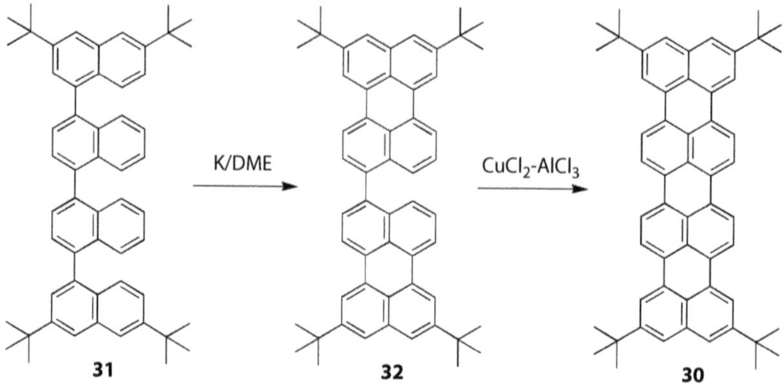

SCHEME 6.9 Synthesis of quaterrylene derivative **30**.

30. Singly linked precursor **31** successively underwent reductive cyclodehydrogenation to generate partially cyclized intermediate **32** and subsequent oxidative cyclodehydrogenation to give final quaterrylene **30** (Koch and Müllen 1991). Four *tert*-butyl groups were designed to resolve the solubility problem resulting from the strong tendency of this ladder-type molecule to form aggregates (Scheme 6.9).

Since higher-order rylenes are electron-rich π-systems and are relatively unstable upon exposure to air, electron-withdrawing groups such as dicarboxylic imide are introduced to the reactive *peri*-positions to improve their chemical and photostability. The rylene bis(dicarboximide)s **33a-f** (Figure 6.5) were found to be more stable than the unmodified rylenes and additional bathochromic shift in absorption was observed due to intramolecular donor–acceptor interaction. The presence of imide moiety also provided opportunities to introduce bulky groups for the purpose of increasing solubility by suppressing aggregation in solution. Furthermore, a more

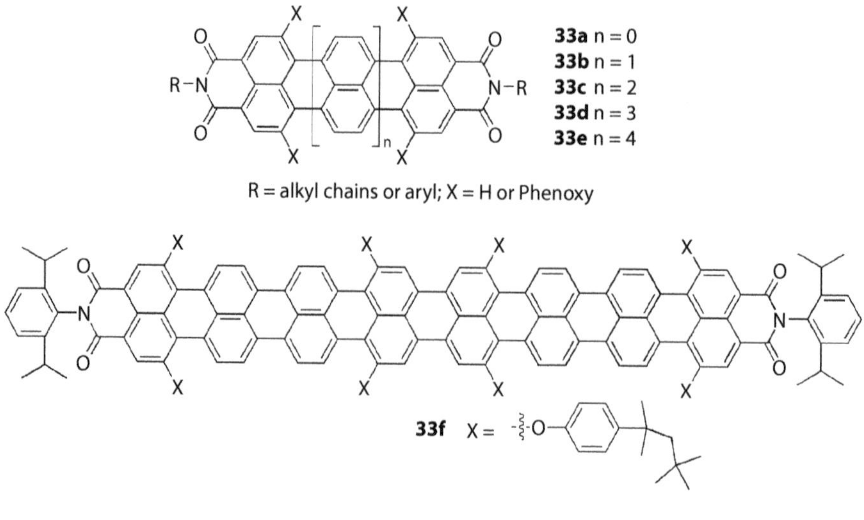

FIGURE 6.5 Structure of rylene bis(dicarboximide)s.

increased solubility can be achieved in the case of *bay*-brominated rylene imide by means of nucleophilic substitution with bulky phenoxy groups. The increase in solubility makes the preparation of higher rylene bisimides, namely pentarylene bisimide, hexarylene bisimide, heptarylene bisimide, and octarylene bisimide, practically possible (Pschirer et al. 2006). In common with rylenes, extension of the conjugation length along the long molecular axis of rylene imide to form higher homologues not only promotes their absorption into longer spectral regions but also significantly enhances their extinction coefficients. The major challenge for the synthesis of rylene imide is the intramolecular ring cyclization of the singly linked precursor, which usually can be prepared by transition metal-catalyzed intermolecular coupling reactions (e.g., Suzuki, Yamamoto couplings) between appropriate building blocks. Different cyclodehydrogenation methods are then required to be used to promote cyclization depending on the electronic properties of respective units in the singly linked precursor. Substitution by bulky phenoxy groups at the bay positions effectively improves solubility and processability of the higher rylene dyes and also results in a moderate bathochromic shift of their absorption spectra. The highest rylene compound reported to date is the octarylene bisimide **33f** (Qu et al. 2008).

A different strategy recently developed to construct rylene backbone is based on fused *N*-annulated perylene analogues (Jiang 2008). Poly(*peri-N*-annulated perylene) **34** (Li and Wang 2009, Y. Li et al. 2010) and its carboximide derivative **35** (Jiao et al. 2009) can be regarded as perfect nanosized graphene ribbons containing nitrogen atoms annulated onto the armchair edge (Figure 6.6). Due to the different electronic properties of building blocks in **34** and **35**, the DDQ/Sc(OTf)$_3$ system has been used in the former to promote the last-step cyclization considering the electron-rich property of the *N*-annulated perylene core. Conversely, a reductive cyclodehydrogenation strategy by mild base was applied in the latter owing to the introduction of electron-withdrawing carboximide groups in **35**, which lowers the HOMO energy level (n = 0) and makes **35** very stable upon exposure to the light as well as oxygen

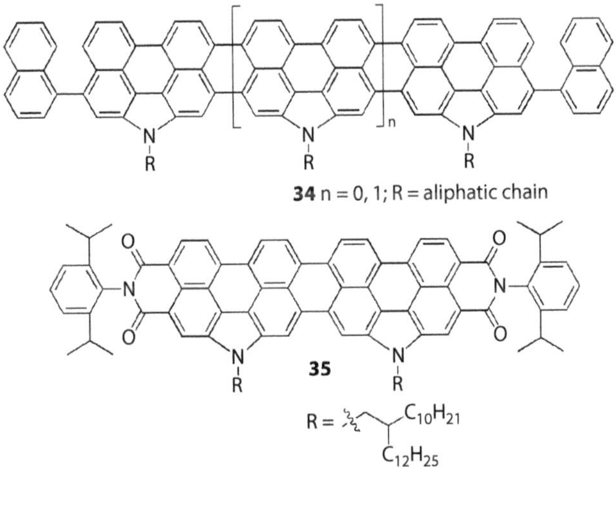

FIGURE 6.6 *N*-annulated rylene and rylene bis(dicarboximide)s.

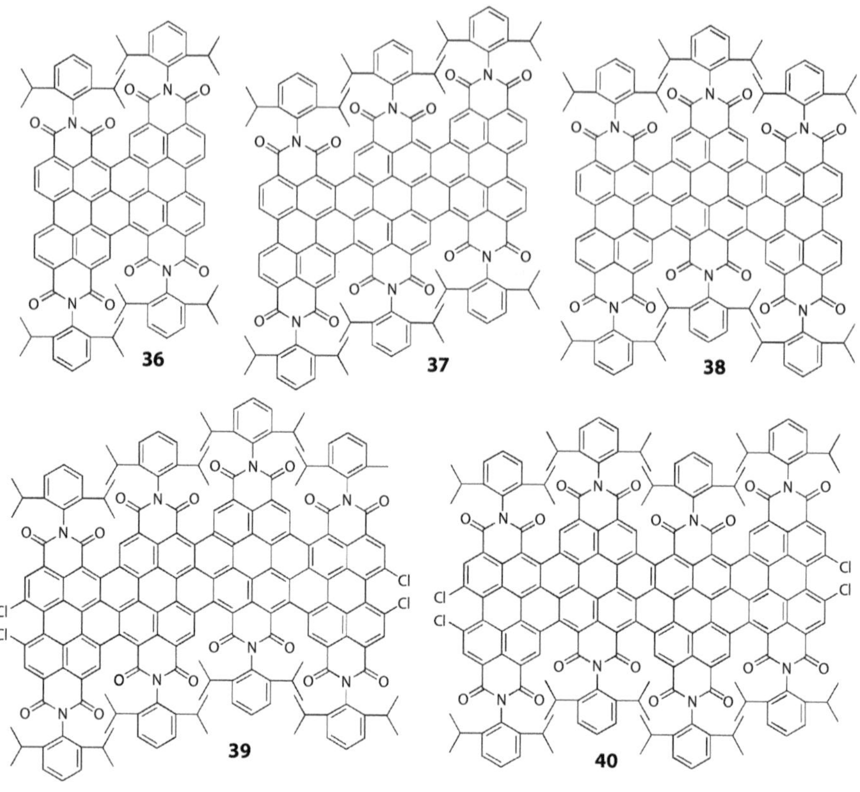

FIGURE 6.7 Triply linked perylene bisimides.

(Jiao et al. 2009). Moreover, compound **35** emits strong fluorescence with quantum yield up to 55% in dichloromethane. Such a high quantum yield for near-infrared (NIR) dyes are remarkable given that many NIR-absorbing dyes usually exhibit low fluorescence quantum yields.

Another family of triply linked oligo-perylene bisimides (Figure 6.7), reported by Wang's group, also can be regarded as expanded rylene derivatives. Related nanosized graphenes **36–40** (Qian et al. 2005, Qian et al. 2007, Zhen 2010) can be obtained by copper-involved condensation of tetrabromo (chloro)-perylene bisimides along the bay region under different conditions. Due to the two possible coupling positions, there are structural isomers for higher analogues. These fully conjugated graphene-type compounds display broad and red-shifted absorption and strong electron-accepting ability.

Higher acenes (e.g., tetracenes) have been incorporated into the rylene skeleton (Figure 6.8). Synthesis of the dibenzopentarylene bis(dicarboximide) **41** also applied the sequential Suzuki cross-coupling reaction and two-step oxidative and reductive cyclodehydrogenation (Avlasevich et al. 2006). It should be noted that the combination of $Pd_2(dba)_3$ with DPEPhos (bis(2-(diphenylphosphino) phenyl)ether) as a ligand was required during the process of Suzuki cross-coupling for the steric reason. Also

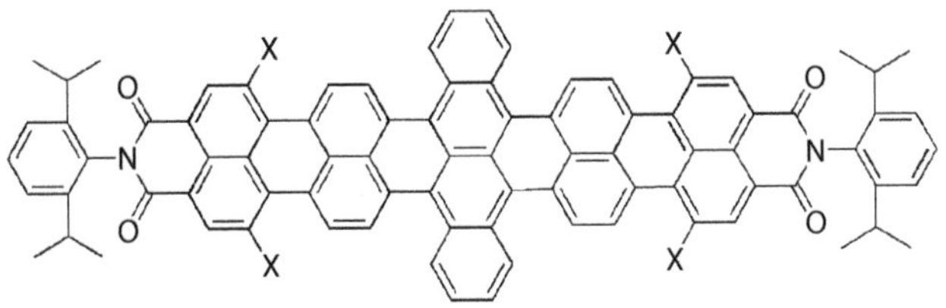

41 X = 4-(*t*-octyl) phenoxy

FIGURE 6.8 Dibenzopentarylene bis(dicarboximide).

noteworthy is that the introduced tetracene subunit on one hand helps to result in remarkably red-shifted absorption bands with absorption maximum at 1037 nm. On the other hand, it has a negative effect on the stability of the obtained dye **41** owing to the twisted structures resulting from steric repulsion between the perylene and tetracene units.

6.2.3.2 Zethrene and Its Derivatives

Zethrenes, which are not as well investigated as the other aromatic compounds such as acenes and rylenes, can be regarded as an interesting class of nanosized graphenes in which the two naphthalene units are fused with one or more hexagonal rings into a Z-shape (Figure 6.9). This type of hydrocarbon has fixed double bonds (highlighted in bold form) between the naphthalene units, and depending on the number of fixed double bonds, the molecules are accordingly called zethrene, heptazethrene, and octazethrene. Theoretical calculations predicted that zethrenes also have diradical character at ground state (Figure 6.9) and they will show interesting nonlinear optical properties and near-infrared absorption (Nakano et al. 2007, Désilets et al. 1995). However, successful syntheses of zethrene and its derivatives were seldom reported due to their low accessibility and high sensitivity in the presence of oxygen and light, especially in a dilute solution.

The first synthesis of zethrene was reported by Clar (Clar et al. 1955), and a more convenient access to zethrene was found accidently by Staab (Staab et al. 1968) and Sondheimer (Mitchell and Sondheimer 1970) during their independent attempts to synthesize tetradehydrodinaphtho[10]annulene, which was highly unstable and could be automatically transformed into zethrene via transannular cyclization. It was not until 2009 that pure tetradehydrodinaphtho[10]annulene (**44**), the

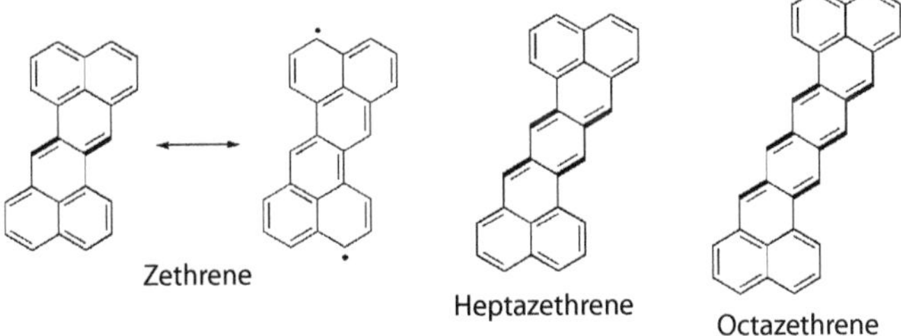

FIGURE 6.9 Structure of zethrene, heptazethrene, and octazethrene.

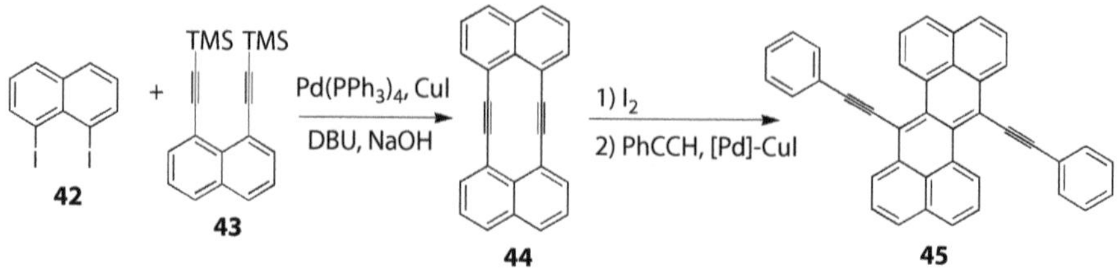

SCHEME 6.10 Synthesis of 7,14-diphenyl zethrene reported by Tobe's group (Umeda et al. 2009).

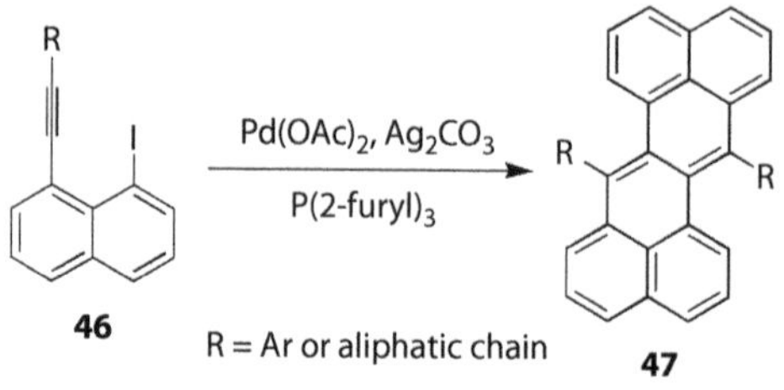

SCHEME 6.11 Synthesis of 7,14-substituted zethrenes by Pd-catalyzed annulation reaction reported by Y.-T. Wu et al. (2010).

precursor to zethrene, was first isolated by Tobe's group (Umeda et al. 2009); they also managed to synthesize stable 7,14-disubstituted zethrene derivatives such as **45** (Scheme 6.10). The synthetic route began with in situ desilylation and Sonogashira-Hagihara coupling reaction between **42** and desilylated **43** to afford intermediate **44** in 22% yield, which further underwent transannular cyclization with iodine followed by Sonogashira-Hagihara coupling reaction with phenylacetylene to give the stable zethrene derivative **45** (Scheme 6.10).

New and exciting progress in the synthesis of zethrene was made in 2010 by Y.-T. Wu's group (Y.-T. Wu et al. 2010). They explored a metal-catalyzed annulation reaction of haloarenes **46** to prepare zethrenes **47** with varied substitutions blocked at 7,14-positions in up to 73% yields (Scheme 6.11).

In parallel to this work, J.-S. Wu's group independently reported dicarboxylic imide groups substituted zethrene derivative **48** (Sun et al. 2010) based on one pot reaction of Stille cross-coupling between **49** and bis(tri-n-butylstannyl)acetylene **50**, followed by simultaneous transannular cyclization of the dehydro[10]annulene intermediate compound (Scheme 6.12). The attachment of electron-withdrawing groups is supposed to improve the chemical and photostability of reactive zethrene species by lowering the HOMO energy level and also red shifting the absorption spectra to the far-red and NIR region due to the acceptor-donor-acceptor structure. As expected, the obtained zethrene bisimide **48** displays advantages over unmodified zethrene, including good chemical stability and photostability with half-life times determined as 4320 minutes under irradiation of ambient light, red-shifted absorption, and enhanced fluorescence quantum yield. Attempted bromination at the 7,14-positions by N-bromosuccinimide (NBS) in dimethylformamide (DMF) gave the oxidized product zethrenequinone **51** due to the butadiene character of the central

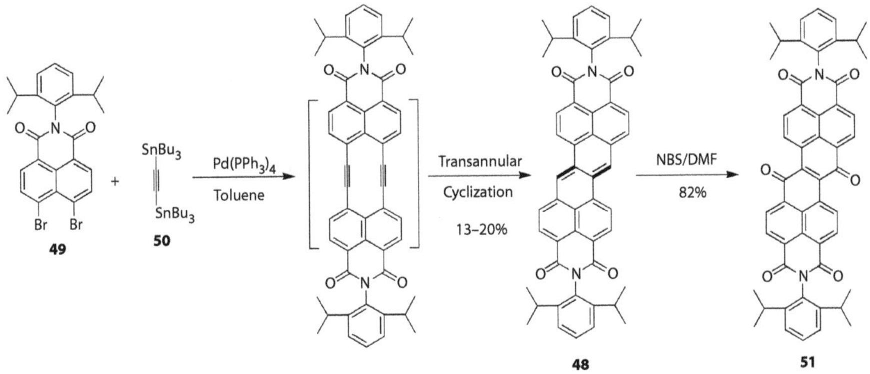

SCHEME 6.12 Synthesis of zethrene bis(dicarboximide) and its quinone reported by Sun et al. (2010).

fixed double bonds. The synthesis of heptazethrene (Clar and Macpherson 1962) and octazethrene (Erünlü 1969) was also attempted. However, they both showed very high reactivity and analytically pure compounds were not obtained and identified.

6.2.3.3 Bisanthene and Teranthene

Extension of conjugation along the long molecular axis of perylene to its higher homologue is well known and such an extension yields the higher-order rylene molecules with longer absorption in wavelength. Theoretical calculations indicate that laterally extended perylene derivatives (also called as *periacene*), namely bisanthene, peritetracene, and peripentacene (Scheme 6.13), would also lead to good candidates for NIR absorbing dyes (Désilets et al. 1995, Zhao et al. 2008). Only two aromatic sextet benzenoid rings can be drawn for periacenes, thus the higher-order periacenes are expected to have a high chemical reactivity.

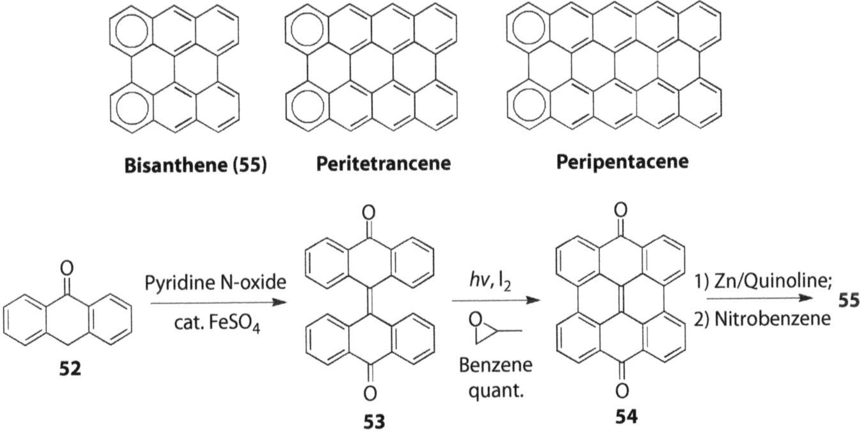

SCHEME 6.13 Structure of bisanthene, peritetracene and peripentacene and the improved synthesis of bisanthene.

The synthesis of bisanthene was first reported by Clar and co-workers (Clar 1948a) and the improved synthesis was recently developed by Bock's and Wu's groups (Saïdi-Besbes et al. 2006, Zhang et al. 2009, J. Li et al. 2010). The synthetic route began with the homocoupling of anthracen-10(9H)-one **52** in the presence of pyridine *N*-oxide and a catalytic amount of $FeSO_4$ to yield bisanthracenequinone **53**, which was subsequently photo-cyclized by UV light irradiation to afford the bisanthenequinone **54** (Scheme 6.13). Treatment of **54** with Zn dust in quinoline followed by oxidative dehydrogenation with nitrobenzene provided the target bisanthene **55**. The parent bisanthene was a blue colored compound with absorption maximum at 662 nm in benzene. However, it showed very poor stability due to its high-lying HOMO energy level, allowing the addition reaction with singlet oxygen at the active *meso*-positions to occur (Arabei and Pavich 2000).

The necessary research to obtain stable bisanthene derivatives was systematically conducted by Wu's group applying different strategies (Yao et al. 2009, Zhang et al. 2009, J. Li et al. 2010). The first method is the attachment of electron-withdrawing dicarboxylic imide groups onto the zigzag edges (Scheme 6.14), which are capable of lowering the relatively high-lying HOMO level in bisanthene (Yao et al. 2009). The key intermediate compound **56** was first prepared from anthracene by a stepwise Friedel-Crafts reaction, oxidation, bromination and imidization reactions. Compound **56** then underwent a $Ni(COD)_2$-mediated Yamamoto homocoupling reaction to give anthracene dicarboxylic imide dimer **57**. After base-promoted cyclization by *t*-BuOK and 1,5-diazabicyclo(4.3.0)non-5-ene (DBN), the fully fused bisanthene bis(dicarboximide) **58** was successfully synthesized in moderate yield (Scheme 6.14). Compared with the parent bisanthene **55**, compound **58** displayed good photostability and no significant change could be observed as a solution of **58** was exposed to air. In addition, a significant red shift of 170 nm at absorption maximum was observed for **58** compared with **55**.

The second method is to introduce aryl or alkyne substituents onto the most reactive *meso*-positions of the bisanthene (Figure 6.10) so as to not only stabilize the

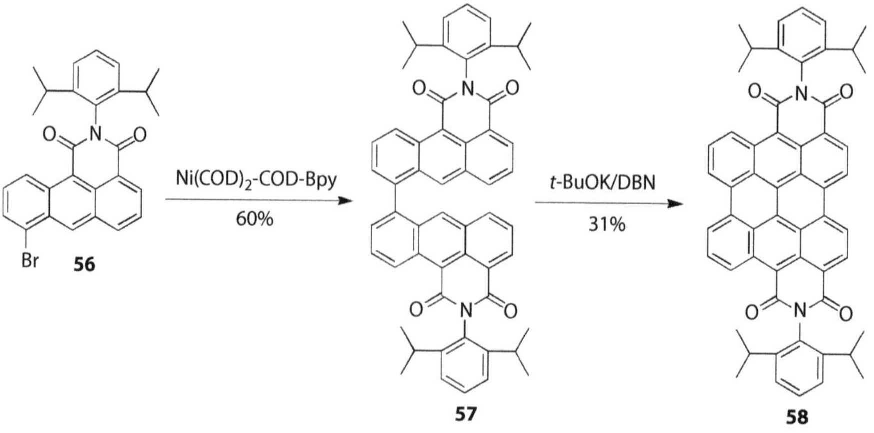

SCHEME 6.14 Synthesis of bisanthene bis(dicarboximide) reported by Wu (Yao et al. 2009)

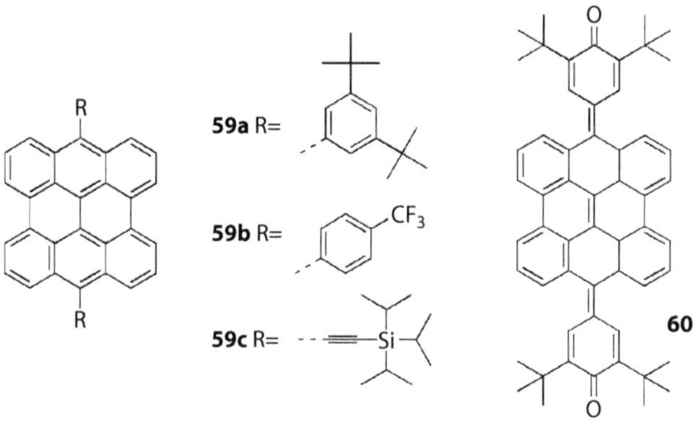

FIGURE 6.10 *Meso*-substituted bisanthene and extended bisanthenequinone.

active bisanthene core but also cause an additional bathochromic shift to the NIR spectral region (J. Li et al. 2010). Based on these considerations, Wu's group synthesized three *meso*-substituted bisanthenes **59a–c** using nucleophilic addition of aryl or alkyne Grignard reagent to the bisanthenequinone **54** followed by reduction/aromatization of the as-formed diol. This synthetic strategy was also used to prepare quinoidal bisanthene **60** (Figure 6.10), which can be regarded as a rare case of soluble and stable nanosized graphene with a quinoidal character (Zhang et al. 2009). Owing to the extended π-conjugation of the bisanthene core through the aryl and triisopropylsilylethynyl (TIPSE) moieties in **59** and the quinoidal character in **60**, the absorption maxima of **59** and **60** more or less shift to longer wavelengths. Furthermore, solutions of **59** and **60** are stable for weeks under ambient conditions, showing higher stability than their parent bisanthene **55**.

When three anthracene units are fused together via three single bonds between neighboring anthryls, the obtained structure is called as teranthene, which is expected to be even less stable than the reactive bisanthene. Very recently, Kubo's group reported the first successful synthesis of a teranthene derivative **66** (Konishi et al. 2010). As shown in Scheme 6.15, nucleophilic addition of lithium reagent **61** to 1,5-dichloroanthraquinone **62** followed by reductive aromatization with NaI/NaH_2PO_2 gave the teranthryl derivative **63**. Cyclization combined with demethylation of **63** was carried out with KOH/quinoline to give the partially ring-closed quinone **64**, which was treated with mesitylmagnesium bromide in the presence of $CeCl_3$ and further underwent reductive aromatization to generate partially cyclized compound **65**. The last cyclization was promoted by DDQ/$Sc(OTf)_3$, followed by quenching the reaction with hydrazine to give **66** as a dark green solid. It is noteworthy that this molecule is considered to possess a prominent singlet biradical character in the ground state due to the stabilization effect of the six aromatic sextet rings (Scheme 6.15).

6.2.3.4 Peritetracene and Peripentacene

Theoretical calculations by Jiang et al. (2007) point out that size is critical, which is correlated to the small HOMO–LUMO gap for the closed-shell state, for periacene

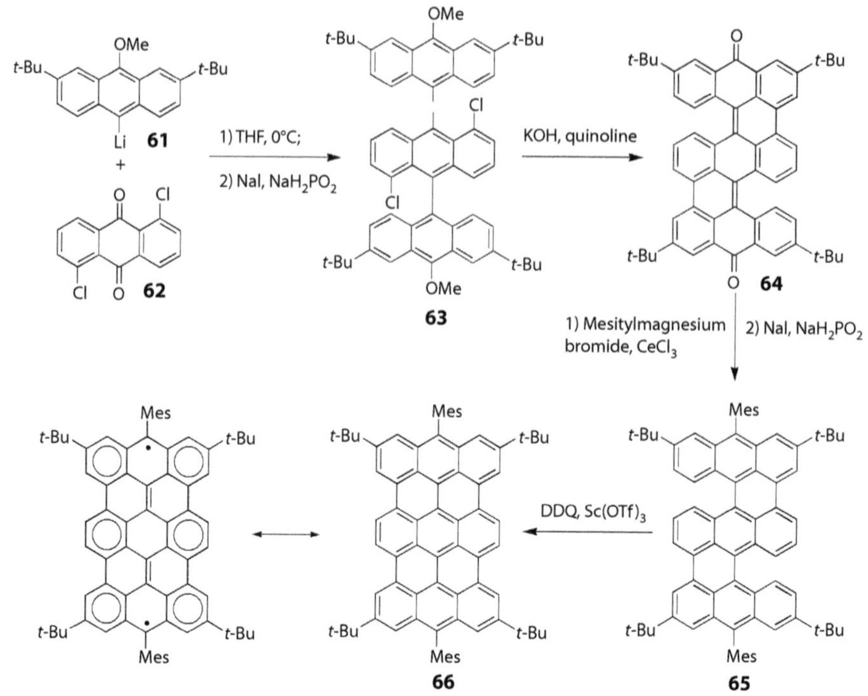

SCHEME 6.15 Synthesis of teranthene derivative.

and peritetracene. Below the critical size, periacenes have a closed-shell nonmagnetic ground state, while beyond the critical size, periacenes have an antiferromagnetic ground state and resemble infinite zigzag-edged graphene nanoribbons. HOMO–LUMO gap changes for the periacenes drop quickly with increases in molecular sizes from 1.87 eV for perylene to 0.08 eV for peripentacene. As a consequence, synthesis of higher periacenes is extremely challenging and little chemistry has been developed for this purpose.

The preliminary studies toward peripentacene were carried out in Wu's group (Zhang et al. 2010a, 2010b). As shown in Scheme 6.16, the bispentacenequinone **68**, obtained from dimerization of pentacenyl monoketone **67**, was subjected to nucleophilic addition of triisopropylsilylethynyl lithium reagent followed by reductive aromatization with NaI/NaH_2PO_2 to give the cruciform 6,6′-dipentacenyl **69** (Zhang et al. 2010a). Compound **69** exhibits two face-to-face π-stacking axes in a single crystal and this allows two-directional isotropic charge transport. Field-effect transistor (FET) mobilities up to 0.11 $cm^2\ V^{-1}\ s^{-1}$ were obtained based on vapor-deposited thin films. In addition, fused bispentacenequinone **70** (Zhang et al. 2010b), which can be regarded as a precursor for synthesis of peripentacene derivatives, was prepared with formation of two new C–C bonds via oxidative photocyclization of **68**. However, the subsequent nucleophilic reaction of compound **70** with excess Grignard reagent of 1-bromo-3,5-di-*tert*-butylbenzene in anhydrous tetrahydrofuran (THF) followed by acidification in air did not generate the desired 1, 2-addition adduct.

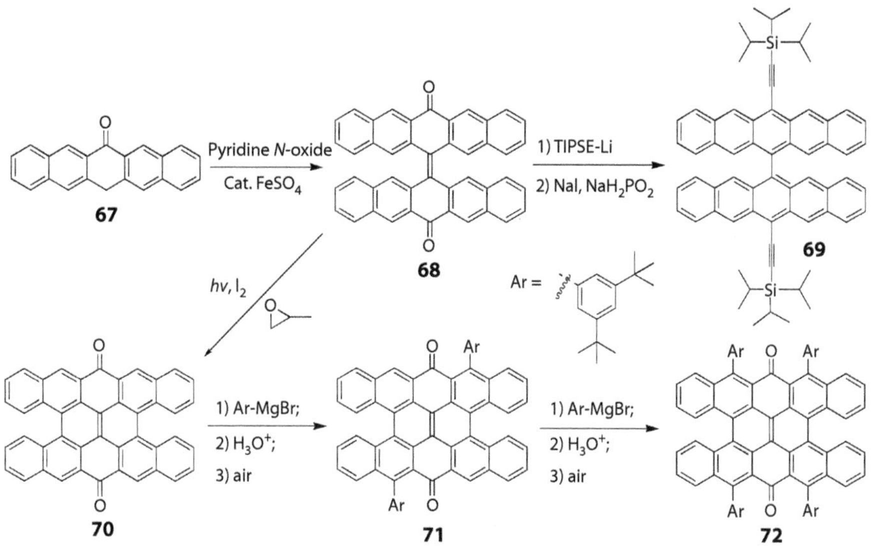

SCHEME 6.16 Synthesis of cruciform 6,6′-dipentacenyl and fused bispentacenequinone.

Alternatively, an unexpected Michael 1,4- addition product **71** was obtained and confirmed by single-crystal analysis. Further treatment of **71** with excess Grignard reagent followed by acidification in air gave the tetraaryl-substituted fused bispentacenequinone **72**. Single-crystal analysis reveals that there are α, β-unsaturated ketone structures in the fused bispentacenequinones **70** and **71**, which may explain the unusual Michael additions. So far, synthesis of peritetracene derivatives has not been reported and only one potential precursor, i.e., monobromo-tetracene dicarboximide, was reported by Wu's group (Yin et al. 2010).

6.2.3.5 Circumacenes

The annellation of two extra benzenoid rings to the bay sides of periacenes leads to another interesting type of nanosized graphene called circumacenes (Figure 6.11). The name comes from the feature that the central acene unit (including the benzene ring) is circularly annellated with benzene rings and accordingly, the molecules are called circumbenzene (i.e., coronene), circumnaphthalene (i.e., ovalene), circumanthracene, and so on. Three aromatic sextet rings can be drawn for coronene and

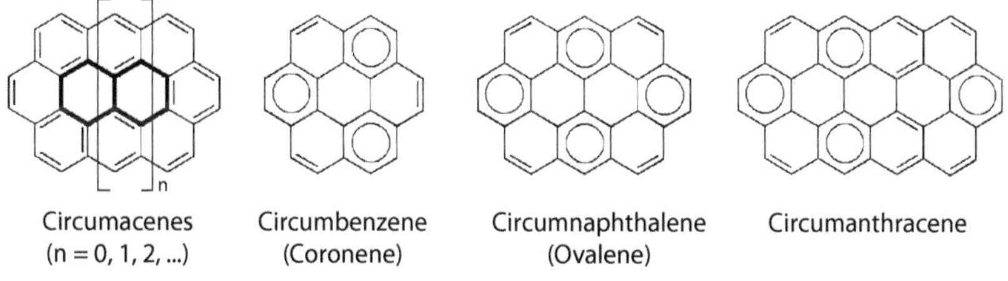

FIGURE 6.11 Structure of circumacenes.

four aromatic sextet rings can be drawn for higher-order circumacenes (Figure 6.11). In contrast, only two aromatic sextet benzenoid rings can be drawn for periacenes (Scheme 6.13). As a result, circumacene usually has a larger HOMO–LUMO energy gap and a higher stability than the respective periacene, which can be explained by Clar's aromatic sextet rule. Some widely investigated members in this family include coronene, ovalene, and circumanthracene.

Coronene is the smallest homologue of benzene with sixfold symmetry, which has a unique electronic structure due to the perfect delocalization of aromaticity between the six outer rings. In the crystalline state, coronene tends to form a bidimensional array of parallel columns. The chemical modifications of coronene core made by attaching different substituents are able to tune its electronic properties and self-assembly. The interest of coronene as a particularly good candidate for obtaining columnar liquid-crystalline self-assembly over wide temperature ranges is, therefore, not surprising. The typical method toward unmodified coronene **75** was developed by Clar (Clar and Zander 1957) as shown in Scheme 6.17. Commercialized perylene **29a** was treated with maleic anhydride under oxidative conditions to obtain the corresponding carboxylic anhydrides **73** followed by a soda-lime ($Ca(OH)_2$–$NaOH$–KOH–H_2O) triggered decarboxylation to yield **74**, which underwent another oxidative Diels-Alder reaction and decarboxylation to form coronene **75**.

Since perylene has a diene character at the *bay*-positions, the oxidative Diels-Alder reaction taking place along the *bay*-positions of perylene by using different electron-withdrawing dienophiles therefore allows easy access to numerous coronene derivatives **76** (Rao and George 2010), **77**, and **78** (Alibert-Fouet et al. 2007) with varied electronic and self-assembling properties (Figure 6.12).

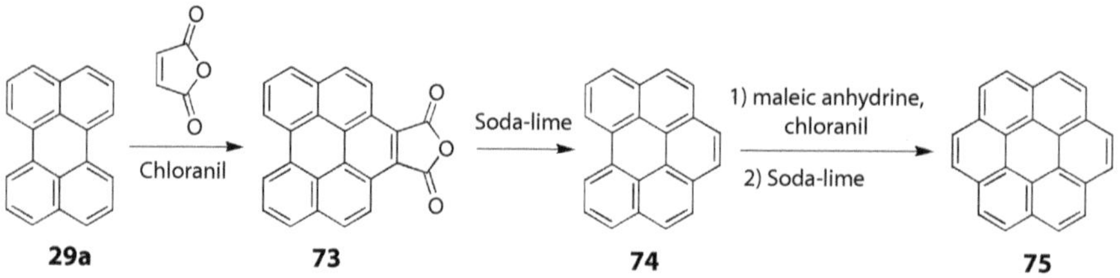

SCHEME 6.17 Synthesis of coronene from perylene.

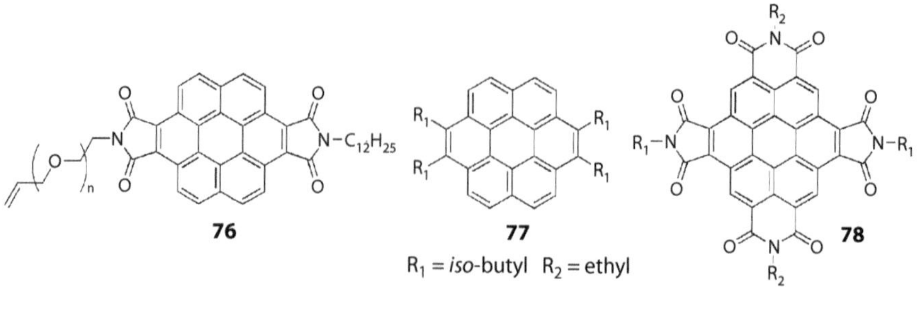

FIGURE 6.12 Coronene derivatives prepared from perylene.

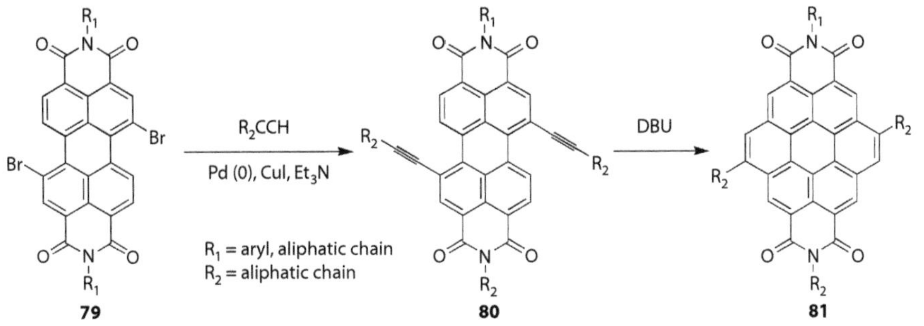

SCHEME 6.18 Synthesis of coronene bisimides by base-mediated cyclization.

An alternative strategy applied to synthesize coronene compounds is based on the Sonogashira-Hagihara coupling reaction of dibrominated perylene bisimide **79** followed by cyclization reaction in the presence of a strong yet nonnucleophilic base such as 1,8-diazabicyclo[5.4.0]undec-7-ene (DBU) (Scheme 6.18) (Rohr et al. 1998). Using this synthetic method, varied substitutions are able to be directly attached to the coronene core (Rohr et al. 2001).

Ovalene, which contains a fully circularly benzannellated naphthalene subunit, is the second member of the circumacene family (Figure 6.13). The first synthesis of ovalene was reported by Clar (1948b) following a strategy similar to that for synthesizing coronene. Twofold Diels-Alder cycloaddition of bisanthene with maleic anhydride followed by soda-lime-triggered decarboxylation gave ovalene **82** in good yield. A series of ovalene derivatives **83** (Saïdi-Besbes 2006) and **84** (Fort et al. 2009) were obtained by twofold oxidative Diels-Alder reactions with different electron-withdrawing dienophiles.

Although in 1956 Clar reported the synthesis of circumanthracene via a "controlled graphitization" process, his group published corrections 25 years later and claimed that the compound prepared earlier was not the desired circumanthracene (Clar et al. 1981). The first successful preparation of circumanthracene was described by Diederich et al. (Broene and Diederich 1991). According to their report, the fourfold photocyclization of **86** to **87** produced an excellent yield under UV light

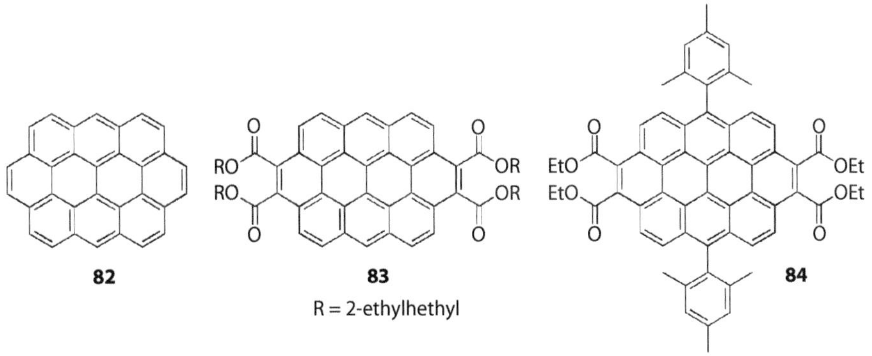

FIGURE 6.13 Ovalene and its derivatives.

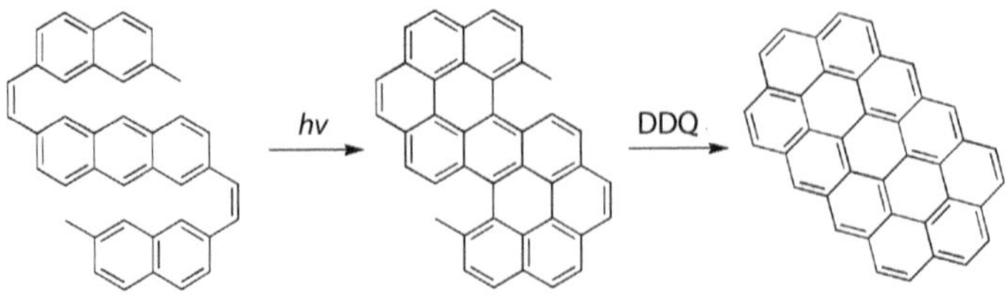

SCHEME 6.19 Synthesis of circumanthracene.

irradiation (Scheme 6.19). Further cyclization triggered by 2,3-dichloro-5,6-dicyano-1,4-benzoquinone (DDQ) in the dark afforded an insoluble crystalline precipitate as the expected circumanthracene **85**.

6.2.4 Nanosized Graphenes with Armchair Edge and All-Benzenoid Character

The previous research disclosed that nanosized graphenes with a fully armchair edge structure usually exhibit high stability but a large band gap (Jiang 2008). Consequently, this type of PAH molecule has an all-benzenoid structure. Among the nanosized graphenes with armchair edges, hexa-*peri*-hexabenzocoronene (HBC) and its derivatives have to be mentioned as the most famous armchair-edged nanosized graphenes. According to Clar's aromatic sextet rule, HBC belongs to an all-benzenoid polycyclic aromatic hydrocarbon because seven aromatic benzenoid aromatic sextet rings can be drawn for HBC and no additional isolated double bond exists in this structure (Figure 6.14). This feature results in high stability compared to other linear and *meta*-annulated aromatics such as acene and phene. In addition, the intrinsic self-organization behavior and outstanding electronic properties of HBC makes it attractive in organic electronic devices (detailed discussion will be given in the next section). A troublesome issue for HBC molecules is their extremely low solubility due to the strong intermolecular π–π interaction. Therefore, the rational design of HBC derivatives with appropriate molecular size and sufficient solubilizing groups is highly desirable. Two comprehensive reviews have covered the syntheses and characterizations of all-benzenoid polycyclic aromatic hydrocarbons with armchair edge structures (Wu and Müllen 2006, Wu et al. 2007) and thus only a brief overview will be given here.

With the development of modern chemistry, synthetic breakthroughs have been achieved that make possible the selective and effective synthesis of a series of HBC derivatives under mild conditions. A widely used approach to HBCs with sixfold symmetry was developed by Müllen's group (Herwig et al. 1996, Wu et al. 2007). Substituted hexaphenyl benzene **89** was synthesized by $Co_2(CO)_8$-promoted cyclotrimerization of R-substituted diphenylacetylene **88**, which subsequently underwent oxidative cyclodehydrogenation to give the fused HBC derivatives **90** in high yields (Scheme 6.20). In principle, three types of reagents can be used in the last step to promote intramolecular ring closure, including $CuCl_2/AlCl_3$ or $Cu(OTf)_2/AlCl_3$ in CS_2, or $FeCl_3$ in nitromethane. Various solubilizing flexible chains are introduced at

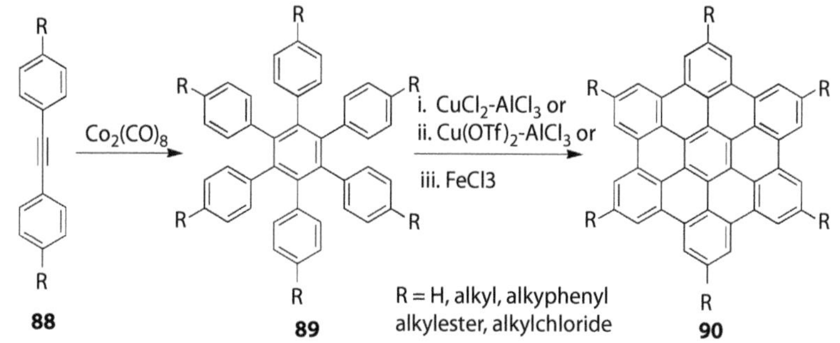

SCHEME 6.20 General synthesis of sixfold symmetric hexa-*peri*-hexabenzocoronene derivatives.

the peripheries of the HBC core to increase its solubility and processability on one hand, and on the other hand to facilitate the formation of a columnar liquid crystalline phase, which is governed by both flexible side chains and the π–π stacked HBC molecules.

The HBC derivatives **94** with low symmetry were synthesized by an alternative route, following a synthetic route containing Diels-Alder cycloaddition between tetraphenylcyclopentadienone derivatives **91** and diphenylacetylene derivatives **92**, and subsequent oxidative cyclodehydrogenation of precursor **93** with $FeCl_3$ (Scheme 6.21). Based on this synthetic strategy, HBCs possessing solubilizing chains and one or several bromine atoms with different symmetries were synthesized (Ito et al. 2000). The presence of bromine atoms allowed further functionalizations by transition metal-catalyzed coupling reactions such as Kumada, Suzuki, Hagihara, Negishi, and Buchwald coupling reactions. The attachment of different functionalities allows more exact control of the order of HBC molecules in the bulk state, the alignment of the discs on the surface, and the intramolecular binary energy/electron transport. These will be discussed in Section 6.3.2.

By employing the two well-known synthetic strategies mentioned previously, a series of large all-benzenoid nanosized graphenes with different molecular sizes, symmetries, and peripheries were obtained. The largest graphene-like molecule to date containing 222 carbons (**99**) and other all-benzenoid nanographenes such as

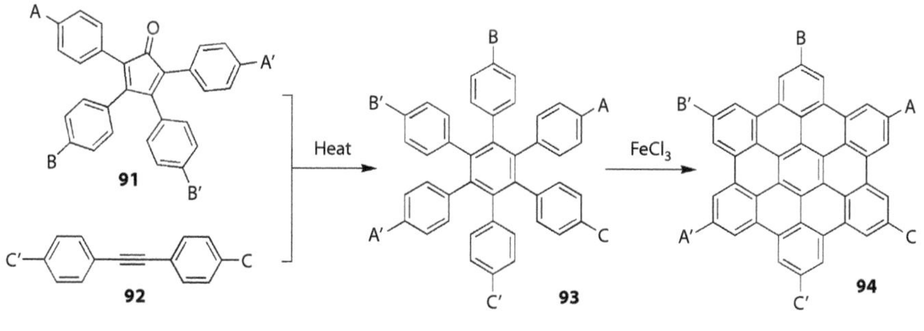

SCHEME 6.21 Synthesis of hexa-*peri*-hexabenzocoronene derivatives with low symmetry.

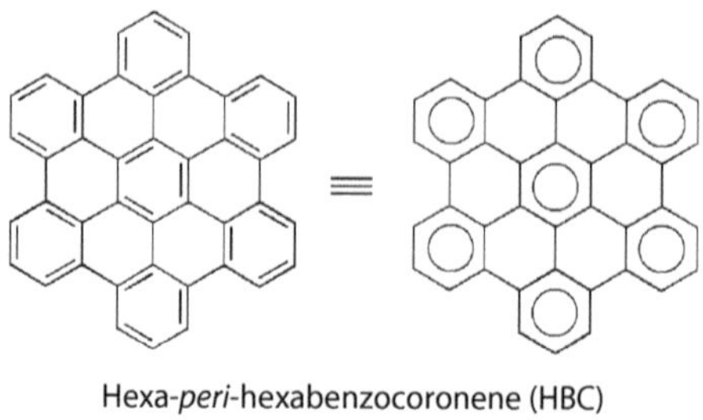

Hexa-*peri*-hexabenzocoronene (HBC)

FIGURE 6.14 Structure of hexa-*peri*-hexabenzocoronene.

R = Alkyl chains

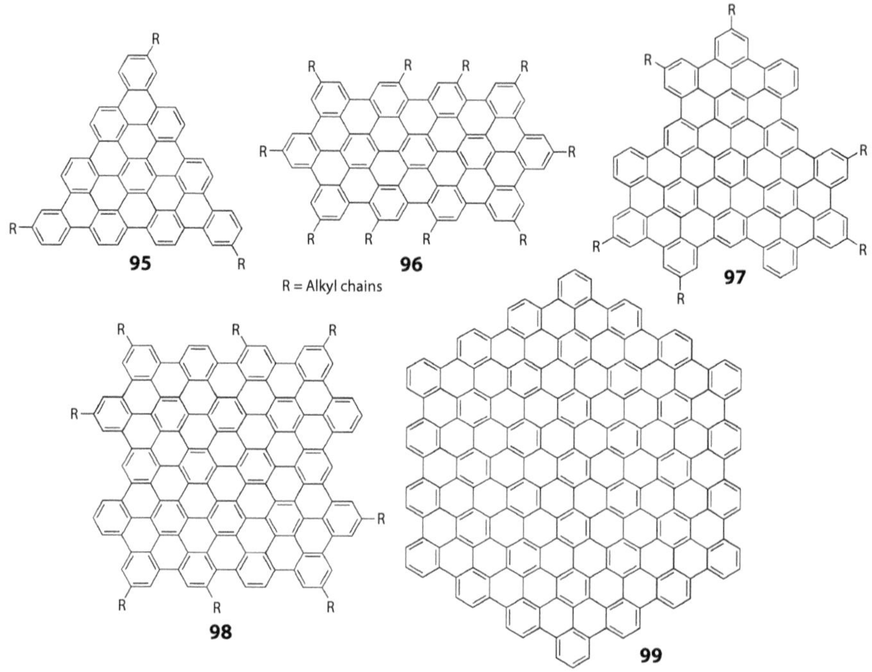

FIGURE 6.15 Various all-benzenoid nanographene molecules with different sizes and symmetries.

95–98 with triangle shape, linear shape, cordate shape, and square shape (Figure 6.15) have been successfully synthesized and successively reported (Wu et al. 2007).

6.3 MATERIAL APPLICATIONS

6.3.1 Organic Dyes

Dye chemistry is believed to be one of the most explored areas in industrial organic chemistry. Recent developments in the field of electronics and in the area of bio-imaging have boosted interest in the development of next-generation functional dyes. There is a very comprehensive range of common dyes (e.g., BODIPYs, fluoresceins, rhodamines) that are now known, some of which have even been commercialized. Certain nanosized graphenes (e.g., rylenes, circumacenes), however, have gained recognition as being some of the most versatile dyes, and the number of them has steadily increased over the last two decades. There is increased interest recently in the design and synthesis of dyes that function in the NIR spectral region due to their promising applications for organic solar cells, bioimaging, and nonlinear optics (Fabian et al. 1992, Qian and Wang 2010, Jiao and Wu 2010).

6.3.1.1 NIR Dyes

Compared with commercialized dyes, nanosized graphenes usually exhibit excellent chemical stability and photostability. However, most nanosized graphenes are only capable of capturing UV or visible light. For practical applications such as solar cells, the materials should have good light-harvesting capability, not only in the UV-Vis spectral range, but also in the NIR range because 50% of the radiation energy in sunlight is in the infrared region. Promotion of the absorption and emission of nanosized graphene-based dyes into the NIR region can normally be achieved by extension of the π-conjugation, or by constructing a push-pull motif, or in rare cases by quinoidization (Jiao and Wu 2010), all of which will produce a decrease in the HOMO-LUMO band gaps, and concomitant bathochromic shift of their absorption and emission bands.

Although the absorption of acenes significantly red shifts with an increase in the number of six-membered rings, they are seldom or never used as dyes because of the poor stabilities of their higher homologues and their relatively low absorption intensities in the long-wavelength region. A representative family in dye chemistry is rylene and its derivatives. For instance, perylene bisimide has a brilliant red color with an absorption maximum at 540 nm and terrylene bisimide was reported to have a green color with absorption maximum at 650 nm (Hortrup et al. 1997). For their higher homologues, the absorptions are bathochromically shifted into the NIR region with large molar extinction coefficients (Quante and Müllen 1995, Pschirer et al. 2006). The successful preparation of octarylene diimide with an absorption maximum at 1066 nm is noteworthy (Qu 2008).

Periacenes were also calculated to be good candidates for near-infrared absorbing dyes. However, structural modification to stabilize these compounds is necessary, which includes attachment of the electron-withdrawing groups and blocking the most active positions as mentioned in Section 6.2.3.3. The recently synthesized bisanthene derivatives **58–60** by Wu's group can serve as stable and soluble NIR dyes (Yao et al. 2009, J. Li 2010, Zang et al. 2009).

6.3.1.2 Bioimaging

In the NIR region, biological samples have low background fluorescence signals, and a concomitant high signal-to-noise ratio. Moreover, NIR light can penetrate deeply into sample matrices due to low light scattering. Thus, in vivo and in vitro imaging of biological samples using NIR dyes are promising and attractive (Kiyose et al. 2008). To take full advantage of organic NIR dyes in the biological field, the design of the dyes is not just directed toward color tuning; photostability, biocompatibility, and sufficient water solubility are also some of the key issues that should be taken into consideration.

Design and synthesis of nanosized graphenes with NIR absorption and emission properties is attainable, and tuning of their photostability and binding ability to a single analyte are also feasible through chemical modifications of the dye molecules. Sufficient water solubility, however, limits the number of biological applications that can involve nanosized graphenes. Nanosized graphenes intrinsically have a hydrophobic nature because of the highly π-conjugated hydrocarbon structures. Gaining solubility in aqueous solutions is usually achieved by linking water-solubilizing groups such as sulfonic acid moieties, quaternized amine groups, crown ethers, polyethylene oxide, and peptide chains to the nanosized graphene core. The presence of highly hydrophilic groups usually makes the dye molecules water soluble, but in many cases these nanosized graphenes show almost no fluorescence in water due to their strong tendency to form aggregates in a polar environment. Consequently, the appropriate choice of water-solubilizing groups while maintaining a high fluorescence quantum yield is of great importance for applications in bioimaging. Exciting progress has been made in the last 6 years based on perylene bisimides **100–101** (Qu et al. 2004, Peneva et al. 2008a) and terrylene bisimide **102–103** (Peneva et al. 2008b, Jung et al. 2009) (Figure 6.16). Through chemical modifications, these rylene dyes combine the exceptional photophysical properties of the rylene bisimide dyes and a recognition unit for site-specific labeling of proteins, and meet all the criteria for bioimaging: (1) good water solubility, which was achieved by the attachment of the solubilizing groups at the *bay*-positions; (2) high fluorescence quantum

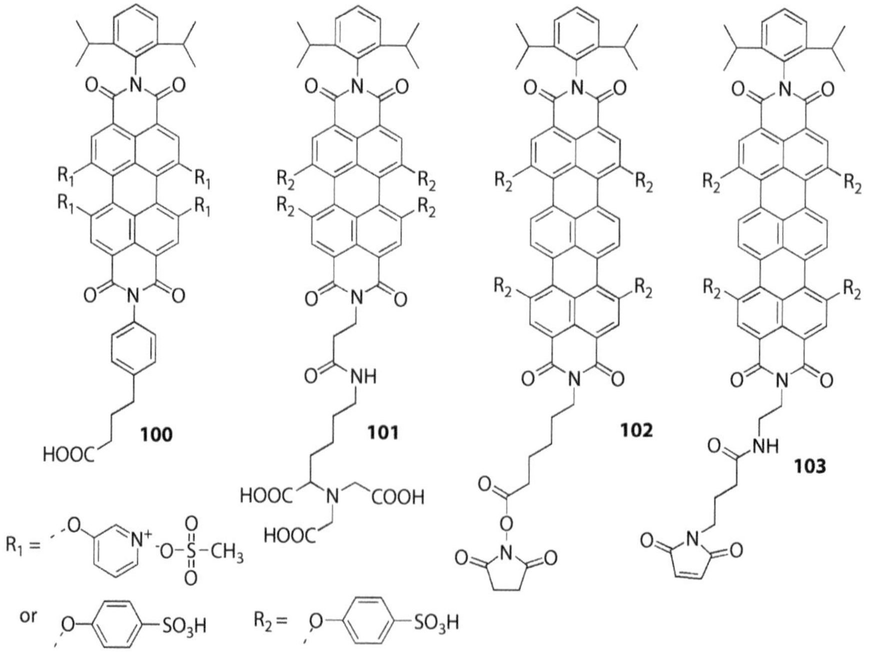

FIGURE 6.16 Some perylene- and terrylene-based water-soluble fluorescent probes.

yield, which was maintained at a high level despite the structural modification; (3) high chemical and photostability, which is considered to be one of the advantages of rylene dyes; (4) nontoxicity; (5) good biocompatibility; and (6) possible commercial viability and scalable production, which are attributed to the high yields in each step.

6.3.2 Charge Transporting Materials

6.3.2.1 Control of Liquid Crystal (LC) Phase

Organic electronic devices have developed rapidly since their discovery in the mid-1980s. In organic electronic devices, the efficiency of the charge carrier transport through the organic molecules is one of the key factors determining device performance. The materials with efficient charge-carrier transport properties are highly desirable in many electronic devices. In principle, charge-carrier transport properties are considered to depend on both the intrinsic electronic properties of the materials and microscopic and macroscopic order of the molecules in solid state. As a result, control of the degree of the molecular order and organization is necessary.

The alkyl chain-substituted nanosized graphenes, particularly the all-benzenoid PAHs, can form self-assembled one-dimensional columnar structures. This is due to strong π–π interactions between the rigid aromatic core and the nanoscale phase separation between the rigid core and the flexible alkyl chains at the periphery. Therefore, the overlap of π-orbitals between neighboring planes is maximized and charge carriers can easily travel along the one-dimensional columns. Furthermore, alkyl chain-substituted nanosized graphenes are able to self-heal structural defects due to their liquid crystalline (LC) character, and can be processed from solution because of their high solubility. All of these advantages inspire researchers to rationally control the LC mesophases of nanosized graphenes. For example, as very promising organic semiconductors in recent years, a great number of HBC derivatives have been developed with different sizes, symmetries, and substitutions. Three major methods have been utilized to tune their LC phase. The first method is a well-developed way in which the length and branching of the alkyl chains are adjusted. The long and branched chains tend to decrease the phase transition temperature. As shown in Figure 6.17, the isotropic temperatures are higher than 450°C for the n-alkyl-substituted HBCs **104a** (Herwig et al. 2006) while the dove-tailed chain-substituted HBC **104b** (Pisula et al. 2006) shows isotropic temperature below 46°C. The second method is to tune the molecular size. A wide-range columnar liquid crystalline phase from room temperature up to >400°C was observed in the dodecylphenyl-substituted HBC **104c** (Fechtenkötter et al. 1999), whereas further increasing the size of the core to **96** and **97** results in broader columnar liquid crystalline phases, which cannot become isotropic when melted below 600°C. The third method is to introduce functional groups or additional noncovalent interactions at the side chains. For instance, additional hydrogen bonding units were introduced in **105**, which significantly enhanced the self-assembling abilities of the HBC and thus endowed it with excellent gelation ability (Dou et al. 2008). Compound **106** is also interesting, in which the local dipoles are presented (Feng et al. 2008). As a result, the solution processing onto a substrate resulted in the formation of exceptionally

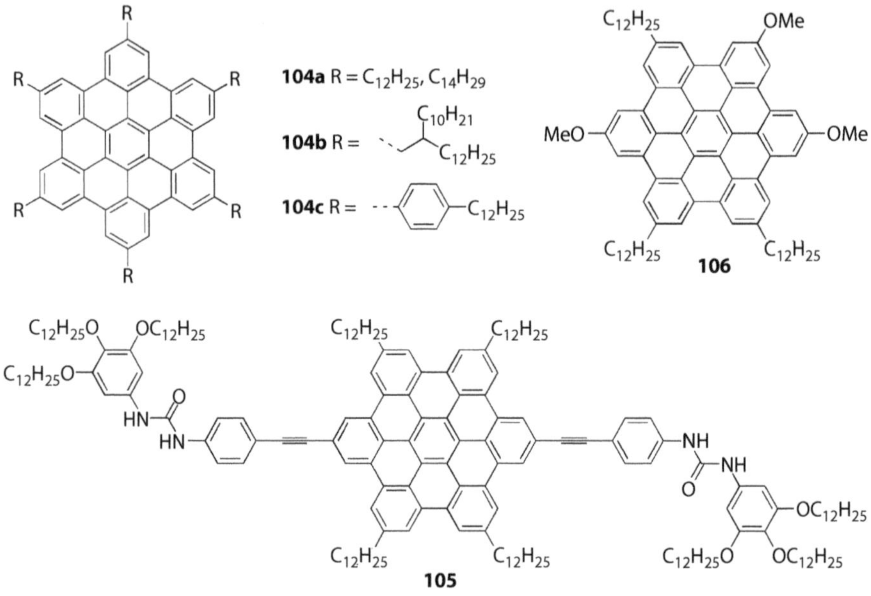

FIGURE 6.17 Some HBC derivatives with controlled liquid crystalline mesophases and self-assembly.

long fibrous microstructures. The rational control of the LC phase is beneficial to enhance the charge transporting properties of materials, qualifying them as promising semiconductors for organic electronic devices such as field-effect transistors.

6.3.2.2 Field-Effect Transistors

Organic semiconductors are normally classified as p-type or n-type, depending on which type of charge carrier is more efficiently transported through the materials, or ambipolar-type, in which both hole and electron can be efficiently transported. Over the past 20 years, organic field-effect transistors (OFETs) based on nanosized graphenes have attracted enormous interest for the realization of organic electronic devices.

A key family of nanosized graphenes for OFETs is acene with pentacene as a representative. Pentacene is a benchmark for FETs exhibiting charge carrier mobility larger than 3 $cm^2\ V^{-1}\ s^{-1}$ (Roberson et al. 2005). Its excellent charge transport is claimed to be due to the extended conjugated system, the efficient intermolecular π–π overlaps, and the appropriate HOMO energy level for hole injection and transport. General problems for pentacenes are their poor stability and solubility, which can be alleviated by the introduction of functional groups at 6, 13-positions. For example, the triisopropylsilylethynyl-substituted pentacene (**5**) exhibited FET hole mobility of 1.42 $cm^2\ V^{-1}\ s^{-1}$ for its single crystal nanowires with enhanced stability (Kim et al. 2007). Apart from pentacene, other acenes also received much interest in OFETs (Wu 2007). For instance, rubrene, a tetracene derivative, still has the highest record for OFETs with mobility approaching 15–40 $cm^2\ V^{-1}\ s^{-1}$ (da Silva Filho et al.

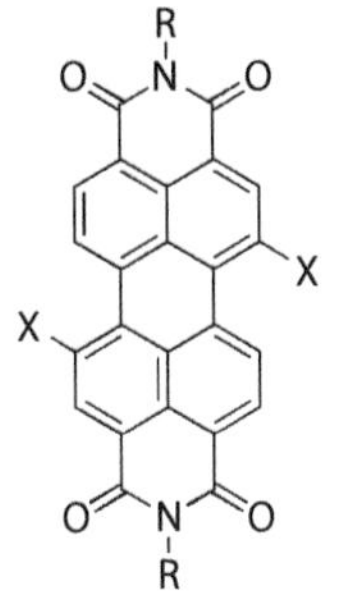

107a: X = H, R = C_6H_5
107b: X = CN, R = cyclohexyl
107c: X = CN, R = $CH_2C_3F_7$

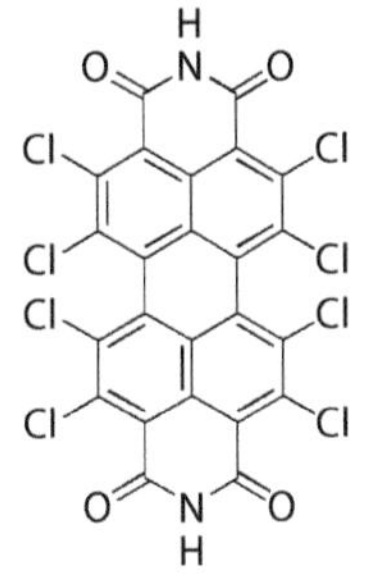

108

FIGURE 6.18 Perylene-based n-type semiconductors.

2005). However, higher acenes such as hexacene and heptacene have not been used for OFETs due to their synthetic difficulties and poor stability.

In the case of HBC-based OFETs, control of the alignment of these discotic materials is the key issue and the discs need to be aligned in the direction that is parallel to the surface. A uniaxial edge-on alignment of the HBC molecules between the source and drain electrodes was achieved by zone-casting techniques or by using preoriented substrates, which allows efficient charge transport in macroscopic thin films. The FET mobility of the oriented films is commonly between 0.0001–0.01 $cm^2\ V^{-1}\ s^{-1}$ (Pisula et al. 2005).

The development of n-type materials is still lagging behind p-type semiconductors due to the low stability of most n-type conductors in the air as well as instability of the electrodes used for n-type materials. One effective method for n-type organic semiconductors is to convert p-type semiconductors into n-type materials by introducing certain strong electron-withdrawing groups. This will lower the LUMO energy level of materials, thus facilitating electron injection and transport. For example, an electron mobility of $10^{-5}\ cm^2\ V^{-1}\ s^{-1}$ was first reported for perylene bisimide **107a** (Figure 6.18) (Horowitz et al. 1996). In fact, perylene bisimide is one of the most promising electron-deficient cores among nanosized graphenes for the design of n-type OFET materials. Further, introduction of electron-withdrawing groups makes it possible to tune their charge transporting properties, the degree of molecular organization, and stability. For instance, cyano group–substituted **107b** has good air stability and exhibits electron mobility of 0.1 $cm^2\ V^{-1}\ s^{-1}$, while **107c** with fluoroalkyl end substitution displays electron mobility as high as 0.64 $cm^2\ V^{-1}\ s^{-1}$ because of the increased intermolecular π–π overlaps induced by the fluoroalkyl substitution in the latter (Jones et al. 2004). The perchlorinated perylenebisimide **108** showed high electron mobility around 0.91 $cm^2\ V^{-1}\ s^{-1}$, although it has a highly twisted structure (Gsänger et al. 2010).

6.3.3 Organic Solar Cells

6.3.3.1 Bulk-Heterojunction (BHJ) Solar Cells

Solar cells, also called photovoltaics, are electronic devices that can directly convert solar energy into electricity. The performance of an organic solar cell is determined by a broad parameter set, such as the HOMO and LUMO energy levels of the donor and acceptor, the absorption behavior and charge transport property of the organic compounds, morphology, and so on. Solar cells are roughly grouped into bilayer solar cells, bulk-heterojunction (BHJ) solar cells, and dye-sensitized solar cells (DSSC) based on the device architecture. Our focus is on the BHJ solar cells and the DSSCs, which usually show higher power conversion efficiency than the bilayer photovoltaics.

To improve the efficiency of bilayer solar cells, one possibility is to enlarge the donor–acceptor interfacial area. In BHJ cells, both donor and acceptor phases are intimately intermixed so that the excitons can readily access the interface and further dissociate to holes and electrons at the donor–acceptor interface. Since the first report on organic solar cells by Tang applying perylene dibenzimidazole as an acceptor (Tang 1986), perylene with strong electron-withdrawing groups has been widely used as an acceptor part in BHJ cells due to the fact that they have high electron affinities and thus can accept electrons from most donor compounds. However, a general drawback of the BHJ cells is that the transport and collection of charges in a disordered nanoscale blend can be hindered by phase boundaries and discontinuities. One way to overcome this drawback is by covalently linking the donor and acceptor in a single polymer chain. For this purpose, perylene derivatives have been linked to other electron-donating units to form donor-accepter co-oligomers or copolymers. For example, Janssen's group first reported two new copolymers **109** and **110** (Figure 6.19) consisting of alternating oligophenylenevinylene donor and perylene bisimide acceptor segments with satisfactory open-circuit voltages around 1–1.2 V (Neuteboom et al. 2003). However, short-circuit current densities were found below 0.012 mA cm^{-2} under AM1.5 conditions, which is due to the fast geminate recombination and poor transport characteristics resulting from face-to-face orientations of oligophenylenevinylene and perylene segments in alternating stacks. Remarkably enhanced power conversion efficiency was achieved by using **111** (Figure 6.19) as

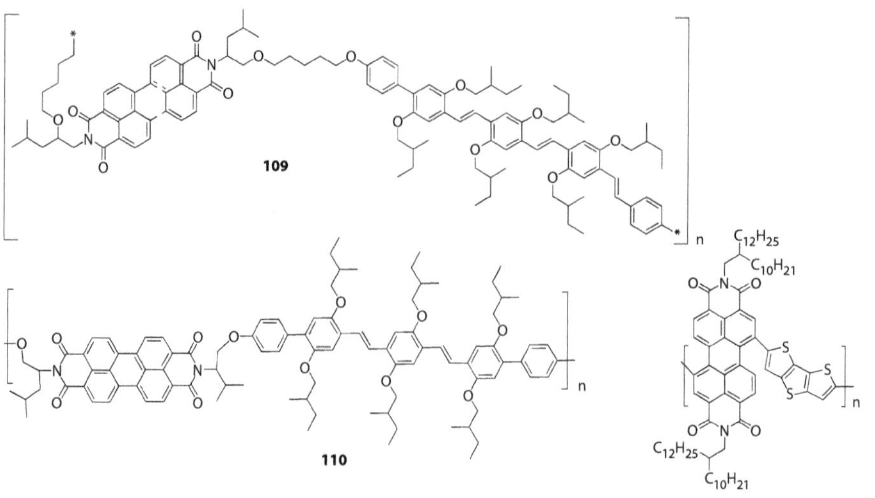

FIGURE 6.19 Perylene bisimide-containing donor–acceptor copolymers for solar cells.

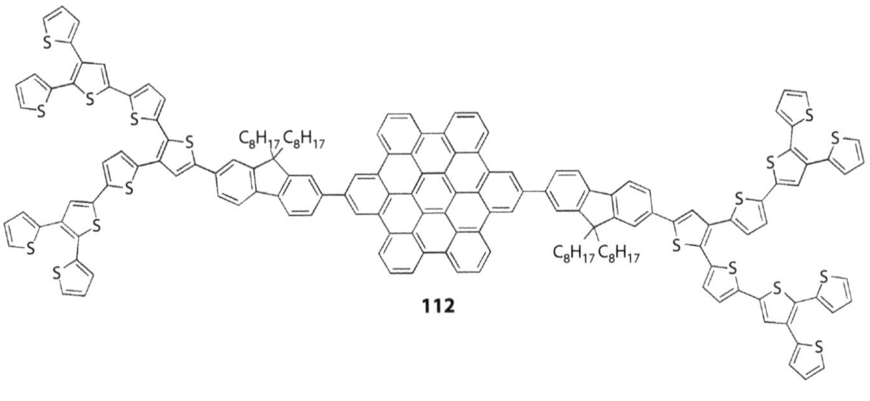

FIGURE 6.20 Thiophene–fluorenene co-oligomers substituted HBC for BHJ cells.

the electron acceptor and another polythiophene polymer as donor to construct BHJ cells (Zhan et al. 2007). In common with low-band polymers, the blend demonstrated very broad absorptions between 250 and 850 nm with power conversion efficiency approaching 1.5%.

Other nanosized graphenes have seldom been used in BHJ cells owing to their limited absorption behavior. A thiophene-fluorene co-oligomer-substituted HBC **112** (Figure 6.20) exhibited broader absorption between 250 and 550 nm and BHJ devices fabricated with **112** as an electron donor and [6,6]-phenyl-C61- butyric acid methyl ester (PCBM) as an electron acceptor showed good performance with power conversion efficiency of 2.5% (Wong et al. 2010). This is attributed to the enhanced light-harvesting property as well as the formation of an ordered morphology in solid state induced by the self-assembly of the HBC molecules.

6.3.3.2 Dye-Sensitized Solar Cell (DSSC)

DSSCs were the first organic photovoltaic products to enter the market because of their high efficiency and stability. In organic DSSCs, photons are collected by organic dyes that are covalently linked to a nanostructured metal oxide electrode (e.g., TiO_2). Improvement of the device's performance has been limited by the light-harvesting capability of state-of-the-art dyes, which lack strong absorption in the far-red/NIR region. Rylene dyes are believed to be ideal candidates as sensitizers in DSSCs due to their excellent light-harvesting properties and available reactive positions, which allow fine chemical modifications to tune the energy level, band gap, and anchoring group. Representative is the perylene monoanhydride **113** (Figure 6.21) with a push-pull motif, showing power conversation efficiency around 3.2% (Edvinsson et al. 2007). A breakthrough was achieved by introducing two phenylthio groups in the 1,6-positions of the perylene core to form dye **114** (Li et al. 2008). This allows fine tuning of the HOMO and LUMO energy levels and the absorption of the dye, giving an efficiency up to 6.8% of standard AM 1.5 solar conditions.

Construction of solid-state DSSCs with a solid organic hole-transporting material has gained considerable attention as an attractive alternative to traditional DSSCs with liquid electrolytes. In this case, some disadvantages such as solvent leakage and

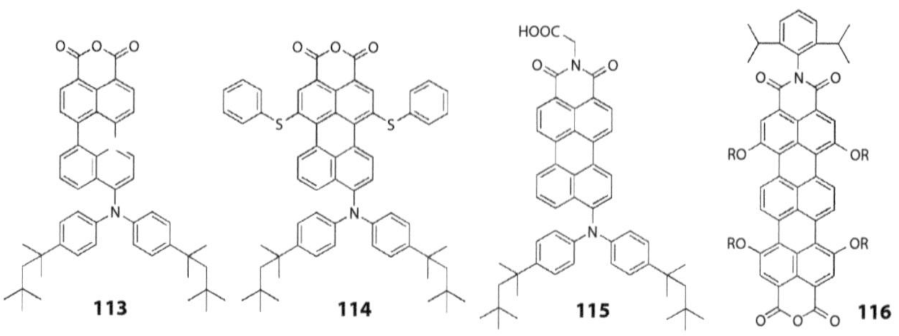

FIGURE 6.21 Perylene- and terrylene-based NIR dyes for DSSCs.

evaporation are prevented. Using dye **115** (Figure 6.21), which possesses a carboxylic acid as the anchoring group, solid-state DSSCs exhibited an efficiency of 3.2%, whereas a poor efficiency as low as 1.2% was obtained for **115** in a liquid-electrolyte-based DSSC (Cappel et al. 2009). Higher rylenes exhibiting a larger π-elongated system and a concomitantly better light-harvesting property, such as terrylene monoimide derivate **116,** have also been tested in DSSCs (Edvinsson et al. 2008). Despite a remarkably broad photocurrent action spectrum arising from its absorption behavior, a moderate efficiency of 2.4% has been achieved, which is attributed to the low voltage caused by the dye's incompatibility with additives.

6.4 CONCLUDING REMARKS

It is obvious that nanosized graphenes play very important roles in materials science. The success of using nanosized graphenes as active materials is ascribed to their unique properties, including good self-assembling and charge transporting properties, and unusual absorption and emission behavior.

Despite encouraging achievements by previous and current contributors, low solubility for larger nanosized graphenes is inherent. Thus, preparation of larger nanosized graphenes still suffers from limitations such as poor solubility. Furthermore, due to the narrow band gap and the relatively high-lying HOMO energy level for some members (e.g., higher-order acenes and periacenes), there is an urgent need to final a way to stabilize them by rational chemical modifications. Synthesis in some cases is also restricted by the multiple steps and overall low yields, which limit their practical applications. Current research into applications that use nanosized graphenes is focused on pentacene, perylene, and HBC derivatives; applying other nanosized graphenes has received less attention. In this respect, attention should be turned toward the other nanosized graphenes and considerable effort should go into the systematic modification of their architectures to endow them with appropriate qualifications.

Specifically, tuning the water solubility and preserving the fluorescence of nanosized graphenes are the key issues for bioimaging applications. One trend is to embed the dye molecules into a hydrophilic shell, which functions as a carrier to deliver the dye into biotargets. This strategy is still in its infancy for nanosized graphene-based dyes; however, it provides an alternative way to exploit applications of nanosized graphenes for biological systems.

Discotic LCs have been used as charge-transporting materials in electronic devices, and the control of the degree of molecular order and organization as well as the alignment of the columnar structure is beneficial to improving device performance. As aspects including design strategies and packing manners are becoming clear, it is conceivable that numerous disc-like nanosized graphenes such as HBC, perylene, coronene, and ovalene have the potential to achieve excellent device performance after further optimization.

In photovoltaics, the exploitation of perylene dyes for organic solar cells has been widely developed. Compared with other organic dyes (e.g., porphyrin in DSSCs, a low-band-gap polymer in BHJ cells), there is still plenty of room for device enhancement. Perylene dyes based on the donor–acceptor concept turn out to be quite successful in the design of highly efficient dyes for photovoltaics. The push-pull effect not only improves the light-harvesting ability of dyes but further facilitates electron injection to the electrode, thus improving the performance of the solar cell device. Furthermore, higher rylenes and other nanosized graphenes with NIR absorptions should receive much more attention in the future due to their better light-harvesting properties.

In short, it seems inevitable that more and more nanosized graphenes will be attainable by state-of-the-art chemistry, and their physical and chemical properties are adjustable for different applications in material science.

ABBREVIATIONS

BHJ: bulk heterojunction
BODIPY: 4,4-difluoro-4-bora-3a,4a-diaza-s-indacene
COD: cis, cis-1,5-cyclooctadiene
DBN: 1,5-diazabicyclo(4.3.0)non-5-ene
DBU: 1,8-diazabicyclo[5.4.0]undec-7-ene
DDQ: 2,3-dichloro-5,6-dicyano-1,4-benzoquinone
DSSC: dye-sensitized solar cell
FETs: field-effect transistors
HBC: hexa-*peri*-hexabenzocoronene
HOMO: highest occupied molecular orbital
LC: liquid crystalline (crystal)
LUMO: lowest unoccupied molecular orbital
NIR: near infrared
NMR: nuclear magnetic resonance
OFETs: organic field-effect transistors
PAHs: polycyclic aromatic hydrocarbons
TIPSE: triisopropylsilylethynyl
Tf: trifluoromethanesulfonyl
Vis: visible
UV: ultraviolet

REFERENCES

Alibert-Fouet, S., Seguy, I., Bobo, J., Destruel, P., and Bock, H. 2007. Liquid-crystalline and electron-deficient coronene oligocarboxylic esters and imides by twofold benzogenic Diels–Alder reactions on perylenes. *Chem. Eur. J.* 13: 1746–1753.

Anthony, J. E. 2006. Functionalized acenes and heteroacenes for organic electronics. *Chem. Rev.* 106: 5028–5048.

Anthony, J. E. 2008. The larger acenes: Versatile organic semiconductors. *Angew. Chem. Int. Ed.* 47: 452–483.

Arabei, S. M., and Pavich, T. A. 2000. Spectral-luminescent properties and photoinduced transformations of bisanthene and bisanthenequinone. *J. Appl. Spectrosc.* 67: 236–244.

Avlasevich, Y., Kohl, C., and Müllen, K. 2006. Facile synthesis of terrylene and its isomer benzoindenoperylene. *J. Mater. Chem.* 16: 1053–1057.

Avlasevich, Y., and Müllen, K. 2006. Dibenzopentarylenebis(dicarboximide)s: Novel near-infrared absorbing dyes. *Chem. Commun.* 4440–4442.

Bailey, W. J., and Madoff, M. 1953. Cyclic dienes. II. A new synthesis of pentacene. *J. Am. Chem. Soc.* 75: 5603–5604.

Bendikov, M., Wudl, F., and Perepichka, D. F. 2004. Tetrathiafulvalenes, oligoacenenes, and their buckminsterfullerene derivatives: The brick and mortar of organic electronics. *Chem. Rev.* 104: 4891–4945.

Bohnen, A., Koch, K. H., Lüttke, W., and Müllen, K. 1990. Oligorylene as a model for "poly(*peri*-naphthalene)." *Angew. Chem. Int. Ed.* 29: 525–527.

Broene, R. D., and Diederich, F. 1991. The synthesis of circumanthracene. *Tetrahedron Lett.* 32: 5227–5230.

Cappel, U. B., Karlsson, M. H., Pschirer, N. G., et al. 2009. A broadly absorbing perylene dye for solid-state dye-sensitized solar cells. *J. Phys. Chem. C* 113: 14595–14597.

Chun, D., Cheng, Y. N., and Wudl, F. 2008. The most stable and fully characterized functionalized heptacene. *Angew. Chem. Int. Ed.* 47: 8380–8385.

Clar, E. 1939. Vorschlage zur nomenklatur kondensierter ringsysteme (aromatische kohlenwasserstoffe, XXVI. Mitteil.). *Chem. Ber.* 75B: 2137–2139.

Clar, E. 1948a. Aromatic hydrocarbons XLII. The condensation principle, a simple new principle in the structure of aromatic hydrocarbon. *Chem. Ber.* 81: 52–63.

Clar, E. 1948b. Synthesis of ovalene. *Nature* 161: 238–239.

Clar, E. 1964. *Polycyclic hydrocarbons*. Vol. I, p. 47. New York: Academic Press.

Clar, E., Lang, K. F., and Schulz-Kiesow, H. 1955. Aromatische kohlenwasserstoffe. *Chem. Ber.* 88: 1520–1527.

Clar, E., and Macpherson, I. A. 1962. The significance of Kekulé structure for the stability of aromatic system-II. *Tetrahedron* 18: 1411–1416.

Clar, E., and Mullen, A. 1968. The non-existence of a threefold aromatic conjugation in linear benzologues of triphenylene (starphenes). *Tetrahedron* 24: 6719–6724.

Clar, E., Robertson, J. M., Schlögl, R., and Schmidt, W. 1981. Photoelectron spectra of polynuclear aromatics. 6. Applications to structural elucidation: "circumanthracene." *J. Am. Chem. Soc.* 103: 1320–1328.

Clar, E., and Zander, M. 1957. Syntheses of coronene and 1,2:7,8-dibenzocoronene. *J. Chem. Soc.* 61: 4616–4619.

da Silva Filho, D. A., Kim, E.-G., and Brédas, J.-L. 2005. Transport properties in the Rubrere crystal: Electronic coupling and vibrational reorganization energy. *Adv. Mater.* 17: 1072–1076.

Désilets, D., Kazmaier, P. M., and Burt, R. A. 1995. Design and synthesis of near-infrared absorbing pigments. I. Use of Pariser-Parr-Pople molecular orbital calculations for the identification of near-infrared absorbing pigment candidates. *Can. J. Chem.* 73: 319–324.

Dou, X., Pisula, W., Wu, J., Bodwell, G. J., and Müllen, K. 2008. Reinforced self-assembly of hexa-*peri*-hexabenzocoronenes by hydrogen bonds: From microscopic aggregates to macroscopic fluorescent organogels. *Chem. Eur. J.* 14: 240–249.

Edvinsson, T., Li, C., Pschirer, N., et al. 2007. Intramolecular charge-transfer tuning of perylenes: Spectroscopic features and performance in dye-sensitized solar cells. *J. Phys. Chem. C* 111: 15137–15140.

Edvinsson, T., Pschirer, N., Schöneboom, J., et al. 2009. Photoinduced electron transfer from a terrylene dye to TiO_2: Quantification of band edge shift effects. *Chem. Phys.* 357: 124–131.

Erünlü R. K. 1969. Octazthren. *Liebigs Ann. Chem.* 721: 43–47.

Fabian, J., Nakanzumi, H., and Matsuoka, M. 1992. Near-infrared absorbing dyes. *Chem. Rev.* 92: 1197–1226.

Fechtenkötter, A, Saalwächter, K., Harbison, M. A., Müllen, K., and Spiess, H. W. 1999. Highly ordered columnar structures from hexa-*peri*-hexabenzocoronenes: Synthesis, X-ray diffraction, and solid-state heteronuclear multiple-quantum NMR Investigations. *Angew. Chem. Int. Ed.* 38: 3039–3042.

Feng, X., Pisula, W., Takase, M., et al. 2008. Synthesis, helical organization, and fibrous formation of C_3 symmetric methoxy-substituted discotic hexa-*peri*-hexabenzocoronene. *Chem. Mater.* 20: 2872–2874.

Fieser, L. F. 1931. Reduction products of naphthacenequinone. *J. Am. Chem. Soc.* 53: 2329–2341.

Fort, E. H., Donovan, P. M., and Scott, L. T. 2009. Diels-Alder reactivity of polycyclic aromatic hydrocarbon bay regions: Implications for metal-free growth of single-chirality carbon nanotubes. *J. Am. Chem. Soc.* 131: 16006–16007.

Gsänger, M., HakOh, J., Könemann, M., et al. 2010. A crystal-engineered hydrogen-bonded octachloroperylene diimide with a twisted core: An n-channel organic semiconductor. *Angew. Chem. Int. Ed.* 49: 740–743.

Herwig, P., Kayser, C. W., Müllen, K., and Spiess, H. W. 1996. Columnar mesophases of alkylated hexa-*peri*-hexabenzocoronenes with remarkably large phase widths. *Adv. Mater.* 8: 510–513.

Horowitz, G., Kouki, F., Spearman, P., et al. 1996. Evidence for n-type conduction in a perylene tetracarboxylic diimide derivative. *Adv. Mater.* 8: 242–245.

Hortrup, F. O., Müller, G. R. J., Quante, H., et al. 1997. Terrylenimides: New NIR fluorescent dyes. *Chem. Eur. J.* 3: 219–225.

Ito, S., Wehmeier, M., Brand, J. D., et al. 2000. Synthesis and self-assembly of functionalized hexa-*peri*-hexabenzocoronenes. *Chem. Eur. J.* 6: 4327–4342.

Jiang, D. 2007a. Unique chemical reactivity of a graphene nanoribbon's zigzag edge. *Chem. Phys.* 126: 134701.

Jiang, D. 2007b. First principles study of magnetism in nanographenes. *Chem. Phys.* 126: 134703.

Jiang, D., and Dai, S. 2008. Circumacenes versus periacenes: HOMO–LUMO gap and transition from nonmagnetic to magnetic ground state with size. *Chem. Phys. Lett.* 466: 72–75.

Jiang, W., Qian, H, Li, Y., and Wang, Z. 2008. Heteroatom-annulated perylenes: Practical synthesis, photophysical properties, and solid-state packing arrangement. *J. Org. Chem.* 73: 7369–7372.

Jiao, C., Huang, K., Luo, J., et al. 2009. Bis-*N*-annulated quaterrylenebis(dicarboximide) as a new soluble and stable near-infrared dye. *Org. Lett.* 11: 4505–4511.

Jiao, C., and Wu, J. 2010. Soluble and stable near-infrared dyes based on polycyclic aromatics. *Curr. Org. Chem.* 14: 2145–2168.

Jones, B. A., Ahrens, M. J., Yoon, M.-H., et al. 2004. High-mobility air-stable n-type semiconductors with processing versatility: dicyanoperylene- 3,4:9,10-bis(dicarboximides). *Angew. Chem. Int. Ed.* 43: 6363–66.

Jung, C., Ruthardt, N., Lewis, R., et al. 2009. Photophysics of new water-soluble terrylenediimide derivatives and applications in biology. *ChemPhysChem* 10: 180–190.

Kaur, I., Jazdzyk, M., Stein, N. N., Prusevich, P., and Miller, G. P. 2010. Design, synthesis, and characterization of a persistent nonacene derivative. *J. Am. Chem. Soc.* 130: 16274–16286.

Kaur, I., Jia, W. L., Kopreski, R. P., et al. 2008, Substituent effects in pentacenes: Gaining control over HOMO-LUMO gaps and photooxidative resistances. *J. Am. Chem. Soc.* 130: 16274–16286.

Kaur, I., Stein, N. N., Kopreski, R. P., and Miller, G. P. 2009. Exploiting substituent effects for the synthesis of a photooxidatively resistant heptacene derivative. *J. Am. Chem. Soc.* 132: 1261–1263.

Kim, D. H., Lee, D. Y., Lee, H. S., et al. 2007. High-mobility organic transistors based on single-crystalline microribbons of triisopropylsilylethynyl pentacene via solution-phase self-assembly. *Adv. Mater.* 19: 678–682.

Kiyose, K., Kojima, H., and Nagano, T. 2008. Functional near-infrared fluorescent probes. *Chem. Asian J.* 3: 506–515.

Koch, K.-H., and Müllen, K. 1991. Polyarylenes and poly(arylenevinylene)s, 5: Synthesis of tetraalkyl-substituted oligo(1,4-naphthylene)s and cyclization to soluble oligo(peri-naphthylene)s. *Chem. Ber.* 124: 2091–2100.

Konishi, A., Hirao, Y., Nakano, M., et al. 2010. Synthesis and characterization of teranthene: A singlet biradical polycyclic aromatic hydrocarbon having Kekulé structures. *J. Am. Chem. Soc.* 132: 11021–11023.

Li, C., Yum, J.-H., Moon, S.-J., et al. 2008. An improved perylene sensitizer for solar cell applications. *ChemSusChem* 1: 615–618.

Li, J., Zhang, K., Zhang, X., et al. 2010. *Meso*-substituted bisanthenes as new soluble and stable near-infrared dyes. *J. Org. Chem.* 75: 856–863.

Li, Y., Gao, J., Motta, S. D., Negri, F., and Wang, Z. 2010. Tri-*N*-annulated hexarylene: An approach to well-defined graphene nanoribbons with large dipoles. *J. Am. Chem. Soc.* 132: 4208–4213.

Li, Y., and Wang Z. 2009. Bis-*N*-annulated quaterrylene: An approach to processable graphene nanoribbons. *Org. Lett.* 11: 1385–1387.

Marschalk, Ch. 1939. Linear hexacenes. *Bull. Soc. Chim.* 6: 1112–1121.

Miao, Q., Chi, X., Xiao, S., et al. 2006. Organization of acenes with a cruciform assembly motif. *J. Am. Chem. S*oc.128: 1340–1345.

Mitchell, R. H., and Sondheimer, F. 1970. The attempted synthesis of a dinaphtha-1,6-bisdehydro[10]annulene. *Tetrahedron* 26: 2141–2150.

Nakano, M., Kishi, R., Takebe, A., et al. 2007. Second hyperpolarizability of zethrenes. *Compt. Lett.* 3: 333–338.

Neuteboom, E. E., Meskers, S. C. J., van Hal, P. A., et al. 2003. Alternating oligo(*p*-phenylene vinylene)-perylene bisimide copolymers: Synthesis, photophysics, and photovoltaic properties of a new class of donor-acceptor materials. *J. Am. Chem. Soc.* 125: 8625–8638.

Payne, M. M., Parkin, S. R., and Anthony, J. E. 2005. Functionalized higher acenes: Hexacene and heptacene. *J. Am. Chem. Soc.* 127: 8028–8029.

Peneva, K., Mihov, G., Herrmann, A., et al. 2008a. Exploiting the nitrilotriacetic acid moiety for biolabeling with ultrastable perylene dyes. *J. Am. Chem. Soc.* 130: 5398–5399.

Peneva, K., Mihov, G., Nolde, F., et al. 2008b. Water-soluble monofunctional perylene and terrylene dyes: Powerful labels for single-enzyme tracking. *Angew. Chem. Int. Ed.* 47: 3372–3375.

Pisula, W., Kastler, M., Wasserfallen, M., et al. 2006. Relation between supramolecular order and charge carrier mobility of branched alkyl hexa-*peri*-hexabenzocoronenes. *Chem. Mater.* 18: 3634–3640.

Pisula, W., Menon, A., Stepputat, M., et al. 2005. A zone-casting technique for device fabrication of field-effect transistors based on discotic hexa-*peri*-hexabenzocoronene. *Adv. Mater.* 17: 684–689.

Pschirer, N. G., Kohl, C., Nolde, F., Qu, J., and Müllen, K. 2006. Pentarylene- and hexarylenebis(dicarboximide)s: Near-infrared-absorbing polyaromatic dyes. *Angew. Chem. Int. Ed.* 45: 1401–1404.

Qian, G., and Wang, Z. 2010. Near-infrared organic compounds and emerging applications. *Chem. Asian J.* 5: 1006–1029.

Qian, H., Negri, F., Wang, C., and Wang, Z. 2005. Fully conjugated tri(perylene bisimides): An approach to the construction of *n*-type graphene nanoribbons. *J. Am. Chem. Soc.* 130: 17970–17976.

Qian, H., Wang, Z., Yue, W., and Zhu, D. 2007. Exceptional coupling of tetrachloroperylene bisimide: Combination of Ullmann reaction and C-H transformation. *J. Am. Chem. Soc.* 129: 10664–10665.

Qu, H., and Chi, C. 2010. A stable heptacene derivative substituted with electron-deficient trifluoromethylphenyl and triisopropylsilylethynyl groups. *Org. Lett.* 12: 3360–3363.

Qu, J., Kohl, C., Pottek, M., and Müllen, K. 2004. Ionic perylenetetracarboxdiimides: Highly fluorescent and water-soluble dyes for biolabeling. *Angew. Chem. Int. Ed.* 43: 1528–1531.

Qu, J., Pschirer, N. G., Koenemann, M., Müllen, K., and Avlasevic, Y. 2008. Heptarylene-and octarylenetetracarboximides and preparation thereof. US2010/0072438.

Quante, H., and Müllen, K. 1995. Quaterrylenebis(dicarboximides). *Angew. Chem. Int. Ed. Engl.* 34: 1323–13225.

Rao, K. V., and George, S. J. 2010. Synthesis and controllable self-assembly of a novel coronene bisimide amphiphile. *Org. Lett.* 12: 2656–2659.

Roberson, L. B. Kowalik, J., Tolbert, L. M., et al. 2005. Pentacene disproportionation during sublimation for field-effect transistors. *J. Am. Chem. Soc.* 127: 3069–3075.

Rohr, U., Kohl, C., Müllen, K., van de Craatsb A., and Warman, J. 2001. Liquid crystalline coronene derivatives. *J. Mater. Chem.* 11: 1789–1799.

Rohr, U., Schlichting, P., and Böhm, A., et al. 1998. Liquid crystalline coronene derivatives with extraordinary fluorescence properties. *Angew. Chem. Int. Ed.* 37: 1434–1437.

Saïdi-Besbes, S., Grelet, É., and Bock, H. 2006. Soluble and liquid-crystalline ovalenes. *Angew. Chem. Int. Ed.* 45: 1783–1786.

Scholl, R., Seer, C., and Weitzenböck, R. 1910. Perylen, ein hoch kondensierter aromatischer kohlenwasserstoff $C_{20}H_{12}$. *Chem. Ber.* 43: 2202–2209.

Staab, H. A., Nissen, A., and Ipaktschi, J. 1968. Attempted preparation of 7,8,15,16-tetradehydrodinaphtho[l,8-ab; 1,8-fs]cyclodecene. *Angew. Chem. Int. Ed. Engl.* 7: 226.

Sun, Z., Huang, K., and Wu, J. 2010. Soluble and stable zethrenebis(dicarboximide) and its quinone. *Org. Lett.* 12: 4690–4693.

Tang, C. W. 1986. Two-layer organic photovoltaic cell. *Appl. Phys. Lett.* 48: 183–185.

Tönshoff, C., and Bettinger, H. F. 2010. Photogeneration of octacene and nonacene. *Angew. Chem. Int. Ed.* 49: 4125–4128.

Umeda, R., Hibi, D., Miki, K., and Tobe, Y. 2009. Tetradehydrodinaphtho[10]annulene: A hitherto unknown dehydroannulene and a viable precursor to stable zethrene derivatives. *Org. Lett.*11: 4104–4106.

Wong, W. W. H., Ma, C.-Q., Pisula, W., et al. 2010. Self-assembling thiophene dendrimers with a hexa-*peri*-hexabenzocoronene core-synthesis, characterization and performance in bulk heterojunction solar cells. *Chem. Mater.* 22: 457–466.

Wu, J. 2007. Polycyclic aromatic compounds as materials for thin-film field-effect transistors. *Curr. Org. Chem.* 11: 1220–1240.

Wu, J., and Müllen, K. 2006. All-benzenoid polycyclic aromatic hydrocarbons: Synthesis, self-assembly and applications in organic electronics. In *Carbon-rich compounds*, ed. Michael M. Haley and Rik R. Tykwinski, 90–139. Weinheim: Wiley-VCH.

Wu, J., Pisula, W., and Müllen, K. 2007. Graphenes as potential material for electronics. *Chem. Rev.* 107: 718–747.

Wu, T.-C., Chen, C.-H., Hibi, D., et al. 2010. Synthesis, structure, and photophysical properties of dibenzo-[de,mn]naphthacenes. *Angew. Chem. Int. Ed.* 49: 7059–7062.

Yao, J., Chi, C., Wu, J., and Loh, K. 2009. Bisanthracene bis(dicarboxylic imide)s as soluble and stable NIR dyes. *Chem. Eur. J.* 15: 9299–9302.

Yin, J., Qu, H., Zhang, K., et al. 2009. Electron-deficient triphenylene and trinaphthylene carboximides. *Org. Lett.* 14: 3028–3031.

Yin, J., Zhang, K., Jiao, C., Li, J., et al. 2010. Synthesis of functionalized tetracene dicarboxylic imides. *Tetrahedron Lett.* 51: 6313–6315.

Zhan, X., Tan, Z., Domercq, B., et al. 2007. A high-mobility electron-transport polymer with broad absorption and its use in field-effect transistors and all-polymer solar cells. *J. Am. Chem. Soc.* 129: 7246–7247.

Zhang, K., Huang, K., Li, J., et al. 2009. A soluble and stable quinoidal bisanthene with NIR absorption and amphoteric redox behavior. *Org. Lett.* 11: 4854–4857.

Zhang, X., Jiang, X., Luo, J., et al. 2010a. A cruciform 6,6′-dipentacenyl: Synthesis, solid-state packing and applications in thin-film transistors. *Chem. Eur. J.* 16: 464–468.

Zhang, X., Li, J., Qu, H., Chi, C., and Wu, J. 2010b. Fused bispentacenequinone and its unexpected Michael addition. *Org. Lett.* 12: 3946–3949.

Zhao, Y., Ren, A.-M., Feng, J.-K., and Sun, C.-C. 2008. Theoretical study of one-photon and two-photon absorption properties of perylene tetracarboxylic derivatives. *J. Chem. Phys.* 129: 014301.

Zhen, Y., Wang, C., and Wang, Z. 2010. Tetrachloro-tetra(perylene bisimides): An approach towards n-type graphene nanoribbons. *Chem. Commun.* 46: 1926–1928.

7 Graphene-Reinforced Ceramic and Metal Matrix Composites

Debrupa Lahiri and Arvind Agarwal

CONTENTS

This chapter deals with graphene-reinforced metals and ceramic nanocomposites, their synthesis techniques, and potential applications. A classification of composite preparation techniques has been made based on the mechanism. Future scope and potential of these nanocomposites is also discussed.

7.1 INTRODUCTION

Graphene is one of the most attractive materials being widely researched in recent times. The increased popularity of graphene originates from its wondrous properties, which include, but are not limited to, excellent electron mobility through its atom-thick sp^2 bonded 2-D structure, its high current density, excellent mechanical properties, optical transmittance, and thermal conductivity [1–5]. Table 7.1 presents the salient properties of graphene. These intrinsic properties of graphene have found potential in myriad applications, for example, electronic devices, field emitters, batteries, solar cells, electronic displays, sensors, thermally and electrically conducting composites, and structural composites with enhanced mechanical properties. Graphene-based composites have thus generated enormous interest. In addition to the excellent properties of graphene, another aspect of its popularity is its easy availability. As has been commented by Professor Kotov, "When carbon fibers just won't do, but nanotubes are too expensive, where can cost-conscious materials scientists go to find a practical conductive composite? The answer could lie with graphene sheets" [6].

TABLE 7.1
Important Physical and Mechanical Properties of Graphene

Property	Graphene	Reference
Electron Mobility	$15000\ cm^2V^{-1}s^{-1}$	[1]
Resistivity	$10^{-6}\ \Omega$-cm	[1]
Thermal Conductivity	$4.84–5.3 \times 10^3\ Wm^{-1}K^{-1}$	[4]
Coefficient of Thermal Expansion	$-6 \times 10^{-6}/K$	[5]
Elastic Modulus	0.5–1 TPa	[2]
Tensile Strength	130 GPa	[4]
Transmittance	>95% for 2 nm thick film >70% for 10 nm thick film	[2]

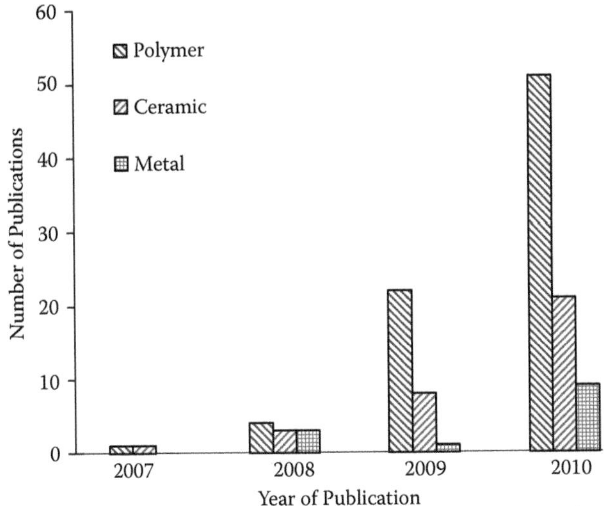

FIGURE 7.1 Publications on graphene-reinforced nanocomposites classified based on the matrix material (metal/ceramic/polymer) and year of publication [Source: www.scopus.com].

Figure 7.1 presents a chronological trend of publications on graphene–polymer, graphene–metal, and graphene–ceramic composites that clearly shows the growing research interest in the field. Closer observation of Figure 7.1 reveals increasing interest for graphene-containing composites of all three types; that is, polymer, ceramic, and metal matrices. But the number of studies is larger for polymer-based composites than for the other two categories. This trend is similar to the initial period of development in carbon nanotube (CNT)-reinforced composites [7–8]. The reason for this trend could be the easy fabrication route of polymer-based composites as they do not generally involve high temperature and pressure. Thus, it is easier to retain the homogeneous dispersion and structural integrity of graphene. On the other hand, the successful processing of metal– and ceramic–graphene composites poses lots of challenges. A couple of detailed review articles have also been published on polymer–graphene composites with comprehensive discussions on their properties and potential applications [4,9].

Polymer-based composites are not suitable for applications requiring higher temperature and strength. This is where metal- or ceramic-based composites become important. Taking into account the reviews already available on polymer–graphene systems, and also considering the potential of metal– and ceramic–graphene composites, the focus of this chapter is graphene-reinforced metal- and ceramic-based composite systems. The composite fabrication processes are classified based on the synthesis mechanism. The proposed applications of metal- and ceramic-based graphene composites, studied to date are in electronic devices, sensors, or similar fields. Since most of these studies are for nonstructural applications, the interface bonding between graphene and the matrix for effective load transfer is not given much importance. However, considering the excellent elastic modulus and tensile strength of graphene (see Table 7.1), the mechanical properties of graphene-reinforced composites could be another interesting area of research to explore their potential for structural applications. To date there is only one publication on ceramic-based composites that has attempted to evaluate mechanical properties of graphene reinforcement [10]. No study exists on metal–graphene composites at the time of this writing. It must be noted that the effect of graphene addition on the mechanical properties of polymers has been addressed [4,9–11].

Carbon nanotubes (CNTs) have been actively researched in the last 15 years as a potential reinforcement for composites. A brief section in this chapter includes a comparison of CNT and graphene as reinforcements used in some ceramic and polymer matrix composite systems [11–15]. Scientific predictions about the effectiveness of graphene reinforcement with respect to CNTs have also been made. A brief discussion is presented on the potential processing routes for synthesizing graphene-reinforced metal and ceramic composites at macroscale for structural applications. In summary, the future of graphene-based nanocomposites has been touched upon with a mention of challenges and areas requiring attention. All the studies available on metal– and ceramic–graphene composites are summarized in Tables 7.2 and 7.3, for the benefit of readers and future researchers.

7.2 CERAMIC-GRAPHENE COMPOSITES

The studies on ceramic–graphene composites have explored a large variety of ceramics including SiO_2 [16], TiO_2 [17–21], NiO [22–24], Co_3O_4 [24–26], ZnO [22,27–29], Al_2O_3 [13,30–31], Cu_2O [32], SnO_2 [14,33–34], MnO_2 [35–37], and other compounds, including Si_3N_4 [15], Si–O–C [38], and $LiFePO_4$ [39]. The proposed applications of these composites are mainly in electronic devices for field emitters [29], Li-ion batteries [14,25,32–35,38–40], solar cells [21], super capacitors [26,28,36,41], and photo catalysts [12,19–20,27]. Graphene is mostly used as reinforcement for the ceramic matrix to enhance its performance in electronic applications. But there are few studies in which graphene has been used as the major element or the matrix of the composite with ceramic particles as the second phase [26,36]. The following subsections focus on the major processing techniques and applications of ceramic–graphene composite systems.

7.2.1 Processing Techniques

Several processing techniques have been adopted for fabricating ceramic–graphene composites with the main aim of having uniform distribution of ceramic and graphene phases. Like CNTs, graphene sheets tend to agglomerate and form clusters. Thus, their uniform distribution in the matrix remains a challenge during composite fabrication. The applications of these composites proposed so far are mainly for electronic devices and not in structural components. Thus, the interfacial bonding between ceramic and graphene leading to strengthening has not been investigated to a significant extent. The following subsections present the classification of major processing techniques to fabricate ceramic–graphene composites.

TABLE 7.2
Summary of Studies of Ceramic–Graphene Composites

Composition	Processing Technique	Mechanical Properties	Other Properties	Potential Applications	Reference
SiO_2–Graphene (6.6 wt. %)–composite film of 20–30 nm thickness	GO exfoliated in water/ethanol; Adding tetramethyl orthosilicate in dispersion to form sol; spin coated on borosilicate glass or silicon; exposed in saturated vapor of hydrazine monohydrate to chemically reduce GO to graphene sheet; thermally treated at 400°C to form consolidated silica films.	NA	Electrical conductivity: 3 wt.% GO: 8×10^{-4} S/cm 11 wt.5 GO: 0.45 S/cm Transmittance: 0 wt.% GO: 0.986 11 wt.% GO 0.94–0.96	Transparent electrical conductor for solar reflective windshield, self-cleaning window, electrostatic charge–dissipating coating, solar cell, sensor devices, etc.	[16]
TiO_2–Graphene	TiO_2 colloidal suspension prepared by adding titanium isopropoxide in ethanol and stirring vigorously; graphite oxide powder is mixed in the suspension; UV-induced photocatalytic reduction of graphene oxide to graphene is carried out by UV irradiation of the suspension while passing nitrogen through it.	NA	Electrical resistivity: Before reduction of GO: 233 KΩ After reduction of GO for 2 h: 30.5 KΩ	Optoelectronic and energy conversion devices	[17]
Ni/NiO–Graphene and ZnO–Graphene composites	Colloidal dispersion of Ni-Zn; anionic clay and a separate dispersion of GO is added in drops in decarbonated water to mix; acid leaching in HCl for 24 h; heat treated for 1 h (30–600°C). 1. Acid leaching (HCl) to form Graphene–ZnO composite with Ni and NiO nanoparticles 2. Base leaching (NaOH) for Graphene–NiO and Graphene–Ni composites	NA	NA	NA	[22]

Continued

TABLE 7.2 (continued)
Summary of Studies of Ceramic–Graphene Composites

Composition	Processing Technique	Mechanical Properties	Other Properties	Potential Applications	Reference
ZnO–Graphene	GO made by Hummer process; reduced using hydrazine monohydrate; pasted on indium tin oxide glass (ITO) substrate; ultrasonic spray pyrolysis at a frequency; 1.65 MHz, 420°C, 5 min.	NA	Specific capacitance ZnO-Graphene (ITO): 11.3 F/g ZnO-(ITO): 0.7 F/g Graphene- (ITO): 3 F/g Promising reversible charge/discharge ability with typical supercapacitor behavior.	Supercapacitor	[28]
Si_3N_4, Al_2O_3, Y_2O_3–Graphene (3 wt. %) ***Compared with CNT-reinforced composites*	Composite power prepared by attritor milling in distilled water; uniaxial dry pressing, 220 MPa; gas pressure sintering, 1700°C, 2 MPa, nitrogen.	NA	Electrical conductivity Si_3N_4 – Graphene – insulator – overloaded (10 M Ω measurement limit Si_3N_4 – MWCNT – 2.95 S/m Si_3N_4 – SWNT – 15.27 S/m	NA	[15]
ZnO nanowire–Graphene Sheet Composite	ZnO nanowire grown on Si substrate by vapor phase deposition; Ni nanoparticle is coated on ZnO nanowire by magnetron sputtering; graphene sheet grown on coated ZnO nanowire in radio frequency PECVD.	NA	Turn on field: ZnO-Graphene: 1.3 V/µm ZnO: 2.5 V/µm Field enhancement factor (β) ZnO–graphene: 1.5×10^4 ZnO: 7.4×10^3	Field Emitters	[29]
TiO_2–Graphene Sheet	Composite synthesized using tetrabutyl titanate and graphene sheet by sol-gel method.	NA	Photocatalytic activity – hydrogen evolution rate by water photo-splitting TiO_2–graphene: 8.6 µmol.h^{-1} TiO_2: 4.5 µmol.h^{-1}		[19]

SiOC–Graphene Nanosheet (4-25 wt.%)	Composite fabricated by dispersing GO powder into polysiloxane (precursor liquid for SiOC) followed by cross linking and pyrolysis; the composite powder is mixed with acetylene black powder and polyfluortetraethylene and ethanol to make slurry; the slurry is spurred on stainless steel surface and rolled to thin disc for fitting in coin type battery.	NA	Initial discharge capacity SiOC–Graphene (25 wt.%): 1141 $mAhg^{-1}$ SiOC: 656 $mAhg^{-1}$ Graphene: 540 $mAhg^{-1}$ Reversible discharge capacity after 20 cycles SiOC–Graphene (25 wt.%): 357 $mAhg^{-1}$ SiOC: 148 $mAhg^{-1}$ Graphene: 350 $mAhg^{-1}$	Li-ion battery anode	[38]
TiO_2–reduced graphene oxide	TiO_2–GO composite prepared by hydrolysis of TiF_4 at 60°C for 24 h in presence of aqueous dispersion of GO; for reduction of GO, hydrazine hydrate is added to TiO_2-GO suspended in DI water; stirred at 100°C for 24 h.	NA	NA	NA	[18]
Cu_2O–Graphene	GO and copper acetate dispersed in ethanol; sonicated, centrifuged, and washed and dried; the mixture is mixed in ethylene glycol, sonicated, and heated up to 160°C for 2 h; centrifuged, washed, and dried; Cu_2O particles form in situ from absorbed copper acetate on GO and reduces GO to graphene.	NA	Tested for Li-ion battery anode No comparison for properties Low specific capacity compared to other anode materials	Li-ion battery anode, hydrogen production, solar energy, catalysis	[32]
Al_2O_3–Graphene (5 vol. %)	Alumina and natural graphite powder ball milled in ethanol (250 rpm, 30 h) ; spark plasma sintering, 1400°C, 3 min soaking, 60 MPa, heating rate; 80–100°C/min, vacuum.	NA	NA	NA	[31]

Continued

TABLE 7.2 (continued)
Summary of Studies of Ceramic–Graphene Composites

Composition	Processing Technique	Mechanical Properties	Other Properties	Potential Applications	Reference
SiOC–Graphene	Polysilane and polystyrene (1:1 weight ratio) dissolved in toluene; pyrolized at 1000°C in Ar for 1 h.	NA	First lithiation and delithiation capacities are 867 and 608 $mAhg^{-1}$, respectively. First Coulombic efficiency: 70%	Li-ion battery anode material	[40]
TiO_2–Graphene Sheet	Graphene sheet suspension synthesized by chemical reduction of exfoliated GO; for TiO_2–graphene composite colloid titanium butoxide is added to graphene sheet suspension; composite film on ITO glass is prepared by electrophoretic deposition.	NA	Electrical resistivity TiO_2–Graphene: $3.6 \pm 1.1 \times 10^2$ Ω.cm TiO_2: 2.1 ± 0.9 X 10^5 Ω.cm For photovoltaic cell Power efficiency TiO_2–Graphene: 1.68% TiO_2: 0.32%	Dye sensitized solar cells	[21]
TiO_2 (P25)–Graphene	TiO_2 (P25) and GO suspension in distilled water and ethanol were treated at 120°C for 3 h to reduce GO and simultaneously deposit TiO_2 on carbon substrate.	NA	Degradation of methyl blue in ~ 1 h TiO_2–Graphene: 85% TiO_2–CNT: 70% TiO_2: 25% Better light absorption range and charge separation and transportation.	Photocatalyst for purification of water and air	[12]
ZnO–Graphene	GO and zinc acetylacetonate dispersed in hydrazine hydrate; heated at 180°C for 16 h in autoclave; product isolated by centrifuging, washed, dried.	NA	Could improve UV-vis absorption of ZnO.	Pollutant decomposition by photocatalysis	[27]

Al_2O_3–Graphene nanosheet	Expanded graphite–Al_2O_3 powder ball milled for 30 h with N-methyl-pyrollidone as media; spark plasma sintered, heating rate 140°C/min, 1300°C, 3 min Dwell, 60 MPa, vacuum.	NA	Electrical conductivity increases with graphene content and also linearly with temperature for same graphene content.	Conductive ceramics	[13]
SnO_2–Graphene (2.4 wt. %)	GO and $SnCl_2.2H_2O$ added in aqueous HCl (35 wt.%) medium; stirred for 2 h; the product obtained by centrifuging, washing, redispersing, and spray drying; spray dried powders annealed at 400°C in Ar for 2 h.	NA	Capacity after 30 cycles SnO_2: 264mAh/g (31%) SnO_2–Graphene: 840 mAh/g (86%)	Li-ion battery anode materials	[33]
MnO_2–Graphene (32 & 80 wt.%)	Redox reaction; graphene in water suspension; $KMnO_4$ added; stirred; heated in microwave (2450 MHz, 700 W) for 5 min; deposit washed in distilled water and alcohol; dried at 100°C, 12 h in vacuum.	NA	Specific capacitance MnO_2: 103 Fg^{-1} Graphene: 104 Fg^{-1} MnO_2–Graphene (32%): 228 Fg^{-1} Capacitance retention ratio at 100 mVs^{-1} MnO_2: 33% MnO_2–Graphene (32%): 88% Capacitance retention ratio is very high at a wide range of scan rates. Capacitance decreases only 4.6% of initial value after 15,000 cycles.		[36]
δ-MnO_2–Graphene	Redox reaction; graphene in water suspension; $KMnO_4$ added; stirred; heated in microwave (2450 MHz, 700 W) for 5 min; deposit washed in distilled water and alcohol; dried at 100°C, 12 h in vacuum.	NA	Ni ion adsorption capacity MnO_2: 30.63 $mg.g^{-1}$ Graphene: 3 $mg.g^{-1}$ MnO_2–Graphene: 46.55 $mg.g^{-1}$ Endothermic reaction; Spontaneous adsorption process and can be reused for 5 times with 91% recovery rate.	Removal of Ni ion from waste water	[37]

Continued

TABLE 7.2 (continued)
Summary of Studies of Ceramic–Graphene Composites

Composition	Processing Technique	Mechanical Properties	Other Properties	Potential Applications	Reference
Co_3O_4–Graphene	GO is suspended in ultrapure water and ammonia; cobalt phthalocyanine molecules are added by ultrasonication for 3 h; hydrazine solution is added at 40°C and stirred for 12 h; precipitate is collected and heated at 400°C in air to form Co_3O_4–Graphene composite.	NA	Reversible capacity – After first cycle Pure Co_3O_4: 900 $mAhg^{-1}$ Commercial Co_3O_4: 791 $mAhg^{-1}$ Graphene: 372 $MAhg^{-1}$ Co_3O_4–Graphene: 754 $mAhg^{-1}$ Reversible capacity – After 20 cycles Co_3O_4: 388 $mAhg^{-1}$ Co_3O_4–Graphene: 760 $mAhg^{-1}$	Anode material for Li-ion battery	[25]
ZnO–Graphene SnO_2–Graphene	Graphene is mixed with ethyl cellulose and terpineol; coated on graphite sheets by screen printing; heated at 100°C for 1 h; ZnO and SnO_2 deposited onto graphene film by ultrasonic spray pyrolysis from 0.3 M zinc/tin acetate aqueous solution, at 430°C for 5 min with carrier gas (air) flow rate of 2 ml/min.	NA	Specific capacitance Graphene: 61.7 F/g ZnO–Graphene: 42.7 F/g SnO_2–Graphene: 38.9 F/g Charge transfer resistance Graphene: 2.5 Ω ZnO–Graphene: 0.6 Ω SnO_2–Graphene: 1.5 Ω Power density Graphene: 2.5 kW/kg ZnO–Graphene: 4.8 kW/kg SnO_2–Graphene: 3.9 kW/kg	Supercapacitor	[41]

Co_3O_4–Graphene (75.6 wt.%)	GO is dispersed in water; cobalt nitrate hexahydrate and urea are added by magnetic stirring; heated in microwave oven (2450 MHz, 700 W) for 10 min; precipitate is filtered and washed with water and absolute alcohol; dried at 100°C for 12 h in vacuum; calcined in a muffle furnace at 320°C for 1 h in air.	NA	Specific capacitance Graphene: 169.3 F/g Co_3O_4–Graphene: 243.2 F/g Excellent long cycle life with 95.6% specific capacitance retained after 2000 cycle tests.	Supercapacitor	[26]
$LiFePO_4$–Graphene	Graphene is suspended in DI water; aqueous solution of $(NH_4)_2Fe(SO_4)_2 6H_2O$ and $NH_4H_2PO_4$ is added; LiOH is added with blowing nitrogen gas; deposit is washed with DI and separated by centrifuging; deposit put in pellets; heated at 120°C for 5 h and at 700°C for 18 h; grinding the resulting product to obtain the composite.	NA	Specific capacity $LiFePO_4$: 113 $mAhg^{-1}$ $LiFePO_4$–Graphene: 160 $mAhg^{-1}$ After 80 cycles, 97% of initial capacity is retained in the composite structure.	Lithium secondary batteries	[39]
TiO_2–Graphene (1, 5 & 10 wt. %)	GO is dissolved in ethanol; sodium borohydride is added to reduce GO to graphene; graphene is washed with DI water and dried; graphene is dispersed in ethanol; tetrabutyl titanate is added; acetic acid glacial and DI water is added; obtained sol is dried at 80°C for 10 h; the precursor is annealed in air/nitrogen at 450°C for 2 h.	NA	Hydrogen evolution rate TiO_2: 4.5 $\mu mol.h^{-1}$ TiO_2–Graphene: 8.6 $\mu mol.h^{-1}$	Hydrogen evolution from water photocatalytic splitting	[20]

Continued

TABLE 7.2 (continued)
Summary of Studies of Ceramic–Graphene Composites

Composition	Processing Technique	Mechanical Properties	Other Properties	Potential Applications	Reference
SnO_2–Graphene (<40 wt.%)	$SnCl_2$. $2H_2O$ was dissolved in HCl and GO dispersed in DI water and both the solutions are mixed with vigorous stirring; the product is separated by centrifuging and washed in distilled water; the precipitate is dried at 80°C overnight; reduction of GO to graphene is done by treating at 300°C for 2 h in Ar.	NA	Specific capacity after first cycle SnO_2: 838 $mAhg^{-1}$ Graphene: 462 $mAhg^{-1}$ SnO_2–MWCNT: 720 $mAhg^{-1}$ SnO_2–Graphene: 786 $mAhg^{-1}$ Specific capacity after 50 cycles SnO_2: 137 $mAhg^{-1}$ Graphene: 197 $mAhg^{-1}$ SnO_2–MWCNT: 440 $mAhg^{-1}$ SnO_2–Graphene: 558 $mAhg^{-1}$ Capacity retention after 50 cycles SnO_2: 18% Graphene: 43% SnO_2–MWCNT: 60% SnO_2–Graphene: 71%	Anode materials for Li-ion batteries	[14]
NiO–Graphene	GO is dispersed in DI water; ammonia is added to adjust pH to 10; nickel chloride and hydrazine hydrate is added with stirring; the product is isolated by centrifuging and washed with water and ethanol; dried in vacuum at 45°C for 24 h; annealed at 500°C for 5 h in nitrogen.	NA	NA	NA	[23]

MnO_2–Graphene (reduced GO)	GO and poly(sodium 4-styrene sulfonate) mixed into distilled water by stirring at 90°C for 5 h; $MnSO_4.H_2O$ is added by stirring for 1 h; NH_3 and H_2O_2 added in sequence; refluxed at boiling point for 6 h; hydrazine hydrate is added with constant stirring; after cooling. suspension is filtered washed with distilled filter and dried in vacuum at 100°C.	NA	MnO_2- reduced GO composite causes oxygen reduction, but only RGO cannot. The reduction of oxygen by MnO_2 –reduced GO is a diffusion controlled process.	Alkaline fuel cells, metal/air batteries	[35]
SnO_2–Graphene	Graphene nanosheets are produced by arc discharge evaporation of graphite in NH_3; He atmosphere; NaOH is added to aqueous solution of $SnCl_4$ under stirring; resulting hydrosol is mixed with graphene dispersed in ethyl alcohol H_2SO_4 is added; the precipitate is separated using centrifuging; heated at 400°C for 2 h in Ar.	NA	Initial capacity: Graphene: 339 $mAhg^{-1}$ SnO_2–Graphene: 673 $mAhg^{-1}$	Anode material for Li-ion batteries	[34]
Al_2O_3–Graphene (2 wt.%)	GO is exfoliated in water; gradually dripped in alumina suspension in water with mechanical stirring; reduced by hydrazine monohydrate at 60°C for 24 h; spark plasma sintering at 1300°C for 3 min, 50 MPa in Ar atmosphere.	Fracture Toughness Al_2O_3 – 3.4 $MPa.m^{0.5}$ Al_2O_3 – Graphene – 5.21 $MPa.m^{0.5}$	Electrical conductivity Al_2O_3–Graphene: 131 $S.m^{-1}$ (13 orders of magnitude higher than pure Al_2O_3)	NA	[30]

Continued

TABLE 7.2 (continued)
Summary of Studies of Ceramic–Graphene Composites

Composition	Processing Technique	Mechanical Properties	Other Properties	Potential Applications	Reference
CoO/Co_3O_4/NiO–Graphene	Cetyltrimethylammonium (CTA)-intercalated GO is prepared by stirring GO in an aqueous solution of cetyltrimethyl ammonium bromide for 3 days; the solid product is washed with water to pH ~7 and dried at 65°C; colloidal dispersion of GO-CTA by sonicating in butanol; monolayer colloidal dispersion of Ni-OH-DS/Co-OH-DS by sonication in butanol for 30 min; the colloidal dispersions of GO-CTA and M-OH-DS are mixed and sonicated for 30 min; slid product (GO-intercalated a-hydroxides) are washed with water and dried; subjected to isothermal heating in air/nitrogen at 300°C for 2 h.	NA	NA	NA	[24]

TABLE 7.3
Summary of Studies of Metal–Graphene Composites

Composition	Processing Technique	Mechanical Properties	Other Properties	Potential Applications	Reference
Pt–Graphene	Chloroplatinic acid hexahydrate is added to aqueous dispersion graphene; methanol is added; sodium carbonate is added to adjust the pH to 7; kept at 80°C for 90 min with constant stirring; few drops of diluted sulfuric acid solution are added to precipitate Pt–Graphene; precipitate rinsed with water and methanol; dried at 70°C for 15 min.	NA	Capacitance Graphene: 14 F/g Pt–Graphene: 269 F/g	Supercapacitor, fuel cell	[52]
Au/Pd/Pt–Graphene	GO is dispersed in water; ethylene glycol and metal precursor ($HAuCl_4.3H_2O/K_2PdCl_4/K_2PtCl_4$) water solution are added by magnetic stirring; heated to 100°C for 6 h; separated from solution by centrifuging and washed with DI water; dried in vacuum at 60°C for 12 h.	NA	Acts as catalyst for methanol oxidation.	Methanol fuel cell	[46]
Pt–Graphene (55 wt.%)	GO is dispersed in water; H_2PtCl_6 is added under stirring; pH is adjusted to 10 with NaOH; $NaBH_4$ is added slowly; collected after washing with DI water and ethanol; vacuum dried at 40°C.	NA	Electrochemically active surface area Pt–Vulcan: 30.1 m^2g^{-1} Pt–Graphene: 44.6 m^2g^{-1} Current density for methanol electro-oxidation Pt–Vulcan: 101.2 $mA.mg^{-1}$ Pt–Graphene: 199.6 $mA.mg^{-1}$	Catalyst carrier in electrocatalysis and fuel cells	[51]

Continued

TABLE 7.3 (continued)
Summary of Studies of Metal–Graphene Composites

Composition	Processing Technique	Mechanical Properties	Other Properties	Potential Applications	Reference
Au–Graphene	NA (mostly MD simulation)	NA	DNA gets adsorbed on Au-Graphene surface much faster than only Au surface	Biosensors, biodevices and DNA sequencing	[54]
Co –Graphene	GO is suspended in ultrapure water and ammonia; cobalt phthalocyanine molecules are added by ultrasonication for 3 h; hydrazine solution is added at 40°C and stirred for 12 h; precipitate is collected and pyrolized at 800°C in Ar.	NA	NA	Anode material for Li-ion battery	[25]
Au–Graphene	Graphene sheets are dispersed in distilled water; cast on glassy carbon (GC) electrode surface and dried to form a film; Au is electrodeposited on the graphene nanosheet (on GC) through electrodeposition from aqueous solution of $HAuCl_4$ precursor.	NA	Au–Graphene composite shows higher peaks in voltammograms of electrocatalytic reduction of oxygen and glucose oxidation than only Au and only graphene.	Fuel cell and bioelectroanalytical chemistry	[50]

Si–Graphene (50 wt. %)	Mixing of nanosized Si and graphene in mortar.	NA	Capacity – first cycle Si: 3026 $mAhg^{-1}$ Si–Graphene: 2158 $mAhg^{-1}$ Capacity after 30 cycles Si: 346 $mAhg^{-1}$ Si–Graphene: 1168 $mAhg^{-1}$ Coulombic efficiency—first cycle Si: 58% Si–Graphene: 73% Coulombic efficiency after 30 cycles Si–Graphene: 93%	Anode material for Li-ion battery	[53]
Au–Graphene-Horseradish peroxidase (HRP)-Chitosan (CS)-	Graphene and HRP are co-immobilized on CS; casted on a glassy carbon (GC) electrode surface; Au is electrodeposited on it.	NA	The largest cathodic peak is observed for composite with graphene. At same concentration of H_2O_2 – Au/Graphene/HRP/CS /GC electrode shows 3 times higher current response than Au/HRP/CS/GC electrode and 20 times higher than HRP/CS/GC electrode.	H_2O_2 biosensor	[49]
Au–Graphene	GO is suspended in water and exfoliated; casted on glassy carbon (GC) electrode; dried under infrared lamp; GO is reduced to ER-GO by electrochemical reduction method; the ER-GO/GC electrode is immersed in $HAuCl_4$ – for absorption of $AuCl_4^-$; reduction is performed by potentiodynamic electrodeposition.	NA	Electroactive surface area ER-GO/GC electrode: 0.109 cm^2 Au/ER-GO/GC electrode: 0.152 cm^2 Electron transfer resistance reduces in Au-ER-GO/GC electrode more than Au/ER-GO electrode.	Electrochemical detection for DNA specific sequence	[47]

Continued

TABLE 7.3 (continued)
Summary of Studies of Metal–Graphene Composites

Composition	Processing Technique	Mechanical Properties	Other Properties	Potential Applications	Reference
Au–Graphene	Polyvinylpyrrolidone (PVP)-protected graphene is stabilized in an aqueous solution of chitosan; the solution is dropped on glassy carbon (GC) electrode surface and dried; cyclic volammetry (CV) scan of the electrode in $KAuCl_4$ solution; rinsing the electrode in water.	NA	Apparent electrode area GC: 0.0164 cm^2 Graphene-Chitosan-GC: 0.149 cm^2 Au-Graphene-Chitosan-GC: 0.737 cm^2 The sensitivity of Au–Graphene composite electrode at lower Hg concentration is 708.3 µA/ppb	Environmental monitoring – detection of mercury	[48]

7.2.1.1 Reduction of GO (Graphene Oxide) to Graphene after Chemical Mixing with Ceramic

Most of the studies on ceramic–graphene composites have followed this route using graphene oxide (GO) as the starting material [12,14,16–18,22–27,30,32–33,35,38,40]. This process is also referred to as the *in-situ composite-forming process* because the reduction of GO to graphene and the formation of metallic oxide from metallic salt take place simultaneously. Graphene oxide is used as the starting material because it is easy to disperse GO uniformly in distilled water, which also helps in maintaining its dispersion in the composite at the final stage. Graphene-containing composites with SiO_2 [16], TiO_2 [12,7–18], NiO [22–23], SiOC [38–40], Cu_2O [32], ZnO [27,41], Co_3O_4 [24–26], SnO_2 [14,33,41], MnO_2 [35], and Al_2O_3 [30] have been processed through this route.

Most of the research studies in this processing category use graphene oxide produced by the Hummers method [42], which is exfoliated in water to produce a stable suspension of individual graphene oxide sheets. The metallic salt is then added to the suspension to form a sol in many of the cases. The final stage of composite formation can be of three different types.

1. The metallic salt is in-situ oxidized by GO to form a composite of graphene and metallic oxide [14,32–33].
2. GO is chemically reduced (mostly using hydrazine monohydrate) to graphene sheets in the composite [16,18,23,25,27,30,35].
3. The composite is pyrolized or decomposed at high temperature to transform GO into graphene [12,22,24,26,38].

The composite is sometimes consolidated by thermal treatment at higher temperature [16]. Watcharotone et al. published the first ever work on ceramic–graphene composite film in 2007 [16]. Tetramethyl orthosilicate was used to form a sol with GO followed by reduction using hydrazine monohydrate and thermal consolidation at 400°C in inert atmosphere. Watcharotone et al. also observed in-plane orientation of graphene oxide sheets in composite, which was attributed to the shear stress generated due to the application of centrifugal force during spin coating of the composite sol on the substrate [16]. But no direct evidence or quantification of the fraction of GO being converted to graphene was presented in this study. Indirect evidence of GO being reduced to graphene in composite is established by an increasing electrical conductivity of the composite, which was nonconductive with GO [16]. In contrast, Ji et al. [38] have shown the presence of graphene in the pyrolized mixture of GO and polysiloxane-forming SiOC–graphene composite as shown in Figure 7.2 [38]. The x-ray diffraction (XRD) pattern of SiOC–graphene nanosheet composite shows absence of a GO peak and a broad graphite (002) peak. The broadness of the peak could be due to the nanosize of graphene in the composite [38]. In another effort, Williams et al. performed UV-irradiation assisted photocatalytic reduction of GO in TiO_2 suspension to form the graphene–TiO_2 composite [17]. The color of the suspension and reduced electrical resistance of the composite are presented as proof of GO being reduced to graphene in the composite [17]. TiO_2–graphene composite creation

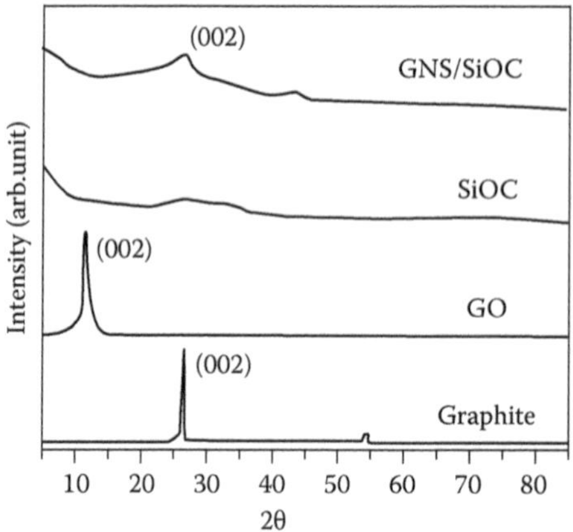

FIGURE 7.2 XRD pattern of graphite, GO, silicon oxycarbide, and graphene nanosheet–silicon oxycarbide composite. (From Ji, F., Feng, J. M., Su, D., Wen, Y. Y., Feng, Y., and Hou, F. 2009. Electrochemical performance of graphene nanosheets and ceramic composites as anodes for lithium batteries. *J. Mater. Chem.* 19: 9063–9067. With permission.)

through GO route is also prepared by thermal decomposition of GO at the final stage [12]. This study has shown uniformly distributed TiO_2 nanoparticles on two-dimensional graphene in transmission electron microscopy (TEM) images (Figure 7.3). Graphene vibration peak in Fourier transform infrared (FTIR) spectra of the TiO_2–graphene composite structure is presented in Figure 7.3 [12]. Microwave-assisted heating is reported to be successful for synthesizing Co_3O_4–graphene composite

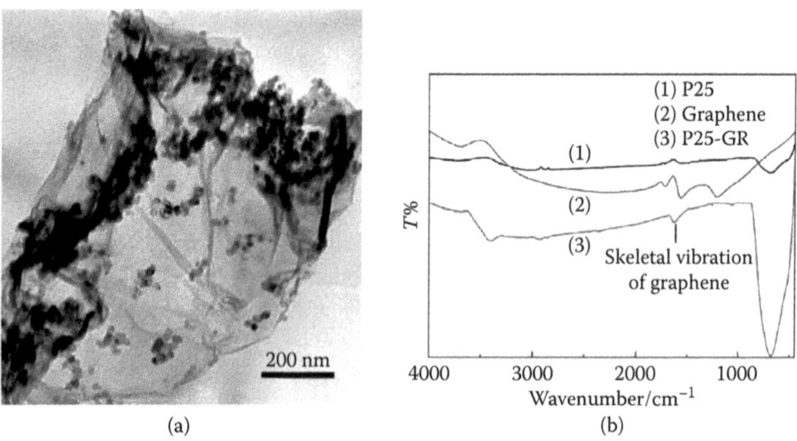

FIGURE 7.3 (a) TEM image of TiO_2–graphene composite with TiO_2 particles on the surface of graphene and concentrating along the wrinkles. (b) Fourier transform infrared (FTIR) spectra of (1) TiO_2, (2) graphene obtained by hydrothermal reduction, and (3) TiO_2-GR in the range of 4000–450 cm^{-1}. (From Zhang, H., Lu, X., Li, Y., Wang, Y., and Li, J. 2010. P25-Graphene composite as a high performance photocatalyst. *ACS Nano* 4: 380–386. With permission.)

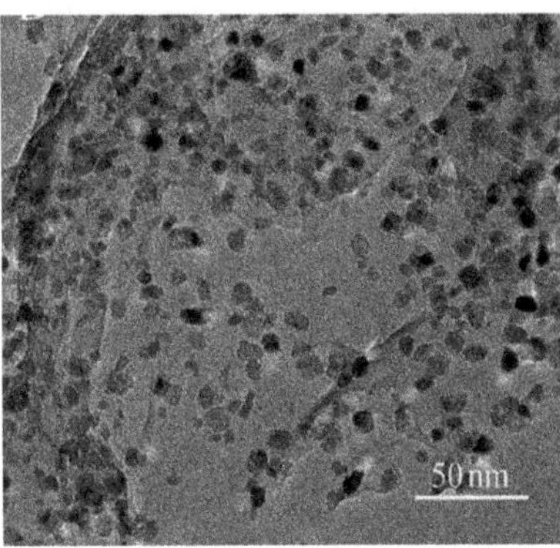

FIGURE 7.4 TEM image of MnO_2-reduced GO composite with MnO_2 particles uniformly dispersed on graphene sheet. (From Qian, Y., Lu, S., and Gao, F. 2010. Synthesis of manganese dioxide/reduced graphene oxide composites with excellent electrocatalytic activity toward reduction of oxygen. *Mater. Lett.* 65: 56–58. With permission.)

from a suspension of GO and cobalt-based salt (cobalt nitrate hexahydrate) [26]. Nethravathi et al. have followed the route of thermal treatment and acid leaching of GO-intercalated nickel–zinc hydroxysalt (anionic clay) to obtain a ceramic–graphene composite [22]. However, a very indirect proof of GO reduction to graphene in composite is provided in this study by presence of NiO and ZnO and a comparatively sharpened graphite peak in the XRD pattern [22]. In another study by the same group on synthesis of NiO– and Co_3O_4–graphene composites, using an identical processing route, the presence of graphene is proven by XRD pattern and IFTR spectrum [24]. But Nethravathi et al., in their later publication, mentioned incomplete reduction of GO with poorly ordered graphene production [24]. Wu et al. have proven the presence of graphene, by absence of a GO peak in the XRD pattern and C–O–H-related peaks of GO in FTIR of a ZnO–graphene composite, fabricated by hydrazine hydrate reduction of GO [27]. A similar processing route, applied for preparing Co_3O_4–graphene, NiO–graphene, and MnO_2–graphene composites, also reveals homogeneously dispersed oxide particles in graphene sheets (Figure 7.4) [23,25,35]. Ji et al. have also observed the absence of C–O and C–OH peaks in FTIR spectrum when GO is reduced to graphene in the NiO–graphene composite [23]. Al_2O_3–graphene composite powder has been prepared by reducing GO through hydrazine monohydrate in composite suspension. The purpose was to obtain homogeneous dispersion of the reinforcement phase in the powder stage [30]. Cu_2O–graphene and SnO_2–graphene nanocomposites are fabricated by in situ reduction of GO to form Cu_2O/SnO_2. Apart from the functionality of the ceramic in specific application, the purpose of these composites is to prevent aggregation of graphene sheets by forming Cu_2O/SnO_2 phase(s) in between [14,32–33].

This processing route is successful in homogeneous dispersion of the constituent in the composite structure. Graphene oxide has a negative charge when dispersed in water. Hence, the electrostatic repulsion between negatively charged graphene oxide sheets generates a stable aqueous suspension [43–44], which in turn helps in maintaining the degree of graphene dispersion in the composite. But the main challenge associated with this process route is the graphene content. The elemental analysis of graphene produced by chemical reduction of GO establishes the presence of oxygen in significant amounts [43]. Theoretical calculations also suggest that it is difficult to reduce GO completely to graphene through this route [43]. It would be even more difficult to reduce GO when it is in the composite. Thus, further research into the modification of this process is required to ensure the purity of graphene in this class of composite.

7.2.1.2 Chemical Mixing of Graphene with Ceramic

Comparatively fewer studies have used graphene sheets instead of GO as a starting material to mix with metallic oxides and salts to fabricate the composite. The graphenes used in these studies are either produced by chemical reduction of GO [19–20,28,36–37,39] or arc-discharge evaporation of graphite [34]. The composites of metallic oxides that have been fabricated by this route include TiO_2 [19], ZnO [28], MnO_2 [36–37], SnO_2 [34], and $LiFePO_4$ [39].

ZnO is deposited using an ultrasonic spray pyrolysis technique onto a graphene film to fabricate a ZnO–graphene composite film to be used in supercapacitors. The authors claimed that graphene was covered with densely packed ZnO grains of irregular shape [28]. The purity of the graphene obtained by reduction of GO is not mentioned in this study [28]. Zhang et al. used graphene and tetrabutyl titanate as a starting material to form TiO_2–graphene composite by sol-gel method [19–20]. MnO_2–graphene composites are fabricated by redox reaction under microwave irradiation of graphene–potassium permanganate suspension. $KMnO_4$ is reduced to MnO_2 by a sacrificial carbon substrate, which in turn is deposited on the surface of graphene sheets. However, the probability of oxidation of graphene in the proposed mechanism has not been addressed [36–37]. Figure 7.5 presents scanning electron microscope (SEM) and TEM images of MnO_2, graphene nanosheet, and the composite fabricated through this process [37]. A $LiFePO_4$–graphene composite was obtained by co-precipitation from an aqueous solution of iron-sulfate salt and LiOH with graphene suspended in it. Figure 7.6 presents the atomic force microscopy (AFM) image of a $LiFePO_4$–graphene composite surface clearly showing uniformly distributed $LiFePO_4$ particles on the graphene sheet [39]. SnO_2 is synthesized by hydrolysis of $SnCl_4$ and loaded on graphene sheets suspended in aqueous solution to form the SnO_2–graphene composite precipitate [34].

Most of the composites fabricated through this process route still use graphene synthesized through chemical reduction of GO. Thus, the ratio of graphene to GO remains questionable due to incomplete reduction. Further, graphene nanosheets have a tendency to form agglomerates, which can restrict their uniform distribution in the matrix [17]. But none of the studies have addressed this matter in greater detail. The agglomeration of graphene becomes more prominent for macroscale composite structures, where graphene is used as reinforcement in the ceramic matrix. But that is not the case for any of the studies presented here, since the majority of the studies employ graphene sheets covered with nanoceramic particles for electronic devices and similar applications. Still, agglomeration of graphene could be detrimental because it can actually form a very thick graphite-like structure, defeating the purpose of having an increased graphene surface area available for the efficiency of the targeted application.

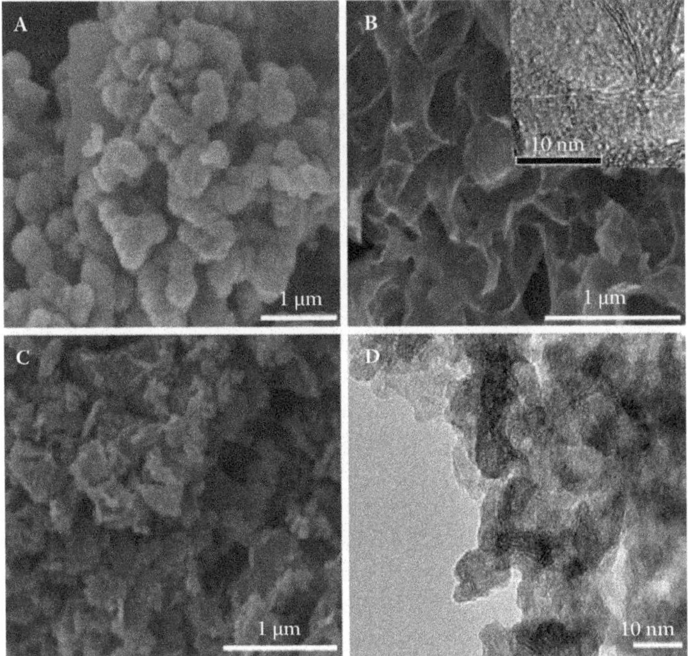

FIGURE 7.5 SEM images of (A) MnO_2, (B) graphene nanosheet, (C) MnO_2–graphene composite, and (D) TEM image of MnO_2–graphene composite. (From Ren, Y., Yan, N., Wen, Q., Fan, Z., Wei, T., Zhang, M., and Ma, J. 2010. Graphene/δ-MnO_2 composite as adsorbent for the removal of nickel ions from wastewater *Chem. Eng. J.* doi:10.1016/j.cej.2010.08.010. With permission.)

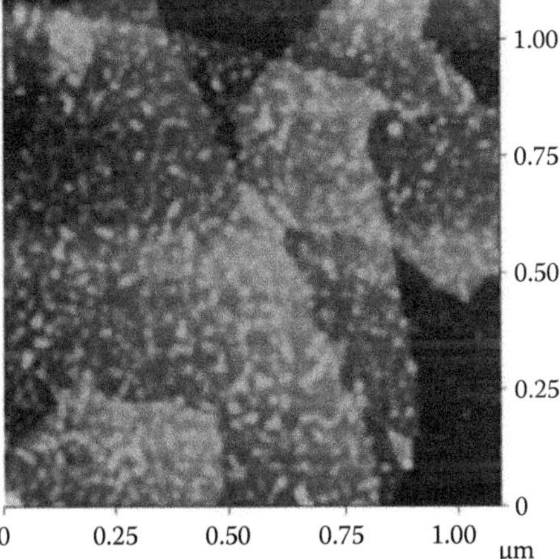

FIGURE 7.6 AFM image revealing morphology of $LiFePO_4$–graphene composites. (From Ding, Y., Jiang, Y., Xu, F., Yin, J., Ren, H., Zhou, Q., Long, Z., and Zhang, P. 2010. Preparation of nano-structured LiFePO4/graphene composites by co-precipitation method. *Electrochem. Commun.* 12: 10–13. With permission.)

7.2.1.3 Mechanical Mixing of Graphene with Ceramic

Ceramic-based composites of graphene at macroscale are mostly synthesized by the powder metallurgy route; that is, preparing the composite powder by mechanical mixing and high temperature consolidation. Few reports are available for composites in this category. Al_2O_3 [13,30–31] and Si_3N_4–Al_2O_3–Y_2O_3 mixture [15] reinforced with graphene have been processed through this route to form a composite. Natural graphite [31] or thermally exfoliated graphite [13] have also been employed as the starting material for Al_2O_3 composite. The graphite is ball milled with Al_2O_3 powder to prepare the composite powder, which is spark plasma sintered to the consolidated structure. He et al. found that the graphene thickness in the composite decreased with milling time, due to milling-induced fracture and deformation [31]. Fan et al. claimed to achieve homogeneous distribution of graphene in an Al_2O_3 matrix by ball milling. Figure 7.7 shows the graphene nanosheets semiwrapping the Al_2O_3 grains and forming a network [13]. Wang et al. have used GO dispersed in water as the starting material to mix with Al_2O_3. The GO is then reduced into graphene to achieve a better dispersion of graphene in the ceramic matrix [30]. The graphene sheets are also reported to inhibit grain growth of alumina during high-temperature consolidation process by restricting the diffusion [30–31]. The Si_3N_4–Al_2O_3–Y_2O_3 matrix composite was prepared by attritor-milled mixing constituent powders and sintering in nitrogen [15]. No observation was made on the distribution of graphene in a ceramic matrix. The effect of high-temperature exposure to graphene during composite consolidation has also not been addressed yet. This category of process route is very promising for preparing bulk graphene-reinforced composites. But ball milling can induce severe damage to graphene sheet structure, which has a deleterious effect on its reinforcing efficiency and needs to be carefully investigated.

7.2.1.4 Electrophoretic Deposition

TiO_2–graphene composite has been synthesized using electrophoretic deposition [21]. The graphene prepared by chemical reduction of GO is suspended in ethanol and reacted with titanium butoxide to attach TiO_2 particles to suspended graphite. This dispersion is then used for synthesis of the composite film on an indium tin oxide (ITO) glass substrate through electrophoretic deposition. The synthesis principle is explained schematically in Figure 7.8(a)–(e) [21]. The dispersion of graphene nanosheets in TiO_2 particles is reveled in the SEM image of the composite film (Figure 7.8[f]) [21]. This process may be a promising one as the suspension is reported to be stable over several weeks [21]. Thus, the uniform distribution of graphene in the matrix is ensured. But the thickness of the composite remains limited in electrodeposited films (< 0.5 mm) and thus may not be very suitable for fabricating bulk structures.

7.2.1.5 Chemical Vapor Growth of Graphene on Ceramic

Direct growth of graphene on ceramic particles by chemical vapor deposition (CVD) is a promising method to ensure uniform dispersion of graphene in the composite. However, the ceramic surface has to be coated with metallic catalyst particles

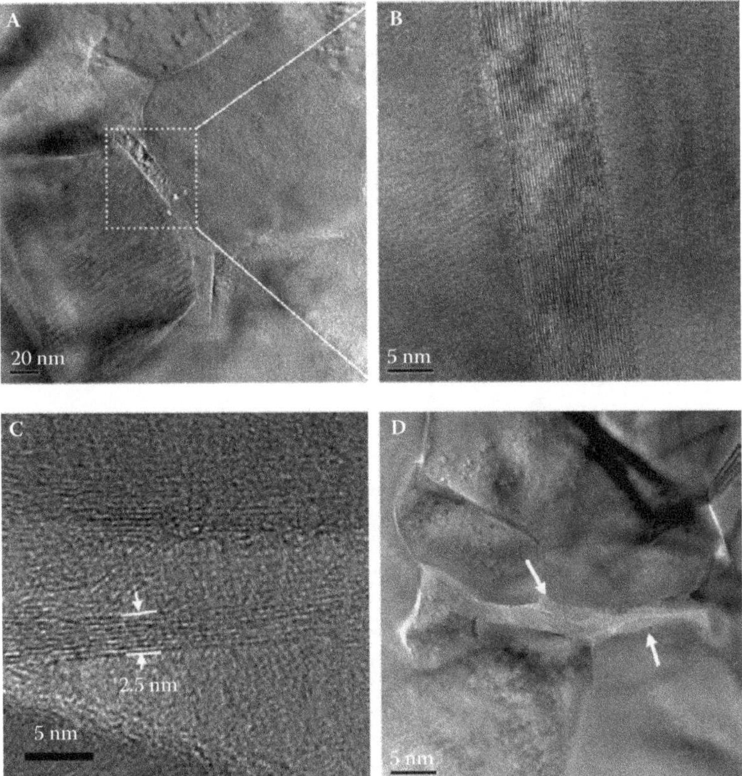

FIGURE 7.7 TEM and HRTEM images of Al_2O_3–graphene composite showing (A) graphene nanosheet surrounding Al_2O_3 nanoparticles, (B) a magnified image of graphene nanosheet with a thickness of about 10 nm in (A), (C) graphene with a thickness of 2.5 nm, and (D) overlap of graphene nanosheet between Al_2O_3 nanoparticles. (From Fan, Y., Wang, L., Li, J., Sun, S., Chen, F., Chen, L., and Jiang, W. 2010. Preparation and electrical properties of graphene nanosheet/Al_2O_3 composites. *Carbon* 48: 1743–1749. With permission.)

(Ni or Cu) to grow graphene on it. To date, only one study has reported using plasma-enhanced chemical vapor deposition (PECVD) to grow graphene directly on ZnO nanowires. Figure 7.9 shows ZnO nanowires uniformly coated by CVD-grown graphene [29]. More studies via this processing route are required to explore its full potential in processing ceramic–graphene composites.

7.2.1.6 Ultrasonic Spray Pyrolysis

Another novel technique for fabricating ceramic–graphene composite adopted is ultrasonic spray pyrolysis [41]. The graphene produced through chemical reduction of GO is screen printed on the graphite substrate. Subsequently, SnO_2 /ZnO particles are deposited using ultrasonic spray pyrolysis to fabricate the composite film. This process is also limited to generating very thin composite films and needs more research to establish its efficiency [41].

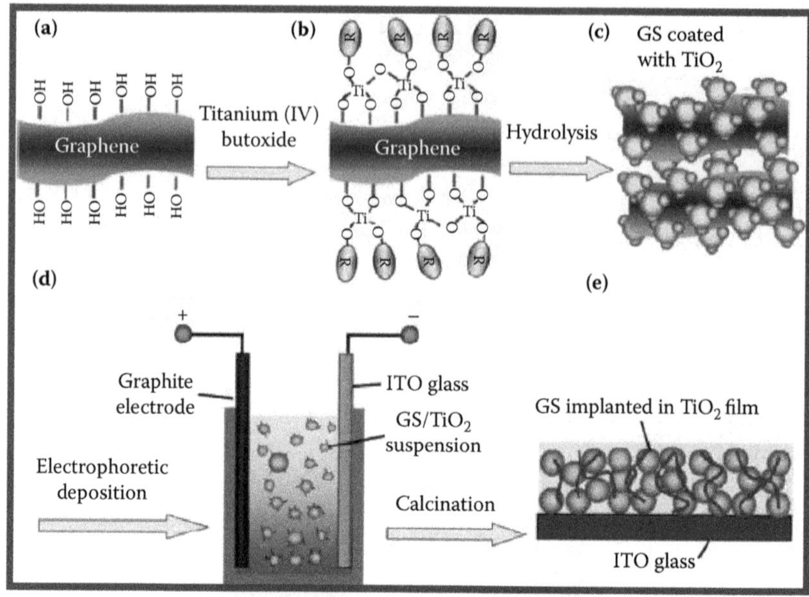

(f)

FIGURE 7.8 Schematic flowchart of in situ incorporation of graphene nanosheet in nanostructured TiO_2 film showing (a) graphene prepared by chemical exfoliation with residual oxygen-containing functional groups, such as hydroxyl; (b) titanium(IV) butoxide grafted on the reduced graphene surfaces by chemisorptions; (c) graphene coated with TiO_2 colloids after hydrolysis; (d) electrophoretic deposition process used to prepare GS/TiO_2 composite films; (e) structure of the graphene–TiO_2 composite film after calcinations; and (f) SEM image of the TiO_2–graphene composite film. (From Tang, Y. B., Lee, C. S., Xu, J., Liu, Z. T., Chen, Z. H., He, Z., Cao, Y. L., Yuan, G., Song, H., Chen, L., Luo, L., Cheng, H. M., Zhang, W. J., Bello, I., and Lee, S. T. 2010. Incorporation of graphenes in nanostructured TiO2 films via molecular grafting for dye-sensitized solar cell application. *ACS Nano* 4: 3482–3488. With permission.)

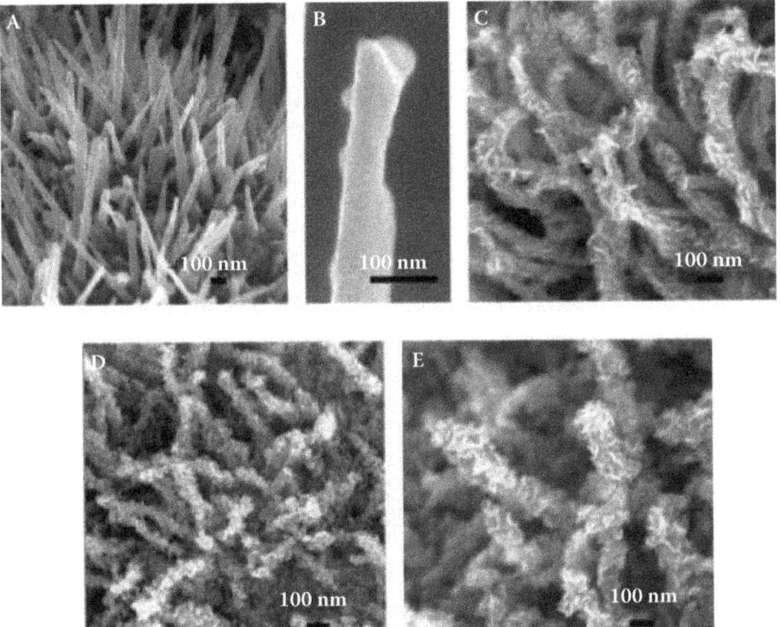

FIGURE 7.9 SEM image of (A) as-grown ZnO nanowires, (B) ZnO nanowire coated with Ni for 5 s, graphene-grown nanowires for (C) 2 min, (D) 5 min, and (E) 10 min after 3 s of coating. (From Zheng, W. T., Ho, Y. M., Tian, H. W., Wen, M., Wen, M., Qi, J. L., and Li, Y. A. 2009. Field emission from a composite of graphene sheets and ZnO nanowires. *J. Phys. Chem. C* 113: 9164–9168. With permission.)

7.2.2 Properties and Applications of Ceramic-Graphene Composites

Ceramic–graphene composites reported to date are mainly proposed and studied for their electronic properties and manifested toward applications in electronic devices. The following subsection presents a brief summary of the major applications of these composites and the effect of graphene addition on the related properties.

7.2.2.1 Electrical Conductivity

The high electron mobility and electrical conductivity of graphene (Table 7.1) have made it an excellent candidate for ceramic-based composites to improve their conductivity in electronic devices. Only a few of the studies have directly reported the electrical conductivity of the composites having SiO_2 [16], TiO_2 [17,21], and Al_2O_3 [13,30] as the ceramic phase. The electrical conductivity of SiO_2–graphene composite increases significantly with 3 orders of magnitude, from 8×10^{-4} Scm^{-1} to 0.45 Scm^{-1}, with graphene content being increased from 3.9 to 11 wt.% [16]. William et al. studied electrical resistance of a TiO_2–graphene system prepared by UV radiation-assisted reduction of GO to graphene in the composite. They found that the high electrical resistance of 233 KΩ in the composite with GO changed to 30.5 KΩ after reduction to graphene [17]. Similar electrical resistivity of 36 KΩ for

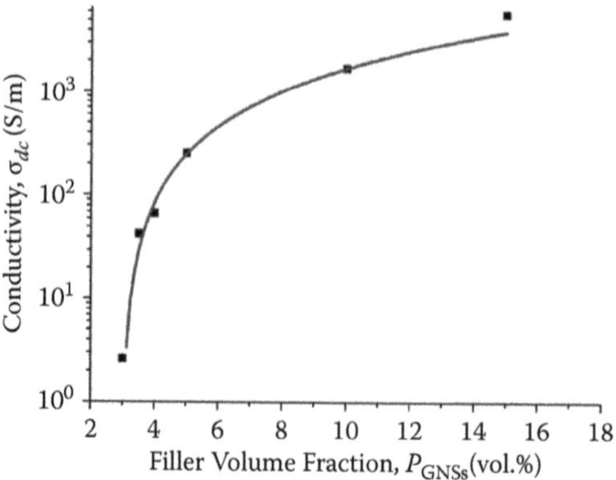

FIGURE 7.10 Electrical conductivities of Al_2O_3–graphene nanosheet composites as a function of filler graphene volume fraction. (From Fan, Y., Wang, L., Li, J., Sun, S., Chen, F., Chen, L., and Jiang, W. 2010. Preparation and electrical properties of graphene nanosheet/Al_2O_3 composites. *Carbon* 48: 1743–1749. With permission.)

a TiO_2–graphene composite is also reported by Tang et al. displaying an impressive improvement over TiO_2 having a resistance of 210 KΩ [21]. Incorporation of graphene in Al_2O_3 is also reported to increase the electrical conductivity of the composite by 170% [13,30]. Figure 7.10 shows the increase in electrical conductivity of Al_2O_3–graphene composites with varying graphene content [13]. The incorporation of graphene in ceramics is capable of forming numerous conductive paths in the composite, which results in easy flow of electric current in ceramics, which are otherwise poor electrical conductors.

7.2.2.2 Supercapacitors

Carbonaceous materials and metal oxides are the two most suitable materials for use in supercapacitor electrodes. In carbon-based materials, the energy is stored due to formation of an electric double layer [28,41]. Graphene is a suitable candidate from the carbon family for this purpose due to its high mobility, chemical and mechanical stability, large surface area, and excellent electrochemical performance over a wide range of voltage scan rates [45]. Metal oxides further improve the capacitance of carbon-based supercapacitors by pseudo capacitive reaction [28,36,41]. It has been interesting to study the performance of metal oxide–graphene composites as capacitor electrodes [26,28,36,41]. Studies on ZnO–SnO_2–MnO_2– Co_3O_4–graphene composites have shown increased specific capacitance as compared to just graphene or the metal oxide (Figure 7.11) [26,28,36,41]. Lu et al. have shown better performance of a ZnO–graphene composite compared to SnO_2–graphene [41]. The increase in the capacitance in metal oxide–graphene composites has been attributed to several factors [26,28,36]. First, the well-dispersed metal oxide particles inhibit stacking and aggregation of graphene nanosheets, providing more surface area for double-layer capacitance. Second, graphene provides a highly conductive network for electron transport during the charge–discharge process. Third, improved interfacial contact between metal oxide particles and graphene and increased surface area improve the

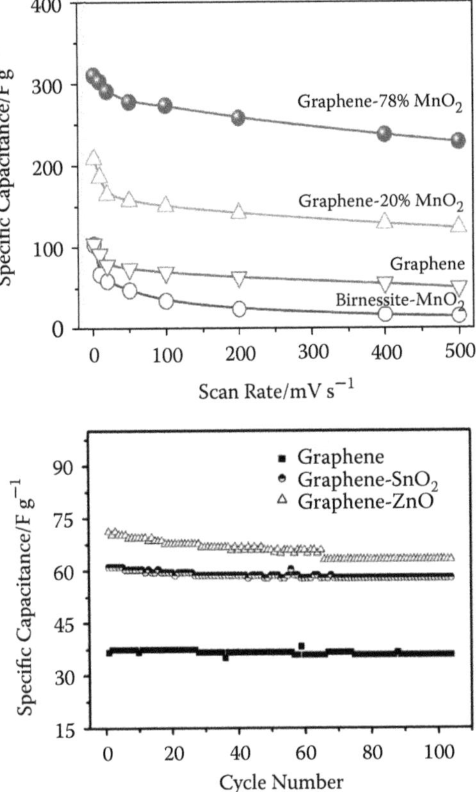

FIGURE 7.11 Specific capacitance of different composite electrodes in comparison to graphene and metal oxide electrodes. (Top from Yan, J., Fan, Z., Wei, T., Qian, W., Zhang, M., and Wei, F. 2010. Fast and reversible surface redox reaction of graphene–MnO_2 composites as supercapacitor electrodes. *Carbon*, 48: 3825–3833; bottom from Lu, T., Zhang, Y., Li, H., Pan, L., Li, Y., and Sun, Z. 2010. Electrochemical behaviors of graphene–ZnO and graphene–SnO_2 composite films for supercapacitors. *Electrochimica Acta* 55: 4170–4173. With permission.)

accessibility of the composite to electrolytes and shorten the diffusion and migration pathway. Thus, the metal oxide–graphene composites are very promising for supercapacitor applications.

7.2.2.3 Field Emitters

Graphene is of great interest in field emitters due to its suitable electronic properties and large surface area with sharp edges, which could be modified to a sharp, needle-type morphology. ZnO-based nanostructures are also considered for the same application due to their high surface-to-volume ratio, thermal stability, and oxidation resistance. Zheng et al. proposed a ZnO–graphene composite nanostructure as a field emitter to tap the advantage of both ZnO and graphene simultaneously, in field-emission devices [29]. Coating of a graphene sheet on Ni-coated ZnO nanowire creates a pyramid-like structure with a much sharper apex than bare ZnO nanowire. As a result, ZnO–graphene composite structure shows a lower turn-on field and much higher field enhancement factor than ZnO, which are characteristics of better field emitters [29].

7.2.2.4 Li-Ion Battery Anode

The anode of a lithium-ion battery is the component that determines its capacity, stability, and performance. Graphite is commercially used as an anode in lithium-ion batteries due to its structure, which provides enough rechargeable sites for Li-ion. But it is limited in its theoretical capacity (372 $mAhg^{-1}$) [33,38]. Thus other materials, such as Si–O–C or SnO_2, have gained more interest due to their higher capacities [33,38,40]. But the reaction during the lithiation–delithiation process generates volume change and stress in these materials causing poor cyclability and large irreversible capacity. A composite of these materials with a carbon nanostructure is often found to be a suitable solution. In this regard, graphene shows a strong potential due to its high surface area, high room temperature carrier mobility, and enhanced discharging capacity due to an increased number of electrochemically active sites [33,38]. Quite a few studies have explored the possibility of a ceramic–graphene composite to be used in the anodes of Li-ion batteries [14,25,32–35,38–39]. SiOC–graphene composites have shown higher initial and reversible discharge capacities than SiOC or graphene alone [38]. SnO_2–graphene and Co_3O_4–graphene composites also show higher capacity and stability for the number of charging/discharging cycles, over metal oxide or graphene alone (Figure 7.12) [14,25,33–34]. Increases in capacities with increasing graphene content in SiOC clearly indicate the contribution from graphene. The main attribute of graphene in performance as anodes in lithium-ion batteries comes from several factors, that is, (1) graphene acts as structural

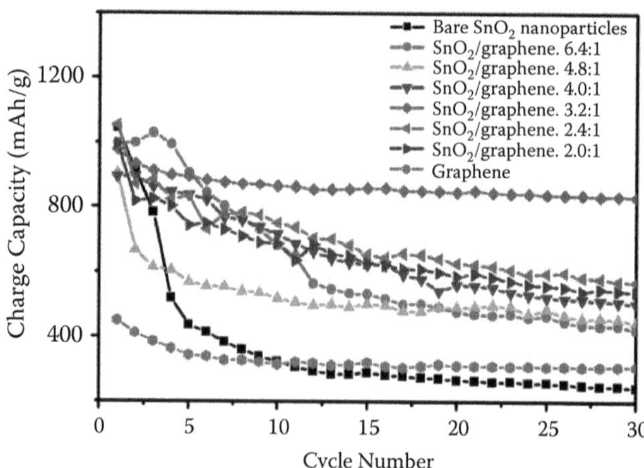

FIGURE 7.12 Cycling performances for bare graphene, SnO_2 nanoparticles, and SnO_2–graphene composite with different ratios. The current density was 67 mA/g. (From Wang, X., Zhou, X., Yao, K., Zhang, J., and Liu, Z. 2011. A SnO_2/graphene composite as a high stability electrode for lithium ion batteries. *Carbon* 49: 133–139. With permission.)

buffer to large volume change in most of the oxides, (2) graphene sheets reduce the contact resistance of oxide particles and provide high electrical conductivity, (3) the inorganic network provided by graphene to oxide particles and the synergistic effect of the interface between two phases that contribute to high electrochemical performance, (4) electroactive component for Li storage in both oxide and graphene contributes to the total capacity of the anode, and (5) formation of a solid electrolyte on the surface of oxide nanoparticles contributes to better cyclic performance of anodes [14,25,33,38–39]. Other composite systems that have shown promise in this application are Cu_2O–graphene [46], $LiFePO_4$–graphene [39], MnO_2–graphene [35].

7.2.2.5 Photocatalytic Activity

The environmental benefits of photodegradation of organic pollutants are attracting a lot of attention from the research community. TiO_2 is a very important material for this purpose due to its long-term thermodynamic stability, strong oxidizing power, and nontoxicity [12,19]. TiO_2–carbon composites have been studied as potential candidates for purification of water and air, but they suffer from a problem involving a decrease in adsorptivity during photodegradation and a lack of reproducibility due to preparation and treatment [12]. Graphene is more suitable for this purpose because of its unique electronic properties and its large surface area for the reaction, and high transparency due to atomic level thickness. Studies on TiO_2–graphene systems have shown increased photocatalytic activities, mostly in terms of the hydrogen evolution rate under UV irradiation [12,19–20]. UV radiation generates photo-induced electron-hole pairs in TiO_2. This photogenerated electron causes splitting of pollutants to evolve hydrogen. The electron and hole recombine very quickly in the presence of TiO_2, which is prevented effectively in the presence of graphene, keeping the electron available for photocatalysis [12,19–20]. Figure 7.13 shows increased photodegradation of methyl blue under UV light by TiO–graphene, compared to only TiO_2

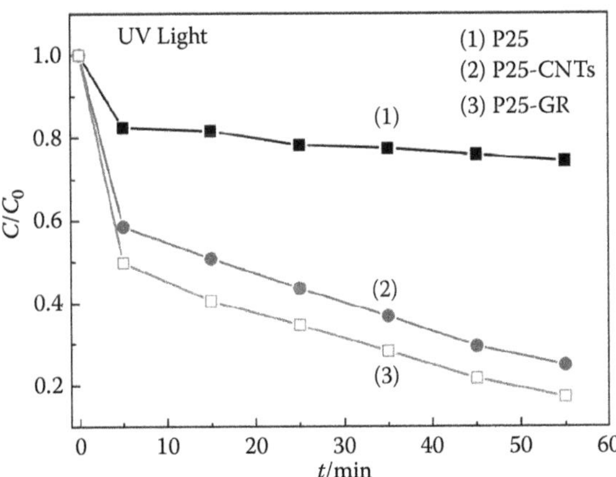

FIGURE 7.13 Photodegradation of methylene blue under UV light over (1) TiO_2, (2) TiO_2–CNTs, and (3) TiO_2–graphene photocatalysts, respectively. (From Zhang, H., Lu, X., Li, Y., Wang, Y., and Li, J. 2010. P25-Graphene composite as a high performance photocatalyst. *ACS Nano* 4: 380–386. With permission.)

and TiO_2–CNT composites. The y-axis of the plot represents normalized temporal concentration changes (C/C_0) [12]. The photocatalytic activity of TiO_2–graphene composite is also taken advantage of in dye-sensitized solar cells (DSSC) [21]. Graphene incorporation improves the conduction pathway of TiO_2, and exhibits excellent photoconversion and photocatalytic properties. The only study in this category reports a decrease in the resistivity of TiO_2–graphene by two orders of magnitude as compared to TiO_2, which helps to enhance cell efficiency up to 320% [21]. In another study on graphene-coated ZnO nanocomposite, Wu et al. commented on the potential application of this composite as a photocatalyst for pollutant decomposition [27].

7.2.2.6 Other Applications

Graphene-incorporated ceramic nanocomposites are also explored for Ni removal from water. MnO_2_graphene composite is studied as an absorbant for removing toxic Ni content from the water through a chemical sorption process. Graphene addition to MnO_2 is reported to increase the adsorption capacity and mechanical intension of the composite, which improves the Ni adsorption efficiency by 52% over MnO_2 alone [37].

7.3 METAL-GRAPHENE COMPOSITE

Research on metal–graphene composites is still in its infancy with very few studies as compared to ceramic–graphene composites. But the growing interest in this field is clearly shown in the significantly increased number of publications, as observed in Figure 7.1. The metal-based composite systems, which have explored the potential of graphene as a second phase includes gold [46–50], platinum [46,51–52], palladium [46], cobalt [25], and silicon [53]. The projected application of these composites is found in energy devices such as Li-ion batteries [53], supercapacitors [52], electrocatalysts [46,50–51], and biosensors [47–49,54]. A brief discussion of different synthesis techniques adopted for fabricating these metal–graphene composites and their expected performance in proposed applications is presented here.

7.3.1 Processing Techniques

The challenge in synthesizing metal–graphene composites remains similar to that for ceramic–graphene systems, which is poor dispersion of graphene in the matrix. An additional challenge for metal–graphene composites could be the reaction at the interface because metals are much more reactive than ceramics. But this problem has not yet been addressed as the composites studied to date are mostly physical mixtures and do not involve high temperatures during processing. The basic principles adopted for synthesizing metal–graphene composites are similar to the ones for ceramic–graphene composites and can be classified into the following categories.

7.3.1.1 Reduction of Graphene Oxide (GO) to Graphene after Chemical Mixing with Metals

This class of processing route involves the mixing of metal particles with GO to form the composite structure, after which GO is reduced to graphene. Composites

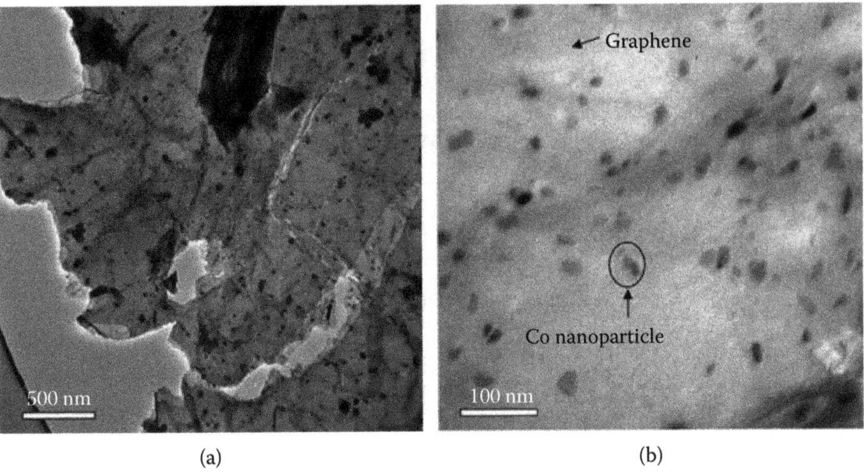

(a) (b)

FIGURE 7.14 (a) Low and (b) high-magnification TEM images of a Co–graphene composite showing the homogeneous dispersion of cobalt nanoparticles in the graphene sheets. (From Yang, S., Cui, G., Pang, S., Cao, Q., Kolb, U., Feng, X., Maier, J., and Mullen, K. 2010. Fabrication of cobalt and cobalt oxide/graphene composites: Towards high-performance anode materials for lithium ion batteries. *Chem. Sus. Chem.* 3: 236–239. With permission.)

containing gold [46], palladium [46], platinum [46,51], and cobalt [25] are fabricated through this route. Xu et al. [46] used a water–ethylene glycol system to disperse GO and metal nanoparticles. The metal nanoparticles absorbed on GO sheets play a major role in the catalytic reduction reaction of GO with ethylene glycol and form the metal–graphene composite. Homogeneous dispersion of GO in water–ethylene glycol systems confirms even distribution of metallic nanoparticles on them. On the other hand, metallic nanoparticles sitting on the graphene sheets prevent their restacking and agglomeration. Platinum–graphene composite has also been prepared by simultaneous reduction of GO and metallic salt (H_2PtCl_6) in water suspension using $NaBH_4$ as the reducing agent [51]. Metallo-organic molecules can also be used with GO to form metal–graphene composites. Cobalt phthalocyanine (CoPc), dispersed homogeneously with GO in water dispersion, is reduced with the hydrazine solution to form a graphene–CoPc aggregate. Upon pyrolysis at 800°C in argon, this aggregate transforms into a Co–graphene composite. The thickness of this composite is reported to be ~7 nm, consisting of several graphene layers, decorated with uniformly distributed Co nanoparticles, as observed in Figure 7.14 [25]. This class of fabrication route assures uniform distribution of phases in metal–graphene composite.

7.3.1.2 Chemical Mixing of Graphene with Metals

This class of processing route uses graphene and metallic salt as the starting materials. The metallic salts are reduced in dispersion and deposited on graphene sheets to form the composite. The aim of fabricating these composite structures is mainly to prevent the aggregation of graphene layers by introducing nanosized metallic particles between the layers. Si et al. deposited platinum nanoparticles on graphene sheets by the chemical reduction of H_2PtCl_6 with methanol [52]. Considering the agglomeration tendency of graphene, Gong et al. used chitosan to form a stable dispersion of graphene in water. After casting and drying this dispersion on glassy carbon electrode, a cyclic voltammetry scan of the electrode is done in $KAuCl_4$ solution to form an Au–graphene composite film on the electrode to be used as a sensor [48]. Like other metal–graphene composite process routes, there are very few reports available in this category and more research is needed for a better understanding and development of this processing route.

7.3.1.3 Mechanical Mixing of Graphene with Metals

Only one report is available on the mechanical mixing of silicon and graphene using mortar [53]. Unlike ceramic–graphene composites, silicon–graphene was not followed by any consolidation technique. This composite is just a physical mixture without any chemical reaction. Silicon nanoparticles were found inside the pores of graphene sheets for reversible Li-ion storage in batteries.

7.3.1.4 Electrodeposition

Electrodeposition of metallic nanoparticles is performed on graphene film–casted electrodes to fabricate the metal–graphene composite film. All three publications available in this category report about synthesis of a gold–graphene composite [47,49,50]. Graphene produced by different methods was dispersed in aqueous solvent and casted on a glassy carbon electrode. Au nanoparticles were deposited from a $HAuCl_4$-containing bath to fabricate the Au–graphene composite film. These composite films are mostly used as biosensors. Figure 7.15 shows a gold nanoparticle–decorated graphene sheet in a Au–graphene composite film [50].

7.3.2 Properties and Applications of Metal-Graphene Composites

The potential applications of metal–graphene composites include anodes for Li-ion batteries, supercapacitors, electrocatalysts, and biosensors.

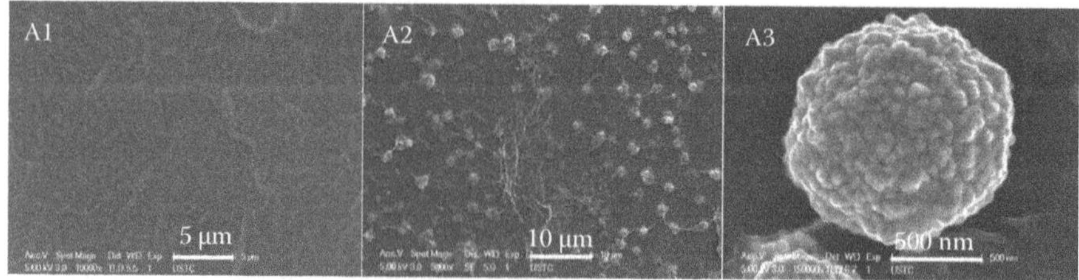

FIGURE 7.15 SEM images of the (A1) graphene film on a glassy carbon electrode surface; (A2–A3) Au nanoparticles electrochemically deposited on the surface of graphene/glassy carbon electrode. (From Hu, Y., Jin, J., Wu, P., Zhang, H., and Cai, C. 2010. Graphene–gold nanostructure composites fabricated by electrodeposition and their electrocatalytic activity toward the oxygen reduction and glucose oxidation. *Electrochimica Acta* 56: 491–500. With permission.)

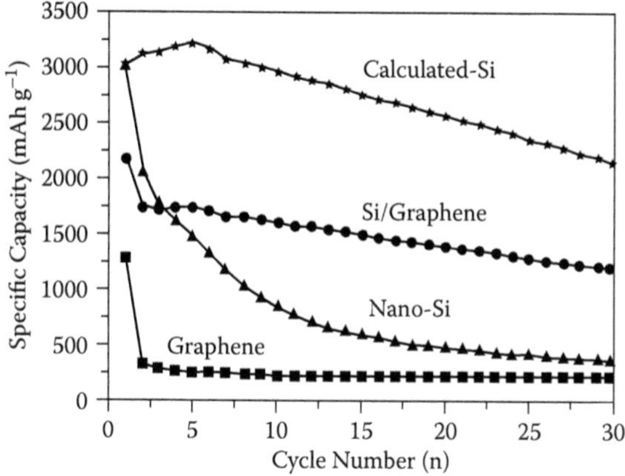

FIGURE 7.16 Cycling stability of graphene, nanosized Si, Si–graphene composite electrodes, and the calculated pure Si contribution. (From Chou, S. L., Wang, J. Z., Choucair, M., Liu, H. K., Stride, J. A., and Dou, S. X. 2010. Enhanced reversible lithium storage in a nanosize silicon/graphene composite. *Electrochem. Commun.* 12: 303–306. With permission.)

7.3.2.1 Anode for Li-Ion Battery

Graphene has been explored as a second-phase addition to silicon for Li-ion battery anode material [53]. Si is an attractive material for Li-ion batteries due to its low discharge potential and high theoretical capacity. But its large volume change associated with the charging–discharging cycle causes loss in electrical contact and capacity fading. Graphene, as an addition to silicon, is a potential solution to this problem because of its good conductivity, chemical stability, and mechanical properties. Chou et al. [53] noted 15% better Coulombic efficiency for Si–graphene composites, though the initial discharge capacity (2158 mAhg^{-1}) was lower than silicon (3026 mAhg^{-1}). But the composite retains 54% of the initial capacity after 30 cycles, whereas Si retains only 11%. Si–graphene composites generate higher specific capacity with an increasing number of cycles than Si or graphene alone (Figure 7.16) [53]. Hence, the presence of graphene in Si improves its suitability as a Li-ion anode material.

7.3.2.2 Supercapacitor

Platinum–graphene composites have been researched for supercapacitor electrode application by Si and Sumulski [52]. The composite structure shows 19 times larger capacitance than graphene alone. The platinum particles act as spacers in the composite to prevent the aggregation of graphene sheets, making it more accessible to electrolytes. Thus, the high surface area of the composite along with the excellent electrical conductivity from both graphene and platinum makes Pt–graphene a very promising material for supercapacitors.

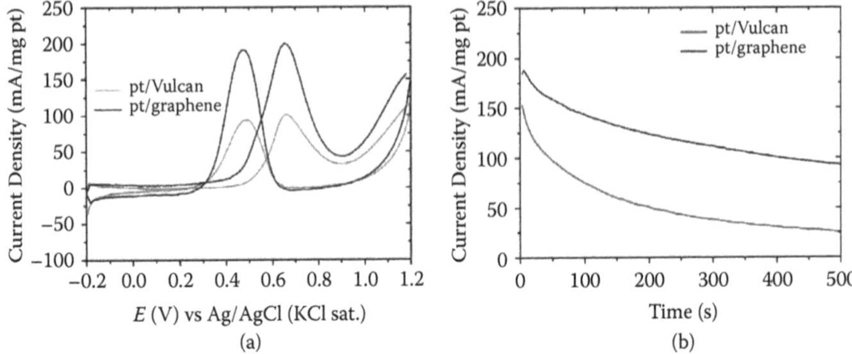

FIGURE 7.17 (a) Cyclic voltammograms of Pt–graphene and Pt–Vulcan in a nitrogen-saturated aqueous solution of 0.5 M H_2SO_4 containing 0.5 M CH_3OH at a scan rate of 50 mVs^{-1}; (b) chronoamperometric curves for Pt–graphene and Pt–Vulcan catalysts in nitrogen-saturated aqueous solution of 0.5 M H_2SO_4 containing 0.5 M CH_3OH at a fixed potential of 0.6 V vs. Ag/AgCl (KCl sat.). (From Li, Y., Tang, L., and Li, J. 2009. Preparation and electrochemical performance for methanol oxidation of pt/graphene nanocomposites. *Electrochem. Commun.* 11: 846–849. With permission.)

7.3.2.3 Electrocatalyst and Methanol Fuel Cell

Noble metal–graphene nanocomposites are very promising for application as electrocatalysts in oxidation reactions. Noble metal particles are of great interest due to their unique properties suitable for electrocatalysis in fuel cells and biosensing. Graphene has created interest in this field due to its higher surface area, excellent conductivity, and good electrocatalytic activity. Pt–graphene nanocomposites are found suitable for methanol oxidation in methanol fuel cells [46,51]. Li et al. reported higher peak current (double) and stability of Pt–graphene composites than Pt–Vulcan, which is the conventional catalyst used in methanol fuel cells (Figure 7.17) [51]. The excellent electrocatalytic performance of Au–graphene nanocomposite has also been applied for glucose oxidation [50].

7.3.2.4 Biosensor and Environmental Monitoring Sensors

Biosensors need electrical conductivity as well as good adsorption of biomolecules to induce strong substrate–molecule coupling. Metallic nanoparticles, mostly used in biosensors, possess excellent electrical conductivity, but have poor biomolecules absorption. Graphene is a very attractive option due to its excellent electrical conductivity along with its special molecular structure, which favors biomolecule adsorption through π-stacking interactions [54]. Thus, metal nanoparticle–graphene composites are suitable candidates for sensor applications. Au–graphene nanocomposites have been used for DNA detection and sequencing [47,54], hydrogen peroxide biosensors [49], and also for mercury detection in water [48]. Metal–graphene composites are found to show high accuracy detection, low detection limits, and increased sensitivity along with a wide linear range and long-term stability [48–49,54].

7.4 EFFECT OF GRAPHENE REINFORCEMENT ON THE MECHANICAL PROPERTIES OF COMPOSITES

The excellent stiffness (0.5–1 TPa) and tensile strength (130 GPa) of graphene (Table 7.1) make it a very efficient reinforcement for strengthening and stiffening of metal, ceramic, or polymer matrix composites. The two most important factors for any second-phase reinforced strengthening of composite structures are: (1) the uniform dispersion of the second phase in the matrix and (2) strong interfacial bonding of the reinforcement phase with the matrix. The uniform distribution of the second phase helps to maintain uniform properties throughout the composite. The interfacial bonding is important for effective load transfer from the matrix to the reinforcement, which is generally the stiffer and stronger phase. Ceramic– and metal–graphene composites, discussed until now, have largely addressed the uniform dispersion of the graphene, but interfacial bonding between graphene and the matrix has not been investigated in detail. The good interfacial bonding in metal- or ceramic-matrix composites requires consolidation using high temperature and/or pressure. Lee et al. [55] performed ab-initio modeling at the Al–graphene interface in a nanocomposite. Their findings show that the Al–graphene interface has substantially higher cohesive energy than typical bulk Al–graphite interface. Also, by controlling the interface characteristics, the chemical type of C–Al bonding is possible at the Al–graphene interface. Thus, it is possible to prepare composites with the desired interfacial strength by a proper choice of synthesis route.

Toughening of the spark plasma-sintered Al_2O_3 matrix using graphene reinforcement has recently been reported [30]. The fracture toughness of Al_2O_3– graphene composites is found to be 5.21 MPa.m$^{0.5}$, which is a 53% improvement over Al_2O_3. Enhancement of mechanical properties in metal– and ceramic–graphene composites are yet to be explored fully, but several studies on polymer-based composites have proven the potential of graphene reinforcement in strengthening the composite structure. Figure 7.18 presents the percentage improvement of elastic modulus in

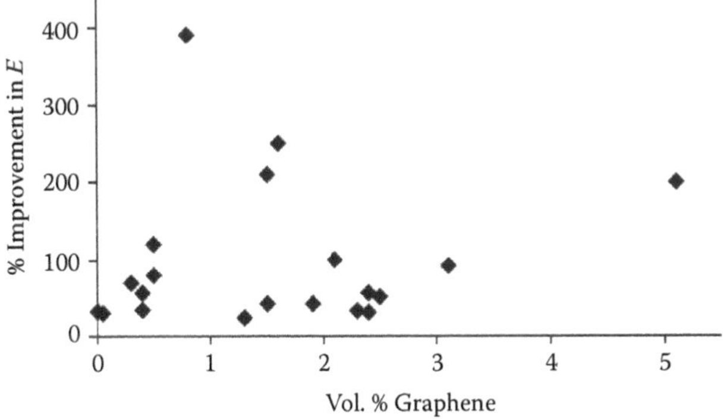

FIGURE 7.18 Improvement in elastic modulus of polymer–graphene composites as a function of graphene content.

different polymer-based composites as a function of graphene content. The reference point for calculation of the percentage improvement is the property of the polymer in the same study without graphene. The data for Figure 7.18 is collected from Kuila et al. [4] and Kim et al. [9]. The polymers studied in this context include epoxy, natural rubber, polycarbonate (PC), polymethylmethacrylate (PMMA), polyvinylidene fluoride (PVF), and polyvinyl alcohol (PVA). A very impressive strengthening is reported for PMMA–(0.01 wt.%) graphene where elastic modulus improved by 33% [10]. A similar observation was made for epoxy–(0.1 wt.%) graphene composite which showed 31% improvement in the elastic modulus [11]. Rafiee et al. attributed this improvement in modulus to the hydrogen bonding interaction of oxygen-functionalized graphene (obtained from thermally reduced GO) with the polymer and also the mechanical interlocking at the wrinkled graphene surface, which may restrict the mobility of polymer chains [11]. Tensile strength of different polymer–graphene composites is reported to increase with graphene reinforcement [4,9,11]. Elongation to fracture generally decreases for polymer–graphene composites [9]. The overall performance of graphene reinforcement in the polymer matrix would encourage its use in ceramic and metal matrixes for similar purposes.

7.5 COMPARISON OF GRAPHENE AND CARBON NANOTUBES AS REINFORCEMENT TO COMPOSITES

Another very attractive carbon nanostructure that has received immense attention as reinforcement for polymers, ceramics, and metals is carbon nanotube (CNT) [7,56–57]. Similar to graphene, the popularity of CNTs is due to its excellent Young's modulus in the range of 200–1000 GPa, tensile strength of 11–63 GPa, and electrical (1850 Scm^{-1}) and thermal (2980 W/mK) conductivities [58–61]. Thus, CNTs are the closest competitor of graphene as a second phase in composite structures for electrical, electronic, sensor, and structural applications. Knowledge about physical and functional properties projects graphene to be better suited in some fields and CNTs in others. For example, the distribution of the second phase in composites is very important for the performance of the structure. Because GO is easily dispersible in water, it is easy to disperse graphene at the GO state in the composite and then reduce it to graphene. But no such method exists to easily disperse CNTs without causing agglomeration in the composite matrix. The 2D structure of graphene with higher surface-to-volume ratio is more favorable for sensor structure and catalytic performance than CNTs [46]. On the other hand, the larger surface area of graphene could be a possible source of problems in metal matrix composites during processing at high temperatures. At high temperatures, metals try to react with the carbon nanostructure. With CNTs (multiwall CNTs), the reaction product formation is less and only on the outermost walls. But the higher surface area of graphene exposes it more to the reaction and to carbide formation. For composite processing techniques that employ high pressure, graphene could be a better option than CNTs. This is due to the hollow tubular structure of CNTs, which can collapse at application of high pressure [62–63], reducing their performance. No such problem would exist in the case of graphene due to its 2D planar structure. CNTs are better in the case of field emitters because of their tubular structure with pointed tips compared to the planar sheet-like structure of graphene with no sharp point suitable for emission. But it is difficult to predict whether graphene- or CNT-containing composites would be better for Li-ion battery applications because the larger surface area of graphene would provide more area for reaction, but the 3D structure of CNT arrays would provide more space for Li-ion storage during cycling than the 2D planar arrangement of graphene. CNTs should act as a better mechanical reinforcement due to their fiber structure, which would give the advantage of short-fiber reinforcement. But during tribological performance, graphene reinforcement may perform better. The lubrication and reduced coefficient of friction (CoF) on CNT-reinforced composites is due to the peeling off of graphene layers from the CNT surface. But application of a load that is less than a critical value does not peel off any graphene layers from the CNT surface and cannot have a positive effect on CoF reduction [62]. Thus, the effect of graphene and CNTs in the composite performance can be judged only on case-by-case basis depending on the intended application. Few studies have reported experimental results on such comparisons, as presented in the following paragraph.

A comparison of graphene- and CNT-reinforced composites in a Si_3N_4–Al_2O_3–Y_2O_3 matrix shows higher electrical conductivity in CNT-containing composites. The authors have attributed the poor electrical conductivity in graphene-reinforced composites to the size and shape of the mixed graphene grains, though no detailed discussion is provided [15]. On the contrary, Fan et al. [13] claimed higher electrical conductivity in a graphene-reinforced Al_2O_3 composite than a CNT-reinforced one, specifically at a higher second-phase content of up to 15 vol.%, which is above the percolation threshold (~3 vol.%). Upon exceeding the percolation threshold, CNTs aggregate and form bundles and contribute very little to electrical conductivity. But this is not the case with graphene sheets. Moreover, the connection between CNTs in composite is mostly point-to-point contact, which incurs high resistance as compared to area-area contact in graphene reinforcement [13]. On the other hand, a TiO_2–graphene composite shows better photocatalytic activity than TiO_2–CNT composites. The better photodegradation of dyes in the presence of graphene is due to its giant π-conjugation system in a 2D planar structure, as compared to CNTs, which causes increased adsorption of dye particles [12]. Zhang et al. [14] have reported high initial reversible capacity and improved cyclic performance of graphene–SnO_2 composite than CNT–SnO_2 composite in Li-ion battery anode material. But the properties could not be compared directly because of compositional differences in the two composite systems. The graphene-containing composite has 60 wt.% SnO_2 compared to only 25 wt.% in CNT-containing composites. Rafiee et al. [11] compared the strengthening of polymers with graphene reinforcement as compared with CNTs. The epoxy-0.1 wt.% graphene composite shows higher elastic modulus, tensile strength, fracture toughness, and fracture energy than epoxy–CNT composites. Better performance of graphene is explained in terms of stronger interfacial bonding between the polymer and graphene, which makes the frictional pull-out of graphene more unlikely. In addition, fracture toughening by the crack deflection process becomes more effective due to planar geometry and the high aspect ratio of graphene, which creates larger polymer–reinforcement interfaces, than in CNTs [11].

7.6 METAL/CERAMIC GRAPHENE COMPOSITE FABRICATION AT MACROSCALE FOR STRUCTURAL APPLICATIONS

Most of the graphene-containing ceramic and metal composites reported to date have been synthesized for electronic, electrochemical, and sensor-related applications. Thus, the requirement was limited to a microscale aggregate without much concern for structural strength. But structural application of graphene-reinforced composites requires fabrication of mechanically strong composites at macroscale length. Preparation of such composites with metal or ceramic as the matrix requires employment of high temperature-assisted and/or pressure-assisted consolidation processes. Few studies on ceramic–graphene composites have adopted such consolidation processes [13,15,30–31], whereas no such report is available on the metal matrix. The major challenges associated are: (1) the distribution of the graphene phase in the ceramic or metal composite without agglomeration and (2) control of reactivity at the interface due to exposure to high temperature and achievement of good interfacial bonding.

GO used as a starting material in composites is an efficient solution to the dispersion problem of graphene in the composite. But the real challenge lies in scaling up this wet chemistry method for bulk production of the composite. CVD growth of graphene on the ceramic or metal parts could be another attractive method to ensure homogeneous distribution of graphene in the matrix. In fact, CVD growth of graphene on thin metal sheets stacked together and consolidated by sintering or the hot/cold deformation process can fabricate laminated composites at macroscale. A similar synthesis technique could be applied for ceramic–graphene composites also. Recently Bakshi et al. [64] observed exfoliation of CNTs to graphene sheets in spark plasma sintering (SPS)-processed TaC composite structure, during application of high pressure and temperature (Figure 7.19). Exfoliation of CNTs could be a possible source of graphene in ceramic matrix composites. However, using CNTs as the starting material for graphene-reinforced composites may not be a cost-effective synthesizing route. Electrodeposition of composite after in-situ reduction of GO into graphene in an electrolytic bath is also an effective method for fabricating composites at macroscale. But electrodeposition is mainly suitable for composite coatings and free-standing thin films. Metal–CNT composites synthesized through electrodeposition are found to be restricted in thickness <200 μm [7].

The other most important factor for structural composites is the strength at the matrix–reinforcement interface. Almost no attention has been given to experimentally analyze the type and nature of the metal– and ceramic–graphene interfaces. The only ab-initio modeling study by Lee et al. barely mentions the possibility of controlling the interfacial strength of aluminum and graphene by synthesis mechanism [55]. But the high temperature reactivity of metals raises further concerns about the fate of the interface due to compound formation. The type of interface adhesion existing in ceramic–graphene composites also needs to be determined, considering the chemically inert nature of that interface. Thus, the metal–graphene and ceramic–graphene interfaces existing in different composites also need to be understood in detail.

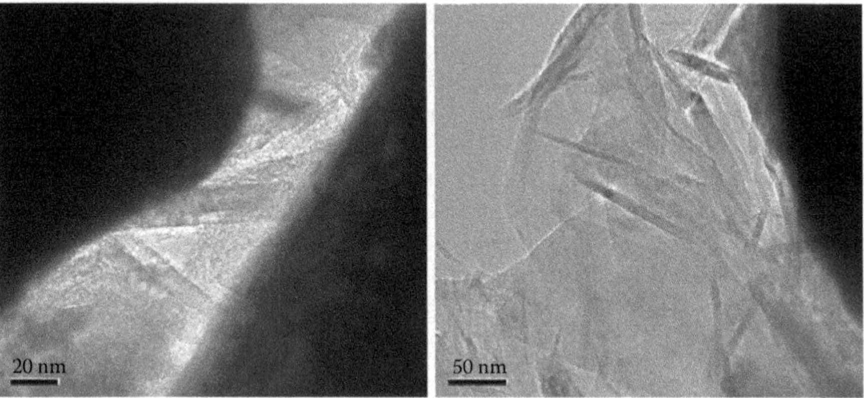

FIGURE 7.19 TEM images of SPS-processed TaC–CNT structure showing presence of carbonaceous structures between TaC grains and damaged CNT structures. (From Bakshi, S. R., Musaramthota, V., Virzi, D. A., Keshri, A. K., Lahiri, D., Singh, V., Seal, S., and Agarwal, A. 2011. Spark plasma sintered tantalum carbide–carbon nanotube composite: Effect of pressure; carbon nanotube length and dispersion technique on microstructure and mechanical properties. *Mater. Sci. Eng. A* 528: 2538–2547. With permission.)

7.7 SUMMARY

Graphene is being explored in many composite structures for its excellent ballistic transportation and mechanical properties. Although most of the studies are focused on polymer matrix composites, the interest is growing steadily and significantly for ceramic and metal matrix composites. Graphene-reinforced metal and ceramic matrix composites explored so far have been mainly for electronic, electrocatalytic, and sensor applications, which mainly employ the electrical conductivity and electrochemical sensitivity of graphene. But the excellent mechanical properties (elastic modulus, tensile strength) of graphene are yet to be taken advantage of in metal and ceramic matrix composites.

The metal and ceramic matrix composites fabricated to date are mostly physical aggregates at smaller scale and volume. Development of a processing route for fabricating structural composites with a consolidated structure is still in its infancy [13,15,30–31]. Fabrication of composite structures at macroscale might suffer from some major challenges. The agglomeration of graphene could be avoided by the GO dissolution method, but controlling the interfacial reactivity to achieve the desired interfacial bonding and strength needs a lot of research.

Apart from the electrical conductivity, electrochemical activity, and mechanical properties, there could be quite a few other attributes of graphene yet to be captured in composite structures. High thermal conductivity [4] and the very low and negative coefficient of thermal expansion [5] of graphene (see Table 7.1) make it a suitable reinforcement for composites to be used in thermal management. Moreover, the low coefficient of thermal expansion and thermal conductivity makes graphene reinforcement very suitable for electronic packaging and interconnects. Graphene shows a strain-dependent conductivity [5], which could be taken advantage of in the piezoelectric composite structure. The lubricating property of graphene is also found effective in reducing wear resistance and increasing load-bearing capacity when used in lubricant [65]. Use of graphene in composites and coatings can also act as a wear-resistance structure. Gradual release of graphene during wear action could provide required lubrication, thus reducing or abolishing the requirement for any lubricant. Graphene-reinforced metal and ceramic composites have a promising future for several applications and require more research to explore their full potential.

REFERENCES

1. Choi, W., Lahiri, I., Seelaboina, R., and Kang, Y. S. 2010. Synthesis of graphene and its applications: A review. *Critical Rev. in Solid State and Mater. Sci.* 35: 52–71.
2. Soldano, C., Mahmood, A., and Dujardin, E. 2010. Production, properties and potential of graphene. *Carbon* 48: 2127–2150.
3. Rao, C. N. R., Sood, A. K., Voggu, R., and Subrahmanyam, K. S. 2010. Some novel attributes of graphene. *J. Phys. Chem. Lett.* 1: 572–580.
4. Kuilla, T., Bhadra, S., Yao, D., Kim, N. H., Bose, S., and Lee, J. H. 2010. Recent advances in graphene based polymer composites. *Prog. Polym. Sci.* 35: 1350–1375.
5. Fuhrer, M. S., Lau, C. N., and MacDonald, A. H. 2010. Graphene: Materially better carbon. *Mater. Res. Soc. Bull.* 35: 289–295.
6. Kotov. N. A. 2006. Materials science: Carbon sheet solutions. *Nature* 442: 254–255.
7. Agarwal, A., Bakshi, S. R., and Lahiri, D. 2010. *Carbon nanotubes: Reinforced metal matrix composites.* Boca Raton, FL: CRC Press.
8. Bakshi, S. R., Lahiri, D., and Agarwal, A. 2010. Carbon nanotube reinforced metal matrix composites: A review. *Int . Mater. Rev.* 55: 41–64.
9. Kim, H., Abdala, A. A., and Macosko, C. W. 2010. Graphene/polymer nanocomposites. *Macromolecules* 43: 6515–6530.
10. Ramanathan, T., Abdala, A. A., Stankovich, S., Dikin, D. A., Herrera-Alonso, M., Piner, R. D., Adamson, D. H., Schniepp, H. C., Chen, X., Rouff, R. S., Nguyen, S. T., Aksay, I. A., Prud'homme, R. K., and Brinson, L. C. 2008. Functionalized graphene sheets for polymer nanocomposites. *Nat. Nanotechnol* 3: 327–331.
11. Rafiee, M. A., Rafiee, J., Wang, Z., Song, H., Yu, Z. Z., and Koratkar, N. 2009. Enhanced mechanical properties of nanocomposites at low graphene content. *ACSNano* 3: 3884–3890.
12. Zhang, H., Lu, X., Li, Y., Wang, Y., and Li, J. 2010. P25-Graphene composite as a high performance photocatalyst. *ACS Nano* 4: 380–386.
13. Fan, Y., Wang, L., Li, J., Sun, S., Chen, F., Chen, L., and Jiang, W. 2010. Preparation and electrical properties of graphene nanosheet/Al2O3 composites. *Carbon* 48: 1743–1749.
14. Zhang, L. S., Jiang, L. Y., Yan, H. J., Wang, W. D., Wang, W., Song, W. G., Guo, Y. G., and Wan, L. J. 2010. Mono dispersed SnO2 nanoparticles on both sides of single layer graphene sheets as anode materials in Li-ion batteries. *J. Mater. Chem.* 20: 5462–5467.
15. Feyni, B., Koszor, O., and Balazsi, C. 2008. Ceramic-based nanocomposites for functional applications. *Nano: Brief Reports and Reviews* 3:323–327.
16. Watcharotone, S., Dikin, D. A., Stankovich, S., Piner, R., Jung, I., Domment, G. H. B., Evmenenko, G., Wu, S. E., Chen, S. F., Liu, C. P., Nguyen, S. B. T., and Ruoff, R. S. 2007. Graphene-silica composite thin films as transparent conductors. *Nano Lett.* 7: 1888–1892.

17. Williams, G., Seger, B., and Kamat, P. V. 2008. TiO_2-graphene nanocomposites. UV-assisted photocatalytic reduction of graphene oxide. *ACS Nano* 2:1487–1491.
18. Lambert, T. N., Chavez, C. A., Harnandez-Sanchez, B., Lu, P., Bell, N. S., Ambrosini, A., Friedman, T., Boyle, T. J., Wheeler, D. R., and Huber, D. L. 2009. Synthesis and characterization of titania-graphene nanocomposites. *J. Phys. Chem. C* 113: 19812–19823.
19. Zhang, X. Y., Li, H. P., and Cui, X. L. 2009. Preparation and photocatalytic activity for hydrogen evolution of TiO_2/graphene sheets composite. *Chinese J. Inorganic Chem.* 25: 1903–1907.
20. Zhang, X. Y., Li, H. P., Cui, X. L., and Lin, Y. 2010. Graphene/TiO_2 nanocomposites: Synthesis characterization and application in hydrogen evolution from water photocatalytic splitting. *J. Mater. Chem.* 20: 2801–2806.
21. Tang, Y. B., Lee, C. S., Xu, J., Liu, Z. T., Chen, Z. H., He, Z., Cao, Y. L., Yuan, G., Song, H., Chen, L., Luo, L., Cheng, H. M., Zhang, W. J., Bello, I., and Lee, S. T. 2010. Incorporation of graphenes in nanostructured TiO_2 films via molecular grafting for dye-sensitized solar cell application. *ACS Nano* 4: 3482–3488.
22. Nethravathi, C., Rajamathi, J. T., Ravishankar, N., Shivkumara, C., and Rajamathi, M. 2008. Graphite oxide-intercalated anionic clay and its decomposition to graphene-inorganic material nanocomposites. *Langmuir* 24: 8240–8244.
23. Ji, Z., Wu, J., Shen, X., Zhou, H., and Xi, H. 2011. Preparation and characterization of graphene/NiO nanocomposites. *J. Mater. Sci.* 46: 1190–1195.
24. Nethravathi, C., Rajamathi, M., Ravishankar, N., Basit, L., and Felser, C. 2010. Synthesis of graphene oxide-intercalated a-hydroxides by metathesis and their decomposition to graphene/metal oxide composites. *Carbon* 48: 4343–4350.
25. Yang, S., Cui, G., Pang, S., Cao, Q., Kolb, U., Feng, X., Maier, J., and Mullen, K. 2010. Fabrication of cobalt and cobalt oxide/graphene composites: Towards high-performance anode materials for lithium ion batteries. *Chem. Sus. Chem.* 3: 236–239.
26. Yan, J., Wei, T., Qiao, W., Shao, B., Zhao, Q., Zhang, L., and Fan, Z. 2010. Rapid microwave-assisted synthesis of graphene nanosheet/Co_3O_4 composite for supercapacitors. *Electrochimica Acta* 55: 6973–6978.
27. Wu, J., Shen, X., Jiang, L., Wang, K., and Chen, K. 2010. Solvothermal synthesis and characterization of sandwich-like graphene/ZnO nanocomposites. *Appl. Surf. Sci.* 256: 2826–2830.
28. Zhang, Y., Li, H., Pan, L., and Sun, Z. 2009. Capacitive behavior of graphene–ZnO composite film for supercapacitors. *J. Electroanalytical Chem.* 634: 68–71.
29. Zheng, W. T., Ho, Y. M., Tian, H. W., Wen, M., Wen, M., Qi, J. L., and Li, Y. A. 2009. Field emission from a composite of graphene sheets and ZnO nanowires. *J. Phys. Chem. C* 113: 9164–9168.
30. Wang, K., Wang, Y., Fan, Z., Yan, J., and Wei, T. 2011. Preparation of graphene nanosheet/alumina composites by spark plasma sintering. *Mater. Res. Bull.* 46: 315–318.
31. He, T., Li, J., Wang, L., Zhu, J., and Jiang, W. 2009. Preparation and consolidation of alumina/graphene composite powders. *Mater. Trans. the Japan Inst. Metal* 50: 749–751.
32. Xu, C., Wang, X., Yang, L., and Wu, Y. 2009. Fabrication of a graphene–cuprous oxide composite. *J. Solid State Chem.* 182: 2486–2490.
33. Wang, X., Zhou, X., Yao, K., Zhang, J., and Liu, Z. 2011. A SnO_2/graphene composite as a high stability electrode for lithium ion batteries. *Carbon* 49: 133–139.
34. Wang, Z., Zhang, H., Li, N., Shi, Z., Gu, Z., and Cao, G. 2010. Laterally confined graphene nanosheets and graphene/SnO_2 batteries. *Nano Res.* 3: 748–756.
35. Qian, Y., Lu, S., and Gao, F. 2010. Synthesis of manganese dioxide/reduced graphene oxide composites with excellent electrocatalytic activity toward reduction of oxygen. *Mater. Lett.* 65: 56–58.

36. Yan, J., Fan, Z., Wei, T., Qian, W., Zhang, M., and Wei, F. 2010. Fast and reversible surface redox reaction of graphene–MnO_2 composites as supercapacitor electrodes. *Carbon*, 48: 3825–3833.
37. Ren, Y., Yan, N., Wen, Q., Fan, Z., Wei, T., Zhang, M., and Ma, J. 2010. Graphene/δ-MnO_2 composite as adsorbent for the removal of nickel ions from wastewater *Chem. Eng. J.* doi:10.1016/j.cej.2010.08.010.
38. Ji, F., Feng, J. M., Su, D., Wen, Y. Y., Feng, Y., and Hou, F. 2009. Electrochemical performance of graphene nanosheets and ceramic composites as anodes for lithium batteries. *J. Mater. Chem.* 19: 9063–9067.
39. Ding, Y., Jiang, Y., Xu, F., Yin, J., Ren, H., Zhou, Q., Long, Z., and Zhang, P. 2010. Preparation of nano-structured $LiFePO_4$/graphene composites by co-precipitation method. *Electrochem. Commun.* 12: 10–13.
40. Fukui, H., Ohsuka, H., Hino, T., and Kanamura, K. 2009. A Si-O-C composite anode: High capability and proposed mechanism of lithium storage associated with microstructural characteristics. *ACS Appl. Mater. Interfaces* 2: 998–1008.
41. Lu, T., Zhang, Y., Li, H., Pan, L., Li, Y., and Sun, Z. 2010. Electrochemical behaviors of graphene–ZnO and graphene–SnO_2 composite films for supercapacitors. *Electrochimica Acta* 55: 4170–4173.
42. Hummers, W. S., Jr., and Offeman, R. E. 1958. Preparation of graphitic oxide. *J. Am. Chem. Soc.* 80: 1339.
43. Park, S., and Ruoff, R. S. 2009. Chemical methods for the production of graphenes. *Nat. Nanotechnol.* 4: 217–224.
44. Li, D., Muller, M. B., Gilje, S., Kaner, R. B., and Wallace, G. G. 2008. Processable aqueous dispersions of graphene nanosheets. *Nat. Nanotechnol.* 3: 101–105.
45. Stoller, M. D., Park, S. J., Zhu, Y. W., An, J. H., and Ruoff, R. S. 2008. Graphene-based ultracapacitors. *Nano Lett.* 8: 3498–3502.
46. Xu, C., Wang, X., and Zhu, J. 2008. Graphene-metal particle nanocomposites. *J. Phys. Chem.* 112: 19841–19845.
47. Du, M., Yang, T., and Jiao, K. 2010. Immobilization-free direct electrochemical detection for DNA specific sequences based on electrochemically converted gold nanoparticle/graphene composite film. *J. Mater. Chem.* 20: 9253–9260.
48. Gong, J., Zhou, T., Song, D., and Zhang., L 2010. Monodispersed Au nanoparticles decorated graphene as an enhanced sensing platform for ultrasensitive stripping voltammetric detection of mercury(II). *Sensors and Actuators B* 150: 491–497.
49. Zhou, K., Zhu, Y., Yang, X., Luo, J., Li, C., and Luan, S. 2010. A novel hydrogen peroxide biosensor based on Au–graphene–HRP–chitosan biocomposites. *Electrochimica Acta* 55: 3055–3060.
50. Hu, Y., Jin, J., Wu, P., Zhang, H., and Cai, C. 2010. Graphene–gold nanostructure composites fabricated by electrodeposition and their electrocatalytic activity toward the oxygen reduction and glucose oxidation. *Electrochimica Acta* 56: 491–500.
51. Li, Y., Tang, L., and Li, J. 2009. Preparation and electrochemical performance for methanol oxidation of pt/graphene nanocomposites. *Electrochem. Commun.* 11: 846–849.
52. Si, Y., and Samulski, E. T. 2008. Exfoliated graphene separated by platinum nanoparticles. *Chem. Mater.* 20: 6792–6797.
53. Chou, S. L., Wang, J. Z., Choucair, M., Liu, H. K., Stride, J. A., and Dou, S. X. 2010. Enhanced reversible lithium storage in a nanosize silicon/graphene composite. *Electrochem. Commun.* 12: 303–306.
54. Song, B., Li, D., Qi, W., Elstner, M., Fan, C., and Fang, H. 2010. Graphene on Au(111): A highly conductive material with excellent adsorption properties for high-resolution bio/nanodetection and identification. *Chem. Phys. Chem.* 11: 585–589.

55. Lee, W., Jang, S., Kim, M. J., and Young, J. M. 2008. Interfacial interactions and dispersion relations in carbon–aluminum nanocomposite systems. *Nanotechnology*. 19: 285701.
56. Tjong, S. C. 2009. *Carbon nanotube reinforced composites: Metal and ceramic matrices*. Weinheim: Wiley-VCH.
57. Mittal, V. 2010. *Polymer nanotube nanocomposites: Synthesis, properties, and applications*. New York: Wiley Publishers.
58. Yu, M. F., Lourie, O., Dyer, M. J., Moloni, K., Kelly, T. F., and Ruoff, R. S. 2000. Strength and breaking mechanism of multiwalled carbon nanotube under tensile load. *Science* 287: 637–640.
59. Singh, S., Pei, Y., Miller, R., and Sundarrajan, P. R. 2003. Long range, entangled carbon nanotube networks in polycarbonate. *Adv. Func. Mater*. 13: 868–872.
60. Che, J., Cagin, I., and Goddard, W. A. III. 2000. Thermal conductivity of carbon nanotubes. *Nanotechnology* 11: 65–69.
61. Ando, Y., Zhao, X., Shimoyama, H., Sakai, G., and Kaneto, K. 1999. Physical properties of multiwalled carbon nanotubes. *Int. J. Inorg. Mater.* 1: 77–82.
62. Lahiri, D., Singh, V., Keshri, A. K., Seal, S., and Agarwal, A. 2010. Carbon nanotube toughened hydroxyapatite by spark plasma sintering: Microstructural evolution and multiscale tribological properties. *Carbon* 48: 3103–3120.
63. Bakshi, S. R., Singh, V., McCartney, D. G., Seal, S., and Agarwal, A. 2008. Deformation and damage mechanisms of multiwalled carbon nanotubes under high-velocity impact. *Scripta Mater.* 59: 499–502.
64. Bakshi, S. R., Musaramthota, V., Virzi, D. A., Keshri, A. K., Lahiri, D., Singh, V., Seal, S., and Agarwal, A. 2011. Spark plasma sintered tantalum carbide–carbon nanotube composite: Effect of pressure; carbon nanotube length and dispersion technique on microstructure and mechanical properties. *Mater. Sci. Eng. A* 528: 2538–2547.
65. Lin, J., Wang, L., and Chen, G. 2011. Modification of graphene platelets and their tribological properties as a lubricant additive. *Tribol. Lett.* 41: 209–215.

8 Graphene-Based Biosensors and Gas Sensors

Subbiah Alwarappan, Shreekumar Pillai, Shree R. Singh, and Ashok Kumar

CONTENTS

8.1 INTRODUCTION

In the field of biosensors and biocatalysts, nanomaterials offer the potential for extremely high surface area-to-volume ratios. This allows the immobilization of large numbers of biomolecules per unit area and results in high-efficiency biosensors with the potential to increase biosensor sensitivity, response time, and selectivity. To date, a wide variety of nanomaterials have been explored for their application in biosensors due to their unique chemical, physical, and optoelectronic properties [1]. Among the various nanomaterials available, carbon nanomaterials have been extensively used during electroanalysis and the most common forms are spherical fullerenes, cylindrical nanotubes, and carbon fibers and carbon blacks. Since the discovery that individual carbon nanotubes (CNTs) can be employed as nanoscale transistors, researchers have recognized their outstanding potential for electronic detection of biomolecules in solution, possibly down to single-molecule sensitivity. In order to detect biologically derived electronic signals, CNTs are often functionalized with linkers, such as proteins and peptides, to interface with soluble biologically relevant targets. For example, incorporation of CNTs and fullerenes has greatly increased the sensitivity of biosensors and their response time possibly due to their high chemical stability, high surface area, and unique electronic properties [2–4]. CNTs have excellent electrocatalytic activities [5] and have been shown to promote electron transfer reactions involving hydrogen peroxide [6], nicotinamide adenine dinucleotide hydrate (NADH) [7,8], cytochrome [9], and ascorbic acid [10]. However, CNTs are expensive, ranging from twenty to hundreds of dollars per gram, and as a result they are often cost-prohibitive for some applications. Exfoliated graphite nanoplatelets or graphene provide a more affordable alternative to CNT. These platelets consist of sp^2 hybridized carbon atoms arranged in a sheetlike structure instead of the cylindrical geometry found in carbon nanotubes. Further, graphene is the "mother" or the basic building block of all other allotropes of carbon. For instance, graphene can be either wrapped up to form a 0D fullerene, or rolled to form 1D carbon nanotubes, or stacked up to yield 3D graphite (Figure 8.1) [11–14]. Graphene has many potential applications due to its excellent mechanical, electrical, thermal, and optical properties, and its large surface-to-weight ratio (for example, 1 g of graphene can cover several football pitches) [15,16]. Some remarkable properties of graphene reported so far include high values of its Young's modulus (~1100 GPa) [17], fracture strength (125 GPa) [17], thermal conductivity (~5000 W $m^{-1}K^{-1}$) [18], mobility of charge carriers (200,000 cm^2 V^{-1} s^{-1}) [19] and specific surface area (calculated value, 2630 m^2 g^{-1}) [20], and fascinating transport phenomena such as the quantum Hall effect [21]. The exceptional thermal, optical, and electrical properties of graphene are the result of its extended π–π conjugation [22]. It is also worth mentioning that graphene possesses a number of surface active functional moieties such as carboxylic, ketonic, quinonic, and C=C. Of these, the carboxylic and ketonic groups

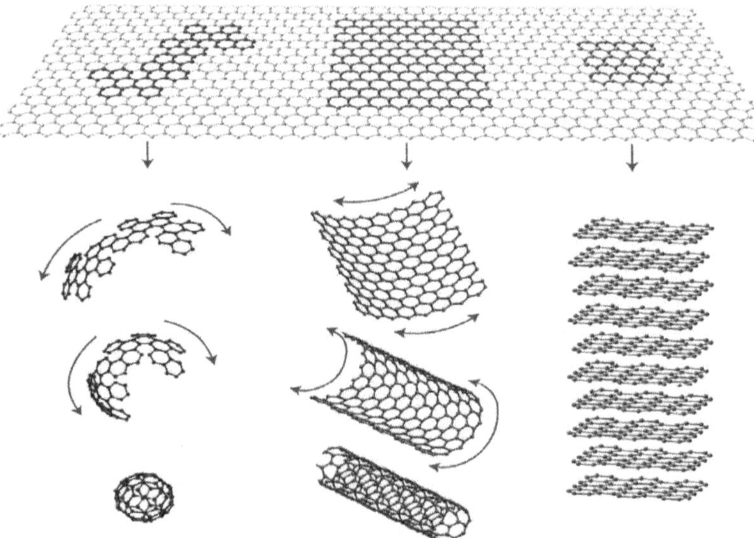

FIGURE 8.1 Scheme showing graphene can be wrapped up to form 0D fullerenes, wrapped up to form 1D CNTs, or stacked up to form 3D graphite. (From A. K. Geim and K. S. Novoselov. 2007. The rise of graphene. *Nat. Mater.* 6, 183–191. With permission..)

are reactive and can easily bind covalently with several biomolecules thereby influencing the possibility of functionalizing graphene with biomolecules for various biosensing applications [23–26].

8.2 ELECTROCHEMISTRY OF GRAPHENE

Graphene is considered as an ideal material for electrochemistry due to its large 2D electrical conductivity, large surface area, and its availability at a cheaper cost. Upon comparison with CNTs, graphene provides two important advantages, and they are: (a) Graphene is free from metallic impurities, whereas CNTs possess numerous metallic impurities. In several cases, such impurities dominate the electrochemistry of CNTs even at <100 ppm levels of impurities in CNTs, thereby leading to misleading conclusions. (b) The production of graphene is cheaper as this synthesis requires only graphite as the starting material.

The latest studies on the electrochemical behavior of graphene indicated that graphene exhibits a wide electrochemical potential window of approximately 2.5 V in 0.1 M PBS (pH 7.0) [25,27,28], which is comparable to that of graphite, glassy carbon (GC), and even boron-doped diamond electrodes [25,27,29], and the charge-transfer resistance on graphene as determined from AC impedance spectra is much lower than that of graphite and GC electrodes [27]. The electron transfer behavior studies of graphene using cyclic voltammetry (CV) of redox couples, such as $[Fe(CN)_6]^{3-/4-}$ and $[Ru(NH_3)_6]^{3+/2+}$ exhibited well-defined redox peaks [30–32]. Both anodic and cathodic peak currents in the CV are linear with the square root of the scan rate, which suggests that the redox processes on graphene-based electrodes are predominantly diffusion controlled [31]. The peak-to-peak potential separations (ΔEp) in CVs for most one-electron-transfer redox couples are quite low, very close to the ideal value of 59 mV. For example, the reported values of graphene electrodes were between 61.5 mV and 73 mV (10 mV/s) for $[Fe(CN)_6]^{3-/4-}$ [30,32–34] and 60–65 mV (100 mV/s) for $[Ru(NH_3)_6]^{3+/2+}$ [32], much smaller than that on GC [29]. The peak-to-peak potential separation is related to the electron transfer (ET) coefficient [35], and a low ΔEp value indicates a fast ET for a single-electron electrochemical reaction [33] on graphene. In order to study the electrochemical response/activity of graphene to different kinds of redox systems, Tang et al. [32] systematically studied three representative redox couples: $[Ru(NH_3)_6]^{3+/2+}$, $[Fe(CN)_6]^{3-/4-}$, and $Fe^{3+/2+}$. As is known, $[Ru(NH_3)_6]^{3+/2+}$ is a nearly ideal outer-sphere redox system that is insensitive to most surface defects or impurities on electrodes and can serve as a useful benchmark in comparing electron transfer of various carbon electrodes; $[Fe(CN)_6]^{3-/4-}$ is *surface sensitive* but not *oxide sensitive*, and $Fe^{3+/2+}$ is both *surface sensitive* and *oxide sensitive* [29]. The apparent electron-transfer rate constants (k_0) calculated from cyclic voltammograms on graphene and GC electrodes are 0.18 cm/s and 0.055 cm/s for $[Ru(NH_3)_6]^{3+/2+}$, respectively [32]. This indicates that the unique electronic structure of graphene, especially the high density of the electronic states over a wide energy range, endows graphene with a fast electron transfer rate [29]. The k_0 for $[Fe(CN)_6]^{3-/4-}$ on graphene and GC were calculated to be 0.49 cm/s and 0.029 cm/s, respectively, and the electron transfer rates for $Fe^{3+/2+}$ at graphene electrode are several orders of magnitude higher than that at GC electrodes [32]. These indicate that the electronic structure and the surface chemistry of graphene are beneficial for electron transfer [32,36,37].

8.3 DIRECT ELECTROCHEMISTRY OF ENZYMES ON GRAPHENE SURFACE

The direct electrochemistry of an enzyme refers to the direct electron communication between the electrode and the active center of the enzyme without the participation of mediators or other reagents [37–39], which is very significant in the development of biosensors, biofuel cells, and biomedical devices [38–42]. However, the realization of direct electrochemistry of redox enzymes on common electrodes is very difficult because the active centers of most redox enzymes are located deep in a hydrophobic cavity of the molecule [40,43]. Carbon nanotubes and metal nanoparticles have exhibited excellent performance in enhancing the direct electron transfer between enzymes and electrodes, and are widely used now [44–47]. Due to its extraordinary electron transport property and high specific surface area [48], functionalized graphene is expected to promote electron transfer between electrode substrates and enzymes [49]. Shan et al. [49] and Kang et al. [50] reported the direct electrochemistry of glucose oxidase (GOD) on graphene. Shan et al. [49] employed chemically reduced graphene oxide (CR-GO) and Kang et al. [50] employed thermally split graphene oxide [51], both of which exhibit similar excellent direct electrochemistry of GOD. Figure 8.2 represents the cyclic voltammograms (CV) of graphene, graphite–GOD, and graphene–GOD modified electrodes in phosphate buffered solution (PBS) [49]. Among all the electrodes employed, only the graphene–GOD electrode exhibited a pair of well-defined redox peaks that correspond to a reversible electron transfer process of a redox active center (flavin adenine dinucleotide, FAD) in GOD, indicating a direct electron transfer of GOD on graphene electrodes. The formal potential calculated by averaging the cathodic and anodic peak potentials was estimated to be −0.43V (*vs.* Ag/AgCl), which is close to the standard electrode potential of FAD/$FADH_2$ [50,51]. The redox peaks of GOD have 69mV peak-to-peak separation and the ratio of cathodic to anodic current intensity was found to be 1 [49]. Furthermore, the peak current densities are in a linear relationship with scan rates [49,50]. These indicate that the redox process of GOD on graphene electrodes is a reversible and surface-confined process [50]. The electron-transfer-rate constant (k_s) of the GOD-on-graphene electrode is 2.83 s^{-1}, much higher than most of the values reported on carbon nanotubes [52–55], indicating that functionalized graphene provides a rapid electron transfer between the redox center of the enzyme and the surface of the electrode [50]. In a recent study, graphene electrodes were found to exhibit a high enzyme loading of about 1.12×10^{-9} mol/cm^2 due to its high surface area [50]. Such a huge amount of enzyme loading is advantageous for increasing the sensitivity of graphene-based biosensors. Furthermore, the direct electron transfer of GOD on graphene is stable (no obvious changes in 15 cycles on the cyclic voltammetric responses of the GOD-graphene-chitosan modified electrode in N_2-saturated PBS) and response retention is above 95% after 1 week of storage [50]. Our group [56] also recently demonstrated the direct electrochemistry of Cytochrome c (Cyt c)

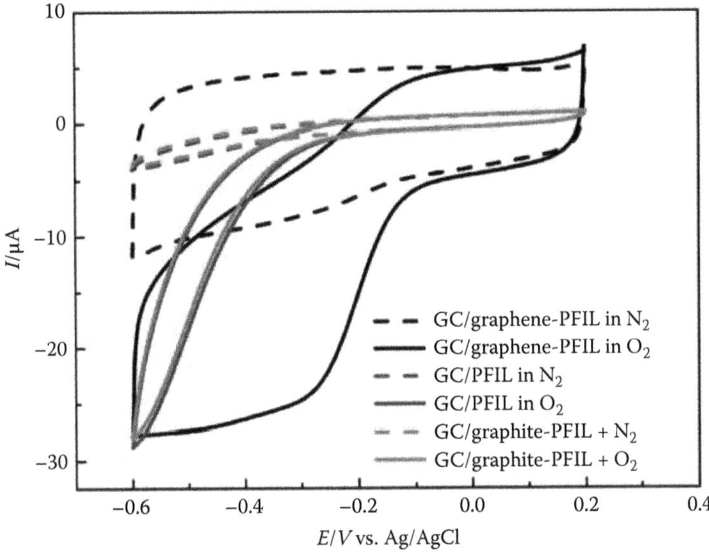

FIGURE 8.2 (*See color insert.*) Cyclic voltammograms at PFIL (blue), graphite-PFIL (magenta), and graphene-PFIL (black) modified electrodes in 0.05 M PBS solution saturated with O_2 (solid) and degassed with pure N_2 (dashed). (From C.S. Shan, H.F. Yang, J.F. Song, D.X. Han, A. Ivaska, and L. Niu. 2009. Direct electrochemistry of glucose oxidase and biosensing for glucose based on graphene. *Anal. Chem.* 81, 2378–2382. With permission.)

at the graphene surface without the need for a co-factor. The excellent performance of graphene with respect to the direct electrochemistry of GOD and Cyt c indicates that graphene is a potential promising material for enzyme electrodes.

8.4 ENZYMATIC ELECTROCHEMICAL BIOSENSORS BASED ON GRAPHENE

Graphene was found to exhibit high electrocatalytic activity to H_2O_2. Further, it serves as an excellent platform for the direct electrochemistry of GOD; thereby, graphene could be an excellent electrode material for oxidase biosensors. For example, there are numerous reports available in the literature about graphene-based glucose biosensors [57–61]. Shan et al. [57] reported the first graphene-based glucose biosensor with a graphene/polyethylenimine functionalized ionic liquid nanocomposite modified electrode that exhibits a wide linear glucose response (2 to 14 mM) together with good reproducibility (relative standard deviation of the current response to 6 mM glucose at −0.5 V was 3.2% for 10 successive measurements) and high stability (response current +4.9% after 1 week) [57]. Recently, Zhou et al. [62] reported a biosensor based on chemically reduced graphene oxide (CR-GO) for the enhanced amperometric detection of glucose. Further, their sensor exhibited a high sensitivity (20.21 μA mM cm^{-2}) and a low detection limit of 2.0 mM (S/N = 3) in the linear range 0.01–10 mM The linear range for glucose detection is wider than that on other carbon materials-based electrodes such as carbon nanotubes [63] and carbon nanofibers

[64]. The detection limit for glucose at the GOD/CR-GO/GC electrode (2.0 mM at −0.20 V) is lower than the reported carbon-based biosensors, such as carbon nanotube paste [65], carbon nanotube nanoelectrodes [66], carbon nanofibers [64], exfoliated graphite nanoplatelets [62], and highly ordered mesoporous carbon [67]. The response at the GOD/CR-GO/GC electrode to glucose is very fast (9±1 s to attain steady-state response) and highly stable (91% signal retention for 5 h), which makes GOD/CR-GO/GC electrodes a rapid and highly stable biosensor to continuously measure plasma glucose level for the diagnosis of diabetes. In another work, graphene dispersed on biocompatible chitosan was also employed to construct glucose biosensors [60]. In this work, it is evident that chitosan helped to form a well-dispersed graphene suspension and immobilize the enzyme molecules, and the graphene-based enzyme sensor exhibited excellent sensitivity (37.93 $mAmM^{-1}cm^{-2}$) and long-term stability for measuring glucose. Graphene/metal nanoparticle (NP)-based biosensors have also been developed. Shan et al. [68] developed a graphene/AuNPs/chitosan composite film–based biosensor that exhibited good electrocatalytic activity to H_2O_2 and O_2. Wu et al. [59] designed a GOD/graphene/PtNPs/chitosan-based glucose biosensor with a detection limit of 0.6 mM glucose. These enhanced performances were attributed to the large surface area and good electrical conductivity of graphene and the synergistic effect of graphene and metal nanoparticles [61,68]. Zhou et al. [62] reported an ethanol biosensor based on graphene-ADH. The ADH-graphene-GC electrode exhibits faster response, wider linear range, and a lower detection limit for ethanol detection compared with ADH-graphite/GC and ADH/GC electrodes. This observed enhanced performance can be explained by the effective transfer of substrate and products through graphene matrixes containing enzymes as well as the inherent biocompatibility of graphene [62].

8.5 GRAPHENE-BASED ELECTROCHEMICAL BIOSENSORS

Hydrogen peroxide is a general enzymatic product of oxidases and a substrate of peroxidases, which is important in biological processes and biosensor development [62]. Hydrogen peroxide is also an essential mediator in food, pharmaceutical, clinical, industrial, and environmental analyses [62]; therefore, it is of great importance to detect H_2O_2. The key point in developing electrodes for detecting H_2O_2 is to decrease the oxidation/reduction overpotentials. Carbon materials, such as carbon nanotubes [44,45,69,70], have been developed in constructing biosensors for detecting H_2O_2. Graphene has shown promise in this respect [31,73]. Zhou et al. [62] studied the electrochemical behavior of hydrogen peroxide on a graphene (chemically reduced graphene oxides, CR-GO) modified electrode, which shows a remarkable increase in electron transfer rate compared with graphite/GC and bare GC electrodes [62]. As shown in the cyclic voltammograms (CVs) in Figure 8.3, the onset potentials of H_2O_2 oxidation/reduction on CR-GO/GC (a1), graphite/GC (b1), and GC electrodes (c1) are 0.20/0.10V, 0.80/−0.35V, and 0.70/−0.25V, respectively, indicating superior electrocatalytic activity of graphene to H_2O_2. The linear relationship of H_2O_2 at −0.2 V on the CR-GO/GC electrode is 0.05–1500 mM, wider than the previously reported results for carbon nanotubes [62]. These can be attributed to the high density of edge-plane-like defective sites on graphene, which might provide many active sites

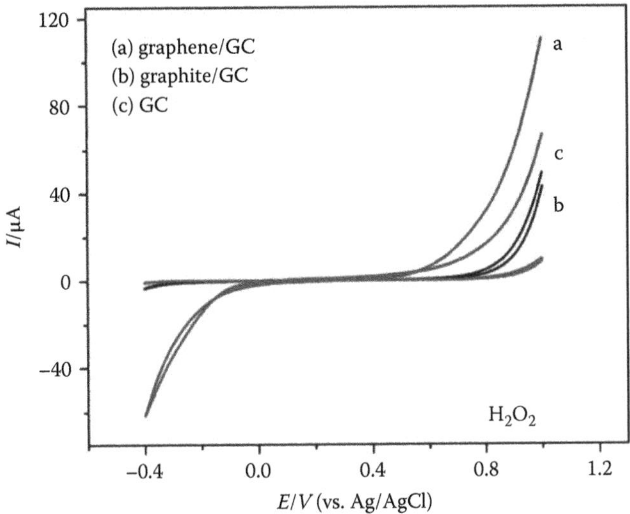

FIGURE 8.3 Background-subtracted CVs (50 mV/s) on (a) graphene/GC, (b) graphite/GC, and (c) GC electrodes in 4 mM H_2O_2 0.1 M PBS (pH 7.0). (From M. Zhou, Y.M. Zhai, and S.J. Dong. 2009. Electrochemical biosensing based on reduced graphene oxide. *Anal. Chem.* 81, 5603–5613. With permission.)

for electron transfer to biological species [29,71–75]. Such significantly enhanced performance on graphene-based electrodes for detecting H_2O_2 may lead to high-selectivity/sensitivity electrochemical sensors.

β-nicotinamide adenine dinucleotide (NAD^+) and its reduced form (NADH) are cofactors of many dehydrogenases [45], which have received considerable interest in developing amperometric biosensors, biofuel cells, and bioelectronic devices associated with NAD^+/NADH-dependent dehydrogenases [70,73]. The oxidation of NADH serves as the anodic signal and regenerates the NAD^+ cofactor, which is of great significance in biosensing important substrates such as lactate, alcohol, or glucose [44]. Problems inherent in such anodic detection are the large overvoltage for NADH oxidation and surface fouling associated with the accumulation of reaction products [44]. Graphene shows promise in addressing these problems. Tang et al. [32] studied the electrochemical behavior of NADH on graphene (chemically reduced graphene oxides, CR-GO) modified electrodes, which show a remarkable increase in electron transfer rate compared with graphite/GC and GC electrodes [32]. The peak potentials of NADH oxidation are shifted from 0.70 V on GC and graphite to 0.40 V on CR-GO (Figure 8.4) [32]. These are attributed to the high density of edge-plane-like defective sites on CR-GO, which provide many active sites for electron transfer to biological species [29,72,73]. Liu et al. [74] reported a further enhanced performance of graphene-based electrodes toward the oxidation of NADH through increasing the dispersity of graphene via noncovalent functionalization of graphene with methylene green (MG). The oxidation of NADH on the MG–graphene electrode takes place at ~0.14 V, which is much lower than that (+0.40 V) for pristine graphene (i.e., without MG functionalization) [74] and carbon nanotube-based biosensors [8,75,76].

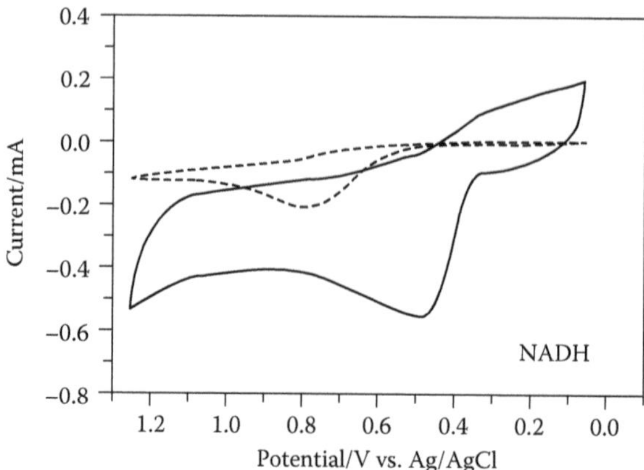

FIGURE 8.4 CV of bare GC (dashed line) and graphene/GC electrodes (solid line) in 0.1 M pH 6.8 PBS containing 1 mM. (From L.H. Tang, Y. Wang, Y.M. Li, H.B. Feng, J. Lu, and J.H. Li. 2009. Preparation, structure and electrochemical properties of reduced graphene sheet films. *Adv. Funct. Mater.* 19, 2782–2789. With permission.)

The enhanced activity of graphene-modified electrodes to NADH oxidation is further confirmed in Lin et al.'s report when compared with bare edge-plane pyrolytic graphite electrodes (EPPGEs), which have many edge-planelike defective sites [31]. It is known that the high density of edge-planelike defective sites, which might provide many active sites for electron transfer to biological species, contribute to the enhanced activity of carbon materials to oxidation/reduction of small biomolecules such as NADH [29,71,72]. The higher activity of functionalized graphene–modified EPPGE compared to bare EPPGE indicates that, in addition to the high density of edge-plane-like defective sites on graphene (chemically reduced graphene oxides), there are other special properties of graphene that contribute to its high activity. The exact mechanisms still need more investigation. Pumera et al.'s report [73] on high-resolution x-ray photoelectron spectroscopy (HRXPS) and ab initio molecular dynamics study of adsorption of NAD^+/NADH (Figure 8.5) might provide preliminary insight into this. It shows that the adsorption of NAD^+ molecules on the edges of graphene is attributed to the interaction with oxygen-containing groups such as carboxylic groups, while graphene edges substituted only with hydrogen are prone to passivation [73]. Therefore, the oxygen-containing groups might play a key role in the enhanced activity of graphene.

Lin et al. [77] recently reported a few-layered, graphene-based electrochemical sensor for the detection of paracetamol. The performance of the senor was demonstrated using cyclic voltammetry and square wave voltammetry. Lin et al.'s [77] sensor exhibited a quasi-reversible response to paracetamol detection. Further, the overpotential of the graphene-based sensor was found to be much smaller than the bare GC electrode.

In another work, Zuo et al. [78] successfully employed GO to demonstrate the possibility of efficient electrical wiring of the redox centers of several metalloproteins containing heme, such as cytochrome c, myoglobin, and horseradish

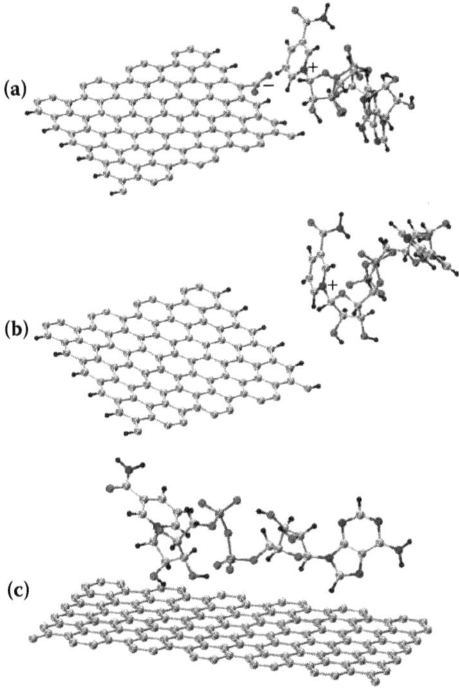

FIGURE 8.5 (*See color insert.*) Model for the adsorption of NAD^+ on a (a) graphene basal plane, (b) graphene edge fully terminated by hydrogen atoms, and (c) graphene edge terminated by hydrogen atoms and containing one $-COO^-$ group *via* Car–Parrinello molecular dynamics. Gray, C; blue, N; red, O; yellow, P; black, H. (From M. Pumera, R. Scipioni, H. Iwai, T. Ohno, Y. Miyahara, and M. Boero. 2009. A mechanism of adsorption of β-nicotinamide adenine dinucleotide on graphene sheets: Experiment and theory. *Chem. Eur. J.* 15, 10851. With permission.)

peroxidase, to the electrode (Figure 8.6). Further, Zuo et al. [78] confirmed that the proteins retained their structural integrity as well as their biological activity when they form binary mixtures with GO [79]. From this observation, it is very promising that GO–protein complexes can be definitely employed for developing a highly sensitive biosensor.

8.6 GRAPHENE-BASED DNA BIOSENSORS

The advantages of electrochemical DNA sensors include their high sensitivity and high selectivity. Additionally, they are cost effective for the detection of selected DNA sequences or mutated genes associated with human disease and promise to provide a simple, accurate, and inexpensive platform for patient diagnosis [80,81]. Electrochemical DNA sensors further allow device miniaturization for samples with tiny volume [25]. Among the various electrochemical DNA sensors, the sensor based on the direct oxidation of DNA is simple and robust [25,81]. Zhou et al. [25] reported an electrochemical DNA sensor based on graphene (chemically reduced graphene

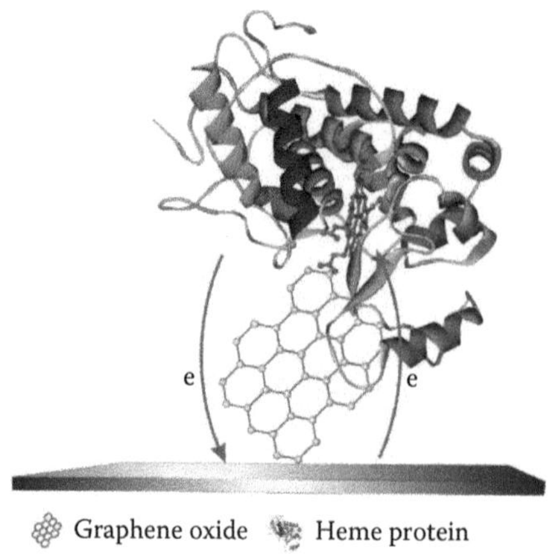

FIGURE 8.6 Scheme showing the protein wiring onto 2D graphene for electrochemical applications. Graphene oxide (GO)-supported heme proteins attached to the glassy carbon electrode surface. (From C. Fu, W. Yang, X. Chen, and D.G. Evans. 2009. Direct electrochemistry of glucose oxidase on a graphite nanosheet–Nafion composite film modified electrode. *Electrochem. Commun.* 11, 997–1000. With permission.)

oxide). As evident from Figure 8.7, the current response of the four free bases of DNA (i.e., guanine [G], adenine [A], thymine [T], and cytosine [C]) on the CR-GO/GC electrode are all separated efficiently, indicating that CR-GO/GC can simultaneously detect four free bases, whereas this is not possible with graphite or glassy carbon. This is attributed to the antifouling properties and the high electron transfer kinetics for base oxidation on CR-GO/GC electrodes [25], which results from the high density of edge-plane-like defects and oxygen-containing functional groups on CR-GO that provide many active sites and promote electron transfer between the electrode and species in solution [29,71,72]. From Figures 8.7(b) and 8.7(c), it is evident that the CR-GO/GC electrode is capable of efficiently separating all four DNA bases in both single-stranded DNA (ssDNA) and double-stranded DNA (dsDNA), which are more difficult to oxidize than free bases at physiological pH without the need of a prehydrolysis step. Further, this electrode provides a single-nucleotide polymorphism (SNP) site for short oligomers with a particular sequence without any hybridization or labeling processes [25]. This is attributed to the unique physicochemical properties of CR-GO (the single sheet nature, high conductivity, large surface area, antifouling properties, high electron transfer kinetics, etc.) [25].

Very recently, Pumera et al. [82] demonstrated that stacked graphene nanofibers (SGNF) exhibited excellent electrochemical sensitivity to the DNA bases, thereby outperforming the sensitivity of CNTs to DNA bases. According to Pumera et al. [82], the observed behavior is due to the presence of an exceptionally large number of accessible graphene-sheet edges on the surface of the nanofibers compared to the CNTs, and it was confirmed by TEM and Raman spectroscopy. Further, Pumera et al. [82] observed approximately four times greater current during the oxidation of adenine, guanine, cytosine, and thymine at the SGNF electrodes than

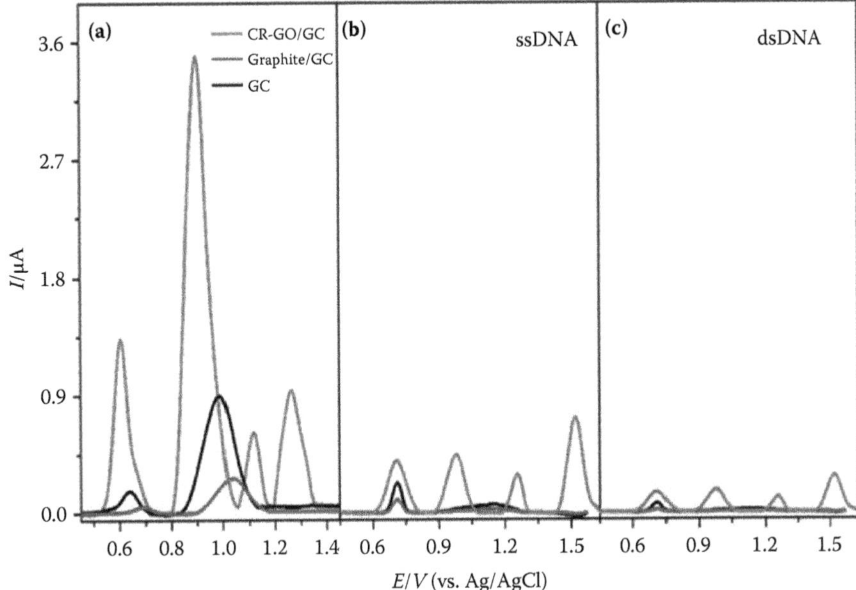

FIGURE 8.7 (*See color insert.*) Differential pulse voltammograms (DPV) for (a) a mixture of DNA free base (G, A, T, and C), (b) ssDNA, and (c) dsDNA in 0.1 M pH 7.0 PBS at graphene/GC (green), graphite/GC (red), and bare GC electrodes (black). Concentrations G, A, T, C, ssDNA, or dsDNA: 10 mg mL−1. (From Y. Liu, D. Yu, C. Zeng, Z-C. Miao, and L. Dai. 2010. Biocompatible graphene oxide-based glucose biosensors. *Langmuir* 26, 6158–6160. With permission.)

on the CNT electrodes. SGNFs also exhibit higher sensitivity than edge-plane pyrolytic graphite electrodes, GC electrodes, or graphite microparticle-based electrodes. Pumera et al. [82,83] also demonstrated that influenza A (H1N1)-related strands can be sensitively oxidized on SGNF-based electrodes, which could therefore be applied to label-free DNA analysis (Figure 8.8).

Recently, Chaniotakis et al. [84,85] employed the SGNFs for the enzymatic detection of glucose. In their work, Chaniotakis et al. [85] adopted the direct immobilization technique to immobilize the enzymes onto a nanofiber surface (see Figure 8.9). From their work, it is evident that surface modification of SGNFs with a biorecognition element will be a highly efficient method for developing a new class of very sensitive, stable, reproducible electrochemical biosensors. Their results suggest that platelet nanofibers are the best materials described to date for developing biosensors that can outperform the existing CNTs or graphite powder.

8.7 GRAPHENE-BASED ELECTROCHEMICAL SENSORS FOR HEAVY METAL ION DETECTION

Graphene-based electrochemical sensors also find potential application in environmental analysis for the detection of heavy metal ions (Pb^{2+} and Cd^{2+}) [86,87]. Li et al.

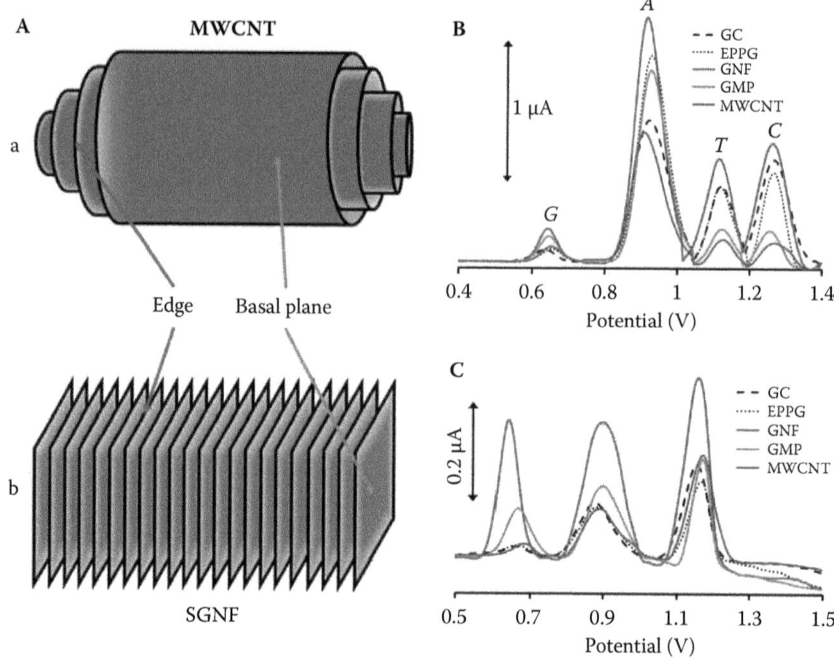

FIGURE 8.8 (*See color insert.*) Stacked graphene nanofibers (SGNFs) outperform carbon nanotubes and graphite in the detection of free DNA bases and A (H1N1) DNA strand. (A) Graphene-sheet orientation in multiwalled carbon nanotubes (MWCNTs) (a) and SGNFs (b). The highly electroactive edge portion of the sheets is represented in yellow. (B) differential pulse voltammograms (DPVs) for a mixture of guanine, adenine, thymine, and cytosine at SGNF (red), graphene platelet (GNP) (green), and MWCNT (blue) electrodes. For comparison, glass-carbon electrode (GCE) (black dashed) and edge-plane pyrolytic graphite (EPPG) (black dotted) electrode signals are also shown. (C) DPVs for ssDNA A (H1N1) at SGNF (red), GNP (green), and MWCNT (blue) electrodes. For comparison, GCE (black dashed) and EPPG (black dotted) electrode signals are also shown. (From A. Ambrosi and M. Pumera. 2010. Stacked graphene nanofibers of electrochemical oxidation of DNA bases. *Phys. Chem. Chem. Phys.* 12, 8943–8947. With permission.)

[86,87] demonstrated that Nafion-graphene composite film-based electrochemical sensors not only exhibit improved sensitivity for the detection of Pb^{2+} and Cd^{2+}, but also alleviate the interferences due to the synergistic effect of graphene nanosheets and Nafion [86]. Further, the stripping current signal is greatly enhanced at the graphene electrodes. It is evident from Figure 8.10 that the stripping current signal is well distinguished. The linear range for the detection of Pb^{2+} and Cd^{2+} is wide (0.5 mgL^{-1} to 50 mgL^{-1} and 1.5 mgL^{-1} to 30 mgL^{-1} for Pb^{2+} and Cd^{2+}, respectively). The detection limits (S/N = 3) are 0.02 mgL^{-1} for both Cd^{2+} and Pb^{2+}, which are more sensitive than that of Nafion film-modified bismuth electrode, [88] and ordered mesoporous carbon-coated GC electrodes [89] and comparable to Nafion/CNT-coated bismuth film electrodes [90]. The enhanced performance is attributed to the unique properties of the graphene (nanosized graphene sheet, nanoscale thickness

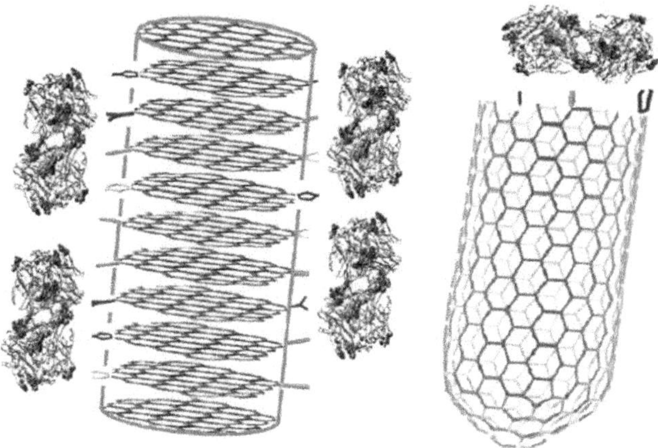

FIGURE 8.9 Immobilization of the model enzyme glucose oxidase on stacked graphene platelet nanofibers and single-walled carbon nanotubes. (From V. Vamvakaki, K. Tsagaraki, and N. Chaniotakis. 2006. Carbon-nanofiber based glucose biosensor. *Anal. Chem.* 78, 5538–5542. With permission.)

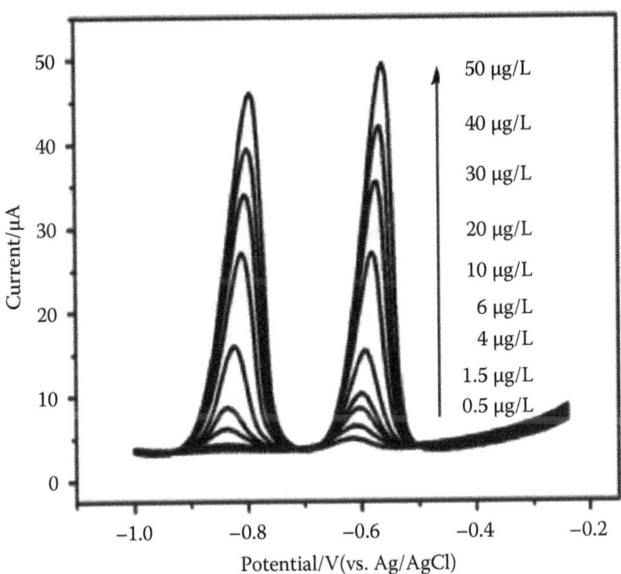

FIGURE 8.10 Striping voltammograms for the different concentrations of Cd^{2+} and Pb^{2+} on an in situ plated Nafion-G-BFE (bismuth film electrode) in solution containing 0.4 mg L^{-1} Bi^{3+}. (From J. Li, S.J. Guo, Y.M. Zhai, and E.K. Wang. 2009. High-sensitivity determination of lead and cadmium based on the Nafion-graphene composite film. *Anal. Chim. Acta* 649, 196–201. With permission.)

of these sheets, and high conductivity), which provided the capability to strongly adsorb target ions, enhanced the surface concentration, improved the sensitivity, and alleviated the fouling effect of surfactants [86,87].

8.8 GRAPHENE-BASED GAS SENSORS

The first experimental study of graphene for gas sensing was performed by Novoselov et al. [11]. In that work, they demonstrated the gas-sensing potential of graphene by showing that graphene devices were doped by exposure to water or ethanol vapors or to ammonia gas. Later, several reports started to appear in experimental and theoretical assessments of graphene's performance in gas sensing (see Table 8.1). All these studies demonstrated the ability of graphene for various gas sensing tasks in a variety of laboratory-based devices, although they are not yet available on the market as commercial sensors. However, it was Schedin et al. [91] who demonstrated that graphene is an ideal material for the high sensitive gas detection. In order to achieve such a high sensitivity toward gas molecules by electrical detection, Schedin et al. [91] fabricated multiterminal Hall bars by conventional lithographic methods from single-layer or few-layer graphene that had been mechanically cleaved from graphite. Adsorption of low concentrations (parts per million) of gases resulted in the concentration-dependent changes in resistivity, after which the sensors were regenerated by annealing at 150°C under vacuum (Figure 8.11[a]). Moreover, the gas-induced changes in resistivity had different magnitudes for different gases and the sign of the change indicated whether the gas was an electron acceptor (e.g., NO_2, H_2O, and I_2) or an electron donor (e.g., CO, ethanol, and NH_3), although interactions between the NH_3 molecules and water adsorbed on the devices probably contributed to the large response shown in Figure 8.11(a) for NH_3 [91]. Based on the low noise level observed in these graphene devices, Schedin et al. [91] confirmed that their sensors had detection limits on the order of one part per billion, the same as that of other existing gas detectors [92–95]. Besides this work, Schedin et al. [91] also performed long-term measurements on extremely diluted NO_2 samples using thoroughly optimized sensors and observed steplike changes in resistivity during adsorption and desorption. Analyzing these quantized data statistically, they interpreted these results as evidence of single-molecule sensitivity as shown in Figure 8.11(b).

Following the success of the early graphene-based sensors, a considerable number of experimental and theoretical reports on graphene-based gas sensors started to appear, several of which sought to understand how the adsorbing molecules alter the conductivity of the graphene. As conductivity is proportional to the product of charge-carrier density and mobility, it is evident that changes in the number density or mobility of carriers, or both, must be responsible for the change in conductivity [91]. However, the relative contributions of these two factors remain uncertain because different proposed mechanisms have yet to be reconciled. For example, Schedin et al. [91] used their Hall-effect measurements as evidence that extra charge carriers were created during gas adsorption on graphene devices. This is a process of chemical doping. Graphene's linear band structure around the Dirac points (the six K points in the Brillouin zone) means that gas adsorption will possibly increase the number of holes if the gas is an acceptor or increase the number of electrons if

TABLE 8.1
Different Types of Important Graphene-Based Gas Sensors

Sensor Material	Method of Detection	Gases or Vapor Detected	Limit of Detection	Reference(s)
Mechanically exfoliated graphene, shaped by lithography	Changes in electrical resistivity	NO_2, H_2O, I_2, NH3, CO, ethanol	~1 ppb	[91]
Mechanically exfoliated graphene, shaped by lithography; measurement made with and without PMMA resist	Changes in electrical current	H_2O, NH_3, nonanal octanoic acid, trimethylamine	Depends on the types of vapor and presence of resist: e.g., «5 ppm for nonanal with resist; >30 ppm without resist	[111]
Hydrazine-reduced graphene oxide	Frequency changes of surface acoustic waves	H_2, CO	~125 ppm CO; < 600 ppm H_2	[112]
Hydrazine-reduced graphene oxide	Reduction changes in electrical resistance	NO_2, NH_3, dinitrotoluene (DNT)	5 ppm NO_2 and NH_3; 28 ppb DNT	[113]
Graphene oxide reduced chemically or thermally or by both approaches	Changes in electrical resistance	H_2O, 0.1 Torr of H_2O vapor		[114]
Low-temperature thermally reduced	Changes in electrical current	NO_2	~2ppm	[115,116]
Hydrazine-reduced graphene oxide	With variable levels of reduction changes in electrical conductance	HCN, chloroethylethyl sulfide (CEES), dimethylmethylphosphonate (DMMP), DNT	70 ppb HCN; 0.5 ppb CEES; 5 ppb DMMP; 0.1 ppb DNT	[117]

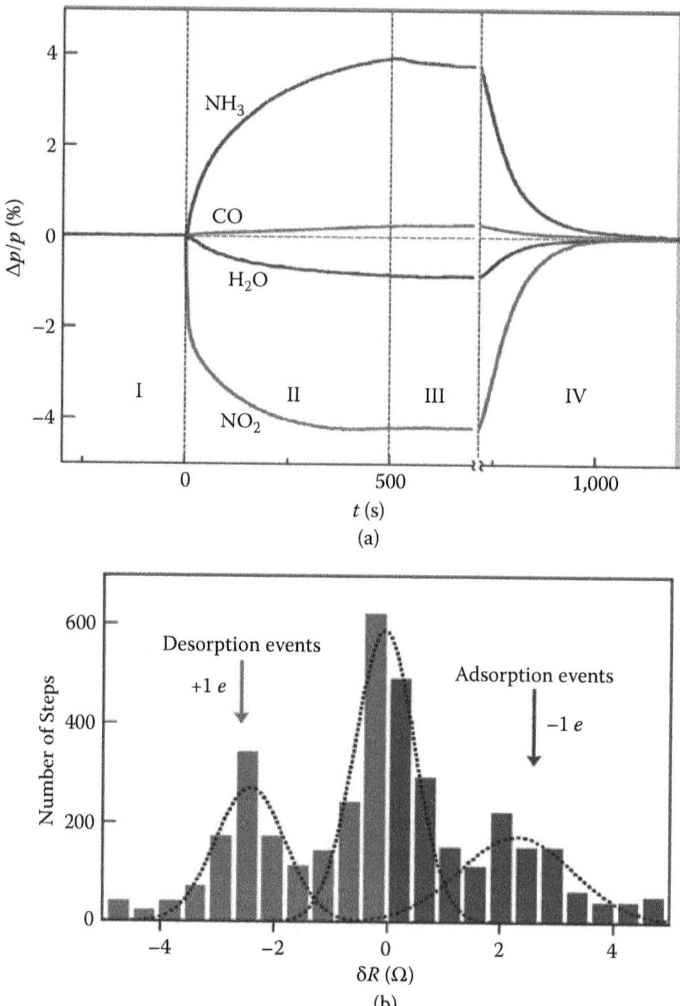

FIGURE 8.11 (a) Changes in resistivity, ΔG/G, of graphene devices during adsorption of different gases, showing rapid responses upon exposure to a fixed volume (5L) of diluent gas that contained 1 ppm of each analyte (Region II). After evacuation (Region III) and annealing at 150°C (Region IV), the sensors returned to their original resistivity. (b) The statistical distribution of step changes in device resistance, δ*R*, during the slow desorption of NO_2. The side peaks around the central "noise peak" were considered evidence for detection of adsorption or desorption of individual gas molecules on the graphene device. (From F. Schedin, A.K. Geim, S.V. Morozov, E.W. Hill, P. Blake, M.I. Katsnelson, and K.S. Novoselov. 2007. Detection of individual gas molecules adsorbed on graphene, *Nat. Mater.* 6, 652–655. Macmillan Publishers Ltd. With permission.)

the gas is a donor, in the chemical equivalent of the ambipolar electric field effect. The charge transfer involved in this process leaves behind a high concentration (exceeding 10^{12} cm^{-2}) of charged surface impurities, and these would be expected to increase scattering, thereby reducing the carrier mobility. Nevertheless, the Schedin [91] team found little change in carrier mobility, as determined from changes in conductivity as a function of changes in carrier density, demonstrating a negligible increase in scattering due to impurities. Based on these facts, they proposed that the ubiquitous layers of water adsorbed on the graphene, or trapped between the graphene and the substrate, provided sufficient dielectric screening of the gas-induced charged impurities to account for the lack of additional scattering with gas adsorption.

On the other hand, Hwang et al. [96] proposed an entirely different mechanism. Hwang et al. [96] employed the resistivity data from some of Schedin's earlier experiments in order to directly calculate the Hall mobility of graphene devices exposed to a constant stream of NH_3 or NO_2. The results showed that mobility rapidly increased and then gradually plateaued, rather than remaining approximately constant. From these observations, Hwang's team [96] suggested an alternate physical model for the changes in graphene's conductivity during gas adsorption. Based on the conductivity experimental measurements in graphene devices and theoretical analysis of possible scattering mechanisms [97,98], they considered that the primary scattering mechanism responsible for limiting carrier mobility near the Dirac point is Coulomb (long-range) scattering due to charged impurities arising from the substrate [96]. This proposed suggestion coincides well with the enhanced mobility of graphene, as high as 230,000 cm^2 V^{-1} s^{-1} [99], when suspended away from the substrate. These suspended mobilities are one order of magnitude larger than the mobilities in typical graphene devices on substrates. Considering Coulomb scattering as the dominant limit on mobility, Hwang et al. [96] proposed that the charged impurities created by gas adsorption on graphene served to partially neutralize the Coulomb scatters induced by the substrate, thereby allowing a rapid increase in electrical mobility. As this "compensation" effect increases with continued gas adsorption, the mobility tends to gradually plateau due to the effects of short-range scattering, possibly caused by defects in the graphene lattice becoming more pronounced.

However, several more detailed experimental studies are still essential to further probe the physical processes involved in gas detection; recent studies started reporting additional insights. Moreover, recognizing that the amount of charge transfer is critical to graphene's sensitivity for gas detection, several teams have modeled the nature of interactions between graphene and different adsorbing gases or vapors [100–117]. Some of the major issues with such work are that (1) different models and approaches generate different adsorption energies and charge transfers, (2) the calculated charge transfer for paramagnetic molecules even depends on the size of the graphene *supercell* that is modeled [102]. For these reasons, this kind of work is limited to providing indicative trends and insights into atomic-scale processes. A recent review of gas adsorption on graphene by Wehling et al. [106] gives a good overview of the present level of understanding. In open-shell adsorbates such as NO_2 or alkali-metal, atoms undergo direct charge transfer to or from graphene, but tend to be weakly bound and relatively mobile at room temperature, unless they form covalent bonds with the graphene, which was observed with the species such as H, F, and

OH. The strong doping by open-shell molecules was made possible by graphene's zero gap electronic structure, which further allows them to undergo charge transfer with any adsorbed species that has even a small chemical-potential mismatch. The predicted phenomenon makes it more sensitive to gas adsorption than semiconductor devices with larger band gaps [104]. However, closed-shell adsorbates such as H_2O and NH_3 do not directly change the band structure of graphene, but rather alter how charges are distributed within graphene or alter the doping of graphene by the supporting substrate [106]. Adsorbed water is a prevalent surface "impurity," and especially when located between the graphene and substrate, can shift the impurity bands of the substrate into the vicinity of graphene's Fermi level and so cause indirect doping of the graphene. Indeed, even if not initially present between the graphene and substrate, water molecules might diffuse into this region upon exposure to humid air. This accumulation of gas molecules at the graphene substrate interface was found in a recent experimental study of NH_3 adsorption and desorption from graphene [108]. Among the observed two-stage desorptions, the fact that one stage was fast and the other slow, was attributed, respectively, to easy desorption of NH_3 from the top surface of the graphene and much slower diffusion of NH_3 from the bottom surface of the graphene. This does not occur through the graphene itself, but involves gas diffusion through the "slit pores" substrate. Other mechanisms have been proposed for water's effect on graphene's properties. One is transfer doping by redox reaction in an analogous manner to the p-type doping of diamond by surface water and oxygen molecules that withdraw electrons from the diamond to form hydroxyl ions [109]. Another is a purely physical effect in which the high dielectric constant of water layers can shield the Coulomb scatterers in the substrate and so alter carrier mobility in graphene [97].

Besides demonstrating the high sensitivity of graphene as an electrical gas sensor and proposing a variety of mechanistic questions about gas sensing, Schedin et al. [91] also found that polymer residues and hydrocarbon contaminants could unintentionally functionalize graphene sensors by providing a surface phase in which the gas molecules can absorb and/or by influencing charge transfer from the gas to the graphene or vice versa. This type of surface layer is particularly pronounced when lithographic resists (typically polymethylmethacrylate [PMMA]) are used. Schedin et al. [91] observed that a 1-nm-thick residue remained on their devices after fabrication, despite thorough cleaning, consistent with a study by Ishigami et al. [110] that showed the difficulty in removing these deposits. The strong influence of residual polymers on electrical sensing by graphene was confirmed by Dan et al. [111]. Further, Dan et al. [111] performed two measurements of the gas-sensing behavior of electrical devices made by electron-beam lithography of mechanically exfoliated graphene. The first set of measurements was done after fabrication of the devices so that the few-layer graphene in the sensor was covered with a PMMA resist. The second set of measurements was done after removal of the resist by heating the devices to 400°C in a reducing H_2/Ar atmosphere for 1 hour. Removing the resist in this way improved the electrical quality of the devices markedly: carrier mobility increased by four times and the doped-carrier density was reduced by nearly 70%. Similarly, cleaning the graphene had a dramatic effect on gas sensing. With the PMMA residue on graphene, the device exhibited ultrafast measurable changes in electrical

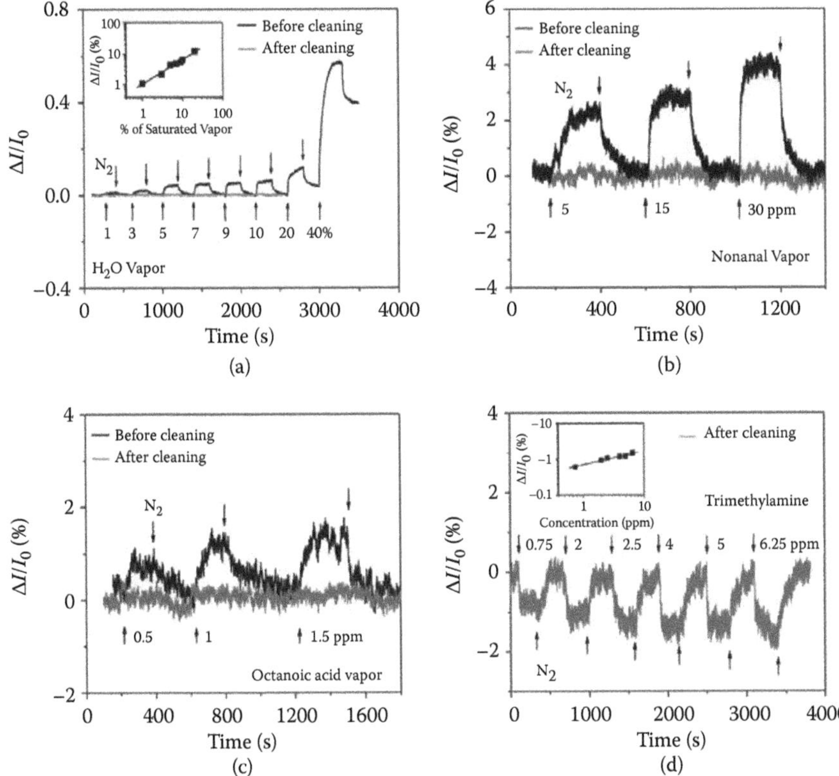

FIGURE 8.12 Gas-adsorption responses of graphene devices with and without a thin polymer layer of lithographic resist during adsorption of (a) water vapor, (b) nonanal vapor, (c) octanoic-acid vapor, and (d) trimethylamine vapor. The black lines (the larger responses) in (a)–(c) are for devices with the resist; i.e., before cleaning. The gray lines (smaller responses) in (a)–(d) are for devices after removal of the resist by annealing; i.e., after cleaning. (From Y.P. Dan, Y. Lu, N.J. Kybert, Z.T. Luo, and A.T.C. Johnson. 2009. Intrinsic response of graphene vapor sensors. *Nano Lett.* 9, 1472–1475. American Chemical Society. With permission.)

response to water and organic vapors in the parts-per-million range or below; after the polymer residue was gone, however, these responses declined by 1 to 2 orders of magnitude, virtually disappearing for some gases (Figure 8.12). Clearly, the high sensitivity to gas adsorption demonstrated by these graphene devices is not intrinsic to graphene, but due to the polymer layer (and any other contaminants, possibly including co-adsorbed water), which absorbs and concentrates the gases near the graphene surface. Despite this, Dan et al. [111] did not dismiss graphene as unsuitable for making sensors. Instead, they pointed out that the inherent two-dimensional nature and electrical properties of graphene have the advantage of low noise, and thus lower detection limits, but that the graphene surface should be cleaned and then deliberately functionalized to further increase sensitivity and/or selectivity.

Recent theoretical studies also supported this conclusion by demonstrating that gas molecules adsorb only weakly on pristine graphene, but adsorb more strongly on doped graphene or defective graphene (i.e., graphene that contains a vacancy) [100,105]. Specifically, defective graphene has the greatest sensitivity for CO, NO, and NO_2, whereas boron-doped (p-type) graphene was found to be sensitive to NH_3. Further, ab-initio study of graphene nanoribbons also implicated dangling edge bonds during strong gas adsorption [101]. However, experimental measurements suggest that such edge effects, though providing energetically favorable sites for gas absorption, contribute only a small part (~2%) of the total electrical response of a typical microscale graphene device [108]. Stronger adsorption, arising from either defects or doping (or edge sites on nanoribbons), generally results in more substantial changes in the density of states and the conductivity of graphene, which were essential for effective sensors. Nevertheless, with these strong interactions, the main problem is that the adsorption is effectively irreversible in realistic time frames and for realistic regeneration temperatures [105]. So there must be a balance between sensitivity and reversibility of the sensing reactions for practical sensors.

8.9 FUNCTIONALIZED GRAPHENE FOR GAS SENSORS

Research work performed on graphene-based gas sensors to date demonstrates that functionalization is the best way to achieve the best sensor performance out of graphene [105,111,118,119]. Functionalization imparts two important things to graphene: (1) it increases the sensitivity of the graphene to the adsorption process and (2) it reduces nonspecific binding and enhances selectivity (or specificity) for the desired analyte. Often, selectivity is equally critical for gas sensors made from other materials and it is a specific challenge for certain classes of sensors such as chemoresistive, field-effect, and piezoelectric gas sensors [93]. On the other hand, some of the functionalization approaches developed for other types of gas sensors, especially for single-walled carbon nanotubes [95], will definitely find applications in graphene-based gas sensors very soon. However, before considering all these possibilities, it is essential to analyze the methods that have been used to functionalize graphene to date. The primary purpose of recent functionalization methods is to increase the ease of exfoliating graphite to produce modified graphene or to make functionalized graphene for other applications, such as polymer nanocomposites. Therefore, many of the chemical-functionalization methods employed to date have used covalent bonding, thereby destroying the sp^2 bonding of graphene's lattice. Methods of non-covalent bonding, however, tend to exploit the extensive opportunities for π-bonding on graphene's basal plane, and are much more amenable than the harsh methods of covalent functionalization, and therefore tend to retain graphene's unique electronic properties for use in sensors.

Very often, chemical exfoliation of graphite is widely employed to synthesize graphene oxide and it results in the covalent functionalization of graphene. The strong oxidation of graphite produces graphene oxide, which incorporates oxygen-containing moieties such as epoxide, hydroxyl, carboxyl, and carbonyl groups. However, a disadvantage of this harsh functionalization process is the formation of many sp^3 bonds and the loss of many of graphene's original properties; but this is not necessarily a problem for all applications. For instance, after rapid heating of graphene oxide, the residual functional groups, as well as the wrinkles in its basal planes due to the changes in bonding, introduce vacancies [120], and it provides strong adhesion with polymers to form superb nanocomposites [121]. Alternatively, the functional groups on graphene oxide provide endless opportunities for further functionalization to create novel carbon sheets such as graphene amine, or new hybrid materials such as DNA tethered to graphene oxide [118]. This approach can be facilitated by chemical-coupling agents such as (O-(7-azabenzotriazol-1-yl)-1,1,3,3-tetramethyluroniumhexa fluorophosphate) or by various other chemical routes. For instance, Shen et al. [122] employed living free-radical polymerization with TEMPO (2,2,6,6-tetramethyl-1-piperidinyloxy) to produce a polystyrene-polyacrylamide block copolymer and covalently attach it to reduced-graphene-oxide sheets to make an amphiphilic form of the modified graphene. Other researchers employed phosgenation of graphene oxide for subsequent covalent attachment of octadecylamine [123] or porphyrin [124] molecules. Still others have used diazonium salts to attach different types of phenyl groups to surfactant-wrapped graphene oxide [125] or to form spontaneous covalent bonds between nitrophenyl and epitaxial graphene [126]. Most of the ongoing approaches to functionalize graphene substantially alter the transport and chemical properties of the graphene (derivatives) that are not suitable to design electrical sensors. A better method is to avoid covalent bonding, thereby reducing the impact of functionalization on graphene's properties [127], and three main approaches have emerged so far. The first avenue for noncovalent functionalization necessitates π-bonding or "π-stacking" between the π-orbitals of graphene's basal plane and those of aromatic functional molecules. For example, this type of bonding, though weaker than covalent bonding, has been used to accomplish the following: functionalize reduced graphene oxide with sulfonated polyaniline [127] or with 1-pyrenebutyrate [128], assemble monolayers of perylene-3,4,9,10-tetracarboxylic dianhydride on epitaxial graphene [129], or load the drug doxorubicin hydrochloride onto graphene oxide for potential drug-delivery applications [130]. The second approach is to simply coat the graphene with an extremely thin polymer layer, although this has largely been used unintentionally as a result of residual lithographic resists. As has been demonstrated by Dan et al. [111], such polymer layers markedly enhance graphene's sensitivity for electrical detection of gas molecules (Figure 8.12); they also might be used to increase selectivity, as is discussed in the following text. The third method draws on electrochemistry to reduce metallic nanoparticles onto graphene derivatives. For instance, Lu et al. [119] have "decorated" expanded graphite nanoplatelets with platinum or palladium nanoparticles through a microwave-heated polyol process (Figure 8.13). In another work, Sundaram et al. [131] employed sheets of reduced graphene oxide as the electrode on which they grew palladium nanoparticles by electrodeposition. However, a considerable amount of work will have to be performed to assess the limits of sensitivity and selectivity imparted to the graphene-based gas sensors by noncovalent methods. Nevertheless, considering the state of the art in functionalizing other types of gas sensors, it is unlikely that such methods will ever create graphene-based devices that are only sensitive to a single gas. Instead, we expect that practical graphene-based gas sensors will follow the trend seen in recent years in which researchers have begun to harness cross-sensitivity in sensors, rather than seeking to obviate it. The basic idea is to design an array of gas sensors, where each element in the array is functionalized differently [93,95]. When exposed to a

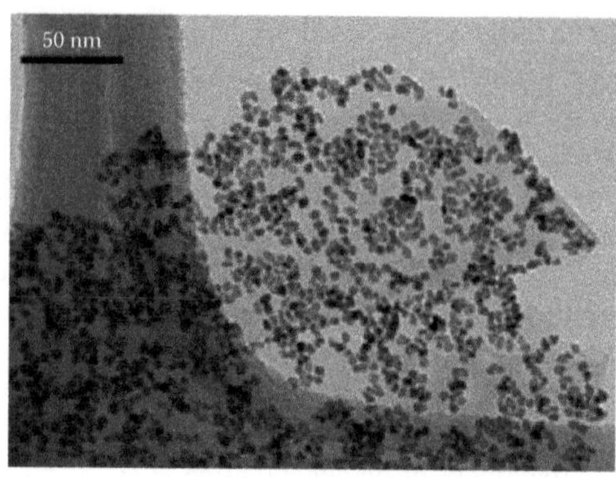

FIGURE 8.13 TEM micrograph of an expanded-graphite nanoplatelet coated with 50 wt % platinum nanoparticles. Darker material in the left and bottom of the micrograph is the sample support. (From Y.F. Xu, Z.B. Liu, X.L. Zhang, Y. Wang, J.G. Tian, Y. Huang, Y.F. Ma, X.Y. Zhang, and Y.S. Chen. 2009. A graphene hybrid material covalently functionalized with porphyrin: Synthesis and optical limiting property. *Adv. Mater.* 21, 1275–1279. American Chemical Society. With permission.)

mixture of gases, the array will produce a matrix of responses, with each sensor's response comprising different contributions from the various gases due to that sensor's particular functionalization. By using multivariate data analysis or *chemometrics* to mathematically extract the different components from across the array for each gas, one can produce a characteristic pattern or "fingerprint" for each type of gas, which can be compared to a database of patterns for gases and vapors. In this way, it becomes possible to identify the types and amounts of mixtures of gases with a single sensor array, which is often called an *electronic nose.* This approach has proven effective in a variety of gas sensors made with different transducer materials and different types of functionalization. One recent example was a tin-oxide semiconductor array, with differences in functionalization achieved by varied levels of palladium doping and different electrode geometries; the eight-element array was used to measure mixtures of CO, CH_4, and H_2O [132]. Another study employed an array of carbon-nanotube field-effect sensors, each of which was decorated with one of four different types of metallic nanoparticles to uniquely identify H_2, H_2S, NH_3, and NO_2 [133]. A third example involved coating surface acoustic wave (SAW) gas sensors with one of five different polymers to produce a sensor array capable of clearly detecting four organic gases commonly used to simulate chemical warfare agents [134]. This sensor-array approach has also been used successfully for optical biosensors built around functionalized nanoparticles [135]. Johnson et al. [136] have also proposed that various base sequences of DNA strands might provide another useful method for variable functionalization of gas-sensor arrays. There are obvious similarities between the basic functionalization approaches employed in the foregoing examples of sensor arrays and the routes already used to functionalize graphene (or its derivatives) nanoparticles, polymers, DNA, and so on. Therefore, it seems reasonable to conclude that the sensor-array approaches could be used, with equal or greater success, on graphene-based gas sensors. Indeed, graphene is likely to prove particularly amenable to creation of sensor arrays because it can be easily integrated into lithographic processes for microfabrication or nanofabrication. This offers potential advantages such as rapid and low-cost production of compact sensors that include large numbers of nanoscale graphene elements; one might even be able to create an array of individual sensor elements from a single, large sheet of graphene, such as that grown by CVD, by means of lithographic patterning and plasma etching. The development and implementation of these kinds of approaches will allow researchers to tailor the size, sensitivity, and selectivity of graphene to create a new generation of gas-sensor arrays that are compact and that can analyze diverse environmental gases and vapors at a parts-per-billon level or perhaps even less.

8.10 CONCLUSION

Graphene, as the "mother" of all other allotropes of carbon, continues to exhibit superior conducting, electronic, and physicochemical properties compared to its counterparts. In addition to these properties, graphene continues to dominate its counterparts (other carbon-based materials) by exhibiting superb performance in direct electrochemistry of enzymes, electrochemical detection of small biomolecules, and in electroanalysis (electrochemical sensors for bioanalysis and environmental analysis). Graphene is also biocompatible and is free from metallic impurities, which often lead us to an incorrect conclusion during electroanalysis using CNTs. On the other hand, better understanding of physics and chemistry at the surface of graphene and interaction of chemicals and biomolecules at the graphene interface will play an important role while employing graphene as a nanoscaffold during chemical and biosensing. For example, the absorption mechanism of molecules on graphene, orientation of biomolecules on the graphene, and how these interactions affect the transport properties of graphene will provide us with a further understanding of graphene and its interaction with molecules, which may lead to great advancements in graphene science and its various applications, such as catalysis and sensors. Nevertheless, several major hurdles have to be resolved prior to employing a truly single sheet of graphene for sensing applications. Further, commercial production of graphene-based gas sensors is still some way off, and some important challenges must be met along the way. Fabrication is one of these challenges, as researchers seek cost-effective production methods that retain graphene's essential properties and can be scaled up. If high-purity graphene is the primary material needed for sensor applications, then, at this stage, CVD or epitaxial growth seem to be the synthetic routes that will likely come to dominate. In conclusion, though there are several reports available on different kinds of biosensing applications of graphene and its derivatives, there is still much room for scientific research and development of graphene-based ultrasensitive sensors for various biosensing applications.

ACKNOWLEDGMENTS

The authors would like to acknowledge the support from National Science Foundation. This work was supported by the following NSF grants: NIRT Number 0404137, CREST Number 0734232, IGERT Number 0221681, and GK12 Number 0638709.

REFERENCES

1. X.L. Luo, A. Morrin, A.J. Killard, and M.R. Smyth. 2006. Application of nanoparticles in electrochemical sensors and biosensors, *Electroanalysis* 18, 319–326.
2. K. Balasubramanian and M. Burghard. 2006. *Anal. Bioanal. Chem.* 385, 452.
3. A. Merkoci. 2006. Carbon nanotubes in analytical sciences. *Microchim. Acta* 152, 157–174.
4. B.S. Sherigara, W. Kutner, and F. D'Souza. 2003. Electrocatalytic properties and sensor applications of fullerenes and carbon nanotubes. *Electroanalysis* 15, 753–772.
5. Y.H. Lin, W. Yantasee, and J. Wang. 2005. Carbon nanotubes for the development of electrochemical biosensors. *Front. Biosci.* 10, 492–505.
6. J. Wang, M. Musameh, and Y.H. Lin. 2003. Solubilization of carbon nanotubes by nafion towards the preparation of amperometric biosensors. *J. Am. Chem. Soc.* 125, 2408–2409.
7. J. Wang and M. Musameh. 2003. Carbon nanotube/Teflon composite electrochemical sensors and biosensors. *Anal. Chem.* 75, 2075–2079.
8. M. Musameh, J. Wang, A. Merkoci, and Y.H. Lin. 2002. Low potential stable NADH detection at carbon nanotube modified glassy carbon electrode. *Electrochem. Commun.* 4, 743–746.
9. J.X. Wang, M.X. Li, Z.J. Shi, N.Q. Li, and Z.N. Gu. 2002. Direct electrochemistry of Cytochrome c at a glassy carbon electrode modified with single-wall carbon nanotubes. *Anal. Chem.* 74, 1993.
10. Z.H. Wang, J. Liu, Q.L. Liang, Y.M. Wang, and G. Luo. 2002. Carbon nanotube modified electrodes for simultaneous detection of dopamine and ascorbic acid. *Analyst* 127, 653–658.
11. K.S. Novoselov, A.K. Geim, S.V. Morozov, D. Jiang, Y. Zhang, S.V. Dubonos, I.V. Grigorieva, and A.A. Firsov. 2004. Electric field effect in atomically thin carbon films. *Science* 306, 666–669.
12. D. Jiang, F. Schedin, T.J. Booth, V.V. Khotkevich, S.V. Morozov, and A.K. Geim. 2005. Two-dimensional atomic crystals. *Proc. Natl Acad. Sci. USA* 102, 10451–10453.
13. K.S. Novoselov, A.K. Geim, S.V. Morozov, D. Jiang, M.I. Katsnelson, I.V. Grigorieva, S.V. Dubonos, and A.A. Firsov. 2005. Two-dimensional gas of massless Dirac fermions in graphene. *Nature* 438, 197–200.
14. A.K. Geim and K.S. Novoselov. 2007. The rise of graphene, *Nat. Mater.* 6, 183–191.
15. G. Brumfiel. 2010. Andre Geim: In praise of graphene. *Nature News*, http://www.nature.com/news/2010/101007/full/news.2010.525.html.
16. S. Mazzocchi. 2010. The big, little substance graphene. PBS.com, http://www.pbs.org/wnet/need-to-know/five-things/the-big-little-substance-graphene/4146/
17. C. Lee, X.Wei, J.W. Kysar, and J. Hone. 2008. Measurement of the elastic properties and intrinsic strength of monolayer graphene. *Science* 321, 385–388.
18. A.A. Balandin, S. Ghosh, W. Bao, I. Calizo, D. Teweldebrhan, F. Miao, and C.N. Lau. 2008. Superior thermal conductivity of single-layer graphene. *Nano Lett.* 8, 902–907.
19. K.I. Bolotin, K.J. Sikes, Z. Jiang, M. Klima, G. Fudenberg, J. Hone, P. Kim, and H.L. Stormer. 2008. Ultrahigh electron mobility in suspended graphene. *Solid State Commun.* 146, 351–355.
20. M.D. Stoller, S. Park, Y. Zhu, J. An, and R.S. Ruoff. 2008. Graphene-based ultracapacitors. *Nano Lett.* 8, 3498–3502.
21. Z. Jiang, Y. Zhanga, Y.W. Tan, H.L. Stormer, and P. Kim. 2007. Quantum Hall effect in graphene. *Solid State Commun.* 143, 14–19.
22. M.J. Allen, V.C. Tung, and R.B. Kaner. 2010. Honey comb graphene: A review of graphene. *Chem. Rev.* 110, 132.

22. S. Alwarappan, A. Erdem, C. Liu, and C-Z. Li. 2009. Probing the electrochemical properties of graphene nanosheets for biosensing applications. *J. Phy. Chem. C* 113, 8853–8857.
23. S. Alwarappan, C. Liu, A. Kumar, and C-Z. Li. 2010. Enzyme-doped graphene nanosheets for enhanced glucose biosensing. *J. Phy. Chem. C* 114, 12920–12924.
24. Y. Liu, D. Yu, C. Zeng, Z-C. Miao, and L. Dai. 2010. Biocompatible graphene oxide-based glucose biosensors. *Langmuir* 26, 6158–6160.
25. M. Zhou, Y.M. Zhai, and S.J. Dong. 2009. Electrochemical biosensing based on reduced graphene oxide. *Anal. Chem.* 81, 5603–5613.
26. D.A. Dikin. 2007. Preparation and characterization of graphene oxide paper. *Nature* 448, 457–460.
27. Y. Shao, J. Wang, H. Wu, J. Liu, I.A. Aksay, and Y. Lin. 2010. Graphene based electrochemical sensors and biosensors: A review. *Electroanalysis* 22, 1027–1036.
28. O. Niwa, J. Jia, Y. Sato, D. Kato, R. Kurita, K. Maruyama, K. Suzuki, and S. Hirono. 2006. Electrochemical performance of angstrom level flat sputtered carbon film consisting of sp^2 and sp^3 mixed bonds. *J. Am. Chem. Soc.* 128, 7144–7145.
29. C.E. Banks, T.J. Davies, G.G. Wildgoose, and R.G. Compton. 2005. Electrocatalysis at graphite and carbon nanotube modified electrodes: Edge plane sites and tube ends are reactive sites. *Chem. Commun.* 829–841.
30. S.L. Yang, D.Y. Guo, L. Su, P. Yu, D. Li, J.S. Ye, and L.Q. Mao. 2009. A facile method for preparation of graphene film electrodes with tailor-made dimensions with Vaseline as the insulating binder. *Electrochem. Commun.* 11, 1912–1915.
31. W.J. Lin, C.S. Liao, J.H. Jhang, and Y.C. Tsai. 2009. Graphene modified basal and edge plane pyrolytic graphite electrodes for electrocatalytic oxidation of hydrogen peroxide and β-nicotinamide adenine dinucleotide. *Electrochem. Commun.* 11, 2153–2156.
32. L.H. Tang, Y. Wang, Y.M. Li, H.B. Feng, J. Lu, and J.H. Li. 2009. Preparation, structure and electrochemical properties of reduced graphene sheet films. *Adv. Funct. Mater.* 19, 2782–2789.
33. N.G. Shang, P. Papakonstantinou, M. McMullan, M. Chu, A. Stamboulis, A. Potenza, S.S. Dhesi, and H. Marchetto. 2008. Catalyst free efficient growth, orientation and biosensing properties of multilayer graphene nanoflake films with sharp edge planes. *Adv. Funct. Mater.* 18, 3506–3514.
34. J.F. Wang, S.L. Yang, D.Y. Guo, P. Yu, D. Li, J.S. Ye, and L.Q. Mao. 2009. Comparative studies on electrochemical activity of graphene nanosheets and carbon nanotubes. *Electrochem. Commun.* 11, 1892.
35. R.S. Nicholson. 1965. Theory and application of cyclic voltammetry for measurement of electrode reaction kinetics. *Anal. Chem.* 37, 1351–1355.
36. A.E. Fischer, Y. Show, and G.M. Swain. 2004. Electrochemical performance of diamond thin film electrodes from different commercial sources. *Anal. Chem.* 76, 2553–2560.
37. R.L. McCreery. 2008. Advanced electrode materials for molecular electrochemistry. *Chem. Rev.* 108, 2646–2687.
38. Y.L. Yao and K.K. Shiu. 2008. Direct electrochemistry of glucose oxidase at carbon nanotube-gold colloid modified electrode with poly(diallyldimethylammonium chloride) coating. *Electroanal.* 20, 1542–1548.
39. C. Leger and P. Bertrand. 2008. Direct electrochemistry of redox enzymes as a tool for mechanistic studies. *Chem. Rev.* 108, 2379–2438.
40. F.A. Armstrong, H.A.O. Hill, and N.J. Walton. 1988. Direct electrochemistry of redox proteins. *Accounts Chem. Res.* 21, 407–413.
41. Y.H. Wu and S.S. Hu. 2007. Biosensor based on direct electron transfer in protein. *Microchim. Acta* 159, 1–17.

42. P.A. Prakash, U. Yogeswaran, and S.M. Chen. 2009. A review on direct electrochemistry of catalase for electrochemical sensors. *Sensors* 9, 1821–1844.
43. A.L. Ghindilis, P. Atanasov, and E. Wilkins. 1997. Enzyme-catalyzed direct electron transfer: Fundamentals and analytical applications. *Electroanalysis* 9, 661–674.
44. J. Wang. 2005. Carbon nanotube based electrochemical biosensors: A review. *Electroanalysis* 17, 7–14.
45. J. Wang and Y.H. Lin. 2008. Functionalized carbon nanotubes and nanofibers for biosensing applications. *Trac-Trends Anal. Chem.* 27, 619.
46. E. Katz and I. Willner. 2004. Integrated nanoparticle-biomolecule hybrid systems: Synthesis, properties and applications. *Angew. Chem. Int. Ed.* 43, 6042–6108.
47. A.K. Sarma, P. Vatsyayan, P. Goswami, and S.D. Minteer. 2009. Recent advances in material science for developing enzyme electrodes. *Biosens. Bioelectron.* 24, 2313–2322.
48. Y. Chen, Y. Li, D. Sun, D. Tian, J. Zhang, and J.-J. Zhu. 2011. Fabrication of gold nanoparticles on bilayer graphene for glucose electrochemical biosensing. *J. Mater. Chem.* 21, 7604–7611.
49. C.S. Shan, H.F. Yang, J.F. Song, D.X. Han, A. Ivaska, and L. Niu. 2009. Direct electrochemistry of glucose oxidase and biosensing for glucose based on graphene. *Anal. Chem.* 81, 2378–2382.
50. X.H. Kang, J. Wang, H. Wu, A.I. Aksay, J. Liu, and Y.H. Lin. 2009. Glucose oxidase-graphene-chitosan modified electrode for direct electrochemistry and glucose sensing. *Biosens. Bioelectron.* 25, 901–905.
51. H.C. Schniepp, J.L. Li, M.J. McAllister, H. Sai, M.H. Alonso, D.H. Adamson, R.K. Prud'homme, R. Car, D.A. Saville, and I.A. Aksay. 2006. Functionalized single graphene sheets derived from splitting graphite oxide. *J. Phys. Chem. B* 110, 8535–8539.
52. Z.H. Dai, J. Ni, X.H. Huang, G.F. Lu, and J.C. Bao. 2007. Direct electrochemistry of glucose oxidase immobilized on a hexagonal mesoporous silica-MCM-41 matrix. *Bioelectrochem.* 70, 250–256.
53. A. Guiseppi-Elie, C.H. Lei, and R.H. Baughman. 2002. Direct electron transfer of glucose oxidase on carbon nanotubes. *Nanotechnology* 13, 559.
54. C.Y. Deng, J.H. Chen, X.L. Chen, C.H. Mao, L.H. Nie, and S.Z. Yao. 2008. *Biosens. Bioelectron.* 23, 1272.
55. C.X. Cai and J. Chen. 2004. Direct electron transfer of glucose oxidase promoted by carbon nanotubes. *Anal. Biochem.* 332, 75–83.
56. S. Alwarappan, R.K. Joshi, M.K. Ram, and A. Kumar. 2010. Electron transfer mechanism of Cytochrome c at graphene electrode. *Appl. Phys. Lett.* 96, 263702.
57. C.S. Shan, H.F. Yang, J.F. Song, D.X. Han, A. Ivaska, and L. Niu. 2009. Direct electrochemistry of glucose oxidase and biosensing for glucose based on graphene. *Anal. Chem.* 81, 2378–2382.
58. Z.J. Wang, X.Z. Zhou, J. Zhang, F. Boey, and H. Zhang. 2009. Direct electrochemical reduction of single-layer graphene oxide and subsequent functionalization with glucose oxidase. *J. Phys. Chem. C.* 113, 14071–14075.
59. H. Wu, J. Wang, X. Kang, C. Wang, D. Wang, J. Liu, I.A. Aksay, and Y. Lin. 2009. Glucose biosensor based on immobilization of glucose oxidase in platinum nanoparticles/graphene/chitosan nanocomposite film. *Talanta* 80, 403–406.
60. X.H. Kang, J.Wang, H. Wu, A.I. Aksay, J. Liu, and Y.H. Lin. 2009. Glucose oxidase–graphene–chitosan modified electrode for direct electrochemistry and glucose sensing. *Biosens. Bioelectron.* 25, 901–905.
61. J. Lu, L.T. Drzal, R.M. Worden, and I. Lee. 2007. Simple fabrication of a highly sensitive glucose biosensor using enzymes immobilized in exfoliated graphite nanoplatelets nafion membrane. *Chem. Mat.* 19, 6240–6246.

62. M. Zhou, Y.M. Zhai, and S.J. Dong. 2009. Electrochemical biosensing based on reduced graphene oxide. *Anal. Chem.* 81, 5603–5613.
63. G.D. Liu and Y.H. Lin. 2006. Amperometric glucose biosensor based on self-assembling glucose oxidase on carbon nanotubes. *Electrochem. Commun.* 8, 251–256.
64. L. Wu, X.J. Zhang, and H.X. Ju. 2007. Amperometric glucose sensor based on catalytic reduction of dissolved oxygen at soluble carbon nanofiber. *Biosens. Bioelectron.* 23, 479–484.
65. M.D. Rubianes and G.A. Rivas. 2003. Carbon nanotubes paste electrode. *Electrochem. Commun.* 5, 689–694.
66. Y.H. Lin, F. Lu, Y. Tu, and Z.F. Ren. 2004. Glucose biosensors based on carbon nanotube nanoelectrode ensembles. *Nano Lett.* 4, 191–195.
67. M. Zhou, L. Shang, B.L. Li, L.J. Huang, and S.J. Dong. 2008. Highly ordered mesoporous carbons as electrode material for the construction of electrochemical dehydrogenase and oxidase based biosensors. *Biosens. Bioelectron.* 24, 442–447.
68. C.S. Shan, H.F. Yang, D.X. Han, Q.X. Zhang, A. Ivaska, and L. Niu. 2009. Graphene/AuNPs/chitosan nanocomposites film for glucose biosensing. *Biosens. Bioelectron.* 25, 1070.
69. J. Wang. 2008. Electrochemical glucose biosensors. *Chem. Rev.* 108, 814–825.
70. J.A. Cracknell, K.A. Vincent, and F.A. Armstrong. 2008. Enzymes as working or inspirational electrocatalysts for fuel cells and electrolysis. *Chem. Rev.* 108, 2439–2461.
71. C.E. Banks, R.R. Moore, T.J. Davies, and R.G. Compton. 2004. Investigation of modified basal plane pyrolytic graphite electrodes: Definitive evidence for the electrocatalytic properties of the ends of carbon nanotubes. *Chem. Commun.* 1804–1805.
72. C.E. Banks and R.G. Compton. 2005. Exploring the electrocatalytic sites of carbon nanotubes for NADH detection: An edge plane pyrolytic graphite electrode study. *Analyst* 130, 1232–1239.
73. M. Pumera, R. Scipioni, H. Iwai, T. Ohno, Y. Miyahara, and M. Boero. 2009. A mechanism of adsorption of β-Nicotinamide adenine dinucleotide on graphene sheets: Experiment and theory. *Chem. Eur. J.* 15, 10851.
74. H. Liu, J. Gao, M.Q. Xue, N. Zhu, M.N. Zhang, and T.B. Cao. 2009. Processing of graphene for electrochemical application: Noncovalently functionalized graphene sheets with water-soluble electroactive methylene green. *Langmuir* 25, 12006–12010.
75. F. Valentini, A. Amine, S. Orlanducci, M.L. Terranova, and G. Palleschi. 2003. Carbon nanotube purification: Preparation and characterization of carbon nanotube paste electrodes. *Anal. Chem.* 75, 5413–5421.
76. M.G. Zhang, A. Smith, and W. Gorski. 2004. Carbon nanotube-chitosan system for electrochemical sensing based on dehydrogenase enzymes. *Anal. Chem.* 76, 5045–5050.
77. X. Kang, J. Wang, H. Wu, J. Liu, I.A. Aksay, and Y. Lin. 2010. A graphene-based electrochemical sensor for sensitive detection of paracetamol. *Talanta* 81, 754–759.
78. X. Zuo, S. He, D. Li, C. Peng, Q. Huang, S. Song, and C. Fan. 2010. Graphene oxide-facilitated electron transfer of metalloproteins at electrode surfaces. *Langmuir* 26, 1936–1939.
79. C. Fu, W. Yang, X. Chen, and D.G. Evans. 2009. Direct electrochemistry of glucose oxidase on a graphite nanosheet–Nafion composite film modified electrode. *Electrochem. Commun.* 11, 997–1000.
80. A. Sassolas, B.D. Leca-Bouvier, and L.J. Blum. 2008. DNA biosensors and microarrays. *Chem. Rev.* 108, 109–139.
81. T.G. Drummond, M.G. Hill, and J.K. Barton. 2003. Electrochemical DNA biosensors. *Nat. Biotechnol.* 21, 1192–1199.
82. A. Ambrosi and M. Pumera. 2010. Stacked graphene nanofibers of electrochemical oxidation of DNA bases. *Phys. Chem. Chem. Phys.* 12, 8943–8947.

83. M. Pumera, A. Ambrosi, A. Bonanni, E.L.K. Chng, and H.L. Poh. 2010. Graphene for electrochemical sensing and biosensing. *Trends in Anal.Chem.* 29, 954–965.
84. V.Vamvakaki, M. Fouskaki, and N.Chaniotakis. 2007. Electrochemical biosensoring system based on carbon nanotubes and carbon nanofibers. *Anal. Lett.* 40, 2271–2287.
85. V. Vamvakaki, K. Tsagaraki, and N. Chaniotakis. 2006. Carbon-nanofiber based glucose biosensor. *Anal. Chem.* 78, 5538–5542.
86. J. Li, S.J. Guo, Y.M. Zhai, and E.K. Wang. 2009. High-sensitivity determination of lead and cadmium based on the Nafion-graphene composite film. *Anal. Chim. Acta* 649, 196–201.
87. J. Li, S.J. Guo, Y.M. Zhai, and E.K. Wang. 2009. Nafion-graphene nanocomposite film as enhanced sensing platform for ultrasensitive determination of cadmium. *Electrochem. Commun.* 11, 1085.
88. G. Kefala, A. Economou, and A. Voulgaropoulos. 2004. A study of Nafion-coated bismuth-film electrodes for the determination of trace metals by anodic stripping voltammetry. *Analyst* 129, 1082–1090.
89. L.D. Zhu, C.Y. Tian, R.L. Yang, and J.L. Zhai. 2008. Anodic stripping determination of lead in tap water at an ordered mesoporous carbon/nafion composite film electrode. *Electroanalysis* 20, 527–533.
90. H. Xu, L.P. Zeng, S.J. Xing, Y.Z. Xian, and G.Y. Shi. 2008. Ultrasensitive voltammetric detection of trace lead (III) and cadmium (III) using MWCNTs-nafion/bismuth composite electrodes. *Electroanalysis* 20, 2655–2662.
91. F. Schedin, A.K. Geim, S.V. Morozov, E.W. Hill, P. Blake, M.I. Katsnelson, and K.S. Novoselov. 2007. Detection of individual gas molecules adsorbed on graphene, *Nat. Mater.* 6, 652–655.
92. K.R. Ratinac, W. Yang, S.P. Ringer, and F. Braet. 2010. Toward ubiquitous environmental gas sensors: Capitalizing on the promise of graphene. *Env. Sci and Tech.* 44, 1167–1176.
93. S. Capone, A. Forleo, L. Francioso, R. Rella, P. Siciliano, J. Spadavecchia, D.S. Presicce, and A.M. Taurino. 2003. Solid-state gas sensors: State of the art and future activities. *J. Optoelect. Adv. Mater.* 5, 1335–1348.
94. C.O. Park, J.W. Fergus, N. Miura, J. Park, and A. Choi. 2009. Solid-state electrochemical gas sensors. *Ionics* 15, 261–284.
95. P. Bondavalli, P. Legagneux, and D. Pribat. 2009. Carbon nanotubes based transistors as gas sensors: State of the art and critical review. *Sens. Actuators B* 140, 304–318.
96. E.H. Hwang, S. Adam, and S. Das Sarma. 2007. Transport in chemically doped graphene in the presence of adsorbed molecules. *Phys. Rev. B* 76, 195421.
97. S. Adam, E.H. Hwang, and S. Das Sarma. 2008. Scattering mechanisms and Boltzmann transport in graphene. *Physica E* 40, 1022–1025.
98. Y.W. Tan, Y. Zhang, K. Bolotin, Y. Zhao, S. Adam, E.H. Hwang, S. Das Sarma, H.L. Stormer, and P. Kim. 2007. Measurement of scattering rate and minimum conductivity in graphene. *Phys. Rev. Lett.* 99, 246803.
99. K.I. Bolotin, K.J. Sikes, Z. Jiang, M. Klima, G. Fudenberg, J. Hone, P. Kim, and H.L. Stormer. 2008. Ultrahigh electron mobility in suspended graphene. *Solid State Commun.* 146, 351–355.
100. Z.M. Ao, J. Yang, S. Li, and Q. Jiang. 2008. Enhancement of CO detection in Al doped graphene. *Chem. Phys. Lett.* 461, 276–279.
101. B. Huang, Z.Y. Li, Z.R. Liu, G. Zhou, S.G. Hao, J. Wu, B.L. Gu, and W.H. Duan. 2008. Adsorption of gas molecules on graphene nanoribbons and its implication for nanoscale molecule sensor. *J. Phys. Chem. C* 112, 13442–13446.
102. O. Leenaerts, B. Partoens, and F.M. Peeters. 2008. Paramagnetic adsorbates on graphene: A charge-transfer analysis. *Appl. Phys. Lett.* 92, 243125.

103. O. Leenaerts, B. Partoens, and F.M. Peeters. 2008. Adsorption of H_2O, NH_3, CO, NO_2, and NO on graphene: A first-principles study. *Phys. Rev. B* 77, 125416.
104. T.O. Wehling, K.S. Novoselov, S.V. Morozov, E.E. Vdovin, M.I. Katsnelson, A.K. Geim, and A.I. Lichtenstein. 2008. Molecular doping of graphene. *Nano Lett.* 8, 173–177.
105. Y.H. Zhang, Y.B. Chen, K.G. Zhou, C.H. Liu, J. Zeng, H.L. Zhang, and Y. Peng. 2009. Improving gas sensing properties of graphene by introducing dopants and defects: A first-principles study. *Nanotechnology* 20, 185504.
106. T.O. Wehling, M.I. Katsnelson, and A.I. Lichtenstein. 2009. Adsorbates on graphene: Impurity states and electron scattering. *Chem. Phys. Lett.* 476, 125–134.
107. M. Chi and Y.P. Zhao. 2009. Adsorption of formaldehyde molecule on the intrinsic and Al-doped graphene: A first-principle study. *Comput. Mater. Sci.* 46, 1085–1090.
108. H.E. Romero, P. Joshi, A.K. Gupta, H.R. Gutierrez, M.W. Cole, S.A. Tadigadapa, and P.C. Eklund. 2009. Adsorption of ammonia on graphene. *Nanotechnology* 20, 245501.
109. S.J. Sque, R. Jones, and P.R. Briddon. 2007. The transfer doping of graphite and graphene. *Phys. Status Solidi A* 204, 3078–3084.
110. M. Ishigami, J.H. Chen, W.G. Cullen, M.S. Fuhrer, and E.D. Williams. 2007. Atomic structure of graphene on SiO_2. *Nano Lett.* 7, 1643–1648.
111. Y.P. Dan, Y. Lu, N.J. Kybert, Z.T. Luo, and A.T.C. Johnson. 2009. Intrinsic response of graphene vapor sensors. *Nano Lett.* 9, 1472–1475.
112. R. Arsat, M. Breedon, M. Shafiei, P.G. Spizziri, S. Gilje, R.B. Kaner, K. Kalantar-Zadeh, and W. Wlodarski. 2009. Graphene-like nanosheets for surface acoustic wave gas sensor applications. *Chem. Phys. Lett.* 46, 344–347.
113. J.D. Fowler, M.J. Allen, V.C. Tung, Y. Yang, R.B. Kaner, and B.H. Weiller. 2009. Practical chemical sensors from chemically derived graphene. *ACS Nano* 3, 301–306.
114. I. Jung, D. Dikin, S. Park, W. Cai, S.L. Mielke, and R.S. Ruoff. 2008. Effect of water vapor on electrical properties of individual reduced graphene oxide sheets. *J. Phys. Chem. C* 112, 20264–20268.
115. G.H. Lu, L.E. Ocola, and J.H. Chen. 2009. Gas detection using low temperature reduced graphene oxide sheets. *Appl. Phys. Lett.* 94, 083111.
116. G.H. Lu, L.E. Ocola, and J.H. Chen. 2009. Reduced graphene oxide for room-temperature gas sensors, *Nanotechnology* 20, 445502.
117. J.T. Robinson, F.K. Perkins, E.S. Snow, Z.Q. Wei, and P.E. Sheehan. 2008. Reduced graphene oxide molecular sensors. *Nano Lett.* 8, 3137–3140.
118. N. Mohanty and V. Berry. 2008. Graphene-based single-bacterium resolution bio-device and DNA transistor: Interfacing graphene derivatives with nanoscale and microscale biocomponents. *Nano Lett.* 8, 4469–4476.
119. J. Lu, I. Do, L.T. Drzal, R.M. Worden, and I. Lee. 2008. Nanometal decorated exfoliated graphite nanoplatelet based glucose biosensors with high sensitivity and fast response. *ACS Nano*, 2, 1825–1832.
120. H.C. Schniepp, J.L. Li, M.J. McAllister, H. Sai, M. Herrera-Alonso, D.H. Adamson, R.K. Prud'homme, R. Car, D.A. Saville, and I.A. Aksay. 2006. Functionalized single graphene sheets derived from splitting graphite oxide. *J. Phys. Chem. B* 110, 8535–8539.
121. T. Ramanathan, A.A. Abdala, S. Stankovich, D.A. Dikin, M. Herrera-Alonso, R.D. Piner, D.H. Adamson, H.C. Schniepp, X. Chen, R.S. Ruoff, S.T. Nguyen, I.A. Aksay, R.K. Prud'homme, and L.C. Brinson. 2008. Functionalized graphene sheets for polymer nanocomposites. *Nat. Nanotechnol.* 3, 327–331.
122. J.F. Shen, Y.H. Hu, C. Li, C. Qin, and M.X. Ye. 2009. Synthesis of amphiphilic graphene nanoplatelets. *Small* 5, 82–85.
123. S. Niyogi, E. Bekyarova, M.E. Itkis, J.L. McWilliams, M.A. Hamon, and R.C. Haddon. 2006. Solution properties of graphite and graphene. *J. Am. Chem. Soc.* 128, 7720–7721.

124. Y.F. Xu, Z.B. Liu, X.L. Zhang, Y. Wang, J.G. Tian, Y. Huang, Y.F. Ma, X.Y. Zhang, and Y.S. Chen. 2009. A graphene hybrid material covalently functionalized with porphyrin: Synthesis and optical limiting property. *Adv. Mater.* 21, 1275–1279.
125. J.R. Lomeda, C.D. Doyle, D.V. Kosynkin, W.F. Hwang, and J.M. Tour. 2008. Diazonium functionalization of surfactant-wrapped chemically converted graphene sheets. *J. Am. Chem. Soc.* 130, 16201–16206.
126. E. Bekyarova, M.E. Itkis, P. Ramesh, C. Berger, M. Sprinkle, W.A. De Heer, and R.C. Haddon. 2009. Chemical modification of epitaxial graphene: Spontaneous grafting of aryl groups. *J. Am. Chem.Soc.* 131, 1336–1337.
127. H. Bai, Y.X. Xu, L. Zhao, C. Li, and G.Q. Shi. 2009. Non-covalent functionalization of graphene sheets by sulfonated polyaniline. *Chem. Commun.* 13, 1667–1669.
128. Y.X. Xu, H. Bai, G.W. Lu, C. Li, and G.Q. Shi. 2008. Flexible graphene films via the filtration of water-soluble non-covalent functionalized graphene sheets. *J. Am. Chem. Soc.* 130, 5856–5857.
129. Q.H. Wang and M.C. Hersam. 2009. Room-temperature molecular resolution characterization of self-assembled organic monolayers on epitaxial graphene. *Nat. Chem.* 1, 206–211.
130. X.Y. Yang, X.Y. Zhang, Z.F. Liu, Y.F. Ma, Y. Huang, and Y. Chen. 2008. High-efficiency loading and controlled release of doxorubicin hydrochloride on graphene oxide. *J. Phys. Chem. C* 112, 17554–17558.
131. R.S. Sundaram, C. Gomez-Navarro, K. Balasubramanian, M. Burghard, and K. Kern. 2008. Electrochemical modification of graphene. *Adv. Mater.* 20, 3050–3053.
132. S. Capone, P. Siciliano, N. Barsan, U. Weimar, and L. Vasanelli. 2001. Analysis of CO and CH_4 gas mixtures by using a micromachined sensor array. *Sens. Actuators B* 78, 40–48.
133. A. Star, V. Joshi, S. Skarupo, D. Thomas, and J.C.P. Gabriel. 2006. Gas sensor array based on metal-decorated carbon nanotubes. *J. Phys. Chem. B* 110, 21014–21020.
134. B.S. Joo, J.S. Huh, and D.D. Lee. 2007. Fabrication of polymer SAW sensor array to classify chemical warfare agents. *Sens. Actuators B* 121, 47–53.
135. M. De, S. Rana, H. Akpinar, O.R. Miranda, R.R. Arvizo, U.H.F. Bunz, and V.M. Rotello. 2009. Sensing of proteins in human serum using conjugates of nanoparticles and green fluorescent protein. *Nat. Chem.* 1, 461–465.
136. A.T.C. Johnson, C. Staii, M. Chen, S. Khamis, R. Johnson, M.L. Klein, and A. Gelperin. 2006. DNA-decorated carbon nanotubes for chemical sensing. *Semicond. Sci. Technol.* 21, S17–S21.

9 Field Emission and Graphene
An Overview of Current Status

Indranil Lahiri and Wonbong Choi

CONTENTS

9.1 INTRODUCTION

In the last five or six years, the truly 2-dimensional material graphene has attracted strong attention from physicists and materials scientists around the world. The awarding of the 2010 Nobel Prize in Physics to Andre Geim and Konstantin Novoselov "for groundbreaking experiments regarding the two-dimensional material graphene" [1,2] has further intensified the enormous interest of the scientific community for this extraordinary material [3–5]. Graphene has been and is being proposed for a wide range of applications, field emission being one of those exciting device applications. Field emitters have found a long list of practical device applications as sources of electrons, starting with electron microscopes and hand-held miniature x-ray sources to high-powered microwave generators and field emission displays [6,7]. This chapter introduces to the reader the basics of the field emission process, followed by a detailed discussion of the progress and prospect of graphene for field emission applications.

Emission of ions or electrons from the surface of energized metals has been known to humankind for more than a century. Emission processes, in a broad sense, can be defined as the process of flow of charge carriers, either ions or electrons, from a highly energized metallic surface to another surface or over some kind of electric potential barrier. Based on the nature of the charge carrier, emission processes can be divided into electron emissions or ionic emissions. On the other hand, depending upon the process of energizing the metal surface, it can be divided into field emissions, thermal emissions, and thermal-field emissions [8].

The process of field emission was first observed by Robert W. Wood in 1897. However, it was due to Walter Schottky in 1923 and R. H. Fowler and L. W. Nordheim in 1928 that an insight into the theory of the process was achieved. In this process, an electric field is applied between two electrodes, kept under vacuum, and the electrons tunnel through the vacuum to emit electrons. This process is generally carried out at room temperature or at slightly elevated temperature. Graphene is being considered as the next-generation electronic material due to its exceptional electronic quality, characterized by high electron mobility, the unique nature of the charge carrier, and the room temperature quantum Hall effect. Since the discovery of few-atom-thick isolated graphene in 2004, much research has been devoted to the theoretical understanding and practical applications of it. Graphene exhibited good field emission behavior and excellent emission stability, comparable to that offered by carbon nanotube (CNT) field emitters. These properties create much interest in its application in future field emission devices. This chapter presents an overview of the application of graphene in field emission devices, along with a very short introduction to the theory of field emission. The theoretical discussion is designed to assist in understanding the performance of practical graphene-based field emitters.

9.1.1 Basics of Field Emission

In order to understand the basic process of field electron emission, it is first necessary to estimate the energy required to emit an electron from a metal surface. The first assumption made in this calculation process is that the metal surface is a semi-infinite plate, having its normal in the z-direction. The surface of the metal is taken as $z = 0$. Among all possible energy terms, the most important is *Fermi energy* (E_F), which is defined as the energy of the highest occupied electron state at absolute zero. Another very important energy term is the *work function* (ϕ) of the metal, which is defined as the minimum amount of energy, at absolute zero, that should be supplied to the metal before an electron can escape from its surface. Another energy term, the *image force*, (given as $-e^2/4z^2$, where the negative sign indicates that the attractive force is inward from the metal surface), is defined as the attraction force that an electron feels toward the plane of a perfect conductor, when situated at a finite distance from it. Adding all these energy terms, the potential energy of an electron on the vacuum side of the metal–vacuum interface is given by [8]

$$V(z) \cong E_F + \phi - e^2/4z^2 \tag{9.1}$$

In the case of a field emission experiment, an external applied field is applied to the surface of the metal. In such cases, the potential energy field seen by an electron is given as

$$V(z) \cong E_F + \phi - e^2/4z^2 - eFz \tag{9.2}$$

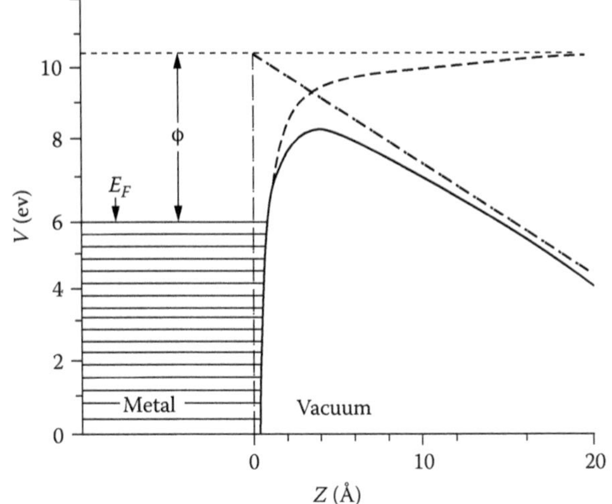

FIGURE 9.1 Surface potential barrier seen by an electron during field electron emission (solid line). Contributions from image potential and applied field are shown by broken and broken-solid line, respectively. (From Modinos, A. 1984. *Field, thermionic and secondary electron emission spectroscopy*. London: Plenum Press. With permission.)

Figure 9.1 schematically shows the potential energy field of an electron during field electron emission. This shows the amount of energy that needs to be supplied to an electron before it can actually escape from the metal surface.

It may be noted from the figure that the potential barrier takes a triangular kind of shape and it is lowest at the highest energy level of electrons. Thus, it is easier for the electrons in higher levels of the conduction band to be emitted easily, at a much lower applied field, than the electrons occupying lower energy levels.

In order to calculate the current density obtained in field electron emission, the first thing to calculate is the number of electrons that hit the surface of the metal, from within the metal. The number of electrons, with normal energy between W and $W + dW$, impinging on the surface of the metal, from within the metal is given as

$$N(W,T)dW = \frac{mdW}{2\pi^2\hbar^3}\int_W^\infty f(E)dE = \frac{mk_BT}{2\pi^2\hbar^3}\ln\left[1+\exp\left(-\frac{W-E_F}{k_BT}\right)\right]dW \quad (9.3)$$

However, all the electrons impinging the surface of the metal cannot emit from the metal surface. It will be decided by the probability $D(W)$, known as the *transmission coefficient*, to be transmitted through the surface potential barrier. Thus, the field emission current density, number of emitted electrons per unit surface area per unit time multiplied by the magnitude of electronic charge, is presented as

$$J(F,T) = e\int_0^\infty N(W,T)D(W)dW \quad (9.4)$$

where T and F denote temperature and applied field, respectively. Different terms of Equation (9.4) can be written as follows:

$$N(W,T) = \frac{mk_BT}{2\pi^2\hbar^3}\ln\left[1+\exp\left(-\frac{W-E_F}{k_BT}\right)\right] \tag{9.5}$$

$$D(W) = \{1+\exp[Q(W)]\}^{-1} \tag{9.6}$$

$$Q(W) \equiv -2i\int_{z1}^{z2}\lambda(z)dz \tag{9.7}$$

$$\lambda(z) = \left[\frac{2m}{\hbar^2}\left(W-E_F-\phi+\frac{e^2}{4z}+eFz\right)\right]^{1/2} \tag{9.8}$$

Using the expressions from Equations (9.5)–(9.8), Equation (9.4) takes the following shape.

$$J(F,T) = \frac{emk_BT}{2\pi^2\hbar^3}\left[\int_0^W \frac{\ln\{1+\exp[-(W-E_F)/k_BT]\}dW}{1+\exp[Q(W)]} + \int_W^\infty \ln[1+\exp\{-(W-E_F)/k_BT\}]dW\right] \tag{9.9}$$

Equation (9.4) is the most generalized form for any kind of emission (field or thermal) from a metal surface. Thus, this generalized form of the equation can be applied to the case of thermionic as well as thermal-field emission. With proper assumptions, Equation (9.4) leads to formulation of the Fowler-Nordheim formula for field emission (at high applied field and low temperature) and to the Richardson and Schottky formulae for thermionic emission (high temperature with weak or no applied field). Since the main focus of this chapter is on field electron emission, it is suitable to mention the Fowler-Nordheim formula at this moment.

$$J(F) = \frac{1.537\times10^{10}F^2}{\phi t^2(3.79F^{0.5}/\phi)}\exp\left[-\frac{0.683\phi^{3/2}}{F}v\left(\frac{3.79F^{0.5}}{\phi}\right)\right]\frac{A}{cm^2} \tag{9.10}$$

Though Equation (9.10), popularly known as the Fowler-Nordheim (or F-N) equation, was originally deduced for metallic electron-emitting surfaces of conventional

3-dimensional materials, it has successfully been used for other types of (nonmetallic and/or non-3-dimensional) materials too.

It may be appreciated at this point that in most of the practical cases, the emitting surface is not flat, but curved and small. Thus, removal of an electron from such a surface leads to consideration of a surface energy term. For such a system, the revised form of Equation (9.2) should take the shape of

$$V(z) \cong E_F + \phi - e^2/4z^2 - eFz - 2\gamma^0/r \tag{9.11}$$

Analysis of Equation (9.11) immediately shows the reason for better performance of nanoemitters. For a nanostructured material, surface energy (γ^0) will be high. If the structure also has a small tip radius (r), then the contribution from the term $2\gamma^0/r$ will be very high, leading to a much smaller value for the potential barrier, $V(z)$. This equation shows clearly how an electron could be extracted more easily from a nanoemitter structure, as compared to a micro- or macro-emitter.

In nano field emitters, it has been observed that the macroscopic field (F_M) applied to it and the local field (F), at the tip of the emitter, is not the same and can be related by the following equation.

$$F = \beta F_M \tag{9.12}$$

where β is known as the *field enhancement factor.*

It is experimentally found that the value of β is very high, in the range of 5000–15,000. The field enhancement factor has also been known to be directly proportional to the aspect ratio (h/r, h being height and r being tip radius) of the emitter. It can immediately be concluded that a wire type of structure will show better enhancement as compared to a platelike structure. It is of no surprise that CNTs, having very high aspect ratios, are known as promising candidates for field emitters. However, as a 2-dimensional material, graphene offers some challenges and a few exciting properties, which are of immense interest in field emission study. While absence of a sharp tip-type edge is a challenge for graphene, the relatively low macroscopic field required for electron emission from graphene-based thin films and their high carrier mobility (~15,000 $cm^2V^{-1}s^{-1}$) combined with low electron mass (~0.007 m_e, m_e is electron mass) are very promising for their application as electron sources. In Section 9.3 of this chapter, a detailed discussion about field emission from graphene and its mechanism is presented.

9.2 MATERIALS USED AS FIELD EMITTERS: PAST AND PRESENT

Field emission has been investigated extensively for almost 50 years. During this time period, many materials have been examined for their field emission response and some of them have been used in practical applications.

Tungsten (W) is the first material to be considered for field emission application. In 1966, Swanson et al. [9] proposed total energy distribution from a

TABLE 9.1
Summary of Materials Used in Field Emitters

Material	Aim of the Study	Reference
Tungsten	Theoretical analysis of total energy distribution of a tungsten field emitter	9
	Emission current and total energy distribution from different crystallographic planes of tungsten, with and without various gas adsorption on them	10–14
	Experimental and theoretical analysis of emission of hot electrons	15
	Field emission from tungsten nanowire	16
	Multistage tungsten-oxide nanowire and its field emission under poor vacuum conditions	17
Carbon	Field emission response from sharpened microsized carbon fiber	18
	Field emission from micro- and nanosized diamond emitter arrays	19–22
	Field emission from single-wall and multiwall carbon nanotubes, in the form of arrays or individual nanotube	6, 7, 23–35
	Field emission from single-layer, multilayer, and thick graphene structures	36–42
Silicon	Large-area arrays of sharply pointed field emitters on Si wafers	43
Molybdenum	Closely packed arrays of microsized Mo cones	44
	Field emission from single crystalline MoO_3 nanobelts	45
Aluminum nitride	Field emission response of AlN nanotubes	46
Boron-Carbon-Nitrogen	Field emission behavior from individual B-C-N nanotube rope, in situ within a low-energy electron microscope	47
Copper oxide	Field emission from aligned cupric oxide nanobelt film, as a function of temperature	48
Tin oxide	Field emission from ~90 μm long SnO_2 nanobelt array, grown on a Si wafer	49
Zinc oxide	Different morphologies of ZnO nanostructure–nanoneedle array, nanowire, nanopencils, nanorod array on different substrates and their field-emission responses	50–53
Lanthanum hexaboride	Microtip and nanostructures of LaB_6 and their field-emission responses	54–56

tungsten field emitter and the research in this field started immediately. Among other materials, LaB_6 has been very popular in practical applications. In the last decade, CNTs were proposed as an excellent field emission material. Table 9.1 summarizes a variety of materials that have been used for field emission [9–56]. The table presents only a fraction of the available literature as references, due to space limitations. The actual amount of published literature on each of these emitter materials is much larger. Figure 9.2 shows the number of publications (decade-wise) for different field emitter materials. It may be observed that carbon nanotubes have attracted the most attention as field emitters in the last decade. The recent trend has been application of graphene and graphene-based materials

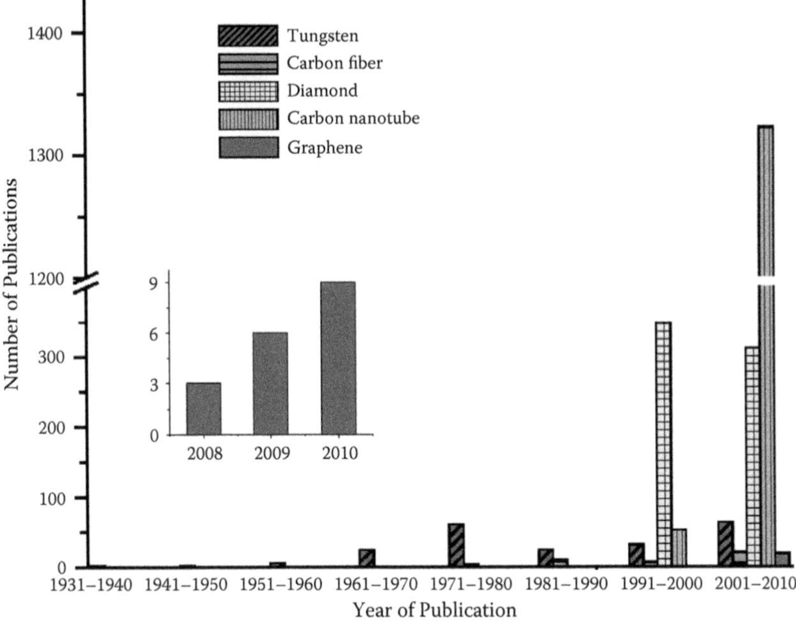

FIGURE 9.2 Research trend in different field emitter materials (decade-wise publication list). (Data compiled using scopus.com.)

for field emission devices. Though application of graphene in a field emission device is still in its preliminary stage, the material has shown enough promise for field emission-related applications. These applications will be discussed in detail in the next section.

9.3 GRAPHENE-BASED STRUCTURES IN FIELD-EMISSION DEVICES

Different carbon-based structures, including diamond, amorphous carbon, carbon nanotubes, and carbon nanosheets have shown good potential as the cathode material for field emission devices [36]. The basic theory of field emission indicates that 1-dimensional nanostructures are the geometrically favored structure for field emission. However, in spite of its 2-dimensional geometry, graphene is considered to be a promising material for field emission [3] applications due to the following reasons:

A. Exceptionally high carrier mobility ($\mu = 15{,}000\ cm^2\ V^{-1}\ s^{-1}$), small electron mass ($0.007m_e$, m_e: free electron mass) [57], and graphene's metallic nature and low contact resistance with metals [58] ensure a very small voltage drop at the graphene–metal interface and also along the graphene surface. A low voltage drop minimizes resistance to the flow of electrons to the emitter.
B. The atomically sharp edges of graphene (thickness 0.34 nm) are expected to offer good field enhancement, thus facilitating easy tunneling of electrons from the edges at a low bias.

C. Further, graphene is known to be chemically stable, has a high melting point, and offers excellent mechanical strength (in fact, defect-free single-layer graphene was shown to be the strongest material known) [59,60]. All these properties are essential for a practical field emitter.

The theory of field emission (see Section 9.1) predicts that a nano-dimensional apex structure (a higher aspect ratio) is the best structure for efficient field emitters. Graphene sheets need to stand erect on their edges, which enables all of their advantageous properties for field emission. However, most of the synthesis processes result in formation of 2-dimensional graphene that has a laterally flat topography. Thus, in spite of having a set of favorable reasons for using graphene, its application as a field emitter has raised debate. It was proposed in the year 2002 that the presence of defects in thin graphitic materials could lead to higher field emission current [61,62]. Introduction of sp^3-like defect sites in an otherwise sp^2-networked structure of graphene (or in any graphitic material) helps lower the work function, thus aiding electrons to easily escape from these sp^3 sites [61]. The dangling bond (DB) states generally dominate as field electron emission sites because the electronic orbitals tend to appear at the edges of graphene sheets and protrude along the direction of the electric field, leading to higher field emission current from these sites [62]. Thus, in order to understand the microscopic mechanisms of field emission from covalently bonded graphitic nanostructures, it seems essential to know about the local electronic properties, the σ or π bonding states, and so on. These initial findings raised some hope that defective graphene edges can really work as a source of field electron emission. After a couple of years (in 2004), a new technique was proposed [63] to synthesize nm-thick free-standing graphitic sheets, erected on the substrate, and the structure was found to offer appreciable field electron emission. This report shows practical application of graphene as a field emission source (Figure 9.3).

These almost vertical carbon nanosheets were synthesized on a variety of substrates (from Si through metals like W, Mo, and so on, to 304 stainless steel and Al_2O_3) without any catalyst or any kind of special surface treatment, using radio frequency plasma-enhanced chemical vapor deposition (rf-PECVD). This structure showed a satisfactory turn-on field (defined as the electric field required to produce 10 $\mu A/cm^2$ emission current density) of 4.7 $V/\mu m$ and maximum emission current density of 0.27 mA/cm^2. Straight line behavior of the Fowler-Nordheim (F-N) plot indicates that emission follows a conventional F-N tunneling mechanism. Interestingly, these so-called carbon nanosheets contained an appreciable number of defects and distortions in the graphene sheets. This nanographitic sheet first proved that edges of a 2-dimensional carbon sheet could act as emission sites and the structure might be considered as a more robust edge emitter than CNTs. These successes in theoretical prediction and experimental evidence of field emission from graphene were good enough to initiate extensive research in this interesting field. However, research in the arena of field emission from graphene did not make any appreciable progress, and even in 2007, field emission applications of graphene were termed a dream [3]. Surprisingly, the dream seemed to come true beginning in 2008 with detailed research efforts in field emission from graphene-based materials and other related

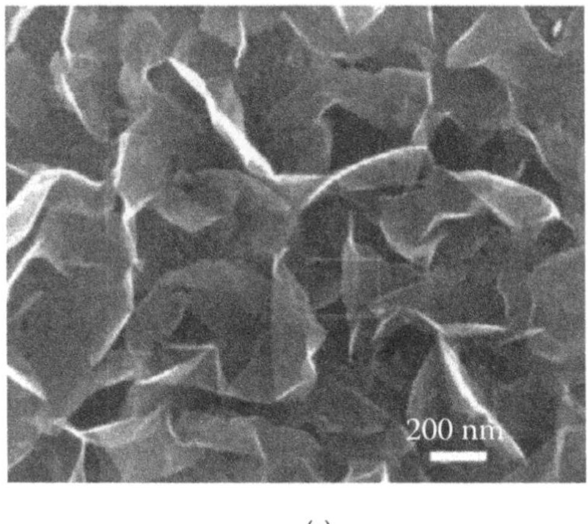

(a)

(b)

FIGURE 9.3 (a) Vertically aligned graphene structure and (b) its field emission response. (From Wang, J.J., Zhu, M.Y., Outlaw, R.A., Zhao, X., Manos, D.M., Holloway, B.C., and V.P. Mammana. 2004. Free-standing subnanometer graphite sheets. *Appl. Phys. Lett.* 85: 1265–1267. Reprinted with permission from AIP.)

and highly interesting applications. Table 9.2 presents an overall summary of important features of the main research work on field emission of graphene [36,37,39–42,63–69]. A detailed discussion about field emission and other related applications of graphene is presented in the following subsections.

9.3.1 Graphene-Based Field Emitters: Experimental Approaches

Experiments involving field emission of graphene evolved in a very systematic manner. After initial experiments with nanocarbon films (discussed above), researchers successfully reported field emission from graphene, specifically few-layer graphene

TABLE 9.2
Summary of Field Emission Studies on Graphene

Structure	Synthesis Method	Field Emission Properties	Reference
Free-standing subnanometer graphite film	Grown on variety of substrates by rf-PECVD, without any catalyst	E_{TO} = 4.7 J = 0.27, at 5.7 V/μm β = NA	63
Nanocrystalline flower-like graphitic films	Grown on ceramic substrate by MPCVD, using Fe-Ni-Cr as catalyst	E_{TO} = 1.26 J = 2.1, at 2.2 V/μm β = 5110	64
Graphene aggregate films	Grown on stainless steel substrate by MPCVD, without any catalyst	E_{TO} = 1.0 J = 2.1, at 2.4 V/μm β = 5110	65
Few-layer graphene nanosheets	Liquid phase exfoliation of graphite and pasting the suspension on Cu substrate	E_{TO} = 1.70 J = 2.4, at 4.5 V/μm β = 7300	66
Vertically aligned few-layer graphene	Grown on Si and Ti by MPECVD	E_{TO} = 1.0 J = 12, at 4.0 V/μm β = 5000	37
Ar plasma-treated graphene paper	Graphene sheet obtained by chemical oxidation; Ar plasma treatment in DC magnetron sputtering	E_{TO} = 1.6 (2.3 for untreated) J = 10, at 3.0 V/μm (same current at 4.4 V/μm for untreated) β = 4015 (3208 for untreated)	67
Ar plasma-treated few-layer graphene	Grown on Si(100) substrate by rf-PECVD	E_{TO} = 2.23 (3.91 for untreated) J = 1.33, at 4.4V/μm (.033, at same field, for untreated) β = 5130 (3188 for untreated)	68
Single-layer graphene film	Prepared from chemical exfoliation of artificial graphite and electrophoretically deposited	E_{TO} = 2.3 J = 10, at 5.2 V/μm β = 3700	39
Individual single-layer graphene	Prepared by mechanical exfoliation of HOPG, on Si/SiO_2	E_{TO} = 12.1 J = 170 × 10^{-6}, at 35 V/μm β = 3519	42
Individual single-layer graphene	Prepared by mechanical exfoliation, on Si/SiO_2	E_{TO} = 150 J = 100 × 10^{-9}, at 250 V/μm β = NA	41
Graphene-polystyrene composite thin film	Graphene oxide prepared by modified Hummers method chemically reduced and suspension spin-coated on degenerately doped Si	E_{TO} = 4.0 J = 1.0, at 14 V/μm β = 1200	36

TABLE 9.2 (continued)
Summary of Field Emission Studies on Graphene

Structure	Synthesis Method	Field Emission Properties	Reference
Graphene sheet on ZnO nanowire	Grown on ZnO nanowire by rf-PECVD, using Ni catalyst	E_{TO} = 1.3 J = 1.4, at 7 V/μm β = 15000	69
Screen-printed graphene film	Graphene prepared by modified Hummers method and suspension screen-printed on Ag-coated glass substrate	E_{TO} = 1.5 J = 2.7, at 3.7 V/μm β = 4539	40

Note: E_{TO} = turn-on field, V/μm; J = emission current density, mA/cm^2; β = field enhancement factor

(FLG), on a variety of substrates. An understanding of the field emission behavior from FLGs led to development of different approaches to further enhance the emission behavior using composite structures or by introducing special processing steps like plasma etching. Another separate field of research concentrated on understanding the basic field emission mechanism from graphene, using single-layer graphene (SLG). This subsection will deal with all of these developments.

Field emission study of graphene started with thicker versions—nanostructured graphitic films [64,65]. The structure can be grown by the microwave plasma chemical vapor deposition (MPCVD) technique on ceramic substrates using Fe-Ni-Cr catalyst or directly on Fe-Ni-Cr alloy substrates. Suitable mixtures of H_2 and CH_4 gas are most often used as the source gas, substrates are heated at ~973 K, and microwave power is used for approximately 10 minutes. The product synthesized by this method looks like petal-like clusters of nanoscale graphitic films. The thickness of such clusters can be in the range of a few tens of nanometers. Essentially, these structures can be considered as thick graphene structures. Even in this thicker state, nanographitic films show appreciable field emission response—a turn-on field of 1.0–1.3 V/μm and emission current density of ~2.1 mA/cm^2, at a reasonably low electric field of 2.2–2.4 V/μm. The field enhancement factor (β) of these emitters is also very good at ~5100. The good emission properties of such nanographitic structures could be due to three reasons. First, some of the graphene sheets were roughly standing vertical to the substrate and along the direction of the applied electric field, which causes high field enhancement. Secondly, the structures contain a sufficiently high number of sharp edges, and electrons could emit from these nanolevel sharp edges much more easily, as compared to other parts of the structure, due to significant field enhancement. Lastly, dangling bonds or sp^3-like defects, present on the edge of graphene sheets, cause local lowering of electron affinity and thus provide a low energy barrier to the flow of electrons.

A proper optimization of the parameters during MPCVD leads to synthesis of high-quality, crystalline, vertically aligned, truly few-layer graphene (FLG) [37]. The typical structure of a FLG and its field emission behavior are shown in

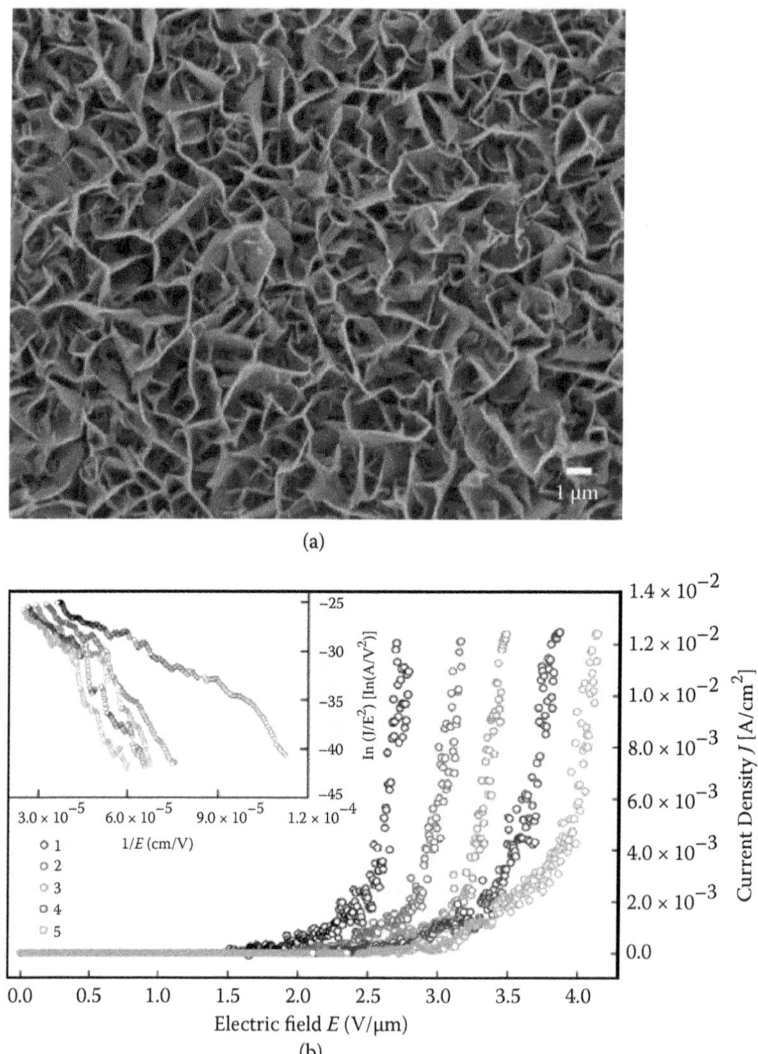

FIGURE 9.4 (a) SEM image (top view) of a typical vertically oriented few-layer graphene (FLG) structure. (b) Field emission responses, i.e., current density-electric field plot and Fowler-Nordheim (F-N) plot (inset) for the same structure, after different time intervals. (From Malesevic, A., Kemps, R., Vanhulsel, A., Chowdhury, M. P., Volodin, A., and Van Haesendonck, C. 2008. Field emission from vertically aligned few-layer graphene. *J. Appl. Phys.* 104: 084301. With permission from AIP.)

Figure 9.4. The height of such FLGs can be limited by controlling the synthesis time. However, to date, a process for directly controlling nucleation sites during graphene synthesis in MPCVD is not very clear. Among other processes, liquid phase exfoliation of graphite is important for FLG production, and the resulting structure shows good field emission response [66]. FLGs show an impressive field emission response: turn-on field of 1.0–1.7 V/μm, emission current density of 2.5–12.0 mA/cm^2 at a sufficiently low electric field of 4.0–4.5 V/μm, and a field enhancement factor in the range of 5000–7300. Though emission properties vary depending upon the synthesis technique followed, FLGs still offer much better performance compared to the thicker version, nanostructured graphitic films.

It should be noted here that the field emission properties shown by FLGs are also comparable to other field emitter structures, such as carbon nanotubes. However, research in this field is still in its infancy and many issues, such as the effects of substrate materials, adsorbed molecules, processing methods on field emission behavior, should be understood clearly for proper exploration of FLGs as field emission sources.

Single-layer graphene (SLG) films show a similar kind of field emission response [39]. Such devices most often use SLGs prepared by mechanical or chemical exfoliation followed by some kind of deposition technique, spin coating or electrophoretic deposition, to form a film on a desired substrate. Such field emitters show appreciably low turn-on and threshold fields, high emission current density (>20 mA/cm^2), and excellent stability of emission (Figure 9.5). It is worth mentioning here that unlike CNTs, which show different field emission behavior in single-wall and multiwall forms, SLGs and FLGs do not show any appreciable difference.

A good understanding of field emission theory and knowledge of field emission responses from graphene led to two possible ways to improve their field emission current. First, graphene edges need to be further sharpened and should be oriented vertically on the substrate, in the direction of the applied electrical field. Second, more defects need to be introduced along the edges. Argon plasma treatment is known to introduce both of these factors in graphene [67,68] (Figure 9.6). Moreover, Ar plasma treatment can also remove folded edges in FLGs, which provides a better geometrical factor and reduces screening effect from neighboring edges. In more planar versions of graphene, such plasma treatment can introduce protrusions, which can enhance field emission. This kind of surface engineering can lead to an almost 50% reduction of turn-on and threshold fields and a significant increase in maximum emission current density, even up to 40 times [68].

Another way to achieve a high aspect ratio is formation of nanoneedles. These structures, which are formed from 2-dimensional graphene sheets, demonstrated stable and high-brightness (10^{12} A sr^{-1} m^{-2}) electron emission, even under poor vacuum conditions (10^{-5} Torr) [57], with an impressive field enhancement factor of 100,000. When used as the electron source in a scanning electron microscope (SEM), nanoneedles produced sharp images and offered a spatial resolution of ~30 nm. Graphene nanoneedles may, therefore, be considered as electron emitters in electron microscopes and handheld x-ray generators. These recent discoveries show that the edge structure, geometry, and chemistry of graphene will play extremely important roles in years to come, for further enhancement of field emission from graphene. However, it is important to note that any surface treatment, be it etching or deposition, needs to be optimized. Otherwise, damaging the surface of graphene sheet (e.g., by laser treatment) or depositing thin films of other materials could significantly reduce the number of emission sites, and hence the total emission current density [69].

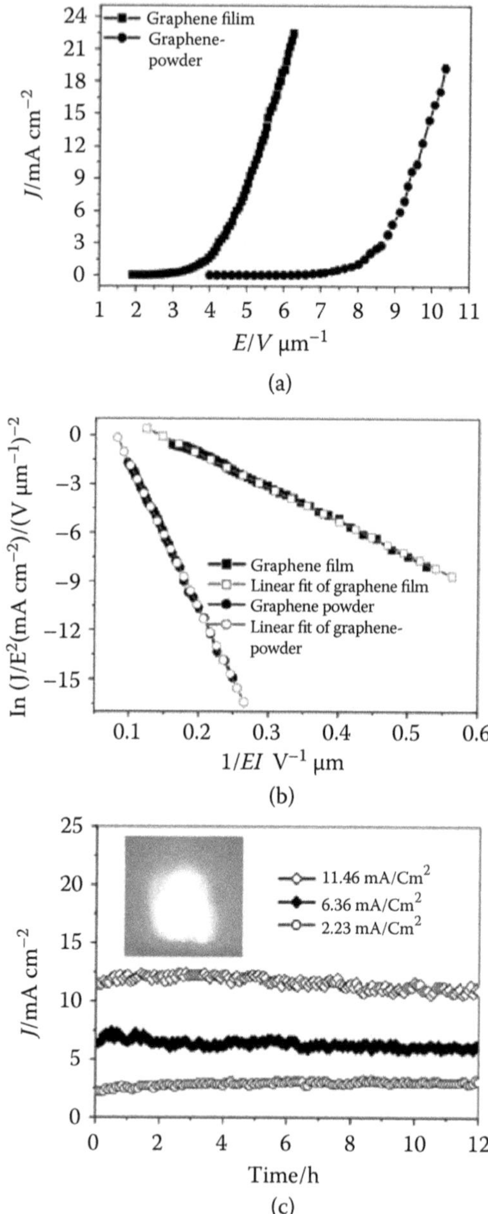

FIGURE 9.5 Comparison of field emission behavior of single-layer graphene (SLG) film and graphene powder coating: (a) current density–applied field plot and (b) F-N plot. (c) Stability of SLG film at different current levels; inset shows field emission image at 3.8 V/μm. (From Wu, Z.-S., Pei, S., Ren, W., Tang, D., Gao, L., Liu, B., Li, F., Liu, C., and H.-M. Cheng. 2009. Field emission of single-layer graphene films prepared by electrophoretic deposition. *Adv. Mater.* 21: 1756–1760. With permission from WILEY-VCH Verlag GmbH & Co.)

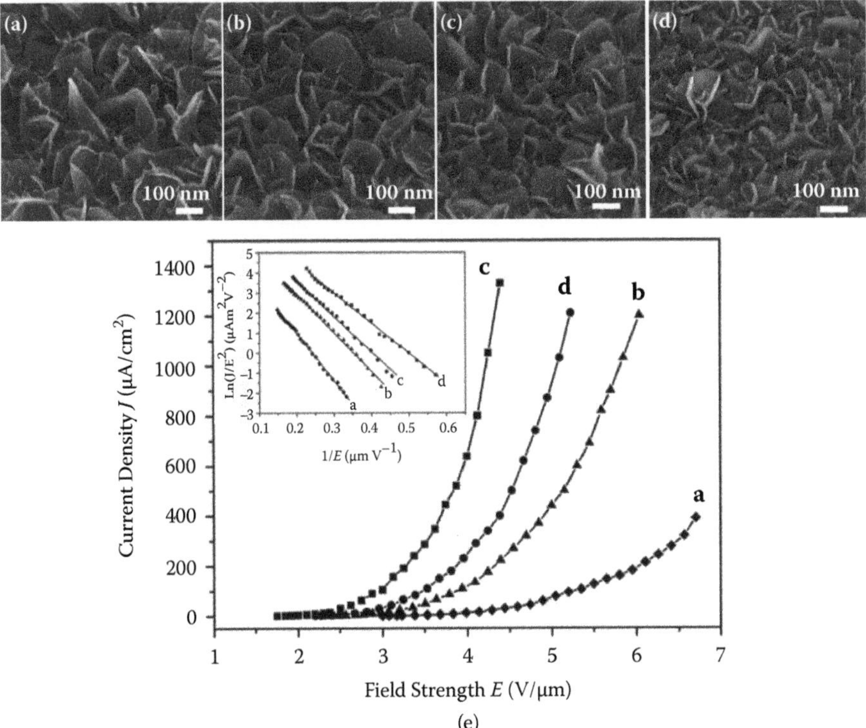

FIGURE 9.6 SEM images of few-layer graphene (FLG) in (a) as-grown state and in Ar plasma treated states, for treatment time of (b) 1 min, (c) 3 min, (d) 5 min. (e) Field emission response of the above-mentioned samples. (From Qi, J.L., Wang, X., Zheng, W.T., Ian, H.W., Hu, C.Q., and Y.S. Peng. 2010. Ar plasma treatment on few layer graphene sheets for enhancing their field emission properties. *J. Phys. D: Appl. Phys.* 43: 055302. With permission from IOP Publishing.)

In addition to enhancing field emission from graphene, understanding the underlying mechanism was stressed. SLGs are the best candidates for such studies [41,42,58]. Two different approaches can be followed for preparing samples for such study. In one approach, a mechanically or chemically exfoliated individual SLG is placed on a Si/SiO_2 substrate and is used as a cathode, while a nanomanipulator-controlled tungsten microtip is used as an anode (Figure 9.7). In the other approach, a nanogap is created by collapsing a single SLG and then the two pieces were used as cathode and anode (Figure 9.8). In the latter approach, electron beam lithography is used to first hold the SLG between two electrode connections, prior to lift-off and breaking the suspended graphene. Since a single SLG is used in all these studies, field emission current varies in the pA–nA range. However, a field enhancement factor as high as 3500 is reported even for a single SLG field emitter. Such high field enhancement shows the potential of graphene as a next-generation field emission device. No emission current could be measured from the flat portions of the SLGs;

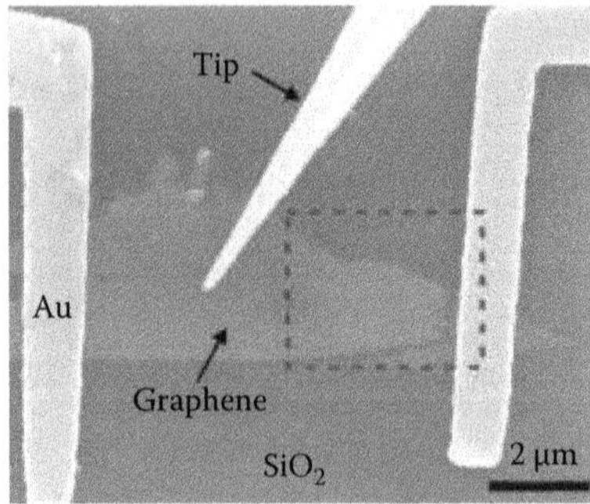

FIGURE 9.7 SEM image of an individual single-layer graphene (SLG), used as the cathode of a field emitter device, along with a tungsten micro-tip as the anode. (From Xiao, Z., She, J., Deng, S., Tang, Z., Li, Z., Lu, J., and N. Xu. 2010. Field electron emission characteristics and physical mechanism of individual single-layer graphene. *ACS Nano* 4: 6332–6336. With permission from the American Chemical Society.)

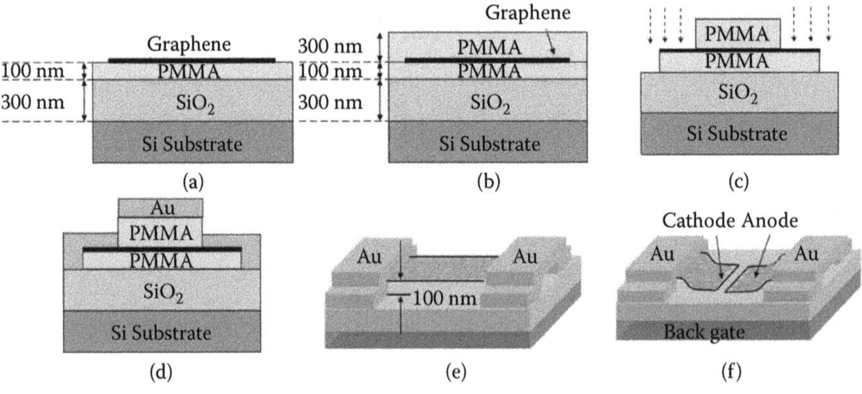

FIGURE 9.8 Schematic diagram of the fabrication process for creating a nanogap in a single SLG. (From Wang, H.M., Zheng, Z., Wang, Y.Y., Qiu, J.J., Guo, Z.B., Shen, Z.X., and T. Yu. 2010. Fabrication of graphene nanogap with crystallographically matching edges and its electron emission properties. *Appl. Phys. Lett.* 96: 023106. With permission from AIP.)

emission current can only be observed from graphene edges. Thus, theoretical explanations of field emission from graphene edges can also be verified experimentally. However, more studies are required to understand the complex electrical field distribution near the edges of SLGs. Simulation and field emission energy distribution studies could be helpful in throwing more light on this issue.

Graphene has also been used as a composite or in hybrid structures for field emission applications. In one such application, a graphene–polystyrene composite was used as the field emitter [36]. During preparation of the cathode for the field emission device, graphene–polystyrene suspension is spin-coated on a Si substrate. Spin-coating speed plays an important role here. At low speed, graphene sheets are

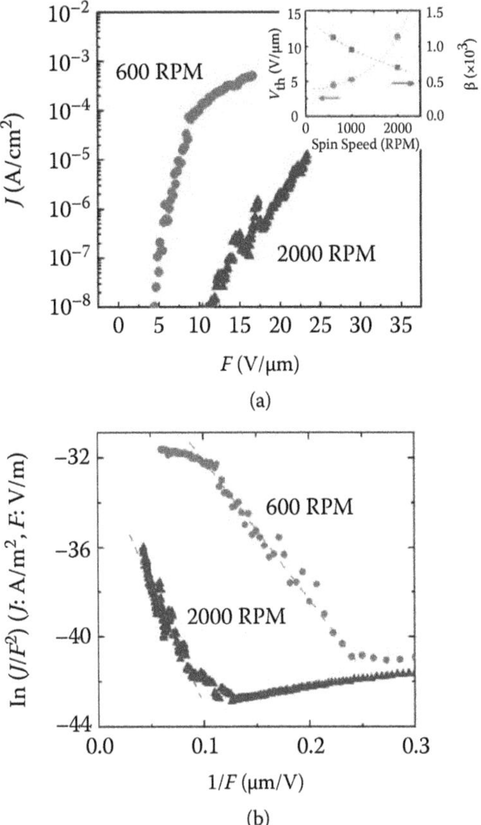

FIGURE 9.9 Field emission response: (a) current density–applied field plot and (b) F-N plot, for graphene-polymer composite, for different spinning speeds. (From Eda, G., Unalan, H.E., Rupesinghe, N., Amaratunga, G.A.J., and M. Chhowalla. 2008. Field emission from graphene based composite thin films. *Appl. Phys. Lett.* 93: 233502. With permission from AIP.)

densely distributed and randomly oriented over the substrate. At such a low speed, shear force is very small and thus it allows random orientation of graphene sheets and the polymer solidifies very quickly before the sheets can lay parallel to the substrate surface. However, at higher speeds, graphene sheets are very sparsely distributed and stay almost parallel to the substrate. As expected from these surface features, a composite formed at a low spin-coating speed shows much better field emission properties than the one prepared at a higher spin-coating speed (Figure 9.9). To date, no other report is available on other types of conventional composites such as metal–graphene and ceramic–graphene composites.

Though no ceramic–graphene structure has been used as a field emitter, a ZnO nanowire–graphene hybrid structure has been used as a field emitter [70]. This structure actually aims to use the high aspect ratio of ZnO nanowires to provide a high apex for graphene sheets, grown on the nanowires by the PECVD method.

FIGURE 9.10 TEM images of ZnO–graphene sheets, grown for (a) 2 min, (b) 5 min, and (c) 10 min. (d) Current density–applied field plots for ZnO (marked A) and ZnO–graphene sheet grown for 2 min (marked B), showing appreciable improvement. Inset shows corresponding F-N plots. (From Zheng, W.T., Ho, Y.N., Tian, H.W., Wen, M., Qi, J.L., and Y.A. Li. 2009. Field emission from a composite of graphene sheets and ZnO nanowires. *J. Phys. Chem. C* 113: 9164–9168. With permission from the American Chemical Society.)

In this device preparation methodology, the thickness of the Ni catalyst (deposited on ZnO nanowires prior to PECVD) and synthesis time during PECVD are two important parameters to control the structure and morphology of graphene sheets (Figures 9.10[a]–[c]). Though ZnO nanowire is also a well-known field emitter material, coating it with graphene clearly enhances the field emission behavior. Compared to a pure ZnO nanowire emitter, this ZnO–graphene hybrid structure has shown almost 50% reduction in turn-on field and 100% increase in field enhancement factor (Figure 9.10 [d]). Thus, similar hybrid structures have good potential for next-generation field emission devices.

For practical application of graphene-based field emitters, it is important to prepare large-scale graphene field emission electrodes. One such technique is screen printing of graphene powder, mixed with a proper solvent, on suitable electrically conducting material, for example, an Ag-coated glass substrate [40]. A field emitter cathode, prepared by this route, showed a very low turn-on field of 1.5 V/μm, good emission current density of 2.7 mA/cm^2, a high field enhancement factor of ~4540, and acceptable stability of emission over a period of 3 hours (Figure 9.11). These types of cathodes are suitable for future practical device applications.

In a different approach, the present authors have demonstrated an all-graphene-based flexible and transparent field emission device for the first time [71]. During this research, chemical vapor deposition (CVD) was used to synthesize graphene on copper foil. Commercially available Cu foil (thickness ~50 μm) was annealed at 1000°C for 1 hour under Ar atmosphere and treated in 1 M acetic acid at 60°C for 10 minutes. The acid treatment is aimed at removing oxides formed on the Cu foil during annealing. After thoroughly washing with deionized water and drying at ambient conditions, Cu foils were inserted in a CVD chamber for graphene synthesis. Graphene on Cu foil was synthesized at 1000°C and atmospheric pressure, using a flow of CH_4 and H_2 gases in 1:4 ratios for 5 minutes. After the growth period, the foil was cooled down to room temperature in an Ar gas environment. Graphene electrodes were prepared by transferring the graphene grown (by thermal CVD) on copper foil onto a polyethylene terephthalate (PET) substrate. A hot press lamination and chemical etching process followed for transferring the graphene onto the transparent flexible substrates. Cu foils with graphene were hot press rolled with a transparent flexible PET film having a thickness of ~50 μm. Concentrated $FeCl_3$ solution was used for complete removal of Cu from the graphene and laminated film. Laminated polymer film with graphene and Cu foil underneath was floated over the $FeCl_3$ acid bath at room temperature. Cu was completely dissolved after 40 min of etching leaving graphene film with the PET substrate. This transparent flexible film was then thoroughly washed with deionized water and dried in air at room temperature before being used as the transparent, flexible electrode. Figures 9.12(A) and 9.12(B) present the size, flexibility, and transparency of the graphene electrode. In a further development, a graphene–CNT hybrid electrode on a transparent, flexible substrate was prepared (Figures 9.12[C]–[E]) [72]. Combined with the transparent, flexible anode, this hybrid graphene–CNT cathode presents a fully transparent, flexible field emission device. A graphene–CNT hybrid cathode was prepared by spin-coating a CNT solution on the graphene electrode. The all-graphene transparent and flexible field emitter structure (Figure 9.12[F]) showed appreciable field emission response. Figure 9.12(G) presents the field emission behavior of this hybrid structure under AC bias, showing appreciable emission current and a high field enhancement factor. The structure also offered good stability in long-term operation, raising hope for application of graphene in future transparent, flexible electronic devices.

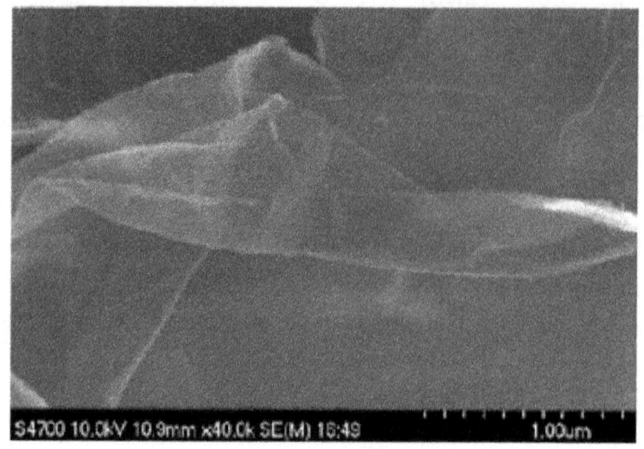
S4700 10.0kV 10.9mm x40.0k SE(M) 16:49
1.00um

(a)

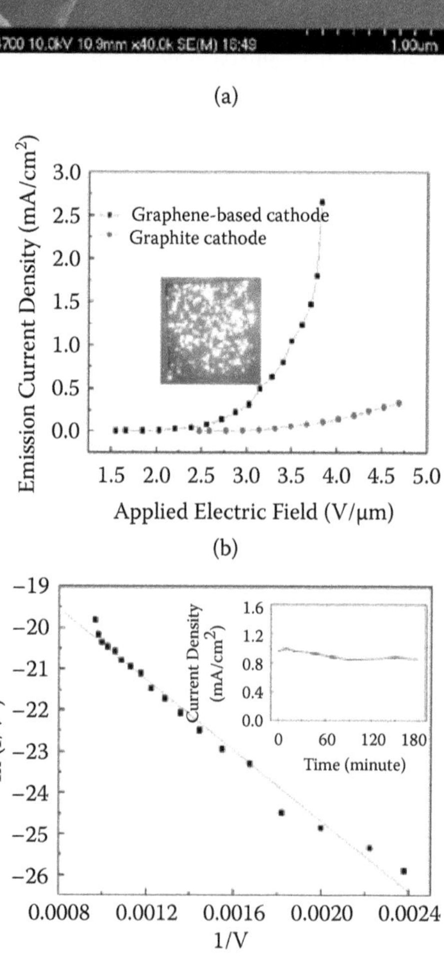
Emission Current Density (mA/cm^2)
3.0
2.5
2.0
1.5
1.0
0.5
0.0
Graphene-based cathode
Graphite cathode
1.5 2.0 2.5 3.0 3.5 4.0 4.5 5.0
Applied Electric Field (V/μm)
(b)
In (I/V^2)
−19
−20
−21
−22
−23
−24
−25
−26
0.0008 0.0012 0.0016 0.0020 0.0024
1/V
Current Density (mA/cm^2)
1.6
1.2
0.8
0.4
0.0
0 60 120 180
Time (minute)
(c)

FIGURE 9.11

9.3.2 Graphene-Based Field Emitters: Theoretical Studies

Along with experimental research regarding field emission from various types of graphene structures, it is also necessary to understand its basic mathematics and physics. Few theoretical studies have been devoted to clarifying these issues. This subsection will discuss developments in the theoretical understanding of field emission from graphene.

Conventionally Fowler-Nordheim plots are used to describe field emission in carbon nanotubes and even in graphene, though the equation was originally developed for 3-dimensional metallic emitters. However, while considering emission from a single SLG, which is a perfect 2-dimensional material, the equation needs to be modified after considering the effect of quantization of motion in the direction of sheet normal [73]. The zero-field barrier, faced by an electron in a state with transverse kinetic energy $W_\perp$, can be written as [41]

$$\phi_s = \phi + W_\perp + W_s$$

where, ϕ is the work function of the material and W_s is the energy difference between the Fermi level and the state. Since longitudinal energy of an electron is very small and the surface barrier is high, the probability of electron tunneling perpendicular to the graphene surface is almost negligible. Thus, electrons can tunnel only from those states where the barrier height is smaller. It is known that edges and corners of a graphene sheet are the most probable sites that can have appreciable contribution to the emission process. Considering the physical properties of graphene, Xiao et al. [41] have shown that line current density of field emission from a single SLG is dependent on the applied field. Two governing equations are proposed for low- and high-field regimes and both of them are a bit different from the conventional F-N equation. For details on the calculation process, readers may refer to the original works by Xiao et al. (55) and Qin et al. [73]. In short, the differences are: (1) the power of the applied field (E) is either 3/2 (at high-field regime) or 3 (at low-field regime), in contrast to the F-N equation where it is 2; (2) the conventional work function of the material (ϕ), as present in the F-N equation, is replaced by ϕ_K (zero-field barrier faced by an electron in the K-state). Thus, for SLGs, the current density–applied field (I-E) relation can be plotted by two different curves: $\ln(1/E^{1/2})$ vs. $1/E$ and $\ln(1/E^3)$ vs. $1/E^2$. While the former plot shows straight line behavior in the high-field region, the latter shows similar behavior in the low-field region.

The field enhancement factor (β) at different locations on a graphene sheet has been shown to be highest at the edges, appreciable along the edges, and almost zero on flat surfaces, far from edges or corners. Thus, emission from a flat graphene surface seems to be negligible. Following a boundary element method (BEM) and

FIGURE 9.11 *(See facing page)* (a) SEM image of graphene sheets on ITO-coated glass substrate. (b) Current density–applied field plot and (c) F-N plot for screen-printed graphene films. Inset of (b) shows emission image of the field emitter device, while inset of (c) presents emission stability of the structure. (From Qian, M., Feng, T., Ding, H., Lin, L., Li, H., Chen, Y., and Z. Sun. 2009. Electron field emission from screen-printed graphene films. *Nanotechnology* 20: 425702. With permission from IOP Publishing.)

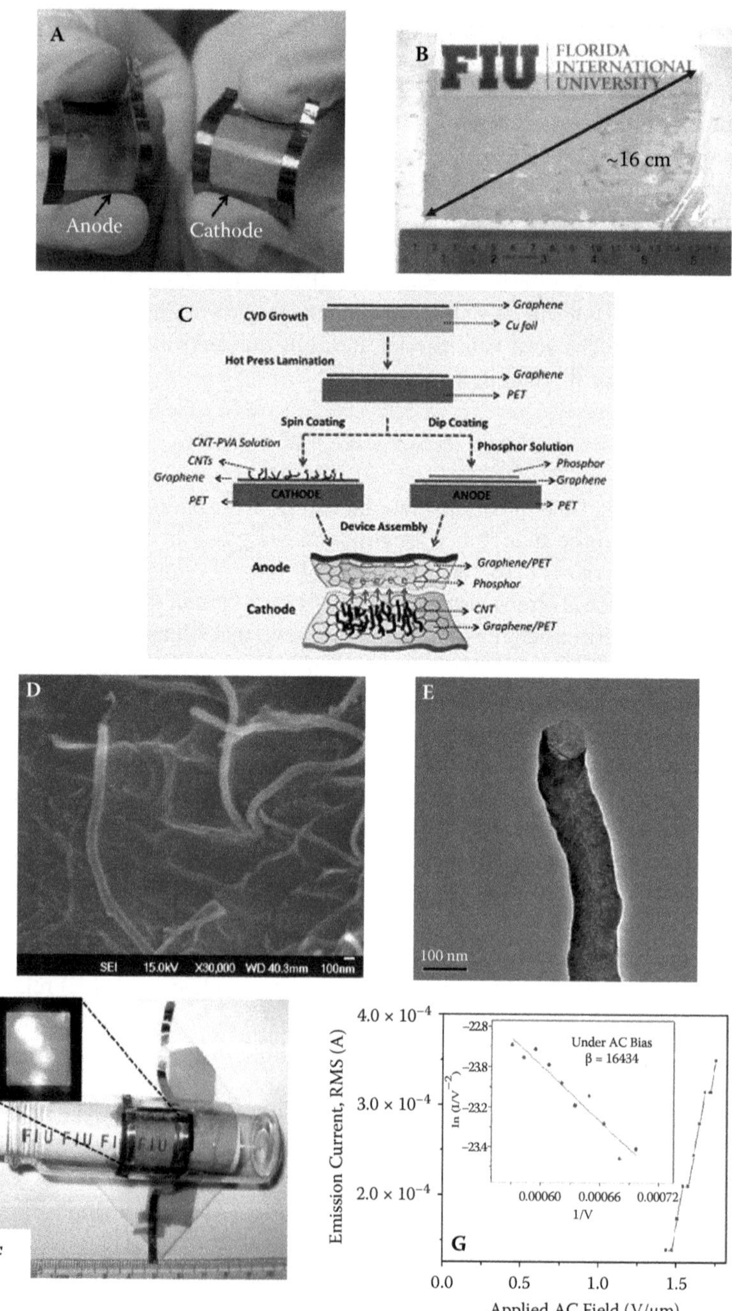
A
Anode
Cathode
B
FIU FLORIDA INTERNATIONAL UNIVERSITY
~16 cm
C
CVD Growth
Graphene
Cu foil
Hot Press Lamination
Graphene
PET
Spin Coating
Dip Coating
CNT-PVA Solution
Phosphor Solution
CNTs
Graphene
PET
CATHODE
ANODE
Phosphor
Graphene
PET
Device Assembly
Anode
Graphene/PET
Phosphor
Cathode
CNT
Graphene/PET
D
SEI 15.0kV X30,000 WD 40.3mm 100nm
E
100 nm
F
G
Under AC Bias
β = 16434
Emission Current, RMS (A)
Applied AC Field (V/μm)
1/V

FIGURE 9.12

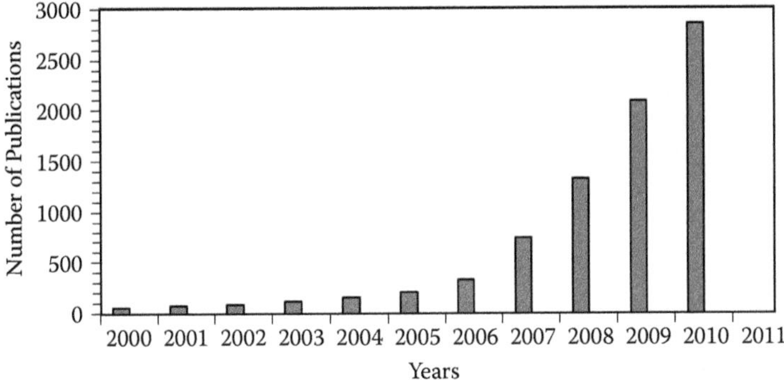

FIGURE 9.13 Graph showing the number of publications on graphene during the last decade. (Data compiled using scopus.com.)

parameterization, Watcharotone et al. [38] derived separate equations for β at corners and on edges. For detailed information on this simulation and parameterization process, readers may refer to the given reference. Though β is smaller on edges than on corners (meeting point of two edges), the edge can have a much higher contribution to the field emission current as the edge area is much higher than the corner areas. Though these initial simulation studies were helpful in understanding the basic field emission mechanism in graphene, still many issues are yet to be understood. In the future, issues like the effective value of β in a graphene field emitter, emission from curved graphene sheets, and so on, need to be explained in much clearer terms.

Apart from graphene-based field emission devices, graphene has found a wide range of applications in new devices. Other chapters of this book have focused on many such applications. All these new device applications of graphene have triggered tremendous excitement in the scientific community and extensive research on graphene is going on worldwide. As can be seen in Figure 9.13, in the last decade, publications on graphene have increased almost exponentially. In the next four to five years, we hope to see detailed research efforts on the basic mechanism and practical device applications of graphene-based materials.

ACKNOWLEDGMENT

I.L. would like to acknowledge financial support from a dissertation year fellowship from Florida International University (FIU).

FIGURE 9.12 *(See facing page)* (A) Transparent, flexible graphene–CNT hybrid electrodes for field emitter device. (B) Large-scale graphene on transparent, flexible substrate. (C) Process flow chart for preparation of graphene–CNT hybrid field emitter. (D) SEM image of PVA-coated CNTs on graphene/PET film. (E) TEM image of a representative CNT structure on graphene. (F) A photograph of the actual field emission experimental setup. (G) Field emission response of the hybrid structure (inset shows F-N plot). (From Lahiri, I., Verma, V.P., and W. Choi. 2011. An all-graphene based transparent and flexible field emission device. *Carbon* 49(5): 1614–1619. Figures (A) and (C)–(G) are reprinted with permission from Elsevier Ltd.)

REFERENCES

1. Nobelprize.org. 2010. The Nobel Prize in Physics. http://nobelprize.org/nobel_prizes/physics/laureates/2010/ (accessed January 23, 2010).
2. Nobelprize.org. 2010. The 2010 Nobel Prize in Physics: Press Release. http://nobelprize.org/nobel_prizes/physics/laureates/2010/press.html (accessed January 23, 2010).
3. Geim, A.K., and K.S. Novoselov. 2007. The rise of graphene. *Nature Mater.* 6: 183–91.
4. Editorial article. 2010. The rise and rise of graphene. *Nature Nanotech* 5: 755.
5. Dresselhaus, M.S., and P.T. Araujo. 2010. Perspectives on the 2010 Nobel prize in physics for graphene. *ACS Nano* 4: 6297–6302.
6. Lahiri, I., Seelaboyina, R., Hwang, J.Y., Banerjee, R., and W. Choi. 2010. Enhanced field emission from multi-walled carbon nanotubes grown on pure copper substrate. *Carbon* 48: 1531–1538.
7. Seelaboyina, R., Lahiri, I., and W. Choi. 2010. Carbon-nanotube-embedded novel three-dimensional alumina microchannel cold cathodes for high electron emission. *Nanotechnology* 21: 145206.
8. Modinos, A. 1984. *Field, thermionic and secondary electron emission spectroscopy.* London: Plenum Press.
9. Swanson, L.W., and L.C. Crouser. 1966. Anomalous total energy distribution for a tungsten field emitter. *Phys. Rev. Lett.* 16: 389–392.
10. Plummer, E.W., and J.W. Gadzuk. 1970. Surface states on tungsten. *Phys. Rev. Lett.* 25: 1493–1495.
11. Ehrlich, C.D., and E.W. Plummer. 1978. Measurement of the absolute tunneling current density in field emission from tungsten. *Phys. Rev. B* 18: 3767–3771.
12. Swanson, L.W. 1978. Current fluctuations from various crystal faces of a clean tungsten field emitter. *Surf. Sci.* 70: 165–180.
13. Engel, T., and R. Gomer. 1969. Adsorption of CO on tungsten: Field emission from single planes. *The J. Chem. Phys.* 50: 2428–2437.
14. Engel, T., and R. Gomer. 1970. Adsorption of oxygen on tungsten: Field emission from single planes. *The J. Chem. Phys.* 52: 1832–1841.
15. Lee, M.J.G. 1973. Field emission of hot electrons from tungsten. *Phys. Rev. Lett.* 30: 1193–1196.
16. Lee, Y.-H., Choi, C.-H., Jang, Y.-T., Kim, E.-K., Ju, B.-K., Min, N.-K., and J.-H. Ahn. 2002. Tungsten nanowires and their field electron emission properties. *App. Phys. Lett.* 81: 745–747.
17. Seelaboyina, R., Huang, J., Park, J., Kang, D.H., and W. Choi. 2006. Multistage field enhancement of tungsten oxide nanowires and its field emission in various vacuum conditions. *Nanotechnology* 17: 4840–4844.
18. Lea, C. 1973. Field emission from carbon fibres. *J. Phys. D: Appl. Phys.* 6: 1105–1114.
19. Kumar, N. 1993. Method of forming field emitter device with diamond emission tips. US Patent No. 5199918, USA.
20. Okano, K., Hoshina, K., Iida, M., Koizumi, S., and T. Inuzuka. 1994. Fabrication of a diamond field emitter array. *Appl. Phys. Lett.* 64: 2742–2744.
21. Zhu, W., Kochanski, G.P., Jin, S., and L. Seibles. 1995. Defect-enhanced electron field emission from chemical vapor deposited diamond. *J. Appl. Phys.* 78: 2707–2711.
22. Talin, A.A., Pan, L.S., McCarty, K.F., Felter, T.E., Doerr, H.J., and R.F. Bunshah. 1996. The relationship between the spatially resolved field emission characteristics and the Raman spectra of a nanocrystalline diamond cold cathode. *Appl. Phys. Lett.* 69: 3842–3844.
23. de Heer, W.A., Châtelain, A., and D. Ugarte. 1995. A carbon nanotube field-emission electron source. *Science* 270: 1179–1180.

24. Choi, W.B., Chung, D.S., Kang, J.H., Kim, H.Y., Jin, Y.W., Han, I.T., Lee, Y.H., Jung, J.E., Lee, N.S., Park, G.S., and J.M. Kim. 1999. Fully sealed, high-brightness carbon-nanotube field-emission display. *Appl. Phys. Lett.* 70: 3129–3131.
25. Xu, X., and G.R. Brandes. 1999. A method for fabricating large-area, patterned, carbon nanotube field emitters. *Appl. Phys. Lett.* 74: 2549–2551.
26. Fransen, M.J., van Rooy, Th. L., and P. Kruit. 1999. Field emission energy distributions from individual multiwalled carbon nanotubes. *Appl. Surf. Sci.* 146: 312–327.
27. Dean, K.A., and B.R. Chalamala. 1999. The environmental stability of field emission from single-walled carbon nanotubes. *Appl. Phys. Lett.* 75: 3017–3019.
28. Saito, Y., and S. Uemura. 2000. Field emission from carbon nanotubes and its application to electron sources. *Carbon* 38: 169–182.
29. Sharma, R.B., Tondare, V.N., Joag, D.S., Govindaraj, A., and C.N.R. Rao. 2001. Field emission from carbon nanotubes grown on a tungsten tip. *Chem. Phys. Lett.* 344: 283–286.
30. Teo, K.B.K., Chhowalla, M., Amaratunga, G.A.J., Milne, W.I., Pirio, G., Legagneux, P., Wyczisk, F., Pribat, D., and D.G. Hasko. 2002. Field emission from dense, sparse, and patterned arrays of carbon nanofibers. *Appl. Phys. Lett.* 80: 2011–2013.
31. Jonge, N. De, Lamy, Y., Schoots, K., and T.H. Oosterkamp. 2002. High brightness electron beam from a multi-walled carbon nanotube. *Nature* 420: 393–396.
32. Liu, D., Zhang, S., Ong, S.-E., Benstetter, G., and H. Du. 2006. Surface and electron emission properties of hydrogen-free diamond-like carbon films investigated by atomic force microscopy. *Mater. Sci. Engg.* 426: 114–120.
33. Seelaboyina, R., Huang, J., and W.B. Choi. 2006. Enhanced field emission of thin multiwall carbon nanotubes by electron multiplication from microchannel plate. *Appl. Phys. Lett.* 88: 194104.
34. Seelaboyina, R., Boddepalli, S., Noh, K., Jeon, M., and W. Choi. 2008. Enhanced field emission from aligned multistage carbon nanotube emitter arrays. *Nanotechnology* 19: 065605.
35. Lahiri, I., Seelaboyina, R., and W. Choi. 2010. Field emission response from multiwall carbon nanotubes grown on different metallic substrates. In *Materials Research Society Symposium Proceedings,* Vol. 1204, ed. Y.K. Yap, K18–21. Boston: Materials Research Society.
36. Eda, G., Unalan, H.E., Rupesinghe, N., Amaratunga, G.A.J., and M. Chhowalla. 2008. Field emission from graphene based composite thin films. *Appl. Phys. Lett.* 93: 233502.
37. Malesevic, A., Kemps, R., Vanhulsel, A., Chowdhury, M.P., Volodin, A., and Van Haesendonck, C. 2008. Field emission from vertically aligned few-layer graphene. *J. Appl. Phys.* 104: 084301.
38. Watcharotone, S., Ruoff, R.S., and F.H. Read. 2008. Possibilities for graphene for field emission: Modeling studies using the BEM. *Physics Procedia* 1: 71–75.
39. Wu, Z.-S., Pei, S., Ren, W., Tang, D., Gao, L., Liu, B., Li, F., Liu, C., and H.-M. Cheng. 2009. Field emission of single-layer graphene films prepared by electrophoretic deposition. *Adv. Mater.* 21: 1756–1760.
40. Qian, M., Feng, T., Ding, H., Lin, L., Li, H., Chen, Y., and Z. Sun. 2009. Electron field emission from screen-printed graphene films. *Nanotechnology* 20: 425702.
41. Xiao, Z., She, J., Deng, S., Tang, Z., Li, Z., Lu, J., and N. Xu. 2010. Field electron emission characteristics and physical mechanism of individual single-layer graphene. *ACS Nano* 4: 6332–6336.
42. Lee, S.W., Lee, S.S., and E.-H. Yang. 2009. A study on field emission characteristics of planar graphene layers obtained from a highly oriented pyrolyzed graphite block. *Nanoscale Res. Lett.* 4: 1218–1221.

43. Thomas, R.N., Wickstrom, R.A., Schroder, D.K., and H.C. Nathanson. 1974. Fabrication and some applications of large-area silicon field emission arrays. *Solid-State Electronics* 17: 155–163.
44. Spindt, C.A., Brodie, I., Humphrey, L., and E.R. Westerberg. 1976. Physical properties of thin-film field emission cathodes with molybdenum cones. *J. Appl. Phys.* 47: 5248–5263.
45. Li, Y.B., Bando, Y., Golberg, D., and K. Kurashima. 2002. Field emission from MoO_3 nanobelts. *Appl. Phys. Lett.* 81: 5048–5050.
46. Tondare, V.N., Balasubramanian, C., Shende, S.V., Joag, D.S., Godbole, V.P., Bhoraskar, S.V., and M. Bhadbhade. 2002. Field emission from open ended aluminum nitride nanotubes. *Appl. Phys. Lett.* 80: 4813–4815.
47. Dorozhkin, P., Golberg, D., Bando, Y., and Z.-C. Dong. 2002. Field emission from individual B-C-N nanotube rope. *Appl. Phys. Lett.* 81: 1083–1085.
48. Chen, J., Deng, S.Z., Xu, N.S., Zhang, W., Wen, X., and S. Yang. 2003. Temperature dependence of field emission from cupric oxide nanobelt films. *Appl. Phys. Lett.* 83: 746–748.
49. Chen, Y.J., Li, Q.H., Liang, Y.X., Wang, T.H., Zhao, Q., and D.P. Yu. 2004. Field-emission from long SnO_2 nanobelt arrays. *Appl. Phys. Lett.* 85: 5682–5684.
50. Zhu, Y.W., Zhang, H.Z., Sun, X.C., Feng, S.Q., Xu, J., Zhao, Q., Xiang, B., Wang, R.M., and D.P. Yu. 2003. Efficient field emission from ZnO nanoneedle arrays. *Appl. Phys. Lett.* 83: 144–146.
51. Jo, S.H., Banerjee, D., and Z.F. Ren. Field emission of zinc oxide nanowires grown on carbon cloth. 2004. *Appl. Phys. Lett.* 85: 1407–1409.
52. Wang, R.C., Liu, C.P., Huang, J.L., Chen, S.-J., Tseng, Y.-K., and S.-C. Kung. 2005. ZnO nanopencils: Efficient field emitters. *Appl. Phys. Lett.* 87: 013110.
53. Zhao, Q., Zhang, H.Z., Zhu, Y.W., Feng, S.Q., Sun, X.C., Xu, J., and D.P. Yu. 2005. Morphological effects on the field emission of ZnO nanorod arrays. *Appl. Phys. Lett.* 86: 203115.
54. Windsor, E.E. 1969. Construction and performance of practical field emitters from lanthanum hexaboride. *Proc. IEEE* 116: 348–350.
55. Qi, K.C., Lin, Z.L., Chen, W.B., Cao, G.C., Cheng, J.B., and X.W. Sun. 2008. Formation of extremely high current density LaB_6 field emission arrays via e-beam deposition. *Appl. Phys. Lett.* 93: 093503.
56. Zhang, H., Tang, J., Yuan, J., Ma, J., Shinya, N., Nakajima, K., Murakami, H., Ohkubo, T., and L.-C. Qin. 2010. Nanostructured LaB_6 field emitter with lowest apical work function. *Nano Lett.* 10: 3539–3544.
57. Matsumoto, T., Neo, Y., Mimura, H., Tomita, M., and N. Minami. 2007. Stabilization of electron emission from nanoneedles with two dimensional graphene sheet structure in a high residual pressure region. *Appl. Phys. Lett.* 90: 103516.
58. Wang, H.M., Zheng, Z., Wang, Y.Y., Qiu, J.J., Guo, Z.B., Shen, Z.X., and T. Yu. 2010. Fabrication of graphene nanogap with crystallographically matching edges and its electron emission properties. *Appl. Phys. Lett.* 96: 023106.
59. Lee, C., Wei, X., Kysar, J.W., and J. Hone. 2008. Measurement of the elastic properties and intrinsic strength of monolayer graphene. *Science* 321: 385–388.
60. Dumé, B. 2010. Graphene has record-breaking strength. *Physics World.* http://physicsworld.com/cws/article/news/35055 (accessed January 27, 2011).
61. Obraztsov, A.N., Volkov, A.P., Boronin, A.I., and S.V. Kosheev. 2002. Defect induced lowering of work function in graphite-like materials. *Diamond Related Mater.* 11: 813–818.
62. Araidai, M., Nakamura, Y., and K. Watanabe. 2004. Field emission mechanisms of graphitic nanostructures. *Phys. Rev. B* 70: 245410.

63. Wang, J.J., Zhu, M.Y., Outlaw, R.A., Zhao, X., Manos, D.M., Holloway, B.C., and V.P. Mammana. 2004. Free-standing subnanometer graphite sheets. *Appl. Phys. Lett.* 85: 1265–1267.
64. Deng, J., Zhang, L., Zhang, B., and N. Yao. 2008. The structure and field emission enhancement properties of nano-structured flower-like graphitic films. *Thin Solid Films* 516: 7685–7688.
65. Lu, Z., Wang, W., Ma, X., Yao, N., Zhang, L., and B. Zhang. 2010. The field emission properties of graphene aggregates films deposited on Fe-Cr-Ni alloy substrates. *J. Nanomaterials* 2010: 148596.
66. Dong, J., Zeng, B., Lan, Y., Tian, S., Shan, Y., Liu, X., Yang, Z., Wang, H., and Z.F. Ren. 2010. Field emission from few-layer graphene nanosheets produced by liquid phase exfoliation of graphite *J. Nanosci. Nanotechnol.* 10: 5051–5055.
67. Liu, J., Zeng, B., Wu, Z., Zhu, J., and X. Liu. 2010. Improved field emission property of graphene paper by plasma treatment. *Appl. Phys. Lett.* 97: 033109.
68. Qi, J.L., Wang, X., Zheng, W.T., Ian, H.W., Hu, C.Q., and Y.S. Peng. 2010. Ar plasma treatment on few layer graphene sheets for enhancing their field emission properties. *J. Phys. D: Appl. Phys.* 43: 055302.
69. Obraztsov, A.N., Gröning, O., Zolotukhin, A.A., Zakhidov, Al. A., and A.P. Volkov. 2006. Correlation of field emission properties with morphology and surface composition of CVD nanocarbon films. *Diamond Related Mater.* 15: 838–841.
70. Zheng, W.T., Ho, Y.N., Tian, H.W., Wen, M., Qi, J.L., and Y.A. Li. 2009. Field emission from a composite of graphene sheets and ZnO nanowires. *J. Phys. Chem. C* 113: 9164–9168.
71. Verma, V.P., Das, S., Lahiri, I., and W. Choi. 2010. Large-area graphene on polymer film for flexible and transparent anode in field emission device. *Appl. Phys. Lett.* 96: 203108.
72. Lahiri, I., Verma, V.P., and W. Choi. 2011. An all-graphene based transparent and flexible field emission device. *Carbon* 49(5): 1614–1619.
73. Qin, X.Z., Wang, W.L., Xu, N.S., Li, Z.B., and R.G. Forbes. 2010. Analytical solution for cold field electron emission from a nanowall emitter. *Proc. Royal Soc. A* DOI: 10.1098/rspa.2010.0460.

10 Graphene and Graphene-Based Materials in Solar Cell Applications

Indranil Lahiri and Wonbong Choi

CONTENTS

10.1 INTRODUCTION

Technological and social developments of humankind have been well supported by the discovery of new sources of energy and its proper application in day-to-day life. Since the first installation of electricity grids by Edison, Tesla, and Westinghouse almost 130 years ago, our society has been driven by this source of energy. Most of this electric energy is produced from coal-fired power generation systems. Another source of energy, mainly used in the transportation industry, comes from burning carbon-based fuels, such as gasoline. Today's modern era, however, demands development of new and more efficient technologies for next-generation electrical grids and shifting of the focus of power generation to cleaner and C-free sources [1]. Both the scarcity of carbon-based fossil fuels and environmental pollution created by the burning of fossil fuels are responsible for this essential shift. Possible sources of such clean and renewable energy are solar, nuclear, wind, and geothermal energy. It is worth mentioning here that the world consumes less energy in one year than the

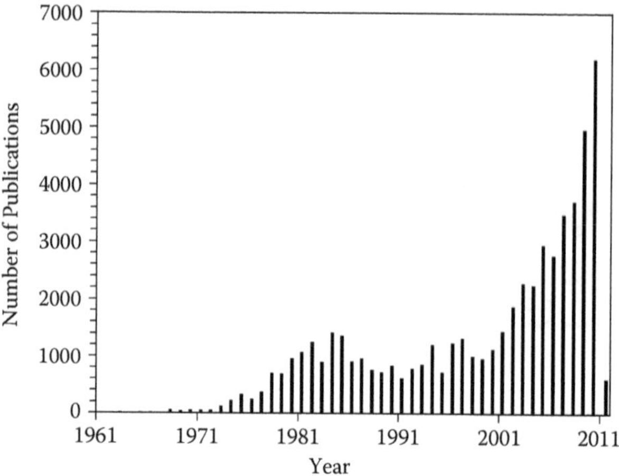

FIGURE 10.1 Research trends in solar cell science and technology. (Data compiled using scopus.com.)

amount of energy we receive from sunlight in one hour (13 terawatts) [2]. It seems obvious from this fact that capturing solar energy is going to be the main thrust for the next few decades [2,3]. In line with this expectation, research on solar cell science and technology has taken a huge jump in recent years (Figure 10.1).

Since the first demonstration of the solar cell in 1954 by Bell Laboratories [4], much progress has been made in this field. The initial cell by Bell was a silicon p-n junction cell. Today, different varieties of solar cells are being studied including amorphous silicon, microcrystalline silicon, polycrystalline silicon, cadmium telluride, and copper indium selenide/sulfide [5]. Demand for next-generation high-efficiency and low-cost solar cells have created excitement in the scientific community to (a) search for new materials that are easily available, low in cost, and nonpolluting and (b) combine the benefits of nanotechnology in photovoltaic systems. In such a context, graphene, due to its exciting electrical and mechanical properties, received immediate attention for possible application as electrodes in future solar cells. As can be seen in Figure 10.2, research on applications of graphene took a sudden jump in 2010 and is expected to continue. The aim of this chapter is to introduce those properties of graphene that are attractive for its application in solar cells, followed by a summary of the actual developments in this research field.

10.2 SOLAR CELL APPLICATIONS: WHY GRAPHENE?

Crystalline silicon solar cells, either mono- or polycrystalline (also known as first-generation solar cells), have a long-standing history of application in real devices, and as of 2008, these cells have captured ~90% of the world's photovoltaic market [6]. While research-grade silicon solar cells offer the highest energy conversion efficiency of 25%, industrial cells are limited to 15–20% efficiency limits [6]. However, Si solar cells suffer from *red losses* and *blue losses*, which could be related to its band

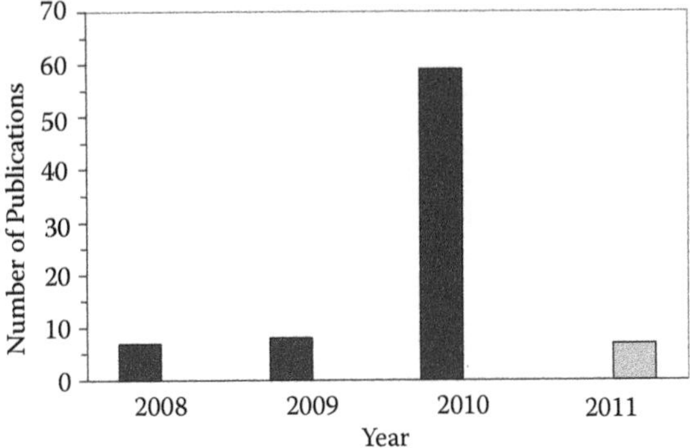

FIGURE 10.2 Application of graphene in solar cell research took a jump in 2010. (Data compiled using scopus.com.)

gap [7]. Hence, in spite of its high efficiency and commercial success, the quest for alternate solar cells was still under way. The driving forces for those research efforts are mainly to obtain higher conversion efficiency of the solar cell through a process that is simple to operate, easy to automate, and capable of bulk production at a much lower cost. To reach electric grid parity and to compete with coal-based or nuclear power, it is necessary to operate solar cells at a total price level of $2 per watt of power produced. Though generating power at such a low cost still remains a big challenge, scientists concentrated their effort on realizing other targets like achieving higher energy efficiency or adding more functionality to the cells, for example, new design concepts to grab sunlight for the full day time, making cells flexible to enable them to be mounted on any surface, and so on. Such continuous research efforts have led to development of new materials like cadmium telluride and copper indium selenide/sulfide thin film solar cells (also known as second-generation solar cells) and new types of solar cells (organic/polymer solar cells, dye sensitized solar cells, multijunction cells, etc., known as third-generation solar cells). Understanding details about each of these systems and the basics of solar cells is beyond the scope of this chapter, but interested readers may refer to relevant sources to gain an overview of each of these topics [6,8–12]. In this section, the focus will be to understand the "how" and "why" of the application of graphene in solar cells.

The advent of nanotechnology has created special interest in the use of nanomaterials in renewable energy sources, such as solar cells. Nanostructures of semiconductors offer quantization effects—the confinement of the electronic particles (electrons and holes) by potential barrier within an extremely small region of space. Confinement is possible in one dimension (for quantum films, 2D materials), two dimensions (quantum wires or rods, 1D materials), and three dimensions (quantum dots, 0D materials). This quantum confinement leads to exciting properties, such as production of more than one electron–hole pair from a single absorbed photon (a process popularly known as *multiple exciton generation*) [13]. Thus, certain nanomaterials that show quantum confinement have the required potential to improve the energy conversion efficiency of solar cells. Fullerenes, carbon nanotubes (CNTs)

[14], and graphene (the result of confinement of Dirac electrons in graphene is known as *Zitterbewegung* or jittery motion of the wave function of the Dirac electrons) [15] are known to offer confinement effects, which has generated interest in their application in solar cells [16]. CNTs were used first in TiO_2-based dye-sensitized solar cells (DSSC); however, the structure resulted in marginal enhancement of efficiency of the solar cell [17,18]. It could be pointed out that a very small point of contact between CNTs and TiO_2 nanospheres restricted electron transport, and hence the efficiency of the cell. At that point in time, it seemed obvious that a 2D material, having electrical properties equal to that of CNTs, could be the best alternative. TiO_2 nanospheres could easily anchor to such 2D materials, accelerate electron flow from the conduction band of TiO_2, and enhance charge separation [19]. Moreover, graphene shows carrier relaxation channels, which bridge the valence and conduction bands leading to carrier multiplication [20,21]. Such multiplication is extremely important for increasing the efficiency of solar cells. Thus, graphene offers a good opportunity for application in solar cells, especially in DSSCs.

Apart from these issues, optical transparency is a critical parameter for the materials to be used in solar cells as conducting electrodes. Two materials, indium tin oxide (ITO) and fluorine tin oxide (FTO), are popularly used in solar cells as the transparent, conducting electrodes [7], ITO being more effective than FTO. For future solar cells, ITO needs to be replaced by another suitable material because ITO (a) is brittle, (b) is unstable in acidic or basic environments, (c) offers poor transparency in the near-infrared region, and (d) is highly expensive due to limited natural resources of indium. In search of a suitable material with high sheet conductivity and transparency, CNT films were again explored [22,23]. However, the performance of these films was found to be inferior to ITO. With its remarkable electrical and thermal properties, chemical stability, mechanical strength, high transparency even in the infrared spectrum, and scalable and low-cost synthesis routes to produce atomic layer films on a variety of substrates, graphene emerged as the best choice for such applications [24]. Graphene is known to have 97.7% optical transparency for single-layer graphene and the transparency is a function of the number of layers in graphene [25,26]; it degrades to almost 88% for just five graphene layers (Figure 10.3). The number of layers of graphene is also known to affect the open circuit voltage (OCV) of graphene-based solar cells [27]. Thus, controlling graphene thickness is of extreme importance. Various techniques for controllable graphene synthesis have already been reported [28].

All the properties discussed previously were the driving force for application of graphene as the transparent, conducting electrode in different varieties of solar cells. However, in DSSCs, large surface area of graphene is expected to generate further interest. The counter electrode is an important part of DSSCs, where tri-iodide ions are reduced to iodide ions. The counter electrode material that is most often used, platinum (Pt), catalyzes this reduction process [29]. The high cost of Pt electrodes demands replacement of this material by an equivalent material. Graphene, with its high charge transport properties and large surface area, seems to be a good candidate for this application.

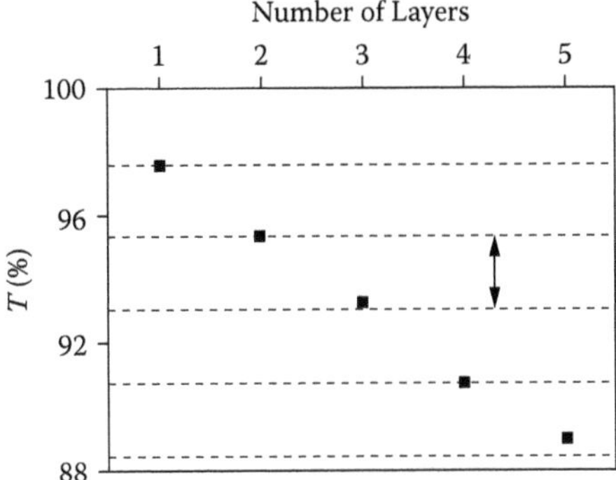

FIGURE 10.3 Transmittance of white light as a function of the number of graphene layers. (From Nair, R.R., Blake, P., Grigorenko, A.N., Novoselov, K.S., Booth, T.J., Stauber, T., Peres, N.M.R., and A.K. Geim. 2008. Fine structure constant defines visual transparency of graphene. *Science* 320: 1308. With permission from AAAS.)

The following sections are dedicated to a discussion of research developments and applications of graphene and graphene-based materials in various types of solar cells.

10.3 GRAPHENE-BASED STRUCTURES IN SOLAR CELL APPLICATIONS

Graphene has been and is being used as a transparent conducting electrode material in various types of inorganic, organic, and dye-sensitized solar cells, and also as counter electrodes in DSSCs (Figure 10.4) [30]. Moreover, functionalized graphene and graphene-based hybrid structures have also been used in first-, second- and third-generation solar cells to enhance their performance. The following subsections will briefly present research developments in each of these categories. It may be mentioned at this point that to date, commercial products are not available for many of these types of solar cells and properties of commercial-grade cells are often lower than those mentioned for research-grade cells.

10.3.1 Graphene-Based Materials as Transparent Conducting Electrodes in Inorganic Solar Cells

Graphene has attracted the most attention as a transparent, conducting electrode in solar cells. For such application, graphene should have good transmittance and low sheet resistance. Figure 10.5 presents a comparison of these properties of graphene with those of other popularly used electrode materials. The figure clearly shows the superior performance of graphene as compared to other materials, as far as these two properties are concerned. It may also be noted that transmittance and sheet resistance is a function of the graphene synthesis technique used, and as of now, chemical vapor deposited (CVD) graphene has shown the best properties, almost comparable to theoretically predicted values. With these excellent properties, graphene takes a lead role in the latest research activities in development of new electrode materials for solar cells [31].

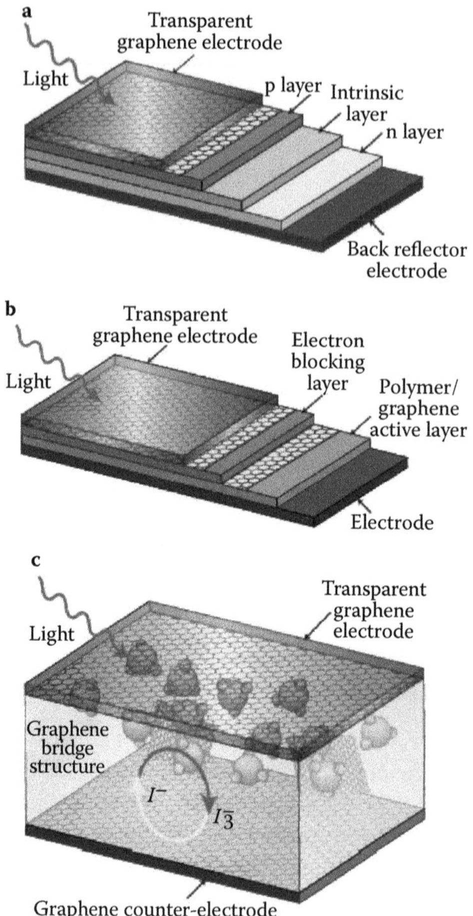

FIGURE 10.4 Schematics of graphene-based solar cells: (a) inorganic, (b) organic, and (c) dye-sensitized solar cells. (From Bonaccorso, F., Sun, Z., Hasan T., and A.C. Ferrari. 2010. Graphene photonics and optoelectronics. *Nature Photonics* 4: 611–622. With permission from Macmillan Publishers Limited.)

FIGURE 10.5 *(See facing page)* Graphene as a transparent conductor. (a) Transmittance for different transparent conductors. (b) Thickness dependence of the sheet resistance. Two limiting lines for GTCFs are also plotted (enclosing the shaded area); calculated from theoretical predictions. (c) Transmittance versus sheet resistance for different transparent conductors; shaded area enclosed by limiting lines for graphene are calculated from theoretical predictions. (d) Transmittance versus sheet resistance for graphene grouped according to production strategies: triangles, CVD; blue rhombuses, micromechanical cleavage (MC); red rhombuses, organic synthesis from polyaromatic hydrocarbons (PAHs); dots, liquid-phase exfoliation (LPE) of pristine graphene; and stars, reduced graphene oxide (RGO). A theoretical line is also plotted for comparison. (From Bonaccorso, F., Sun, Z., Hasan T., and A.C. Ferrari. 2010. Graphene photonics and optoelectronics. *Nature Photonics* 4: 611–622. With permission from Macmillan Publishers Limited.)

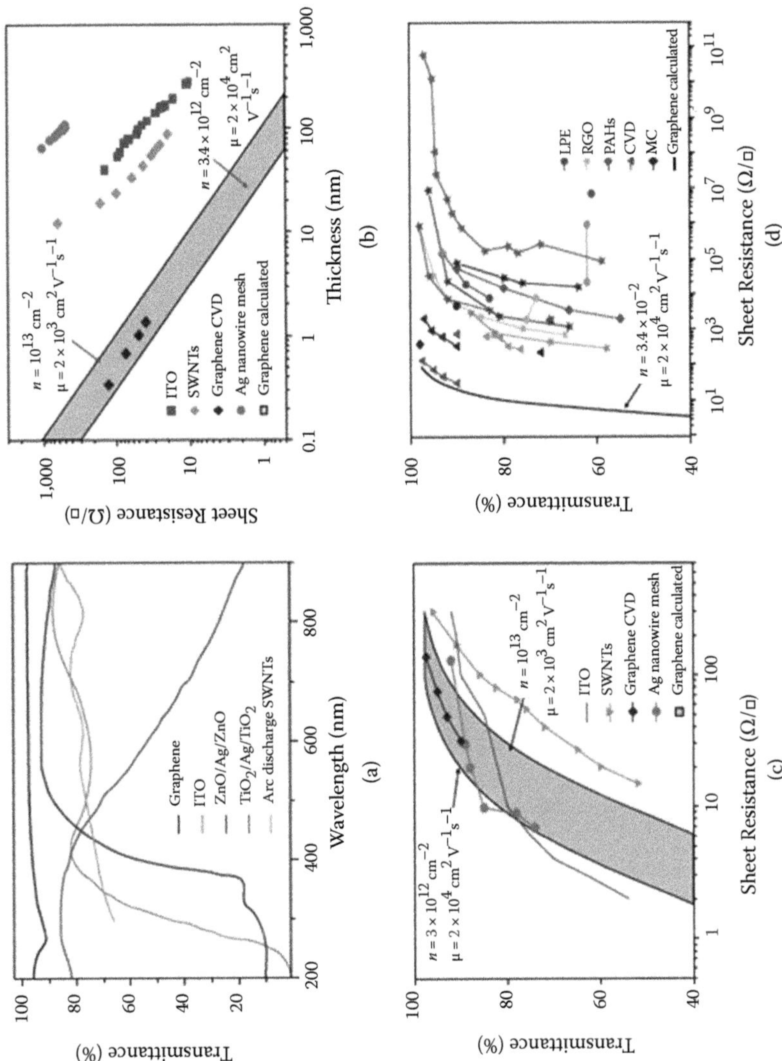
Transmittance (%)
Wavelength (nm)
(a)
Graphene
ITO
ZnO/Ag/ZnO
TiO2/Ag/TiO2
Arc discharge SWNTs
Sheet Resistance (Ω/□)
Thickness (nm)
(b)
ITO
SWNTs
Graphene CVD
Ag nanowire mesh
Graphene calculated
n = 10^13 cm^−2
μ = 2 × 10^3 cm^2 V^−1 s^−1
n = 3.4 × 10^12 cm^−2
μ = 2 × 10^4 cm^2 V^−1 s^−1
Transmittance (%)
Sheet Resistance (Ω/□)
(c)
n = 3 × 10^12 cm^−2
μ = 2 × 10^4 cm^2 V^−1 s^−1
n = 10^13 cm^−2
μ = 2 × 10^3 cm^2 V^−1 s^−1
ITO
SWNTs
Graphene CVD
Ag nanowire mesh
Graphene calculated
Transmittance (%)
Sheet Resistance (Ω/□)
(d)
n = 3.4 × 10^−2
μ = 2 × 10^4 cm^2 V^−1 s^−1
LPE
RGO
PAHs
CVD
MC
Graphene calculated

FIGURE 10.5

One of the initial efforts in this field involved application of water-processable single-layer graphene oxide (GO) films on quartz surfaces [32]. Though unreduced film has shown sufficient conductivity and high transmittance, a reduction procedure was followed to enhance its properties further. A proper selection of the reduction process is expected to improve film conductivity as well as its transmittance. Thickness of the film also showed a strong relationship to transmittance; while 6-nm film showed ~88% transmittance, an 8-nm film offered only ~60% transmittance. These experimental results indicate the importance of layer-controlled synthesis of graphene film for solar cell application. However, the solution processing of the film reduces the production cost by a huge margin. Following this trend, it was later shown that a controlled chemical reduction of exfoliated GO can produce a so-called chemically converted graphene (CCG) [33]. This CCG film demonstrated sufficiently high bulk conductivity (10^2 S/cm) and 80% transparency at 550 nm. A thermal annealing of the CCG film was found to restore the sp^2 network in the graphene film, enhancing the charge transport in individual sheets. Furthermore, the thermal annealing treatment also helped in reducing the interlayer distance, which aided in charge transport across CCG sheets. Though the power conversion efficiency of this electrode was almost half of that offered by the ITO electrode, its low processing cost and higher fill factor (as compared to other carbon film electrodes reported earlier) showed promise.

A practical solar cell electrode will, however, need large-area films, which is another challenging task related to graphene. Recently it was reported that a modified chemical exfoliation technique could be used to prepare large-area GO films (up to 40,000 μm^2 area) and a novel low-temperature (<100°C) fast reduction by HI acid could lead to graphene films of similar area [34]. The larger the area of the film, the lower the sheet resistance, which could be related to a decrease in the number of intersheet tunneling barriers for large-area graphene sheets. Resistance of such large-area graphene films could be reduced by p-type doping by HNO_3 [35]. Nitric acid is a well-known p-type dopant for graphitic materials, transferring one electron from graphene to nitric acid and forming a charge-transfer complex according to the reaction

$$6HNO_3 + 25C \rightarrow C_{25}^{+}NO_3^{-}.4HNO_3 + NO_2 + H_2O$$

This doping leads to a shift in Fermi level, and increases the carrier concentration (and hence, conductivity) of the graphene layers (Figure 10.6). Two ways of doping could be adopted—last-layer doping (doping done after all the layers are stacked) and interlayer doping (doping done after stacking each layer). This p-type doping of the graphene layer could reduce the sheet resistance of graphene up to a factor of 3, thus enhancing its performance in solar cells.

Another school of scientists opted to experiment with mixing graphene with other forms of carbon (carbon nanotubes or amorphous carbon films) and analyzing their properties for possible application in solar cells. In one such effort, graphene–CNT hybrid thin films were deposited on a polyethylene terephthalate (PET) substrate and the films were doped using HNO_3 [36]. While graphene–CNT hybrid film showed almost an order of magnitude reduction in resistance (at similar transmittance) as compared to graphene films, both films have shown positive effects of doping.

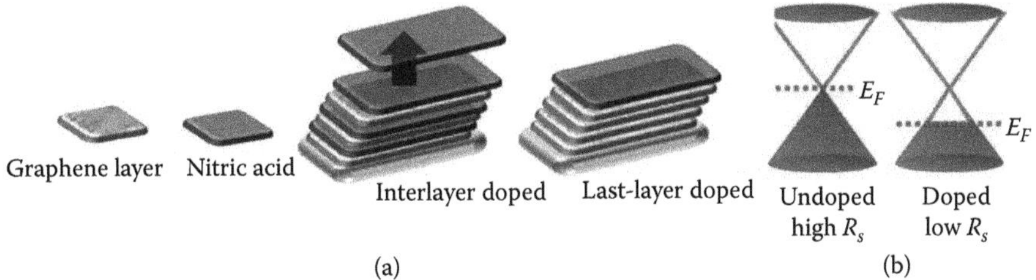

FIGURE 10.6 (a) Schematic illustrating the two different doping methods pursued here. In the interlayer-doped case, the sample is exposed to nitric acid after each layer is stacked, whereas in the last-layer-doped case, the film is exposed to nitric acid after the final layer is stacked. (b) Illustration of the graphene band structure, showing the change in the Fermi level due to chemical p-type doping. (From Kasry, A., Kuroda, M.A., Martyna, G. J., Tulevski, G. S., and A. A. Bol. 2010. Chemical doping of large-area stacked graphene films for use as transparent, conducting electrodes. *ACS Nano* 4 (7): 3839–3844. With permission from the American Chemical Society.)

Furthermore, the films have shown excellent integrity (compared to brittle ITO coatings) after several bending cycles, indicating their promise to be used as transparent, conducting, flexible electrodes in next-generation flexible solar cells. Such exciting findings encourage researchers to explore more and more important applications of graphene-based materials in solar cells.

10.3.2 Graphene-Based Materials as Electrodes in Dye-Sensitized Solar Cells

Photoelectrochemistry has opened up a new direction of converting solar energy into electrical energy by photoelectrochemical cells. The most common form of this type of cell is dye-sensitized solar cells (DSSCs). DSSCs are composed of two electrodes, one of which has a porous layer of particulates covered with a dye sensitizer that absorbs sunlight, and is immersed in an electrolyte solution. In a typical DSSC, the dye molecules become photo-excited upon light illumination and rapidly inject electrons into the conduction band of the semiconductor TiO_2 nanocrystalline films, followed by transportation of the separated holes to the counter electrode by means of a red-ox process. However, even though the conventional TiO_2 DSSCs can reach power conversion efficiency up to 13.4%, the performance of DSSCs is expected to be enhanced further by adapting new materials, a new design of energy band gap alignment, and a new method of photo excitation. Apart from offering good energy conversion efficiency, DSSCs show good transparency and flexibility, at a much lower cost (as compared to industrial solar cells). Recent research focus in DSSCs is centered on applications of n-type semiconducting nanostructured electrodes, ionic liquid electrolytes, and graphene and carbon nanotube electrodes [37–39]. This subsection will present an overview of the developments in the use of graphene in DSSCs.

It is important to mention here that maximum attainable solar-to-electrical power conversion efficiency of an ideal single-junction device is almost 30% [40]. First-generation single-crystal silicon solar cells have shown efficiencies up to 25%, which is very close to the theoretical limit [41]. However, DSSCs are not ideal cells. With the currently available technology (which results in a loss in potential of 0.75 eV), DSSCs can show a maximum power conversion efficiency of 13.4%. If the loss in potential can be reduced to 0.40 eV, then the maximum efficiency could be as high as 20.25%, with an optical band gap of 1.31 eV (940 nm) [42]. One of the ways to reduce loss in potential is to lessen the disorder at the interface and the heterogeneity of the electron transfer process. Graphene is expected to play an important role in this respect.

Initial efforts to incorporate graphene in DSSCs occurred in 2008, when stable aqueous dispersions of graphene sheets were prepared by noncovalent functionalization with 1-pyrenebutyrate (PB^-) and spin-coated on ITO substrates to form DSSC counter electrodes [43]. However, poor quality and the relatively low conductivity (200–300 S/m) of the graphene film restricts efficiency of this cell to an unsatisfactory level of 2.2%. Polystyrene sulfonate-doped poly(3,4-ethylenedioxythiophene) (PEDOT–PSS) has a higher conductivity and much better film-forming ability. Thus, a graphene–PEDOT-PSS electrode is expected to perform better, and in reality, such a solar cell was found to show high transparency (>80%) in the visible wavelength range and a higher energy conversion efficiency of 4.5% [44]. Although performance of this cell is also far from theoretical expectations, still it shows that with proper combination of materials, graphene has the potential to be used in DSSCs. Another well-studied material system as the electrode material for DSSCs is graphene–TiO_2 [45–48] (Figure 10.7). Electrostatic attractive force between these two materials is favorable to bind TiO_2 nanoparticles to the surface of graphene. Thus, this material combination can offer good structural stability. Different methods could be followed to produce this hybrid material—exfoliation of graphene oxide followed by thermal reduction [45], heterogeneous coagulation between graphene and TiO_2 nanoparticles [46], and molecular grafting of chemically exfoliated graphene sheets into TiO_2 nanoparticle film [47]. Out of all these methods, the first method (exfoliation of graphene oxide followed by thermal reduction) showed poor performance as compared to the control FTO electrode, due to series resistances in the cell. However, the other two methods have shown appreciable performances. While the structure produced by the coagulation method has shown very high short-circuit photo-current density (J_{sc}) and overall energy conversion efficiency (η) of 8.38 mA/cm^2 and 4.28%, respectively, these values were actually 66% and 59% more than the control samples (without graphene). The molecular grafted structure showed lower absolute values of J_{SC} and η – 6.67 mA/cm^2 and 1.68%, respectively. These values are, however, 3.5 and 5 times higher than the corresponding values of the control sample (without graphene). Analyzing all these results, it seems that structural integrity of the electrode is very important to reduce cell resistances and the synthesis route plays an extremely important role in determining the final properties of the electrode. Properties of graphene sheets, if annealed, depend strongly on the annealing cycles, too; it was found that graphene nanosheets that are annealed at higher temperatures offered higher J_{SC} and η [49]. This observation is also related to the morphology of the graphene. Positive effects of graphene addition in TiO_2 can be manifold: (a) lower charge-transfer resistance in the hybrid films creating a continuous electron conducting network, which facilitates transport of electrons (significantly longer electron

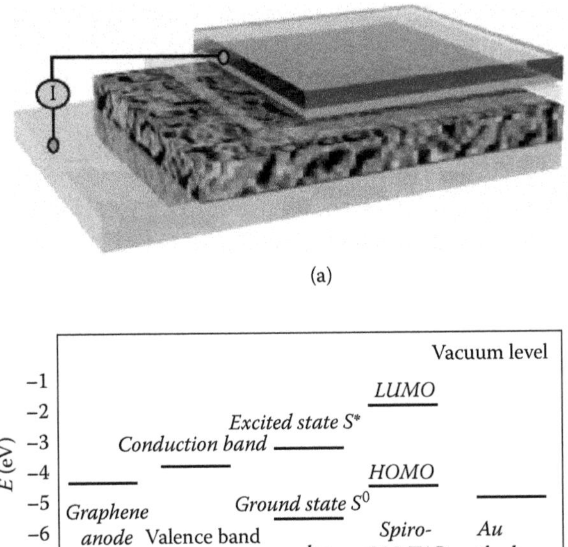

(a)

(b)

FIGURE 10.7 Schematics of solar cell based on graphene electrodes. (a) DSSCs using graphene film as electrode; the four layers from bottom to top are Au, dye-sensitized heterojunction, compact TiO_2, and graphene film. (b) Energy level diagram of graphene/TiO_2/dye/spiro-OMeTAD/Au device. (From Wang, X., Zhi, L., and K. Mullen. 2008. Transparent, conductive graphene electrodes for dye-sensitized solar cells. *Nano Lett.* 8 (1): 323–327. With permission from the American Chemical Society.)

lifetime) and reduces the probability of recombination [50], (b) graphene introduces unique surface morphologies that enhance absorption of more dye molecules, leading to more light harvesting, and finally, generation of more photo-induced electrons.

Graphene has another strong contribution in DSSCs; it shows catalytic activity to I^-/I_3^-. Both graphene [51] and partially reduced graphene oxide (GO) [52] show this type of electrocatalytic activity. Cyclic voltammetry (CV) and electrical impedance spectroscopy (EIS) are the best experimental tools to analyze the catalytic effects. It was found that the synthesis techniques have a strong effect on the electrocatalytic properties. The degree of reduction of a reduced GO sheet was found to affect catalytic reaction; a lower degree of reduction causes sluggish catalytic activity. Further, defect density of the graphene also influences catalytic activity. Graphene with higher defects shows improved electrocatalytic properties. This indicates the possibility of tuning the catalytic activity on the graphene surface by functionalization or modification [53]. In a recent study, a functionalized graphene with oxygen-containing sites was observed to offer very high catalytic activity and overall conversion efficiency comparable to that of a Pt electrode [54].

Success of graphene as an electrode material in DSSCs has initiated research in various related directions. These include synthesis of graphene at lower temperatures or room temperature through simple and cost-effective techniques like solvent-based routes [55], application of different types of graphene-based hybrid materials, especially with other C-based nanostructures like CNTs [56,57], development of new morphologies of graphene to further enhance its activities [58,59], and so on. These new structures have often shown interesting properties, including efficiencies higher than 5%. However, multiwalled carbon nanotubes (MWCNT) are not expected to offer any enhancement in solar cell applications, as its inner walls do not contribute to the electrochemical activities, but absorb light. Other graphene-based composite structures have good potential to replace conventional Pt and FTO in DSSCs.

Very recently, other graphene-based composite structures—polystyrene (PS) and polyvinyl chloride (PVC) fibers incorporated into TiO_2 nanoparticles and graphene nanoflakes—have shown tunable superhydrophobicity and electrochemical properties [60]. Such materials can provide a new dimension to DSSCs by offering self-cleaning and anti-icing surfaces. Detailed characterization of such materials for their suitability in solar cell application is yet to be performed, but with such new and exciting discoveries, the future of DSSCs looks really promising.

10.3.3 Graphene-Based Materials as Electrodes in Organic Solar Cells

Recently, organic–polymer solar cells have attracted immense attention due to their capability to offer a new dimension to conventional solar cells—flexibility. Moreover, materials for this type of cell are often solution-processable, lightweight, and low cost, making the cell attractive for practical applications. Graphene has also shown a capability to maintain structural integrity and sheet resistance under bent state, creating more enthusiasm for application of graphene as electrodes in organic solar cells. Though many polymer-based graphene materials have found application in DSSCs, in organic solar cells, graphene is mainly added as an electron acceptor [61]. To understand its function in organic solar cells, a glimpse into the basic mechanism of organic cells [7] is necessary.

When a photon is absorbed by a conjugated polymer having highly conjugated π-systems, an electron is excited from the highest occupied molecular orbital (HOMO) to the lowest unoccupied molecular orbital (LUMO), leading to a strongly bound electron–hole pair (exciton). This pair can be separated by creating a heterojunction with an acceptor material. The acceptor material should have an electron affinity larger than that of the polymer, but lower than its ionization potential. Moreover, its HOMO level should be at a level lower than that of the polymer. These properties ensure that the photo-excited polymer transfers the electron to the acceptor material, but retains the hole in its valence band. The most widely used heterojunction is bulk heterojunction (BHJ). Many polymers and their composites have been used so far in BHJ materials. The most successful materials have been soluble poly(3-hexylthiophene) (P3HT) and/or poly(3-octythiophene) (P3OT) as the donor and [6,6]-phenyl C61-butyric acid methyl ester (PCBM) (this material is a fullerene derivative) as the electron acceptor. Since efficiency of organic photovoltaic (OPV) devices using these materials is still low, it is necessary to look for new materials. In this context, graphene is being explored for its function as an electron acceptor material.

Simulation studies have predicted efficiencies for graphene-based organic photovoltaic devices: 12% for single cells and up to 24% for stacked cells [62]. Experiments regarding application of graphene in such cells are thus aimed at achieving this efficiency level. Initial efforts of using graphene in organic solar cells were not very successful; these cells have shown lower short-circuit photo-current density (J_{sc}) and lower energy conversion efficiency (η) [63,64], compared to their control devices. Such inferior properties were related to higher sheet resistance of graphene and unoptimized process parameters during synthesis of graphene. In a little different approach, chemically converted graphene–CNT composite films were prepared in a solution-based low-temperature process without using surfactant and later used as a platform for the fabrication of P3HT:PCBM photovoltaic devices [65]. Though this hybrid material has shown better properties compared to its ITO-based control device, especially after bending, the overall J_{SC} and η were low, 3.45 mA/cm^2 and 0.85%, respectively. Electrodes prepared from graphene could be bent to 138° without any degradation, while ITO electrodes start cracking at a 60° bending angle, leading to irreversible failure [66]. Although this proof-of-concept device failed to provide excellent results, it showed promise of graphene-based structures in OPV devices.

Continuity of graphene film is one of the important issues. Compared to stacked graphene flakes, a continuous graphene sheet (of similar transparency) shows much lower sheet resistance and higher efficiency of the cell [66]. This finding poses a challenge to scientists to synthesize larger-area continuous graphene films and to successfully transfer them onto polymeric substrates. Such large-area graphene could be produced by chemical vapor deposition (CVD) [66] and a solution-processing route [67]. Apart from controlling the morphology of the graphene sheet, another way to improve efficiency of graphene-based OPVs is to functionalize the graphene. Different organic and inorganic materials such as $AuCl_3$ [68], butylamine [69], pyrenebutyrate [70], P3HT [71], have been used to functionalize graphene before using it in electrodes of BHJ solar cells. Among different varieties of functionalized graphene, the electrode in which CH_2OH-terminated regioregular poly(3-hexylthiophene) (P3HT) was chemically grafted onto carboxylic groups of graphene oxide (GO) via esterification reaction [71] (Figure 10.8) was found to offer a 200% increase in power conversion efficiency over the control sample. Though the actual efficiency was still very low (0.61%), proper functionalization of graphene could be a feasible route to enhance the efficiencies of organic solar cells.

Composites of graphene, especially with other carbon-based nanostructures, could be treated as alternate options. Transparent graphene constructed carbon films (TGF), when used in organic solar cells, have shown appreciable (81%) transparency at 550 nm, good conductivity, and minimum leakage current [72]. On the other hand, a composite electrode of solution-processable functionalized graphene (SPFGraphene) and functionalized multiwalled carbon nanotubes (f-MWCNT) also showed improved performance in a heterojunction solar cell [73]. This composite structure offers good carrier transport, better exciton separation, and suppresses charge recombination, all of which lead to better photovoltaic performance. A recent report has proposed a new kind of composite structure using graphene as an atomic

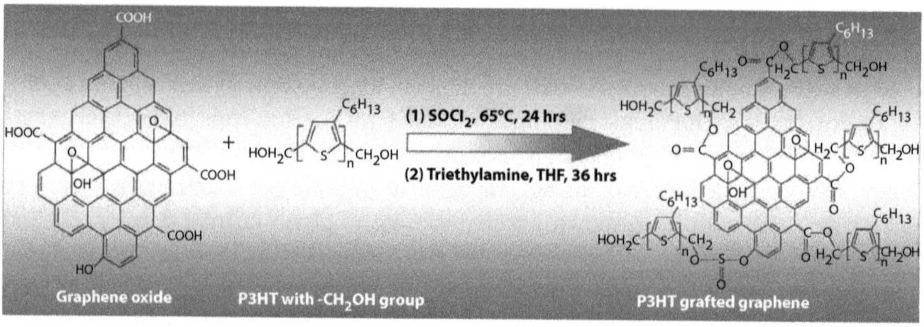

FIGURE 10.8 Schematic process chart of synthesis for chemical grafting of CH_2OH-terminated P3HT chains onto graphene, which involves the $SOCl_2$ treatment of GO (step 1) and the esterification reaction between acyl-chloride-functionalized GO and MeOH-terminated P3HT (step 2). (From Yu, D., Yang, Y., Durstock, M., Baek, J.-B., and L. Dai. 2010. Soluble P3HT-grafted graphene for efficient bilayer-heterojunction photovoltaic devices. *ACS Nano* 4(10): 5633–5640. With permission from the American Chemical Society.)

template and structural scaffold on which to assemble 1D organic nanostructures [74]. This novel graphene-organic hybrid structure has generated interest in application of novel nanostructures in solar cells.

Graphene oxide (GO) thin films can also be used in organic solar cells as a hole transport and electron blocking layer [75–77]. This layer is generally deposited between the photoactive P3HT:PCBM and transparent, conducting ITO layers. Presence of this GO layer inhibits electron–hole recombination, decreases leakage current, and enhances efficiency of the OPV cell. The function of the GO film is again strongly dependent on the film thickness [76] (optimum at 2–3 nm thickness), functionalizing, and annealing effects [78]. Ongoing and future research on this topic will be designed to optimize these parameters to extract the best achievable properties from these materials.

Through the analysis of extensive research done in the last few years involving the application of graphene in organic solar cells, it seems that a single layer of graphene is not sufficient to offer the high sheet conductivity required for such application. Thus, doping and stacking multiple layers of graphene become necessary. During stacking, interface coupling between different layers becomes an important issue. Recently it was demonstrated that a direct layer-by-layer transfer of graphene without any impurities can achieve 83.3% of the power conversion efficiency of the ITO-based control samples [79]. There could be other ways to control the interface. Jo et al. [80] used a work-function engineered multilayered graphene structure in an organic solar cell. Basically, an interfacial dipole layer has been applied to increase the built-in potential and improve charge extraction. Interface engineering seems to play a significant role in device design and more detailed studies are required to optimize the interface. A different way to achieve better performance from graphene-based solar cells could be through band-gap engineering. In order to effectively tune the band gap, 2D graphene sheets are converted to 0D graphene quantum dots [81]. Graphene quantum dots (QDs) were prepared through an electrochemical route and the process was performed in 0.1 M phosphate buffer solution (PBS) with a filtration-formed graphene film as a working electrode. As-prepared graphene quantum dots were mono-dispersed in a uniform diameter of ~3–5 nm. The solution remained homogeneous even after 3 months at room temperature without any visible change (e.g., aggregation or color change). These graphene QDs offer unique electron transport along with new quantum confinement and edge effects. Moreover, its shape and size control can lead to realization of new properties.

10.3.4 Graphene-Based Materials as Electrodes in Other Solar Cells

DSSCs and organic–polymer solar cells have shown definite promise as next-generation solar cells and application of graphene in these types of cells has been very exciting so far. Apart from these two types of cells, graphene has also been used either individually or as a hybrid material in many other types of solar cells. The silicon Schottky junction solar cell is one such cell. A graphene-on-silicon solar cell, which was created by depositing a graphene sheet on an n-type Si wafer and then constructing the Schottky junction cell (Figure 10.9), has shown an energy conversion efficiency of ~1.65%, a fill factor of 0.56, open circuit voltage of 0.48 V, and short circuit current of 6.5 mA [82,83], measured in well-known "air mass 1.5 global illumination" conditions. Though efficiency of this cell is low, it still represents a new type of photovoltaic device. Similar attempts to use graphene nanowhiskers on

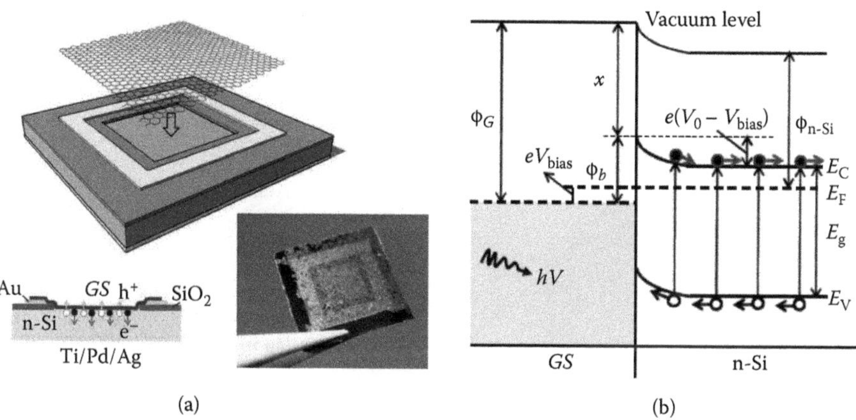

FIGURE 10.9 Graphene-silicon Schottky junction cell. (a) Schematic of the device configuration. Bottom-left inset: cross-sectional view, photogenerated holes (h^+) and electrons (e^-) are driven into graphene sheet (GS) and n-Si, respectively, by the built-in electric field. Bottom-right inset: photograph of a GS/n-Si Schottky cell with a 0.1 cm^2 junction area. (b) Energy diagram of the forward-biased GS/n-Si Schottky junction upon illumination. Φ_G (4.8 ≈ 5.0 eV), $\Phi_{n\text{-}Si}$ (4.25 eV) is the work function of GS and n-Si, respectively. V_0 is the built-in potential. Φ b is the barrier height. χ is the electron affinity of silicon (4.05 eV). E_g is the band gap of silicon (1.12 eV), and E_F is the energy of the Fermi level. V_{bias} is the applied voltage. The depth of the Fermi level below the Si conduction band edge ($E_C - E_F$) is ≈0.25 eV for the n-Si used in this work. (From Li, X., Zhu, H., Wang, K., Cao, A., Wei, J., Li, C., Jia, Y., Li, Z., Li, X., and D. Wu. 2010. Graphene-on-silicon Schottky junction solar cells. *Adv. Mater.* 22: 2743–2748. © 2010 Wiley-VCH Verlag GmbH & Co. KGaA, Weinheim. With permission.)

amorphous carbon [84] or CNT films and graphene [85] in silicon Schottky junction solar cells have resulted in almost comparable properties. However, when a CNT network patched with graphene sheets was prepared by a solid-phase layer stacking with ethanol wetting, the electrodes showed high flexibility, transparency (90% transmittance at 550 nm), low sheet resistance (735 Ω/sq), and good power conversion efficiency of up to 5.2% (under AM1.5 illumination) [86]. Properties of this film are much better than other similar types of electrodes, pointing to the importance of the morphology of the structure and the preparation techniques on the final achievable properties of the electrode. Overall, these studies show the potential of graphene in Si-based solar cells.

Graphene has also shown good potential for application in CdS quantum dot (QD) solar cells [87]. A layer-by-layer composite of graphene and CdS QDs (Figure 10.10) showed very exciting properties—6% incident photon-to-charge carrier generation efficiency (ICPE) [88]. This enhancement is more than three times that of any other carbon QD solar cell. The performance of this cell is probably attributable to single layers of graphene, probably enhancing light absorption and charge transport, and the favorable energy band match of graphene and CdS. This success inspired further efforts in different directions to better understand and develop this field of research. While some research aims to develop new synthesis techniques to prepare the graphene–CdS QD structure in a straightforward one-step process, like direct synthesis from graphene oxide by an easy one-step reaction [89], many other efforts are concentrated on understanding the basic mechanism of these types of solar cells through application of new characterization techniques, such as scanning photocurrent microscopy (SPCM) [90]. In a variation of graphene in CdS QD solar cells,

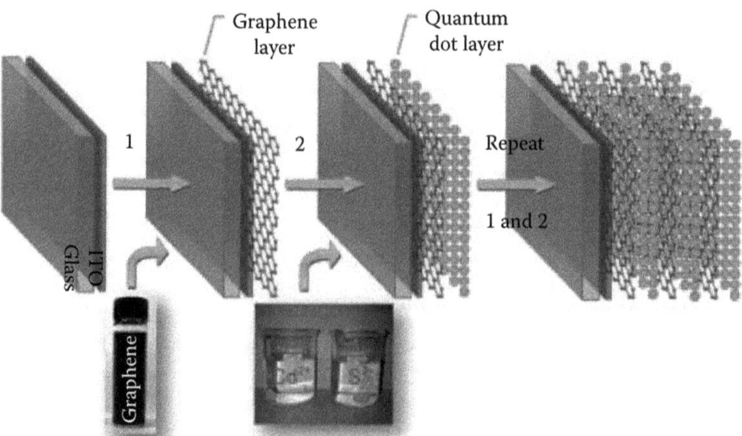

FIGURE 10.10 Schematic of fabrication of graphene-CdS quantum dot layered composite on ITO glass for solar cell application. (From Guo, C.X., Yang, H.B., Sheng, Z.M., Lu, Z.S., Song, Q.L., and C.M. Li. 2010. Layered graphene/quantum dots for photovoltaic devices. *Angew. Chem. Int. Ed.* 49: 3014–3017. © 2010 Wiley-VCH Verlag GmbH & Co. KGaA, Weinheim. With permission.)

new materials were also introduced to demonstrate the potential of such new materials systems and to enhance the performance of QD-sensitized solar cells. TiO_2 is a popular material in solar cells, and a combination of graphene, TiO_2, and CdS QDs was used for a solar cell [91], which showed 56% higher power conversion efficiency as compared to cells prepared without graphene. In such cells, graphene contributes significantly to enhancing electron transport within the cells. Following this trend of using graphene-based hybrid materials in solar cells, graphene has also been used in CdTe QD-sensitized solar cells [92]. Though this very recent effort failed to show any impressive power conversion efficiency, still it clearly demonstrated that the hybrid material performs much better than materials with graphene or QD alone. All such developments show the potential of graphene in various types of solar cells. The most important point is that all such exciting developments took place in the last couple of years, so it would be reasonable to predict that further exciting results might be reported in the near future. Further investigation is required to improve performance of all these graphene-based hybrid materials in solar cell applications.

In a recent and significantly different approach [93], monocrystalline ZnO nanorods (NR) (ZnO has high donor concentration) were electrochemically deposited on highly conductive reduced graphene oxide (rGO) films on quartz. This hybrid material was further used to fabricate inorganic–organic hybrid solar cells with a layered structure of quartz/rGO/ZnO NR/poly(3-hexylthiophene)/poly(3,4-ethylenedioxythiophene): poly(styrenesulfonate) (P3HT/PEDOT:PSS)/Au. Though the initial power conversion efficiency reported from this structure is not very high, this study opens up a new horizon of materials application in solar cell research.

10.4 SUMMARY

In a nutshell, exciting mechanical, electronic, and chemical properties, along with transparency and flexibility, suggest that graphene will be introduced in a wide variety of solar cells. Graphene, graphene oxide, functionalized graphene, and graphene-based hybrid structures have been used in many solar cells and most of those applications have shown easy processability and better performance. New ideas like interface control, band-gap engineering, and graphene-based organic nanowire structures show good potential for the next generation of solar cells. The future of graphene in solar cell applications seems to be highly promising, though much research work is needed to fully explore the benefits of this "wonder" material and convert these research efforts into industrial solar cells.

ACKNOWLEDGMENT

I.L. would like to acknowledge financial support through a dissertation year fellowship from Florida International University (FIU). This work was, in part, supported by the World Class University (WCU) program (R31-2008-000-10092) for W.C.'s sabbatical leave.

REFERENCES

1. Robinson, L. 2010. Shaping what's in store for the next-generation electrical grid. *JOM* 62 (11): 13–16.
2. The American Ceramic Society, Association for Iron & Steel Technology, ASM International, Materials Research Society, The Minerals, Metals, & Materials Society. 2010. *Advanced materials for our energy future*. Warrendale, PA: Materials Research Society.
3. The Minerals, Metals, & Materials Society. 2010. *Linking transformational materials and processing for an energy efficient and low-carbon economy: Creating the vision and accelerating realization*. Warrendale, PA: The Minerals, Metals, & Materials Society.
4. Perlin, J. 2002. *From space to Earth: The story of solar electricity*. Cambridge, MA: Harvard University Press.
5. Jacobson, M.Z. 2009. Review of solutions to global warming, air pollution and energy security. *Energy Environ. Sci.* 2: 148–173.
6. Saga, T. 2010. Advances in crystalline silicon solar cell technology for industrial mass production. *NPG Asia Mater.* 2(3): 96–102.
7. Hu, Y.H.. Wang, H., and B. Hu. 2010. Thinnest two-dimensional nanomaterial–graphene for solar energy. *Chem. Sus. Chem.* 3: 782–796.
8. Hoppe, H., and N.S. Sariciftci. 2004. Organic solar cells: An overview. *J. Mater. Res.* 19(7): 1924–1945.
9. Gratzel, M. 2003. Dye-sensitized solar cells. *J. Photochem. Photobiology C: Photochem. Rev.* 4: 145–153.
10. Meyer, G.J. 2010. The 2010 millennium technology grand prize: Dye-sensitized solar cells. *ACS Nano* 4(8): 4337–4343.
11. Nicholson, P.G., and F.A. Castro. 2010. Organic photovoltaics: Principles and techniques for nanometre scale characterization. *Nanotechnology* 21: 492001.
12. Nelson, J. 2003. *The physics of solar cells*. London: Imperial College Press.
13. Nozik, A.J. 2010. Nanoscience and nanostructures for photovoltaics and solar fuels. *Nano Lett.* 10: 2735–2741.
14. Saito, R., Dresselhaus, G., and M.S. Dresselhaus. 1998. *Physical properies of carbon nanotubes*. London: Imperial College Press.
15. Castro Neto, A.H., Guinea, F., Peres, N.M.R., Novoselov, K.S., and A.K. Geim. 2009. The electronic properties of graphene. *Rev. Modern Phys.* 81: 109–162.
16. Guldi, D.M., and V. Sgobba. 2011. Carbon nanostructures for solar energy conversion schemes. *Chem. Commun.* 47: 606–610.
17. Kongkanand, A., Martine-Dominguez, R., and P.V. Kamat. 2007. Single-wall carbon nanotube scaffolds for photoelectrochemical solar cells. Capture and transport of photogenerated electrons. *Nano Lett.* 7 (3): 676–680.
18. Brown, P., Takechi, K., and P.V. Kamat. 2008. Single-walled carbon nanotube scaffolds for dye-sensitized solar cells. *J. Phys. Chem. C* 112(12): 4776–4782.
19. Yang, N., Zhai, J., Wang, D., Chen, Y., and L. Jiang. 2010. Two-dimensional graphene bridges enhanced photoinduced charge transport in dye-sensitized solar cells. *ACS Nano* 4 (2): 887–894.
20. Winzer, T., Knorr, A., and E. Malic. 2010. Carrier multiplication in graphene. *Nano Lett.* 10: 4839–4843.
21. Mueller, M.L., Yan, X., McGuire, J.A., and L.-S. Li. 2010. Triplet states and electronic relaxation in photoexcited graphene quantum dots. *Nano Lett.* 10: 2679–2682.
22. Hu, L., Hecht, D.S., and G. Gruner. 2004. Percolation in transparent and conducting carbon nanotube networks. *Nano Lett.* 4: 2513–2517.

23. Wu, Z.C., Chen, Z.H., Du, X., Logan, J.M., Sippel, J., Nikolou, M., Kamaras, K., Reynolds, J.R., Tanner, D.B., Hebard, A.F., and A.G. Rinzler. 2004. Transparent, conductive carbon nanotube films. *Science* 305: 1273–1276.
24. Choi, W., Lahiri, I., and R. Seelaboyina. 2010. Synthesis of graphene and its applications: A review. *Critical Rev. Solid State and Mater. Sci.* 35(1): 52–71.
25. Ludwig, A.W.W., Fisher, M.P.A., Shankar, R., and G. Grinstein. 1994. Integer quantum Hall transition: An alternative approach and exact results. *Phys. Rev. B* 50(11): 7526–7552.
26. Nair, R.R., Blake, P., Grigorenko, A.N., Novoselov, K.S., Booth, T.J., Stauber, T., Peres, N.M.R., and A.K. Geim. 2008. Fine structure constant defines visual transparency of graphene. *Science* 320: 1308.
27. Ihm, K., Lim, J.T., Lee, K.-J., Kwon, J.W., Kang, T.-H., Chung, S., Bae, S., Kim, J.H., Hong, B.H., and G.Y. Yeom. 2010. Number of graphene layers as a modulator of the open-circuit voltage of graphene-based solar cell. *Appl. Phys. Lett.* 97: 032113.
28. Wei, D., and Y. Liu. 2010. Controllable synthesis of graphene and its applications. *Adv. Mater.* 22: 3225–3241.
29. Oregan, B., and M. Gratzel. 1991. A low-cost, high efficiency solar cell based on dye-sensitized colloidal TiO_2 films. *Nature* 353: 737–740.
30. Bonaccorso, F., Sun, Z., Hasan T., and A.C. Ferrari. 2010. Graphene photonics and optoelectronics. *Nature Photonics* 4: 611–622.
31. Liang, M., Luo, B., and L. Zhi. 2009. Application of graphene and graphene-based materials in clean energy-related devices. *Int. J. Energy Res.* 33: 1161–1170.
32. Becerril, H.A., Mao, J., Liu, Z., Stoltenberg, R.M., Bao, Z., and Y. Chen. 2008. Evaluation of solution-processed reduced graphene oxide films as transparent conductors. *ACS Nano* 2 (3): 463–470.
33. Geng, J., Liu, L., Yang, S.B., Youn, S.-C., Kim, D.W., Lee, J.-S., Choi, J.-K., and H.-T. Jung. 2010. A simple approach for preparing transparent conductive graphene films using the controlled chemical reduction of exfoliated graphene oxide in an aqueous suspension. *J. Phys. Chem. C* 114: 14433–14440.
34. Zhao, J., Pei, S., Ren, W., Gao, L., and H.-M. Cheng. 2010. Efficient preparation of large-area graphene oxide sheets for transparent conductive films. *ACS Nano*, 4 (9): 5245–5252.
35. Kasry, A., Kuroda, M.A., Martyna, G.J., Tulevski, G.S., and A.A. Bol. 2010. Chemical doping of large-area stacked graphene films for use as transparent, conducting electrodes. *ACS Nano* 4 (7): 3839–3844.
36. Xin, G., Hwang, W., Kim, N., Cho, S.M., and H. Chae. 2010. A graphene sheet exfoliated with microwave irradiation and interlinked by carbon nanotubes for high-performance transparent flexible electrodes. *Nanotechnology* 21: 405201.
37. Wei, D., Andrew, P., and T. Ryhanen. 2010. Electrochemical photovoltaic cells: Review of recent developments. *J. Chem. Technol. Biotechnol.* 85: 1547–1552.
38. Wei, D. 2010. Dye sensitized solar cells. *Int. J. Mol. Sci.* 11: 1103–1113.
39. Calandra, P., Calogero, G., Sinopoli, A., and P.G. Gucciardi. 2010. Metal nanoparticles and carbon-based nanostructures as advanced materials for cathode application in dye-sensitized solar cells. *Int. J. Photoenergy* 2010: 109495.
40. Shockley, W., and H.J. Queisser. 1961. Detailed balance limit of efficiency of p-n junction solar cells. *J. Appl. Phys.* 32 (3): 510–519.
41. Green, M.A. 2009. The path to 25% silicon solar cell efficiency: History of silicon cell evolution. *Prog. Photovolt: Res. Appl.* 17: 183–189.
42. Snaith, H.J. 2010. Estimating the maximum attainable efficiency in dye-sensitized solar cells. *Adv. Func. Mater.* 20 (1): 13–19.

43. Xu, Y., Bai, H., Lu, G., Li, C., and G. Shi. 2008. Flexible graphene films via the filtration of water-soluble noncovalent functionalized graphene sheets. *J. Am. Chem. Soc.* 130 (18): 5856–5857.
44. Hong, W., Xu, Y., Lu, G., Li, C., and G. Shi. 2008. Transparent graphene/PEDOT–PSS composite films as counter electrodes of dye-sensitized solar cells. *Electrochem. Commun.* 10: 1555–1558.
45. Wang, X., Zhi, L., and K. Mullen. 2008. Transparent, conductive graphene electrodes for dye-sensitized solar cells. *Nano Lett.* 8 (1): 323–327.
46. Sun, S., Gao, L., and Y. Liu. 2010. Enhanced dye-sensitized solar cell using graphene–TiO_2 photoanode prepared by heterogeneous coagulation. *Appl. Phys. Lett.* 96: 083113.
47. Tang, Y.-B., Lee, C.-S., Xu, J., Liu, Z.-T., Chen, Z.-H., He, Z., Cao, Y.-L., Yuan, G., Song, H., Chen, L., Luo, L., Cheng, H.-M., Zhang, W.-J., Bello, I., and S.-T. Lee. 2010. Incorporation of graphenes in nanostructured TiO_2 films via molecular grafting for dye-sensitized solar cell application. *ACS Nano* 4 (6): 3482–3488.
48. Zhu, G., Lv, T., Xu, T., Pan, L., and Z. Sun. 2010. Graphene-incorporated nanocrystalline TiO_2 films for dye-sensitized solar cells. In *Proceedings of the 2010 8th International Vacuum Electron Sources Conference and Nanocarbon (IVESC)*, 370–371. Nanjing, China: IEEE.
49. Zhang, D.W., Li, X.D., Chen, S., Li, H.B., Sun, Z., Yin, X.J., and S.M. Huang. 2010. Graphene nanosheet counter-electrodes for dye-sensitized solar cells. In *Proceedings of the 3rd International Nanoelectronics Conference (INEC)*, 610–611. Hong Kong: IEEE.
50. Kim, S.R., Parvez, Md. K., and M. Chhowalla. 2009. UV-reduction of graphene oxide and its application as an interfacial layer to reduce the back-transport reactions in dye-sensitized solar cells. *Chem. Phys. Lett.* 483: 124–127.
51. Hasin, P., Alpuche-Aviles, M.A., and Y. Wu. 2010. Electrocatalytic activity of graphene multilayers toward I^-/I^{3-}: Conditions and polyelectrolyte modification. *J. Phys. Chem. C* 114: 15857–15861.
52. Choi, S.-H., Ju, H.-M., and S.H. Huh. 2010. A catalytic graphene oxide film for a dye-sensitized solar cell. *J. Korean Phys. Soc.* 57(6): 1653–1656.
53. Das, S., Sudhagar, P., Song, D.H., Eto, E., Lee, S.Y., Kang, Y.S., and W. Choi. 2011. Amplifying charge transfer characteristics of graphene for triiodide reduction in dye-sensitized solar cells. *Advanced Functional Materials* (accepted for publication).
54. Roy-Mayhew, J.D., Bozym, D.J., Punckt, C., and I.A. Aksay. 2010. Functionalized graphene as a catalytic counter electrode in dye-sensitized solar cells. *ACS Nano* 4(10): 6203–6211.
55. Wan, L., Wang, S., Wang, X., Dong, B., Xu, Z., Zhang, X., Yang, B., Peng, S., Wang, J., and C. Xu. 2011. Room-temperature fabrication of graphene films on variable substrates and its use as counter electrodes for dye-sensitized solar cells. *Solid State Sciences* 13: 468–475.
56. Choi, H., Kim, H., Hwang, S., Choi, W., and M. Jeon. 2011. Dye-sensitized solar cells using graphene-based carbon nanocomposite as counter electrode. *Solar Energy Materials & Solar Cells* 95: 323–325.
57. Choi, H., Kim, H., Hwang, S., Kang, M., Jung. D.-W., and M. Jeon. 2011. Electrochemical electrodes of graphene-based carbon nanotubes grown by chemical vapor deposition. *Scripta Mater.* 64: 601–604.
58. Yang, N., Zhai, J., Wang, D., Chen, Y., and L. Jiang. 2010. Two-dimensional graphene bridges enhanced photoinduced charge transport in dye-sensitized solar cells. *ACS Nano* 4(2) 887–894.
59. Kavan, L., Yum, J.H., and M. Gratzel. 2011. Optically transparent cathode for dye-sensitized solar cells based on graphene nanoplatelets. *ACS Nano* 5(1): 165–172.
60. Asmatulu, R., Ceylan, M., and N. Nuraje. 2011. Study of superhydrophobic electrospun nanocomposite fibers for energy systems. *Langmuir* 27(2): 504–507.

61. Li, C., and G. Shi. 2011. Synthesis and electrochemical applications of the composites of conducting polymers and chemically converted graphene. *Electrochimica Acta* DOI: 10.1016/j.electacta.2010.12.081.
62. Yong, V., and J.M. Tour. 2010. Theoretical efficiency of nanostructured graphene-based photovoltaics. *Small* 6 (2): 313–318.
63. Wu, J., Becerril, H.A., Bao, Z., Liu, Z., Chen, Y., and P. Peumans. 2008. Organic solar cells with solution-processed graphene transparent electrodes. *Appl. Phys. Lett.* 92: 263302.
64. Wang, X., Zhi, L., Tsao, N., Tomovic, Z., Li, J., and K. Mullen. 2008. Transparent carbon films as electrodes in organic solar cells. *Angew. Chem. Int. Ed.* 47: 2990–2992.
65. Tung, V.C., Chen, L.-M., Allen, M.J., J. Wassei, K., Nelson, K., Kaner, R.B., and Y. Yang. 2009. Low-temperature solution processing of graphene-carbon nanotube hybrid materials for high-performance transparent conductors. *Nano Lett.* 9(5): 1949–1955.
66. De Arco, L.G., Zhang, Y., Schlenker, C.W., Ryu, K., Thompson, M.E., and C. Zhou. 2010. Continuous, highly flexible, and transparent graphene films by chemical vapor deposition for organic photovoltaics. *ACS Nano* 4(5): 2865–2873.
67. Yan, X., Cui, X., Li, B., and L.-S. Li. 2010. Large, solution-processable graphene quantum dots as light absorbers for photovoltaics. *Nano Lett.* 10: 1869–1873.
68. Park, H., Rowehl, J.A., Kim, K.K., Bulovic, V., and J. Kong. 2010. Doped graphene electrodes for organic solar cells. *Nanotechnology* 21: 505204.
69. Valentini, L., Cardinali, M., Bon, S.B., Bagnis, D., Verdejo, R., Lopez-Manchado, M.A., and J.M. Kenny. 2010. Use of butylamine modified graphene sheets in polymer solar cells. J. *Mater. Chem.* 20: 995–1000.
70. Chang, H., Liu, Y., Zhang, H., and J. Li. 2010. Pyrenebutyrate-functionalized graphene/poly(3-octyl-thiophene) nanocomposites based photoelectrochemical cell. *J. Electroanal. Chem.* DOI:10.1016/j.jelechem.2010.10.015.
71. Yu, D., Yang, Y., Durstock, M., Baek, J.-B., and L. Dai. 2010. Soluble P3HT-grafted graphene for efficient bilayer-heterojunction photovoltaic devices. *ACS Nano* 4(10): 5633–5640.
72. Kalita, G., Matsushima, M., Uchida, H., Wakita, K., and M. Umeno. 2010. Graphene constructed carbon thin films as transparent electrodes for solar cell applications. *J. Mater. Chem.* 20: 9713–9717.
73. Liu, Z., He, D., Wang, Y., Wu, H., Wang, J., and H. Wang. 2010. Improving photovoltaic properties by incorporating both SPFGraphene and functionalized multiwalled carbon nanotubes. *Solar Energy Mater. Solar Cells* 94: 2148–2153.
74. Wang, S., Goh, B.M., Manga, K.K., Bao, Q., Yang, P., and K.P. Loh. 2010. Graphene as atomic template and structural scaffold in the synthesis of graphene organic hybrid wire with photovoltaic properties. *ACS Nano* 4(10): 6180–6186.
75. Li, S.-S., Tu, K.-H., Lin, C.-C., Chen, C.-W., and M. Chhowalla. 2010. Solution-processable graphene oxide as an efficient hole transport layer in polymer solar cells. *ACS Nano* 4(6): 3169–3174.
76. Gao, Y., Yip, H.-L., Hau, S.K., O'Malley, K.M., Cho, N.C., Chen, H., and A.K.-Y. Jen. 2010. Anode modification of inverted polymer solar cells using graphene oxide. *Appl. Phys. Lett.* 97: 203306.
77. Yin, B., Liu, Q., Yang, L., Wu, X., Liu, Z., Hua, Y., Yin, S., and Y. Chen. 2010. Buffer layer of PEDOT:PSS/graphene composite for polymer solar cells. *J. Nanosci. Nanotechnol.* 10: 1934–1938.
78. Wang, J., Wang, Y., He, D., Liu, Z., Wu, H., Wang, H., Zhao, Y., Zhang, H., and B. Yang. 2010. Composition and annealing effects in solution-processable functionalized graphene oxide/P3HT based solar cells. *Syn. Met.* 160: 2494–2500.

79. Wang, Y., Tong, S.W., Xu, X.F., Özyilmaz, B., and K.P. Loh. 2011. Interface engineering of layer-by-layer stacked graphene anodes for high-performance organic solar cells. *Adv. Mater.* DOI: 10.1002/adma.201003673.
80. Jo, G., Na, S.-I., Oh, S.-H., Lee, S., Kim, T.-S., Wang, G., Choe, M., Park, W., Yoon, J., Kim, D.-Y., Kahng, Y.H., and T. Lee. 2010. Tuning of a graphene-electrode work function to enhance the efficiency of organic bulk heterojunction photovoltaic cells with an inverted structure. *Appl. Phys. Lett.* 97: 213301.
81. Li, Y., Hu, Y., Zhao, Y., Shi, G., Deng, L., Hou, Y., and L. Qu. 2011. An electrochemical avenue to green-luminescent graphene quantum dots as potential electron-acceptors for photovoltaics. *Adv. Mater.* 23: 776–780.
82. Li, X., Zhu, H., Wang, K., Cao, A., Wei, J., Li, C., Jia, Y., Li, Z., Li, X., and D. Wu. 2010. Graphene-on-silicon Schottky junction solar cells. *Adv. Mater.* 22: 2743–2748.
83. Won, R. 2010. Graphene-silicon solar cells. *Nature Photonics* 4: 411.
84. Li, X., Li, C., Zhu, H., Wang, K., Wei, J., Li, X., Xu, E., Li, Z., Luo, S., Lei, Y., and D. Wu. 2010. Hybrid thin films of graphene nanowhiskers and amorphous carbon as transparent conductors. *Chem. Commun.* 46: 3502–3504.
85. Schriver, M., Regan, W., Loster, M., and A. Zettl. 2010. Carbon nanostructure-aSi:H photovoltaic cells with high open-circuit voltage fabricated without dopants. *Solid State Commun.* 150: 561–563.
86. Li, C., Li, Z., Zhu, H., Wang, K., Wei, J., Li, X., Sun, P., Zhang, H., and D. Wu. 2010. Graphene nano-"patches" on a carbon nanotube network for highly transparent/conductive thin film applications. *J. Phys. Chem. C* 114: 14008–14012.
87. Dai, L. 2010. Layered graphene/quantum dots: Nanoassemblies for highly efficient solar cells. *Chem. Sus. Chem.* 3: 797–799.
88. Guo, C.X., Yang, H.B., Sheng, Z.M., Lu, Z.S., Song, Q.L., and C.M. Li. 2010. Layered graphene/quantum dots for photovoltaic devices. *Angew. Chem. Int. Ed.* 49: 3014–3017.
89. Cao, A., Liu, Z., Wu, M., Ye, Z., Cai, Z., Chang, Y., and Y. Liu. 2010. Synthesis of single-layer graphene-quantum dots nanocomposite directly from graphene oxide. In *Proceedings of the 3rd International Nanoelectronics Conference (INEC)*, 87–88. Hong Kong: IEEE.
90. Dufaux, T., Boettcher, J., Burghard, M., and K. Kern. 2010. Photocurrent distribution in graphene-CdS nanowire devices. *Small* 6(17): 1868–1872.
91. Zhu, G., Xu, T., Lv, T., Pan, L., Zhao, Q., and Z. Sun. 2011. Graphene-incorporated nanocrystalline TiO_2 films for CdS quantum dot-sensitized solar cells. *J. Electroanal. Chem.* 650: 248–251.
92. Lu, Z., Guo, C.X., Yang, H.B., Qiao, Y., Guo, J., and C.M. Li. 2011. One-step aqueous synthesis of graphene–CdTe quantum dot-composed nanosheet and its enhanced photoresponses. *J. Colloid Interface Sci.* 353: 588–592.
93. Yin, Z., Wu, S., Zhou, X., Huang, X., Zhang, Q., Boey, F., and H. Zhang. 2010. Electrochemical deposition of ZnO nanorods on transparent reduced graphene oxide electrodes for hybrid solar cells. *Small* 6(2): 307–412.

11 Graphene

Thermal and Thermoelectric Properties

Suchismita Ghosh and Alexander A. Balandin

CONTENTS

11.1 INTRODUCTION

There has been increasing interest in thermal properties of materials in the scientific and industrial research communities in recent years. This has been mostly motivated by the heat removal and thermal management issues in the electronic industry and also by the need to understand heat conduction at nanoscale from a fundamental science perspective [1–3]. A material's ability to conduct heat is ingrained in its atomic structure and knowledge of thermal conductivity can shed light on many other materials' properties. Thermal transport in two-dimensional (2-D) and one-dimensional (1-D) systems presents a particularly interesting scientific problem. Materials with very high or very low thermal conductivities attract attention due to possible applications for either heat removal or thermal insulation.

11.1.1 Thermal Transport at Nanoscale

Rapid advancement in the synthesis and processing of structures in the nanometer length scale has advocated the scientific understanding of thermal transport in nanostructured materials, devices, and individual nanostructures. We are mostly concerned about thermal transport in nonmetallic, semiconductor systems in which heat is transported by phonons through lattice vibrations [4–6]. Phonons cover a large range in frequency and an even larger range in their mean free path values. However, most of the heat is carried by phonons having a large wave vector and a mean free path (MFP) between 1 and 100 nm at room temperature. Hence, in many systems and devices of our present interest, the scale of the micro/nanostructure is of the same scale as the phonon mean free path and sometimes comparable to the phonon wavelength. The room temperature MFP of bulk silicon ranges from 40–300 nm and the transistor gate length is 20–40 nm. Table 11.1 (adapted from Balandin [7]) will elucidate the different phonon dispersion and scattering mechanisms that have dominant effects in various scales of length, L. In the table, the phonon thermal wavelength has been symbolized as λ. Cahill et al. [8] pointed out that the definition of temperature becomes very important in the scale of phonon mean free path and wavelength. Two regions of space that have a difference in temperature will also have a difference in the distribution of phonons. Scattering of phonons changes their distribution. The anharmonic scattering process occurs on the length of mean free path.

A local region with a particular temperature should therefore be larger than the phonon scattering distance. It is quite difficult to define temperature within the scale of average phonon mean free path. The low-frequency phonons have long mean free paths and vice versa for high-frequency phonons. Hence, for phonons, which carry the bulk portion of the heat, one can think of an average mean free path.

TABLE 11.1
Scattering Process and Length Scale Considerations in Phonon Transport Regime

Length Scale	Phonon Dispersion	Dominant Phonon Scattering Process
$L \gg$ MFP	Bulk dispersion	3-phonon Umklapp, point defect scattering
$\lambda \ll L \leq$ MFP	Bulk dispersion	3-phonon Umklapp, point defect, boundary scattering
$\lambda \leq L \ll$ MFP	Modified dispersion, many phonon branches populated	3-phonon Umklapp, point defect, boundary scattering
$L < \lambda$	Modified dispersion, lowest number of phonon branches populated	Ballistic mode of transport

Source: Adapted from A. A. Balandin. 2004. Thermal conductivity of semiconductor nanostructures. In *Encyclopedia of Nanoscience and Nanotechnology*, edited by H. S. Nalwa, vol. 10, 425–445. Stevenson Ranch, CA: American Scientific Publishers.

This theory might not apply across grain boundaries, since the boundary provides a natural limit to the temperature region. Wave interference is also quite significant in nanoscale devices. In most solids at room temperature, all the phonon states in the Brillouin zone are engaged in the transport mechanism. But since the wavelength of the heat-carrier phonons becomes comparable to the length scale of nanostructures, the traditional Boltzmann equation cannot explain all phenomena. Other theoretical approaches come into play at this point. The phonon confinement effect has a strong influence on phonon relaxation rates, modifies the group velocity, and, in turn, thermal properties. In low-dimensional materials like quantum wells and thin films, phonon scattering from boundaries, alteration of the phonon dispersion, and anharmonic interactions can decrease lattice thermal conductivity and thereby enhance thermoelectric properties [9–11].

11.1.2 Thermal Management in Scaled-Down Devices and Circuits

With the advent of transistors and integrated circuits, there has been a tremendous growth in the technological sphere. Gordon Moore predicted that the number of transistors on a chip would double every two years [12]. Known widely as Moore's Law, this observation has driven the semiconductor industry and the development of silicon-based integrated circuit design. This device miniaturization trend has led to shrinking transistor gate length, wafer size growth, and defect density reduction. This has led to an unprecedented revolution in microprocessor architecture, and the device design methods have increased the performance–cost ratio, integration density, and portability with a generation of huge computing power [13].

In order to achieve the projected performance gain, the threshold leakage current in devices had to be relaxed and the power density became a cause of concern in high-speed microprocessors. Before long, the power density on chips had increased tremendously, and at present this value is on the order of 100 W/cm^2, which is almost equivalent to that in a nuclear reactor, and it may rise even further as seen in Figure 11.1. As the device feature size is approaching 10 nm and beyond, the increased power densities and high chip temperatures will hinder reliable performance of integrated circuits [14,15]. While this is an issue at the chip level and poses a problem for circuit designers, device designers have started facing problems within individual transistors, which gives rise to thermal management issues. During the operation of devices, self-heating might occur as a result of interaction between current carrier electrons and phonons, which are the units of lattice vibration and carriers of heat. Complex device structure and the choice of materials also have significant roles to play in this increased power dissipation. Materials with lower thermal conductivity than silicon, like low *K*-dielectrics, when used in a silicon-on-insulator structure, cause increased thermal resistance due to the presence of interconnects. For a wide range of devices like complementary metal-oxide semiconductor (CMOS), high electron mobility transistors, and quantum cascade lasers, excessive heating severely impedes operations. For photonic devices, this might create a severe problem as heat generation reaches a million watts per unit square area. One of the main reasons

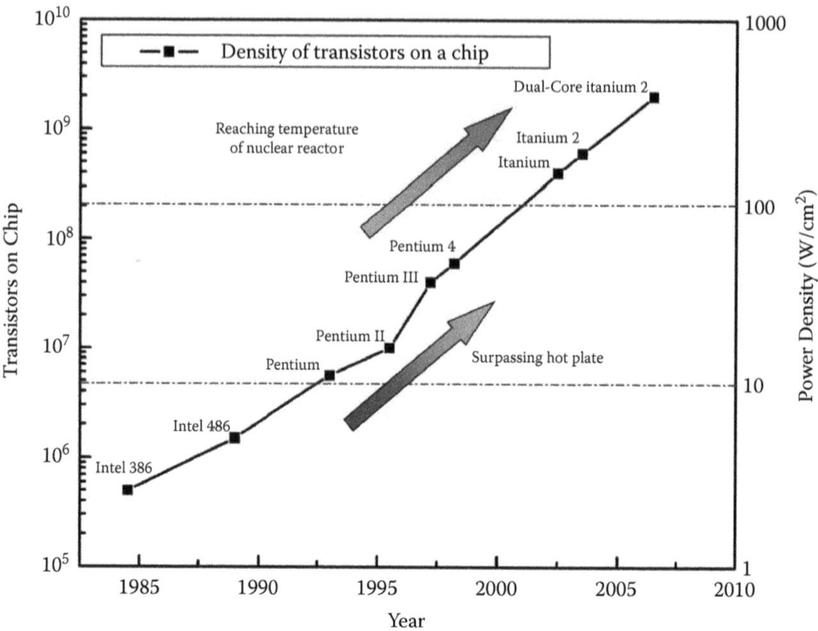

FIGURE 11.1 Transistor density on a chip and corresponding power density over the past 20 years. (Based on data from P. P. Gelsinger. 2001. Microprocessors for the new millennium: Challenges, opportunities, and new frontiers. *Solid-State Circuits Conference, Digest of Technical Papers, ISSCC, 2001 IEEE International*, 22 and from P. P. Gelsinger's lecture "The Era of Tera," Intel Developer Forum, San Francisco, CA, Spring 2004.)

behind this is the nanoscale device feature size approaching the phonon MFP; at such a length scale, phonon boundary scattering starts dominating three-phonon Umklapp scattering. Particularly in the ballistic regime, when device feature size is much less than the phonon MFP, there is a nonequilibrium state in the electron–phonon interaction and the phonons generated have diverse contributions to the thermal transport in devices. Acoustic phonons with large group velocities contribute more to thermal conductivity, as opposed to optical phonons with smaller group velocities. In a CMOS device, the total power loss can be vastly attributed to the switching power loss and the device leakage power. Techniques like transistor gating, power gating, and low-power designs are often implemented to reduce power dissipation, though at the cost of performance and allowable noise limits [16]. Scaling of the supply voltage is also limited by voltage fluctuations [17]. When conventional design is power constrained, in order to maintain optimum device performance, one has to take into account engineering of material parameters or structural geometry so that heat can be removed efficiently. With downscaling of devices and circuits and the increasing problem of heat dissipation, one possible approach to solving this thermal issue is to find a material with very high thermal conductivity so that it can be integrated with Si-based complementary metal-oxide-semiconductor (CMOS) technology.

11.2 THERMAL PROPERTIES OF GRAPHENE

Graphene is a single isolated layer of sp^2 hybridized carbon atoms packed in the form of a honeycomb lattice. Graphene was isolated in its free state, only a few years ago, by micromechanical cleavage of bulk graphite by Novoselov et al. [18,19]. Graphene exhibits many interesting physical properties. This 2-D material forms a unique building block for other graphitic materials; it can be enfolded into 0-D large fullerenes, rolled into 1-D carbon nanotubes, or can also be stacked up to form 3-D bulk graphite. Figure 11.2 shows a unit cell of graphene.

Graphene has a rather unusual energy dispersion relation; the low-lying electrons in single-layer graphene behave like massless relativistic Dirac fermions [19].This gives rise to unique phenomena such as quantum spin Hall effect [20–22], enhanced Coulomb interaction [22–24], suppression of the weak localization [19], and deviation from the adiabatic Born-Oppenheimer approximation [25]. Graphene also has an extraordinarily high room temperature (RT) carrier mobility exceeding 15,000 $cm^2V^{-1}s^{-1}$ and the electric field effect is tunable with either electrons or holes as charger carriers with a concentration of ~10^{13} cm^{-2} [18,26]. In addition, conductance quantization possibilities of inducing a band gap through lateral quantum confinement effects [24,26] and prospects of epitaxial growth [27] make graphene a promising material for future electronic circuits.

There have been theoretical suggestions that graphene has an unusually high thermal conductivity. Ballistic thermal conductance of a two-dimensional system was calculated directly by using the dispersion relations of electrons and phonons, and this method was applied to graphene. Dependence of thermal conductance on temperature and Fermi energy was studied by Saito et al. [28]. Peres and team have used a semiclassical approach and experimental data to study the dependence of thermal conductivity, K, and thermo-power of doped graphene on electron density [29]. Mingo et al. calculated quantum mechanical ballistic thermal conductivity for carbon nanotubes (CNTs), graphene, and graphite [30]. However, before 2008 there was no experimental measurement reported to support the claim of these theoretical works. In order to understand the unique thermal properties of graphene, it

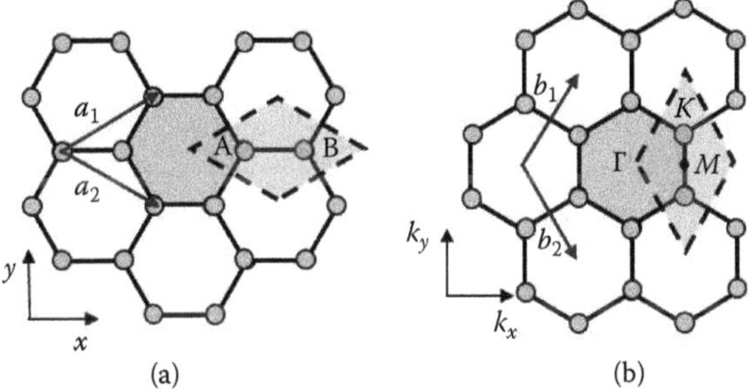

FIGURE 11.2 Unit cell of graphene lattice in (a) real and (b) reciprocal space shown as dotted rhombus. The shaded hexagon is (a) the Wigner Seitz cell in real space and (b) the Brillouin zone of graphene in reciprocal space.

would be useful to start with a review of thermal transport in graphite and CNTs, which are essentially precursors to graphene in terms of advancement in research of carbon-based materials.

11.2.1 Thermal Properties of Graphite and Carbon Nanotubes

In-plane graphite has been known to have one of the highest thermal conductivities (2000 W/mK) among carbon-based materials for a long time [31]. Researchers have also investigated c-axis thermal conductivity of graphite using the Debye model [32,33]. These values are up to two-orders of magnitude smaller than the values extracted for graphite basal planes. The calculations were done assuming that graphite consists of a few monolayers; defect and boundary scattering were not taken into consideration.

After discovery of CNTs, which are unique one-dimensional forms of carbon [34], there were suggestions that this material would have thermal conductivity even higher than crystal graphite [35]. Theoretical studies of thermal conductivity of infinitely long nanotubes have yielded high values comparable to or higher than those of graphite [36]. Single-walled carbon nanotubes (SWCNT) grown in crystalline bundles were studied by Hone et al. [37]. The tubes were a few microns in length with a diameter around 1.4 nm. Room temperature (RT) values of K for a single rope of CNTs were found to be 1750–5800 W/mK. Though graphite and CNTs are both made of graphene sheets, in the case of highly ordered pyrolytic graphite (HOPG), the *ab*-plane thermal conductivity, dominated by acoustic phonons, varies as T^{2-3} up to around 150 K. At higher temperatures, the 3-phonon Umklapp scattering process causes K to decrease with an increase in temperature, T. Interplanar vibrations in graphite give rise to extra phonon modes, which are absent in a single nanotube. The tubular shape of CNTs affects the phonon spectrum and scattering times significantly [38]. Thermal conductivity and thermoelectric power of an isolated multiwalled CNT (MWCNT) were measured and the value of K was found to be ~3000 W/mK around RT [39]. The measured thermoelectric power shows a linear temperature dependence with a value of 80 μV/K at room temperature. This was subsequently followed by thermal conductance analysis of a suspended metallic SWCNT of 1.7 nm diameter and K was found to be 3500 W/mK at RT [40]. Recent work by Aliev et al. [41] shows that the K value for a single MWCNT is 600±100 W/mK , bundled MWCNT has 150 W/mK, and that of an aligned, free-standing MWCNT sheet is 50 W/mK. A self-heating, 3-ω technique was used for this purpose, and this gradual decrease in K was attributed to the quenching of phonon modes in CNTs. A theoretical work by Berber et al. [42] uses molecular dynamic (MD) simulations to determine thermal conductivity of an isolated nanotube at room temperature and indicates an extremely high value, $K \approx 6600$ W/mK. The same calculations suggested that the thermal conductivity of graphene, a single planar layer of carbon atoms, would be even higher.

11.2.2 Thermal Conductivity of Graphene

In spite of the importance of the knowledge of thermal conduction in graphene and theoretical predictions of a very high thermal conductivity, there were no experimental studies on this subject prior to the pioneering work at the University of California-Riverside. This is explained by the fact that conventional techniques like 3-ω, laser flash, thermal bridge, or transient plane source techniques are not well suited for single-layer graphene. In 2007, through direct measurements, it was discovered experimentally that graphene indeed reveals an extremely high thermal conductivity [43,44]. The measurements by Balandin et al. [43] and Ghosh et al. [44] were performed using Raman spectroscopy, where the local temperature rise due to the laser heating was determined through independently measured temperature coefficients of the peaks in the graphene Raman spectrum [45,46]. It was found that the near-RT thermal conductivity of partially suspended single-layer graphene (SLG) is in the range of $K \sim 3000-5300$ W/mK depending on the graphene flake size. These experiments motivated theoretical work on the subject. Nika et al. [47] performed a detailed numerical study of the thermal conductivity of graphene using the phonon dispersion obtained from the valence force field (VFF) method, and treating the three-phonon Umklapp scattering directly considering all phonon relaxation channels allowed in graphene's 2-D Brillouin zone (BZ) [47]. The authors also proposed a simple model for the lattice thermal conductivity of graphene [48] showing that the Umklapp-limited thermal conductivity of graphene increases with the linear dimensions of graphene flakes. The results are in agreement with experimental data from Balandin et al. [43] and Ghosh et al. [44]. This experimental technique was then extended to few-layer graphene (FLG) by Ghosh et al. [49], which allowed the study of dimensional crossover of thermal transport. It was shown that as the number of atomic planes increases from 2 to 4, thermal conductivity changes from ~2800 to ~1300 W/mK, approaching bulk graphite limit.

Reports of the initial experimental and theoretical studies of thermal conduction in graphene stimulated several other interesting research studies in this field. Jiang et al. [50] calculated the thermal conductance of graphene in the pure ballistic limit obtaining a very high value, which is expected for the ballistic regime when no scattering is present. Jauregui et al. [51] reported thermal conduction in chemical vapor deposition (CVD)–grown graphene samples both suspended and SiO_2/Si substrate-supported. The measurement technique was a combination of Raman spectroscopy for thermometry and electrical transport for Joule's heating so that both thermal conductivity and graphene–substrate interface thermal resistance could be calculated. The thermal conductivity of suspended graphene was 1500–5000 W/mK. As a comparison, using thermal coefficients from Calizo et al. [45,46] provided similar results in the range of ~1500–4000 W/mK. A couple of very important results were reported by Cai et al. [52] and Chen et al. [53] on thermal conductivity values of CVD-grown suspended and supported monolayer graphene samples. For supported graphene, the RT thermal conductivity was 370 + 650/−320 W/mK. For graphene suspended over a 3.8-μm-diameter hole, K decreases from 2500 ± 1100 W/mK at 350 K to 1400 ± 500 W/mK at 500 K. Figure 11.3 shows these thermal conductivity values as compared to previously reported values of basal planes of pyrolytic graphite [54,55].

Although these values were slightly less than the previous K values [43,44], the error is comparable to uncertainties caused by Raman measurements and optical absorption. The K of graphene is still higher than that of pyrolytic graphite. Another

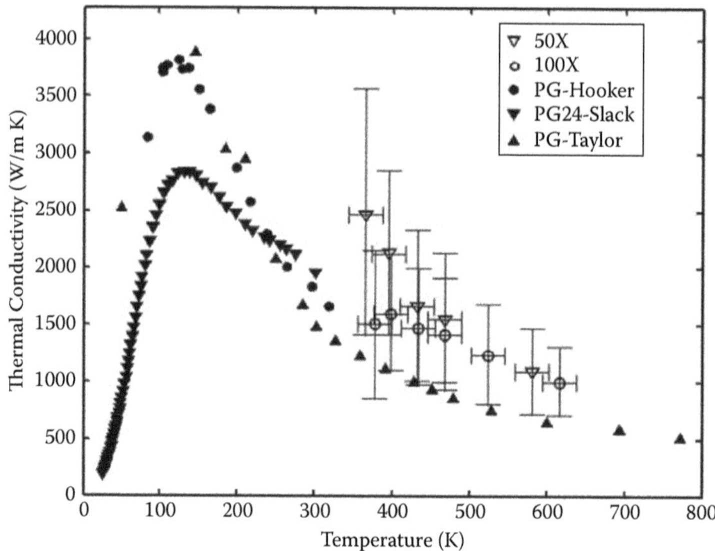

FIGURE 11.3 Thermal conductivity of the suspended CVD graphene measured using the 100X and 50X objective lens as a function of the measured graphene temperature. Also shown in comparison are the literature thermal conductivity data of pyrolytic graphite samples as a function of temperature [54, 55]. (From Cai, W., A. L. Moore, Y. Zhu, X. Li, S. Chen, L. Shi, and R. S. Ruoff. 2010, Thermal transport in suspended and supported monolayer graphene grown by chemical vapor deposition. *Nano Letters*, 10:1645–1651. © 2010 American Chemical Society. With permission.)

research group that repeated the measurements using the micro-Raman spectroscopy-based technique found thermal conductivity of 630 W/mK for suspended graphene at $T \approx 600$ K [56]. Although the measurements were conducted at ambient, the graphene membrane was heated to $T = 660$ K in the center and above ~500 K over most of its area. Since the thermal conductivity decreases with temperature due to Umklapp phonon scattering, this fact can explain the difference with Balandin et al. [43] and Ghosh et al. [44], which reported thermal conductivity near room temperature. Differences in strain distribution in the suspended graphene of various sizes and geometries may also affect the results. In another work by Seol et al. [57], experimental observations suggest that K of single-layer graphene exfoliated on a SiO_2 support is ~600 W/mK, which is still high as compared to metals such as copper. This value is lower than that of freely suspended graphene owing to phonon leakage across the graphene–substrate interface and strong interface scattering of the phonon modes. A study of heat flow across Au/Ti/graphene/SiO_2 interfaces shows that thermal conductance is ~25 $MWm^{-2}K^{-1}$ at RT, which is almost 4 times less as compared to Au/Ti/SiO_2 interface [58]. This shows that heat conduction across metal/graphene/oxide interfaces is restricted by a finite phonon transmission across metal and graphene. For effective use of graphene as a thermal management material in devices, it is important to choose metals with high Debye temperatures for a better energy match between the phonon modes in the graphene and the metals.

11.2.2.1 Thermal Conductivity of Graphene Nanoribbons

With increasing interest in thermal conductivity studies in graphene, there was also a parallel interest in graphene nanoribbons (GNRs). These are graphene strips of width <20 nm. GNRs are similar to CNTs yet have a much simpler fabrication process [59]. Lan et al. [60] determined the thermal conductivity of graphene nanoribbons combining the tight-binding approach and the phonon nonequilibrium Green's function method. The authors found a thermal conductivity of $K = 3410$ W/mK [60], which is clearly above the bulk graphite limit of 2000 W/mK and in agreement with the first experiments [43,44]. A strong edge effect was also revealed by the numerical data. Murali et al. [61] found experimentally that for 20-nm-wide graphene nanoribbons, the value of K was ~1000 W/mK. The value is less than those reported for SLG [42,43] because the nanoribbons are a few nm wide whereas the SLG was ~20 μm in size. This shows that K is indeed dependent on flake size as was reported earlier [47,48]. Another group reported calculation of K ~2000 W/mK for symmetric GNRs using MD studies [62]. Nanoribbons with zigzag edges had larger K than those with armchair edges, and thermal rectification was observed for asymmetric nanoribbons. A strong dependence of thermal conductivity of GNR on shapes and edges was reported by Guo et al. [63], similar to the size dependence reported by Nika et al. [47,48], though the absolute values in this case were much smaller at ~400–500 W/mK due to the smaller size of the structure. Tensile uniaxial strain was found to distinctly reduce the thermal conductivity of GNRs. These works were followed by a few more on GNR thermal conductance. Equilibrium MD simulations were used to compare thermal conductivity of GNRs with smooth and rough edges [64]. Figure 11.4 shows K data for different types of GNRs.

The smooth edges led to much higher K values than rough edges; also, in case of the latter, thermal conductivity was strongly dependent on the width. While smooth-edged GNRs of ~20 Å width without H-termination showed K ~ 3000 W/mK; the GNRs with rough edges with similar attributes had K ~ 500 W/mK.

11.2.3 Thermoelectric Properties of Graphene

Thermoelectric power (TEP) has always been a very powerful tool for probing transport mechanisms in metals and semiconductors. Often the measurement of conductivity (or resistivity) might not be sufficient to distinguish among various scattering mechanisms, but the TEP can be used to probe transport properties since it also provides corresponding knowledge about resistivity. At low temperature, the thermopower is inversely proportional to the impurity potential and density. The thermopower can provide important information about impurity scattering in graphene [65]. This can be either measured directly or extracted via application of Mott's relation to graphene. The latter can also be used to analyze charge conductivity. There has been other theoretical investigation of the thermoelectric properties

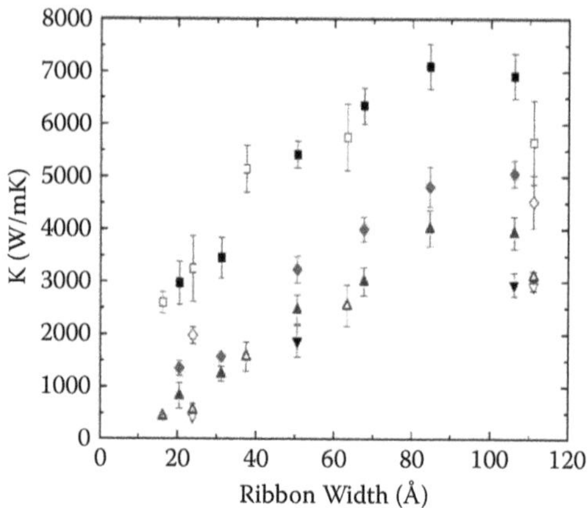

FIGURE 11.4 Comparison of computed thermal conductivity for armchair and zigzag GNRs with smooth, rough, and H-terminated edges as a function of ribbon width. Armchair edge data is displaced by +5Å for clarity. Chart symbols are as follows: ■ κ smooth zigzag, ▲ κ rough zigzag, □ κ smooth armchair, Δ κ rough armchair, ◆ κ smooth+C13 zigzag, ▼ κ rough+C13 zigzag, ◊ κ smooth+C13 armchair, ▿ κ rough+C13 armchair. (From Evans, W. J., L. Hu, and P. Keblinski. 2010. Thermal conductivity of graphene ribbons from equilibrium molecular dynamics: Effect of ribbon width, edge roughness, and hydrogen termination. *Applied Physics Letters*, 96:203103. © 2010 American Institute of Physics. With permission.)

of graphene [66]. The TEP is extremely sensitive to the particle–hole asymmetry of a system and hence it helps in understanding the electronic transport [67]. A recent work by Zuev et al. [68] reports measurement of both the conductance and TEP of graphene using a microfabricated heater with thermometer electrodes. As the majority carrier density switches from electron to hole, the TEP changes from positive to negative across the charge neutrality point (CNP). This is clearly shown in Figure 11.5 where the electrical conductivity and TEP are measured as a function of applied gate voltage over a temperature range of 10–300 K. For the device under consideration, the electrical conductivity reaches a minimum value at the CNP. At ~300 K, the TEP reaches around 80 μV/K.

Graphene's carrier density–dependent thermoelectric coefficient TEP scales linearly with temperature. Under the application of high magnetic field, in the quantum Hall regime, all the TEP-tensor components are quantized in nature, but there were strong deviations from the Mott's relation in the CNP. Similar results were reported by Wei et al. [69]. This involved a study of thermoelectric properties of graphene under zero and applied magnetic fields. The Seebeck coefficient was found to diverge and vary inversely with the square root of carrier density. Experimental observations showed a very large Nernst signal of 50 μV/K at the Dirac point when a magnetic field of 8T was applied. These features can be mostly attributed to the massless particles in graphene. These reports were followed by experimental work of Checkelsky and Ong [70]. The authors aimed to elucidate the anomalous nature of the off-diagonal thermoelectric conductivity peak near the Dirac point. In the thermoelectric response of a quantum Hall regime, the thermopower displays a large

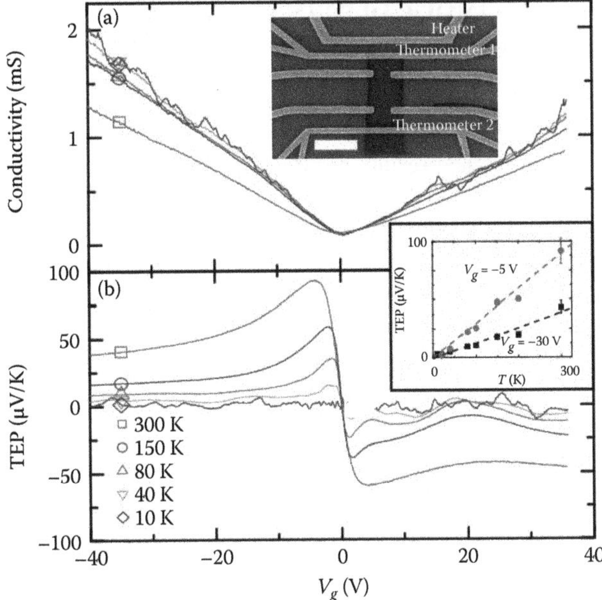

FIGURE 11.5 (a) Conductivity and (b) TEP of a graphene sample as a function of V_g for $T = 10\ K$ to $300\ K$. Upper inset: Scanning electron microscopy image of a typical device with scale bar = 2 μm. Lower inset: TEP values Vs T, with $V_g = -30$V and −5V. (From Zuev, Y. M., W. Chang, and P. Kim. 2009. Thermoelectric and magnetothermoelectric transport measurements of graphene. *Physical Review Letters* 102:096807. © 2009 American Physical Society. With permission.)

peak at each Landau level index $n \neq 0$ and at $n = 0$, and the peak of the thermoelectric conductivity is anomalously narrow by a factor ~4 as compared to its neighbors. The peak magnitude is almost equal in each case, but the area enclosed under the curve is anomalously small. Also, a work by Wang et al. [71] addressed the same issue in the case of bilayer graphene. Electric and thermoelectric transport measurements were performed in a magnetic field up to 15T. Mott's relation fails near the CNP as in single-layer graphene. There was also anomalous behavior of the transverse thermoelectric conductivity and Seebeck coefficient near the CNP and the semiclassical theory cannot explain this phenomenon satisfactorily. This possibly implies the existence of a unique phase of counter-propagating edge channels with opposite spins in the spin-polarized quantum Hall regime. This promises potential application in spin-electronics. Based on the experimental works described previously [68–70], there was an interesting theoretical calculation of graphene thermopower taking into account the energy dependence of various transport scattering times [72]. The calculations were in agreement with the experimental thermopower data and showed that the main scattering mechanism in graphene layers is screened Coulomb scattering by charged impurities. The sign change of the thermopower in the low-density regime, near the Dirac point, was explained using effective-medium theory.

11.3 THERMAL CONDUCTION IN SUSPENDED LAYERS OF GRAPHENE

The first experimental investigations of thermal conduction in graphene were reported by Balandin et al. [43] and Ghosh et al. [44]. In this section we review our experimental and theoretical results pertinent to thermal conduction in suspended graphene layers. The term *suspended* is used to emphasize that the experimental results were achieved for partially suspended graphene flakes not covered by any cap or insulating layer, which might have reduced the heat conduction through graphene. In the theoretical sense, the term suspended indicates that the theory does not include any graphene–substrate interaction, for example, the derivation is performed for graphene in free space. In Sections 11.3.1 and 11.3.2, the experimental setup and the data extraction procedure are explained in detail along with a physical interpretation of the heat conduction in graphene. This is followed by a formal theory in Section 11.3.3, which is an intricate approach for treating the three-phonon Umklapp processes in graphene accurately. A comparatively simple theoretical model is described in Section 11.3.4, which helps in explaining the higher values of thermal conductivity in suspended graphene as compared to basal planes of bulk graphite. This model can also be used for quick estimates of the thermal conductivity in graphene flakes of different sizes. The thermal conduction in few-layer graphene is explained in Section 11.3.5.

11.3.1 EXPERIMENTAL INVESTIGATION OF HEAT CONDUCTION IN GRAPHENE

In order to measure the thermal conductivity of suspended graphene layers, a unique noncontact optical approach was developed based on Raman spectroscopy. It was already known that graphene has distinctive signatures in Raman spectra with a clear *G* peak and *2D* band [73,74]. Moreover, it was also found that the *G* peak of graphene's Raman spectra exhibits strong temperature dependence [45,46]. The latter means that the shift in the position of the *G* peak in response to laser heating can be used for measuring the local temperature rise. The correlation between the temperature rise and the amount of power dissipated in graphene, for the sample with given geometry and proper heat sinks, can provide the value of the thermal conductivity *K*. The schematic for the experiment is shown in Figure 11.6.

Even a small amount of power dissipated in single-layer graphene (SLG) can be sufficient for inducing a measurable shift in the *G* peak position due to the extremely small thickness of the material—one atomic layer. The suspended portion of graphene served several essential functions: (1) accurately determining the amount of power absorbed by graphene through the calibration procedure, (2) forming a two-dimensional in-plane heat front propagating toward the heat sinks, and (3) reducing the thermal coupling to the substrate through the increased micro- and nanoscale corrugations as seen in Figure 11.7.

The thermal coupling between the graphene and Si substrate is small due to very low thermal conductivity of the oxide (~1 W/mK) and large thermal interface resistance of the partially suspended flake. Microscale and nanoscale corrugations of partially suspended graphene reduce the thermal coupling. As a result, the heat wave

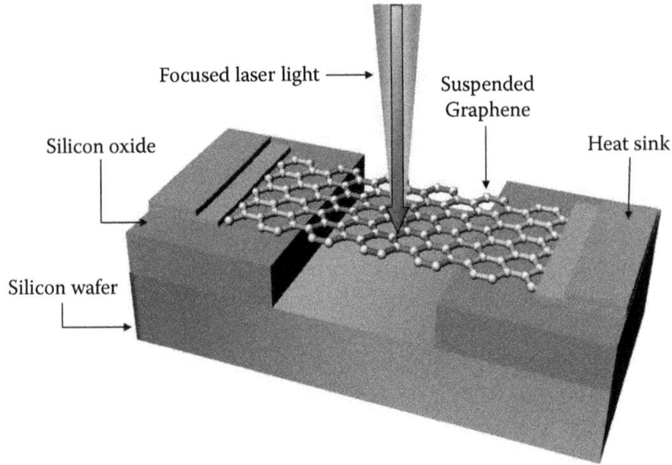

FIGURE 11.6 Schematic of the experimental setup with the excitation laser light focused on graphene suspended across a trench in Si wafer. Laser power absorbed in graphene induces a local hot spot and generates heat wave propagating toward the heat sinks.

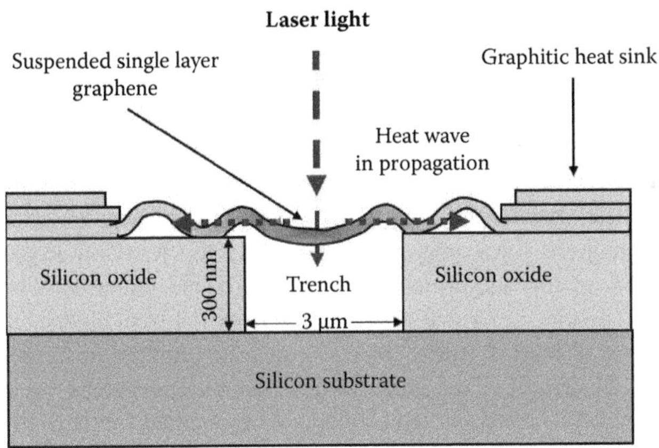

FIGURE 11.7 Illustration of the micro- and nanoscale corrugation formed in the suspended flake, which further reduces the thermal coupling to the substrate. The depicted experimental technique allows one for the steady-state non-contact direct measurement of the thermal conductivity. (Adapted from Ghosh, S., D. L. Nika, E. P. Pokatilov, and A. A. Balandin. 2009. Heat conduction in graphene: Experimental study and theoretical interpretation. *New Journal of Physics*, 11:095012.)

generated over the suspended portion of graphene continues to propagate all the way to the heat sink. From the available excitation wavelengths, 488-nm laser light was chosen for this experiment. The shorter excitation wavelength in the ultraviolet range (e.g., 325 nm) is strongly absorbed and good for heating the sample surface but it does not provide clear Raman signatures for graphene. The longer wavelength (e.g., 633 nm) excites informative scattering spectra from graphene but does not produce local heating as efficiently as 488-nm laser light.

The first step in the measurements was determining the temperature coefficient χ_G for the G peak. To accomplish this task, the laser excitation power was kept at a minimal level and the temperature of the graphene flake was changed externally through the hot–cold cell [45,46]. After the change in the temperature ΔT has been correlated with the change in the $\Delta\omega$ peak position, the micro-Raman spectrometer can be used as a thermometer. During the measurement of the thermal conductivity, the excitation power is intentionally increased to induce local heating. The local temperature rise is determined through the expression $\Delta T = \Delta\omega_G/\chi_G$. It is important to mention here that the measurement technique is steady state. Each data point in the thermal conductivity measurement, that is, recording the G peak position as a function of the excitation power, takes sufficient time (several minutes) for achieving the steady state. The energy deposited by the laser light to the electron gas in graphene is being transferred to phonons very fast. The time constant for the energy transfer from the electrons to acoustic phonons in graphene is on the order of several picoseconds [75–77]. Thus, for the large graphene flakes utilized in our experiments (tens of microns), the changes in the induced hot spot due to the finite thermalization time are small and can be neglected. From the other side, our measurement time was small compared to hours, which are required in order to induce damage or surface modification in graphene by laser light [78].

The long graphene flakes for these measurements were produced using the standard technique of mechanical exfoliation of bulk Kish and highly oriented pyrolytic graphite (HOPG) [18,19]. The trenches were fabricated using reactive ion etching. The width of these trenches ranged from 1 μm to 5 μm with the nominal depth of 300 nm. In the first set of measurements we selected graphene flakes of approximately rectangular shape connected to large graphitic pieces, which acted as heat sinks. The rectangular shape was selected in order to use a simple data extraction procedure based on the one-dimensional heat diffusion equation. These graphitic pieces were at a distance of a few micrometers from the trench edges to ensure that the transport was at least partially diffusive and the phonon mean free path is not limited just by the length of the flake. In the later measurements we utilized well-defined massive metal heat sinks and an elaborate procedure for thermal conductivity extraction based on the numerical solution of the heat diffusion equation. The single-layer graphene flakes were selected using micro Raman spectroscopy by checking the intensity ratio of the G and $2D$ peaks and by $2D$ band deconvolution [73,74]. The combination of these two Raman techniques with atomic force microscopy (AFM) and scanning electron microscopy (SEM) allowed us to verify the number of atomic planes and flake uniformity with a high degree of accuracy.

For the plane-wave heat front propagating in two opposite directions from the middle of the SLG, the expression for K can also be written as $K = (L/2a_G W)(\Delta P_G/\Delta T_G)$, where L is the distance from the middle of the suspended SLG to the heat sink, W is the width, and a_G is the thickness of the flake. Here, ΔT_G is the change in the temperature in the suspended portion of graphene flake due to the change in the power ΔP_G dissipated in graphene. Finally, using the temperature coefficient, χ_G the thermal conductivity can be determined as

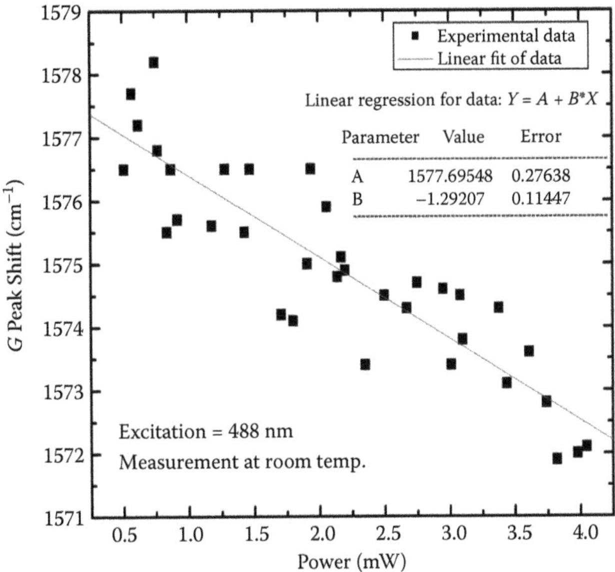

FIGURE 11.8 Shift in G peak spectral position with change in total dissipated power. The spectra are excited at 488 nm and recorded at room temperature in the backscattering configuration.

$$K = (L/2a_G W)\chi_G(\Delta\omega/\Delta P_G)^{-1}. \tag{11.1}$$

The increase in the excitation power led to the increase in the intensity count and red shift of the *G* mode peak. The *G* peak position dependence on the dissipated power for a high-quality suspended SLG is shown in Figure 11.8. The red shift indicates a rise in the local temperature in the middle of the suspended graphene. The extracted slope is $\Delta\omega/\Delta P_G \approx -1.29$ cm^{-1}/mW.

11.3.2 Data Extraction and Power Calibration

The challenge in the measurement of the thermal conductivity with the described optical technique is in accurately determining the power absorbed in the graphene. Only a fraction P_G of the laser light focused on the graphene flake will actually be dissipated in the graphene. Most of the light will be reflected after the light travels through the flake to the trench bottom and is reflected back. The power, which is measured by the detector placed at the position of the flake, is the total power P_D, part of which goes into the graphene flake after two transmissions (incident pass and reflected pass) and the rest is lost in the silicon wafer P_{Si}. It is now known that the fraction of the power absorbed by graphene is 2.3% per layer for light wavelength $\lambda >$ 500 nm. Our measurements were performed at a smaller wavelength (λ = 488 nm) where the absorption is enhanced [79,80,81]. Thus, it was important to determine the absorbed power in the specific conditions of our experiment. The power P_G has

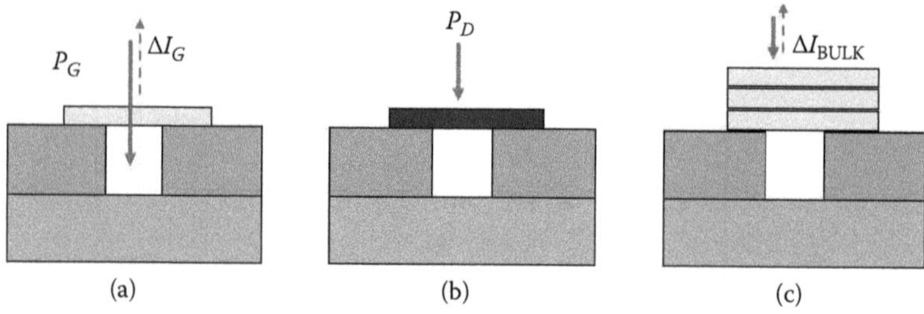

FIGURE 11.9 Illustration of the measurement and calibration procedure for power dissipation in SLG. (a) Integrated Raman intensity is related to the power absorbed in graphene through the scattering cross section and absorption coefficient. (b) Detector is placed at the sample location to measure the power at the surface. (c) Bulk graphite is used as reference for the calibration procedure. The power absorbed in graphene is determined through the ratio of the integrated Raman intensities of graphene and bulk graphite. (From Ghosh, S., D. L. Nika, E. P. Pokatilov, and A. A. Balandin. 2009. Heat conduction in graphene: Experimental study and theoretical interpretation. *New Journal of Physics*, 11:095012. © 2009 IOP Publishing Limited. With permission.)

been measured through the calibration procedure with the bulk graphite serving as a reference as in Figure 11.9. It is based on a comparison of the experimentally determined integrated Raman intensity for the *G* peak from the single-layer graphene and bulk graphite [44]. Following is a derivation of the calibration formula for single-layer graphene.

In order to determine the fraction of light power absorbed in graphene, we closely follow the work reported by Ghosh et al. [44,82]. The intensity of Raman scattering in graphene is given by

$$\Delta I_G = N\sigma_G I_0, \tag{11.2}$$

where N is the number of scattering atoms on the illuminated surface of the cross section A, I_0 is the laser intensity, and σ_G is the Raman scattering cross section. The total power absorbed in graphene can be written as

$$P_G = \alpha_G a_G (1 + R_{Si}) I_o A. \tag{11.3}$$

Here, R_{Si} is the reflection coefficient of silicon, a_G is the thickness of the SLG, and α_G is the absorption coefficient of graphene. We take into account the power that is reflected from the Si trench and absorbed by the suspended portion of graphene. The reflection from the SLG is assumed to be negligible. Expressing I_0 through P_G we have

$$I_0 = \frac{P_G}{\alpha_G a_G (1 + R_{Si}) A}. \tag{11.4}$$

Plugging Equation (11.4) into Equation (11.2) we obtain the following expression for the integrated intensity of Raman light reflected from graphene:

$$\Delta I_G = \frac{N}{A}\sigma_G \frac{P_G}{\alpha_G a_G(1+R_{Si})}. \tag{11.5}$$

Let us now consider the light absorption and Raman scattering in bulk graphite. Highly oriented crystalline graphite consists of atomic planes of graphene, which are bound by the weak Van der Waals interaction. For light with energy above 0.5 eV, graphite behaves essentially as a collection of independent graphene layers [81]. It is also confirmed by the results of measurements of the light absorption in graphene multilayers, which proved that for a wide range of wavelengths, the absorption is constant per atomic layer [79,80]. In the analysis we take into account that the absorption coefficient and Raman scattering cross section defined per layer are the same for graphene and for the single atomic plane in bulk graphite. The integrated Raman scattered intensity from bulk graphite (or HOPG) is obtained by summation over all the n atomic planes stacked together, that is,

$$\Delta I_H = N\sigma_H I_0 \sum_{n=1}^{\infty} e^{-2\alpha_H a_H n} \tag{11.6}$$

where α_H is the absorption coefficient and a_H is the thickness of each layer. The details of this derivation can be found in Ghosh et al. [82]. This leads to

$$\Delta I_H = N\frac{\sigma_H}{2a_H\alpha_H} I_0 .$$

Taking into account that a part of the incident power is reflected from graphite and does not contribute to the Raman scattered intensity, one obtains

$$\Delta I_H = (1/2)(N/A)(\sigma_H/\alpha_H a_H)P_D(1-R_H). \tag{11.7}$$

Combining Equations (11.5) and (11.7) and introducing the ratio $\varsigma = \overline{\Delta I_G}/\overline{\Delta I_H}$, which has to be determined experimentally, we obtain the final expression for the power absorbed in graphene:

$$P_G = (\varsigma/2)[\sigma_H\alpha_G a_G/\sigma_G\alpha_H a_H](1+R_{Si})(1-R_H)P_D. \tag{11.8}$$

Equation (11.8) allowed measurement of the heating power dissipated in the suspended portion of graphene under the specific conditions of the experiment. The ratio $\varsigma = \overline{\Delta I_G}/\overline{\Delta I_H}$, measured for graphene and reference bulk graphite for the G peak and the same frequency interval, stays nearly constant over the excitation power range used in the experiments. The reflection coefficients for silicon and bulk graphite are tabulated quantities but can also be measured directly, while P_D is the actual reading of the power detector taken at the position of the sample. Since the microscopic in-plane Raman cross sections and absorption coefficients are the same for graphene and bulk graphite, the square bracket term in Equation (11.8) is close to one.

With the calibration procedure in place for converting P_D to P_G, the measurement of the thermal conductivity of suspended graphene reduces to measuring the Raman shift $\Delta\omega_G$ as the function of the heating power P_D determined by the detector. The measured slope $\Delta\omega/\Delta P_D$, ratio of the integrated intensities ζ, and the temperature coefficient χ_G give the value of the thermal conductivity of graphene. Initially, the thermal data extraction was accomplished with the simple one-dimensional model, and later improved by utilizing the numeric solution of the heat diffusion equation for a given flake shape. It has been found that the near–RT thermal conductivity of the single-layer suspended graphene is in the range from ~3000 W/mK to 5300 W/mK depending on the lateral size or width of the graphene flakes.

11.3.3 Theory of Heat Conduction in Graphene

In this section we outline the theory of heat conduction in graphene. The description is based on derivations of Nika et al. [47]. The heat flux along a graphene atomic plane can be calculated according to the expression [5,83]

$$\vec{W} = \sum_{s,\vec{q}} \vec{v}(s,\vec{q})\hbar\omega_s(\vec{q})N(\vec{q},\omega_s(\vec{q})) = \sum_{s,\vec{q}} \vec{v}(s,\vec{q})\hbar\omega_s(\vec{q})n(\vec{q},\omega_s), \tag{11.9}$$

where $\vec{v}(s,\vec{q})\hbar\omega_s(\vec{q})$ is the energy carried by one phonon, $\vec{v}(s,\vec{q}) = d\omega_s/dq$ is the phonon group velocity, and $N(\omega,\vec{q}) = N_0(\omega,\vec{q}) + n(\omega,\vec{q})$ is the number of phonons in the flux. Here, N_0 is the Bose-Einstein distribution function and $n = -\tau_{tot}(\vec{v}\nabla T)\,\partial N_0/\partial T$ is the nonequilibrium part of the phonon distribution function N, where τ_{tot} the total phonon relaxation time and T is the absolute temperature.

Comparing the microscopic expression

$$\vec{W} = -\sum_{\beta}(\nabla T)_\beta \sum_{s,\vec{q}} \tau_{tot}(s,\vec{q})v_\beta(s,\vec{q})\frac{\partial N_0(\omega_s)}{\partial T}\vec{v}(s,\vec{q})\hbar\omega_s(\vec{q}) \tag{11.10}$$

with the macroscopic definition of the thermal conductivity

$$W_\alpha = -\kappa_{\alpha\beta}(\nabla T)_\beta hL_xL_y, \tag{11.11}$$

we obtain the following expression for the thermal conductivity tensor

$$\kappa_{\alpha\beta} = \frac{1}{hL_xL_y}\sum_{s,\vec{q}} \tau_{tot}(s,\vec{q})v_\alpha(s,\vec{q})v_\beta(s,\vec{q})\frac{\partial N_0(\omega_s)}{\partial T}\hbar\omega_s(\vec{q}). \tag{11.12}$$

Here, $L_x = d$ is the sample width (graphene flake width), L_y is the sample length, and $h = 0.35$ nm is the thickness of graphene. The diagonal element of the thermal conductivity tensor, which corresponds to the phonon flux along the temperature gradient, is given by

$$\kappa_{\alpha\alpha} = \frac{1}{hL_xL_y}\sum_{s,\vec{q}}\tau_{tot}(s,\vec{q})v^2(s,\vec{q})\cos^2\varphi\frac{\partial N_0(\omega_s)}{\partial T}\hbar\omega_s(\vec{q}). \tag{11.13}$$

Finally, making a transition from the summation to integration and taking into account the two-dimensional density of phonon states, we obtain the expression for the scalar thermal conductivity

$$K = \frac{1}{4\pi k_B T^2 h}\sum_s\int_0^{q_{\max}}\left\{\left[\hbar\omega_s(q)\frac{d\omega_s(q)}{dq}\right]^2\tau_{tot}(s,q)\frac{\exp[\hbar\omega_s(q)/kT]}{\left[\exp[\hbar\omega_s(q)/kT]-1\right]^2}q\right\}dq. \tag{11.14}$$

The detailed description of the theoretical formalism for the phonon heat conduction in graphene was recently reported by some of us elsewhere [47]. Our theoretical approach utilized an original *phase-diagram technique* to account for all three-phonon Umklapp scattering channels allowed by the energy and momentum conservation. We consider two types of three-phonon Umklapp scattering processes [83]. The first type is the scattering when a phonon with the wave vector $\vec{q}(\omega)$ absorbs another phonon from the heat flux with the wave vector $\vec{q}'(\omega')$, that is, the phonon leaves the state $\vec{q}$. For this type of scattering process, the momentum and energy conservation laws are written as

$$\begin{gathered}\vec{q}+\vec{q}' = \vec{b}_i+\vec{q}''\\ \omega+\omega' = \omega''\end{gathered} \tag{11.15}$$

where $\vec{b}_i$, i = 1, 2, 3 is one of the vectors of the reciprocal lattice. The processes of the second type are those when the phonons $\vec{q}$ of the heat flux decay into two phonons with the wave vectors $\vec{q}'$ and $\vec{q}''$ leaving the state $\vec{q}$, or, alternatively, two phonons $\vec{q}'(\omega')$ and $\vec{q}''(\omega'')$ merge together forming a phonon with the wave vector $\vec{q}(\omega)$, which corresponds to the phonon coming to the state $\vec{q}(\omega)$. The conservation laws for this type are given by

$$\begin{gathered}\vec{q}+\vec{b}_i = \vec{q}'+\vec{q}'',\quad i=4,5,6\\ \omega = \omega'+\omega''.\end{gathered} \tag{11.16}$$

To find all possible three-phonon processes, we used a fine mesh $q_j = (j-1)\Delta q$ (j = 1, …, 1001) with the step $\Delta q = q_{\max}/1000 \approx 0.015$ nm^{-1}. For each phonon mode (q_i, s), we found all pairs of the phonon modes ($\vec{q}'$, s') and ($\vec{q}''$, s'') such that the conditions of Equations (11.15) and (11.16) are met. The latter can be done with the help of the ($\vec{q}'$)-space *phase diagrams* technique, which we introduced in reference [47].

Using the general expression for a matrix element of the three-phonon interaction [83] and taking into account all relevant phonon branches and their dispersion as well as all unit vectors of the reciprocal lattice $\vec{b}_1......\vec{b}_6$, directed from the Γ point to the centers of the neighboring unit cells, we obtain for the Umklapp scattering rates

$$\frac{1}{\tau_U^{(I),(II)}(s,\vec{q})} = \frac{\hbar\gamma_s^2(\vec{q})}{3\pi\rho v_s^2(\vec{q})} \sum_{s's'';\vec{b}_i} \iint \omega_s(\vec{q})\omega'_{s'}(\vec{q}')\omega''_{s''}(\vec{q}'') \times$$
$$\left\{ N_0[\omega'_{s'}(\vec{q}')] \mp N_0[\omega''_{s''}(\vec{q}'')] + \frac{1}{2} \mp \frac{1}{2} \right\} \times \delta[\omega_s(\vec{q}) \pm \omega'_{s'}(\vec{q}') - \omega''_{s''}(\vec{q}'')] dq'_l dq'_\perp. \tag{11.17}$$

Here, q'_l and $q'_\perp$ are the components of the vector $\vec{q}'$ parallel or perpendicular to the lines defined by Equations (11.15) and (11.16), correspondingly, $\gamma_s(\vec{q})$ is the mode-dependent Gruneisen parameter, which is determined for each phonon wave vector and polarization branch and ρ is the surface mass density. In Equation (11.17), the upper signs correspond to the processes of the first type while the lower signs correspond to those of the second type. Integrating along $q_\perp$, one obtains the line integral

$$\frac{1}{\tau_U^{(I),(II)}(s,\vec{q})} = \frac{\hbar\gamma_s^2(\vec{q})\omega_s(\vec{q})}{3\pi\rho v_s^2(\vec{q})} \sum_{s's'';\vec{b}} \int_l \frac{\pm(\omega''_{s''} - \omega_s)\omega''_{s''}}{v_\perp(\omega'_{s'})} \left(N'_0 \mp N''_0 + \frac{1}{2} \mp \frac{1}{2} \right) dq'_l. \tag{11.18}$$

The combined scattering rate in both types of the three-phonon Umklapp processes for a phonon in the state $(s,\vec{q})$ is a sum of the first and second types of Umklapp processes. One should note here that for the small phonon wave vectors (long wavelength), $q \to 0$, the Umklapp limited phonon lifetime $\tau_U \to \infty$. For this reason, the calculation of the intrinsic thermal conductivity with only Umklapp scattering is not possible without an arbitrary truncation procedure.

To avoid the unphysical assumptions about the limits of the integration in the thermal conductivity integral, one can include the phonon scattering on boundaries. In the case of graphene, the boundary scattering term corresponds to scattering from the rough edges of graphene flakes. No scattering happens from the top and bottom sides of graphene flake since it is only one atomic layer thick and the phonon flux is parallel to the graphene plane. One can evaluate the rough-edge scattering using a well-known equation [4]

$$\frac{1}{\tau_B(s,q)} = \frac{v_s(\omega_s)}{L}\frac{1-p}{1+p}. \tag{11.19}$$

Here p is the specularity parameter, which depends on the roughness at the graphene edges and L is the width of the graphene flake. The total phonon relaxation rate is found as a sum of the Umklapp and edge scattering. It is important to understand that the thermal conductivity of a two-dimensional system such as graphene cannot be determined without the restriction on the phonon MFP in the long wavelength limit. The phonon scattering on the edges restricts the MFP in the formal theory of thermal conductivity. In this sense, the thermal conductivity limited by the Umklapp and boundary scattering can be considered as an *intrinsic* property of a graphene flake of a particular size. The extrinsic effects, which reduce thermal conductivity, such as phonon scattering on defects, impurities, and grain boundaries are not included in the consideration.

11.3.4 Simple Theory of Thermal Conductivity of Graphene

The calculation of the thermal conductivity within the formal theory outlined in the previous section is a rather complicated procedure. In this section, we describe a simple model of the thermal conductivity of graphene. The model is based on derivation reported by Nika et al. [48]. This model uses a rather general expression for thermal conductivity with the two Gruneisen parameters γ_s obtained independently for each of the heat-conducting phonon polarization branches s, and separate phonon velocities and cutoff frequencies for each phonon branch. The model includes the specifics of phonon dispersion in graphene. The effective parameters γ_s are computed by averaging the phonon mode-dependent $\gamma_s(\vec{q})$ for all relevant phonons (here, $\vec{q}$ is the phonon wave vector). The phonon branches, which carry heat, are longitudinal acoustic (LA) and transverse acoustic (TA). The out-of-plane transverse acoustic phonons (ZA) do not make contributions to heat conduction due to their low group velocity and high $\gamma_s(\vec{q})$.

There is a clear difference in the heat transport in the basal planes of bulk graphite and in single-layer graphene [84,85]. In the former, the heat transport is approximately two-dimensional only until some low-bound cutoff frequency ω_C. Below ω_C, there appears strong coupling with the cross-plane phonon modes and heat starts to propagate in all directions, which reduces the contributions of these low-energy modes to heat transport along the basal planes to negligible values. In bulk graphite there is a physically reasonable reference point for the onset of cross-plane coupling, which is the ZO′ phonon branch near ~4 THz observed in the spectrum of bulk graphite. The presence of the ZO′ branch and corresponding ω_C allows one to avoid the logarithmic divergence in the Umklapp-limited thermal conductivity integral and calculate it without considering other scattering mechanisms. The physics of heat conduction is principally different in graphene where the phonon transport is purely two-dimensional all the way to zero phonon frequency $\omega(q = 0) = 0$. There is no onset of cross-plane heat transport at the long-wavelength limit in the system, which consists of only one atomic plane. There is no ZO′ branch in the phonon dispersion of graphene. Thus, the cutoff frequency for Umklapp processes cannot be introduced by analogy with bulk graphite.

Using an expression for the three-phonon Umklapp scattering from references [84,85] but introducing separate lifetimes for LA and TA phonons, one can write

$$\tau_{U,s}^{K} = \frac{1}{\gamma_s^2}\frac{M\upsilon_s^{\;2}}{k_B T}\frac{\omega_{s,\max}}{\omega^2}, \tag{11.20}$$

where $s = TA, LA$, υ_s is the average phonon velocity for a given branch, T is the absolute temperature, k_B is the Boltzmann constant, $\omega_{s,\max}$ is the maximum cutoff frequency for a given branch, and M is the mass of an atom. To determine γ_s, we averaged $\gamma_s(q)$ obtained from the accurate phonon dispersion calculated using the VFF method [47] and ab initio theory [86]. Substituting $\tau_{tot} = \tau_{U,s}^K$ in Equation (11.14) in the previous section, one can obtain the following formula for intrinsic thermal conductivity in graphene

$$K_U = \frac{1}{4\pi k_B T^2 h} \sum_{s=TA,LA} \int_{q_{min}}^{q_{max}} \left\{ \left[\hbar\omega_s(q) \frac{d\omega_s(q)}{dq} \right]^2 \tau_{U,s}^K(q) \frac{exp[\hbar\omega_s(q)/kT]}{[exp[\hbar\omega_s(q)/kT]-1]^2} q \right\} dq. \tag{11.21}$$

Equation (11.21) can be used to calculate the thermal conductivity with the actual dependence of the phonon frequency $\omega_s(q)$ and the phonon velocity $d\omega_s(q)/dq$ on the phonon wave number. To simplify the model further, one can use the linear dispersion $\omega_s(q) = \upsilon_s q$ and rewrite it as

$$K_U = \frac{\hbar^2}{4\pi k_B T^2 h} \sum_{s=TA,LA} \int_{\omega_{min}}^{\omega_{max}} \left\{ \omega^3 \tau_{U,s}^K(\omega) \frac{exp[\hbar\omega/kT]}{[exp[\hbar\omega/kT]-1]^2} \right\} d\omega. \tag{11.22}$$

Substituting Equation (11.20) to Equation (11.21) and performing integration, one obtains

$$K_U = \frac{M}{4\pi T h} \sum_{s=TA,LA} \frac{\omega_{s,\max}\upsilon_s^2}{\gamma_s^2} F(\omega_{s,\min}, \omega_{s,\max}), \tag{11.23}$$

where

$$F(\omega_{s,\min}, \omega_{s,\max}) = \int_{\hbar\omega_{s,\min}/k_BT}^{\hbar\omega_{s,\max}/k_BT} \xi \frac{exp(\xi)}{[exp(\xi)-1]^2} d\xi = \left[ln\{exp(\xi)-1\} + \frac{\xi}{1-exp(\xi)} - \xi \right] \Big|_{\hbar\omega_{s,\min}/k_BT}^{\hbar\omega_{s,\max}/k_BT} \tag{11.24}$$

In Equation (11.24), $\xi = \hbar\omega/k_BT$, and the upper cutoff frequencies $\omega_{s,\max}$ are defined from the actual phonon dispersion in graphene, calculated using the VFF model [47]: $\omega_{LA,\max}$ = 104.4 THz, $\omega_{TA,\max}$ = 78 THz. The low-bound cutoff frequencies $\omega_{s,\min}$ for each s are determined from the condition that the phonon MFP cannot exceed the physical size L of the flake, that is,

$$\omega_{s,\min} = \frac{\upsilon_s}{\gamma_s} \sqrt{\frac{M\upsilon_s}{k_BT} \frac{\omega_{s,\max}}{L}}. \tag{11.25}$$

The integrand in Equation (11.24) can be further simplified near RT when $\hbar\omega_{s,\max} > k_BT$, and it can be expressed as

$$F(\omega_{s,\min}) \approx -ln\left\{\left|exp(\hbar\omega_{s,\min}/k_BT)-1\right|\right\} + \frac{\hbar\omega_{s,\min}}{k_BT}\frac{exp(\hbar\omega_{s,\min}/k_BT)}{exp(\hbar\omega_{s,\min}/k_BT)-1}. \quad (11.26)$$

The obtained Equations (11.25) and (11.26) constitute a simple analytical model for calculation of the thermal conductivity of graphene layer, which retains such important features of graphene phonon spectra as different υ_s and γ_s for *LA* and *TA* branches. The model also reflects the two-dimensional nature of heat transport in graphene all the way down to zero phonon frequency. Equation (11.26) reduces to the Klemens formula for graphene [85] in the limit $\xi \rightarrow 0$ ($\hbar\omega << k_BT$) and additional assumption of the same γ_s and υ_s for *LA* and *TA* phonons.

Using the equations summarized in Section 11.3.4, we calculated the Umklapp-limited *intrinsic* thermal conductivity of graphene as a function of temperature. The results are shown in Figure 11.10 for several lateral sizes of graphene flakes. The Gruneisen parameters used in this calculation, γ_{LA} = 1.8 and γ_{TA} = 0.75, were obtained by averaging of $\gamma_s(q)$ [48]. An experimental data point after Balandin et al. [43,44] is also shown for comparison. There is good agreement between our model predictions and the available experimental data. One should note that a very different temperature dependence was obtained within this theoretical model for relatively small (or narrow) graphene flakes (5 μm) as compared to large flakes (100 μm). In very narrow graphene flakes and nanoribbons, thermal conductivity increases with temperature, which is related to the size (edge) effect on the phonon MFP.

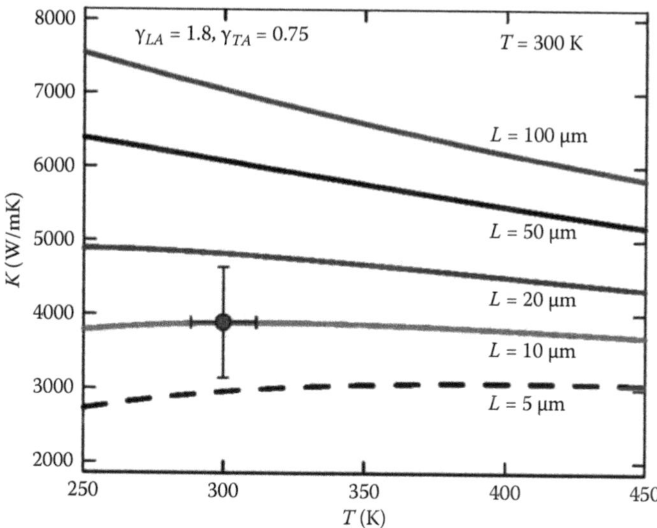

FIGURE 11.10 Thermal conductivity of graphene flake as a function of temperature for several linear dimensions *L* of the flake. (From Nika, D. L., S. Ghosh, E. P. Pokatilov, and A. A. Balandin. 2009. Lattice thermal conductivity of graphene flakes: Comparison with bulk graphite. *Applied Physics Letters*, 94:203103. © 2009 American Institute of Physics. With permission.)

Our model predictions are also in agreement with the tight-binding and nonequilibrium Green's function calculation of thermal transport in graphene nanoribbons reported by Lan et al. [60]. Lan et al. [60] obtained a very strong width dependence of the thermal conductance for narrow graphene ribbons with the width of about 2–20 carbon atoms. In their calculations, the thermal conductance increases with increasing width and with temperature. This is similar to our results for small (narrow) graphene flakes (see the curve for $L = 5$ μm in Figure 11.10). The RT value of the thermal conductivity calculated by Lan et al. [60], K = 3410 W/mK, is clearly above the bulk graphite limit of 2000 W/mK and in agreement with the experiments of Balandin et al. [43,44].

In Figure 11.11 we present the calculated RT thermal conductivity as a function of the flake lateral size, L. The data is presented for the averaged values $\gamma_{LA} = 1.8$ and $\gamma_{TA} = 0.75$ obtained from ab initio calculations, and for several other close sets of $\gamma_{LA,TA}$ to illustrate the sensitivity of the result to Gruneisen parameters. For small graphene flakes, K dependence on L is rather strong. It weakens for flakes with $L \geq 10$ μm. The calculated values are in agreement with the experiment [43,44]. The horizontal dotted line is the experimental thermal conductivity for bulk graphite, which is lower than the theoretical intrinsic limit and is exceeded by graphene's thermal conductivity at smaller L. The Klemens formula [84,85] gives similar dependence but with different absolute values of K due to overestimated γ in his calculations and a few other simplifying assumptions. It should be noted that the calculated thermal conductivity is an *intrinsic* quantity limited by the three-phonon Umklapp scattering only. But it is determined for a specific graphene flake size since L defines the low-bound (long-wavelength) cutoff frequency in Umklapp scattering through Equation (11.25).

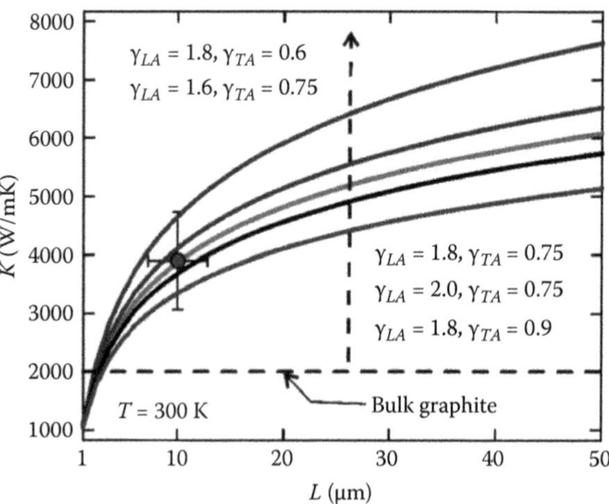

FIGURE 11.11 Thermal conductivity of graphene as a function of the graphene flake size L. It is to be noted that the thermal conductivity of graphene exceeds that of basal planes of graphite when the flake size is larger than a few micrometers. (From Nika, D. L., S. Ghosh, E. P. Pokatilov, and A. A. Balandin. 2009. Lattice thermal conductivity of graphene flakes: Comparison with bulk graphite. *Applied Physics Letters*, 94:203103. © 2009 American Institute of Physics. With permission.)

In experimental conductions, the thermal conductivity will also be limited by extrinsic factors (defects, impurities, grain size), which prevent the growth of the thermal conductivity for very large flakes. With the decreasing specularity parameter (more diffuse scattering), the thermal conductivity calculated with the boundary scattering term approaches the result obtained with the simple model. This is because in the simple model we neglect phonons with frequency $\omega < \omega_{min,s}$ by completely restricting the phonon MFP to the lateral sizes of the flake. The latter corresponds to the perfectly diffusive scattering case. The described models are in good agreement with the measurements and shed light on the heat conduction properties of graphene. Similar dependences were obtained from both approaches: the *intrinsic* thermal conductivity of graphene grows with the increasing linear size of the graphene flake. This is a manifestation of the 2D-nature of the phonon transport in graphene.

11.3.5 Study of Thermal Conduction in Few-Layer Graphene

The evolution of heat conduction as one goes from 2-D graphene, to few-layer graphene, and then to 3-D bulk, is of great interest for both fundamental science and practical applications. This question was addressed by experimentally measuring thermal conductivity of few-layer graphene (FLG) as the number of atomic planes changes from $n = 2$ to $n \approx 10$. The exact mechanisms behind a material's intrinsic ability to conduct heat as its dimensionality changes from two to three dimensions have remained elusive. A large number of FLG samples were prepared by the standard mechanical exfoliation of bulk graphite [18] and suspended across trenches in Si/SiO_2 wafers. The depth of the trenches was ~300 nm while the trench width varied in the range 1–5 μm. The trenches in Si/SiO_2 wafers were made with reactive ion etching (STS). Metal heat sinks were fabricated at distances 1–5 μm from trench edges by electron-beam lithography (SUPRA, Leo) followed by metal deposition. The width of the flakes was from $W \approx 5$ to 16 μm. The metal pads ensured proper thermal contact with the flakes and constant temperature during the measurements. Also, better flow of heat was expected in this setup. The number of atomic layers in graphene flakes was determined with micro-Raman spectroscopy (InVia, Renishaw) through well-established decomposition of the 2-D band in graphene's spectra, as has been mentioned in references [43,45,46,73].

The measurements of thermal conductivity were performed using a direct steady-state noncontact optical technique that was developed on the basis of micro-Raman spectroscopy and has been used previously for suspended SLG flake [43,44]. The size of the laser spot was determined to be 0.5–1 μm. The diameter of the strongly heated region on the flake was somewhat larger due to the indirect nature of energy transfer from light to phonons. FLG samples were heated with a 488-nm laser (Argon Ion) in the middle of the suspended part. The shift of the temperature-sensitive Raman *G* peak in the graphene spectrum [46] defined the local temperature rise in the suspended portion of FLG in response to heating by an excitation laser in the middle of the suspended portion of the flakes. Thermal conductivity was extracted from the power dissipated in the FLG, resulting temperature rise, and flake geometry through a finite-element method solution of the heat diffusion equation [49,87]. It was not possible to mechanically exfoliate

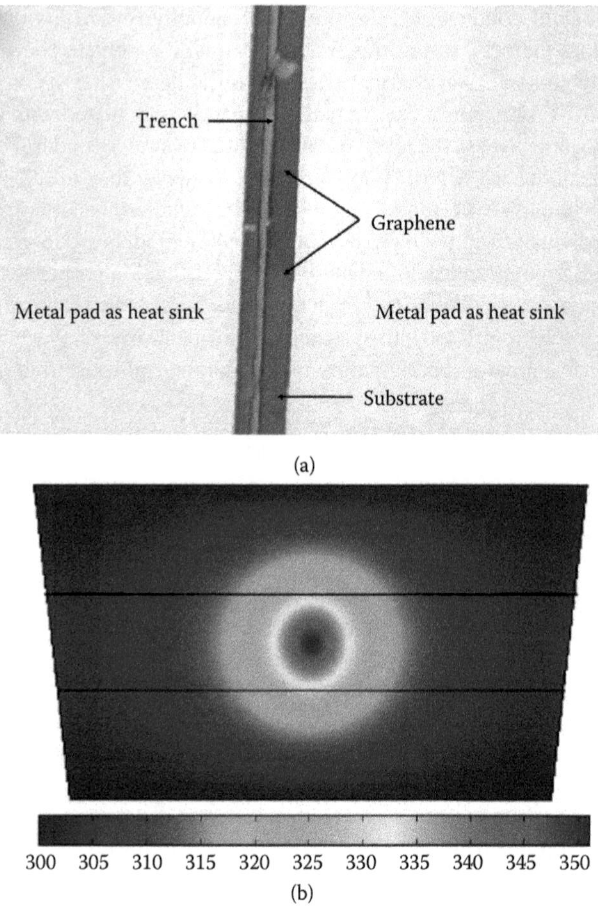

FIGURE 11.12 (a) Optical microscopy images of FLG attached to metal heat sinks. (b) Heat distribution in a nonrectangular-shaped sample using finite element modeling (FEM). The heat source is modeled using a Gaussian distribution.

FLG flakes with a different number of atomic planes n and the same geometry. To avoid damage to graphene the obtained flakes were not cut to the same shape. Instead, the heat diffusion equation was solved numerically for each sample shape to extract its thermal conductivity. A typical FLG flake and its heat distribution are shown in Figures 11.12a and 11.12b, respectively. The heat source is modeled by the Gaussian distribution.

The errors associated with the laser spot size and intensity variation were ~8%, that is, smaller than the error associated with the local temperature measurement by Raman spectrometers (~10–13%). The results were cross-checked with the assumption of a disk-shaped source to take into account local hot spots that might be formed. The calibration procedure for power absorption in graphene is based on a comparison

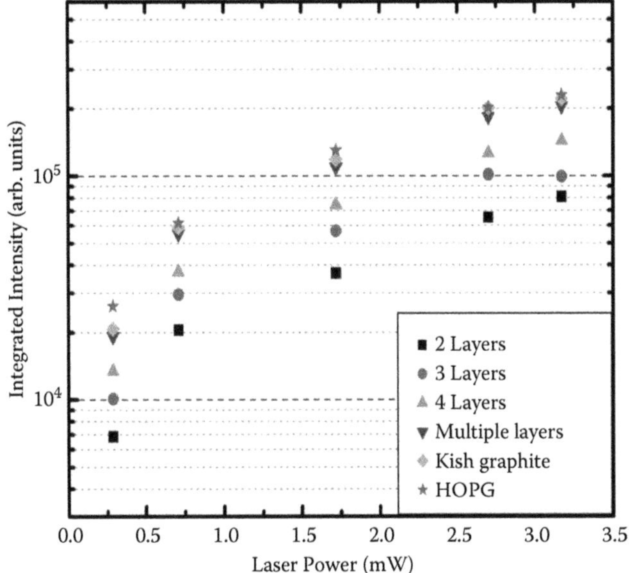

FIGURE 11.13 Integrated Raman intensity of the *G* peak as a function of the laser power at the sample surface for FLG and reference bulk graphite (Kish and HOPG). The data was used to determine the fraction of power absorbed by the flakes. (Adapted from Ghosh, S., W. Bao, D. L. Nika, S. Subrina, E. P. Pokatilov, C. N. Lau, and A. A. Balandin. 2010. Dimensional crossover of thermal transport in few-layer graphene. *Nature Materials* 9:555–558.)

of the integrated Raman intensity of FLG's *G* peak $\overline{I_{FLG}^{G}}$ and that of reference bulk graphite $\overline{I_{BULK}^{G}}$. Figure 11.13 shows measured data for FLG with $n = 2, 3, 4, \sim 8$ and reference graphite. An addition of each atomic plane leads to $\overline{I_{FLG}^{G}}$ increase and convergence with the graphite, while the ratio $\varsigma = \overline{I_{FLG}^{G}}/\overline{I_{BULK}^{G}}$ stays approximately independent of excitation power, indicating proper calibration. This ratio was used to estimate the amount of power absorbed in FLG and was subsequently used in the thermal conductivity equation.

It was found that the near-RT thermal conductivity changes from $K \sim 2800$ W/mK to 1300 W/mK as the number of atomic plains n increases from $n = 2$ to $n = 4$ [49]. Since thermal conductivity of graphene depends on the width of the flakes [47,48] the data for FLG is normalized to the width $W = 5$ μm to allow for direct comparison. At fixed W, the changes in the K value with n are mostly due to modification of the three-phonon Umklapp scattering. The thermal transport in this experiment is in a diffusive regime since L is larger than the phonon MFP in graphene, which was measured to be around ~800 nm near RT [44]. Thus, a gradual change in K was noticed for different numbers of layers from $n = 2$ to $n = 4$ and a heat conduction crossover was actually observed from 2-D graphene to 3-D graphite. This is shown in Figure 11.14. The fact that addition of a number of layers quenches thermal conductivity in graphene is in agreement with the theoretical work of Berber et al. [42]. The bulk value of K along the basal planes is recovered at $n \approx 8$. The ambiguity of

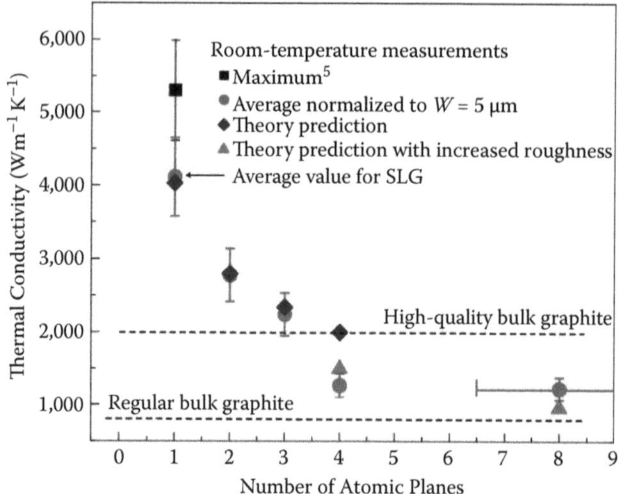

FIGURE 11.14 Thermal conductivity as a function of the number of atomic planes in FLG. Dashed straight lines indicate the range of bulk graphite thermal conductivities. The blue diamonds were obtained from the first-principles theory of thermal conduction in FLG based on the actual phonon dispersion and accounting for all allowed three-phonon Umklapp scattering channels. The green triangles are Callaway-Klemens model calculations, which include extrinsic effects characteristic of thicker films. (From Ghosh, S., W. Bao, D. L. Nika, S. Subrina, E. P. Pokatilov, C. N. Lau, and A. A. Balandin. 2010. Dimensional crossover of thermal transport in few-layer graphene. *Nature Materials* 9:555–558. © 2010 Nature Publishing Group. With permission.)

the data point for $n \approx 8$ is explained by the fact that n in FLG can be determined accurately from Raman spectrum only if $n \leq 6$ [45,46].

This evolution of thermal transport from 2-D graphene to bulk can be explained by the cross-plane coupling of the low-energy phonons and the change in phonon Umklapp scattering. In a bilayer graphene (BLG), though the number of available phonon branches doubles, and the number of conduction channels increases, K decreases because the phase space available for Umklapp scattering increases even more than in SLG. In a single layer of graphene, phonon Umklapp scattering is quenched and the heat transport is limited by the edge or in-plane boundary scattering. This agrees with Klemens's approximation, which explains the higher thermal conductivity in graphene as compared to graphite [85]. The lower K in $n = 4$ can be explained by stronger extrinsic effects, which result from nonuniform thickness of samples.

This theoretical data point at which the crossover of thermal conduction takes place may be related to the AB Bernal stacking and the graphite unit cell. The position of the crossover is also controlled by the width and quality of the FLG flake and the strength of coupling to the substrate. This study helps in understanding the fundamental properties of 2-D systems. The obtained results are important for the proposed graphene and FLG applications in lateral heat spreaders for nanoelectronics.

11.4 COMPARISON OF THERMAL CONDUCTION IN CARBON ALLOTROPES

Carbon materials occupy a unique place in terms of their ability to conduct heat. Different allotropes of carbon span a huge range of thermal conductivity from the lowest value ~ 0.1 W/mK reported for some forms of diamond-like carbons [88] to almost ~3000–5300 W/mK reported for graphene [43,44]. The values for graphene depend on the width (lateral size) of the flake. Essentially, carbon materials can serve both as thermal insulators (diamond-like carbon), as well as superconductors of heat, such as graphene. Microcrystalline diamond (MCD), nanocrystalline diamond (NCD), ultra-nanocrystalline diamond (UNCD), tetrahedral amorphous carbon (ta-C), and carbon-based films occupy all ranges between these two extremes [88–91]. Diamond, which was known to have the highest thermal conductivity among bulk materials, has K values ranging from 800–2000 W/mK [92,93]. In Table 11.2,

TABLE 11.2
Thermal Conductivity of Carbon Materials

Sample	K (W/mK)	Method	Comments	Reference
Graphene	~3080–5300	Optical	Single layer, mechanical exfoliation	Balandin et al., Ghosh et al. [43, 44]
2-Layer Graphene	~2800	Optical	Normalized width 5 μm	Ghosh et al. [49]
3-Layer Graphene	~2250	Optical	Normalized width 5 μm	Ghosh et al. [49]
Graphene	1400–3600	Optical	CVD grown	Cai et al. [52]
Supported graphene	600	Electrical	Exfoliated on SiO_2 substrate	Seol et al. [57]
GNR—Smooth edge	6000	Theoretical, MD simulations	100 Å width, no *H*-termination	Evans et al.[64]
GNR—Rough edge	3500	Theoretical, MD simulations	100 Å width, no *H*-termination	Evans et al.[64]
MW-CNT	>3000	Electrical	Individual	Kim et al. [39]
SW-CNT	~3500	Electrical	Individual	Pop et al. [40]
SW-CNT	1750–5800	Thermocouples	Bundles	Hone et al. [37]
Graphite	~2000	Variety	In-plane	Klemens [84,85]
DLCH	~0.6–0.7	3-omega	H: ~20–35%	Shamsa et al. [88]
NCD	~16	3-omega	Grain size: 22 nm	Liu et al. [89]
UNCD	~6–17	3-omega	Grain size: < 26 nm	Shamsa et al. [91]
ta-C	3.5	3-omega	sp^3: ~ 90%	Shamsa et al. [88]
ta-C	1.4	3-omega	sp^3 : ~ 60%	Balandin et al. [90]
Diamond	800–2000	Laser flash	In-plane	Sukhadolau et al. [92]

thermal conductivities of various carbon-based materials have been summarized. These include single-walled carbon nanotubes (SWCNTs), multiwalled carbon nanotubes (MWCNTs), as well as bulk carbon materials and thin films. The data on NCD, UNCD, ta-C, and hydrogenated diamond-like carbon (DLC) is based on the experimental results obtained [88–91]. The measurements were performed using a variety of techniques including the transient planar source (TPS) "hot-disk," 3-omega, and laser-flash techniques. Detailed analysis of the effects of the intrinsic atomic structure of carbon materials, for example, sp^2 versus sp^3 bonding, cluster formation, and so on; from the extrinsic, for example, phonon scattering on interfaces, on the thermal conductivity of DLC, ta-C, and NCD can be found in references [88–91]. It is interesting to note that there is a wide range of data for the thermal conductivity of CNTs. The conventionally accepted values for CNTs are ~ 3000–3500 W/mK. Thus, graphene can outperform CNTs as heat conductors. Also it was seen that few-layer graphene shows a quenching of thermal conductivity with an increase in the number of layers approaching the limit of bulk graphite [49]. It still remains unclear, though, how the thermal conductivity of graphene will be affected when it is embedded inside a device structure. Future studies might include investigation of in-plane thermal conductivity of graphene and the contact resistance when it is embedded inside a device structure or a bulk material. The theoretical models described in this review shed light on the differences in heat conduction in carbon materials. They can be incorporated into the simulation software for analysis of heat conduction in graphene layers and graphene devices [87,94]. In case of CVD-grown graphene suspended over a 3.8-μm-diameter hole, K was observed to be (2500 ± 1100) W/mK at 350 [52]. Thermal conductivity of single-layer graphene exfoliated on a SiO_2 support is ~600 W/mK, which is still high as compared to metals such as copper [57]. Some interesting works on graphene nanoribbons show that the thermal conductivity is higher for smooth edges than rough edges, and the values might vary widely ranging from 7000 W/mK to 500 W/mK depending on the width of the ribbon and edge characteristics [60,64]. The excellent thermal properties of graphene and the possible applications are covered in reference [95]. Owing to its planar geometry, graphene may have potential for lateral heat spreading. The superior thermal conductivity of graphene is beneficial for all of its proposed device applications, such as low-noise transistors, sensors, and interconnects [96,97].

11.5 CONCLUSIONS

In this chapter, we have reviewed the results of the experimental and theoretical investigation of heat conduction in graphene layers. We have also discussed the thermoelectric properties of graphene. The enhanced thermal conductivity of graphene as compared to that of bulk graphite basal planes can be explained by the 2-D nature of thermal transport in graphene over the whole range of phonon frequencies. The thermal conductivity of graphene was compared with that of other carbon materials. The superior thermal properties of graphene are beneficial for the proposed device applications and thermal management of nanoelectronic chips. Thermoelectric applications of graphene can become possible if its high intrinsic thermal conductivity is suppressed either by controlled introduction of defects and disorder or the use of narrow graphene ribbons.

ACKNOWLEDGMENTS

The work in Balandin Group was supported, in part, by DARPA–SRC through the FCRP Center on Functional Engineered Nano Architectonics (FENA) and Interconnect Focus Center (IFC). The authors are indebted to Prof. E.P. Pokatilov and Prof. D. Nika for illuminating discussions on the theory of heat conduction in graphene. The authors are thankful to the current and former members of the Nano-Device Laboratory who contributed to this investigation.

REFERENCES

1. Tien, C. L., and G. Chen. 1994. Challenges in microscale conductive and radiative heat transfer. *ASME Journal of Heat Transfer* 116:799–807. ASME Publishing, New York.
2. Chen, G. 2004. Nanoscale heat transfer and nanostructured thermoelectric. *2004 Inter Society Conference on Thermal Phenomena* 8–17. IEEE, Piscataway, NJ.
3. Kim, W., S. Singer, and A. Majumder. 2005. Role of nanostructures in reducing thermal conductivity below alloy limit in crystalline solids. *2005 International Conference on Thermoelectric* 9–12. IEEE, Piscataway, NJ.
4. Ziman, J. M. 1963. *Electrons and Phonons.* New York: Oxford University Press.
5. Bhandari, C. M., and D. M. Rowe. 1988. *Thermal conduction in semiconductors.* New York: John Wiley & Sons, Inc.
6. Goldsmid, H. J. 1964. *Thermoelectric refrigeration.* New York: Plenum Press.
7. Balandin, A. A. 2004. Thermal conductivity of semiconductor nanostructures. In *Encyclopedia of nanoscience and nanotechnology*, ed. H. S. Nalwa, 10:425–445. Stevenson Ranch, CA: American Scientific Publishers.
8. Cahill, D. G., W. K. Ford, K. E. Goodson, G. D. Mahan, A. Majumdar, H. J. Maris, R. Merlin, and S. R. Phillpot, 2003. Nanoscale thermal transport. *Journal of Applied Physics* 93:793–818.
9. Balandin, A. A., and K. L. Wang. 1998. Effect of phonon confinement on the thermoelectric figure of merit of quantum wells. *Journal of Applied Physics* 84:6149–6153.
10. Hicks, L. D., and M. S. Dresselhaus. 1993. Thermoelectric figure of merit of a one-dimensional conductor. *Physical Review B* 47:16631–16634.
11. Goodson, K. E., and Y. S. Ju. 1999. Heat conduction in novel electronic films. *Annual Review of Materials Science* 29:261–293.
12. Moore, G. E. 1965. Cramming more components onto integrated circuits. *Electronics* 38:114–117.
13. Gelsinger, P. P. 2001. Microprocessors for the new millennium: Challenges, opportunities, and new frontiers. *Solid-State Circuits Conference, Digest of Technical Papers*, ISSCC, 2001 IEEE International, 22.
14. Haensch, W., E. J. Nowak, R. H. Dennard, P. M. Solomon, A. Bryant, O. H. Dokumaci, A. Kumar, X. Wang, et al. 2006. Silicon CMOS devices beyond scaling. *IBM Journal of Research and Development* 50:339–361.
15. Pop, E., S. Sinha, and K. E. Goodson. 2006. Heat generation and transport in nanometer scale transistors. *Proceedings of the IEEE* 94:1587–1601.
16. Vasudev, P. K. 1996. CMOS scaling and interconnect technology enhancements for low power/low voltage applications. *Solid-State Electronics* 39:481–488.

17. Mutoh, S., T. Douseki, Y. Matsuya, T. Aoki, S. Shigematsu, and J. Yamada. 1995. 1-V power supply high-speed digital circuit technology with multithreshold-voltage CMOS. *IEEE Journal of Solid-State Circuits* 30:847–854.
18. Novoselov, K. S., A. K. Geim, S. V. Morozov, D. Jiang, D. Zhang, S. V. Dubonos, I. V. Grigorieva, and A. A, Firsov. 2004. Electric field effect in atomically thin carbon films. *Science* 306:666–669.
19. Novoselov, K. S., A. K. Geim, S. V. Morozov, D. Jiang, M. I. Katsnelson, I. V. Grigorieva, S. V. Dubonos, and A. A. Firsov. 2005. Two-dimensional gas of massless Dirac fermions in graphene. *Nature* 438:197197–197200.
20. Zhang, Y. B., Y. W. Tan, H. L. Stormer, and P. Kim. 2005. Experimental observation of the quantum Hall effect and Berry's phase in graphene. *Nature* 438:201–204.
21. Abanin, D. A., P. A. Lee, and L. S. Levitov. 2007. Randomness-induced XY ordering in a graphene quantum hall ferromagnet. *Physical Review Letters* 98:156801.
22. Kane, C. L., and E. J. Mele. 2005. Quantum spin Hall effect on graphene. *Physical Review Letters* 74: 161402.
23. Miao, F., S. Wijeratne, Y. Zhang, U. C. Coskun, W. Bao, and C. N. Lau. 2007. Phase coherent transport in graphene quantum billiards. *Science* 317:1530–1533.
24. Peres, N. M. R., A. H. Castro Neto, and F. Guinea. 2006. Conductance quantization in mesoscopic graphene. *Physical Review B* 73:195411.
25. Pisana, S., M. Lazzeri, C. Casiraghi, K. S. Novoselov, A. K. Geim, A. C. Ferrari, and F. Mauri. 2007. Breakdown of the adiabatic Born-Oppenheimer approximation in graphene. *Nature Materials* 6:198–201.
26. Geim, A. K., and K. S. Novoselov. 2007. The rise of graphene. *Nature Materials* 6:183–191.
27. Hass, J., R. Feng, T. Li, X. Li, Z. Zong, W. A. de Heer, P. N. First, E. H. Conrad, et al. 2006. Highly ordered graphene for two dimensional electronics. *Applied Physics Letters* 89:143106.
28. Saito, K., J. Nakamura, and A. Natori. 2007. Ballistic thermal conductance of a graphene sheet. *Physical Review B* 76:115409.
29. Peres, N. M. R., J. dos Santos, and T. Stauber. 2007. Phenomenological study of the electronic transport coefficients of graphene. *Physical Review B* 76:073412.
30. Mingo, N., and D. A. Broido. 2005. Carbon nanotube ballistic thermal conductance and its limits. *Physical Review Letters* 95:096105.
31. Kelly, B. T. 1986. *Physics of graphite.* London: Applied Science Publishers.
32. Sun, K., M. A. Stroscio, and M. Dutta. 2009. Graphite C-axis thermal conductivity. *Superlattices and Microstructures* 45:60–64.
33. Klemens, P. G. 2004. Unusually high thermal conductivity in carbon nanotubes. In *Proceedings of the Twenty-Sixth International Thermal Conductivity Conference, in: Thermal Conductivity*, ed. Ralph B. Dinwiddie, 26:48–57. Lancaster, PA: Destech Publications.
34. Ijima, S. 1991. Helical microtubules of graphitic carbon. *Nature* 354:56–68.
35. Ruoff, R. S., and D. C. Lorents. 1995. Mechanical and thermal properties of carbon nanotubes. *Carbon* 33:925–930.
36. Osman, M. A., and D. Srivastava. 2001. Temperature dependence of the thermal conductivity of single-wall carbon nanotubes. *Nanotechnology* 12:21–24.
37. Hone, J., M. Whitney, C. Piskoti, and A. Zettl. 1999. Thermal conductivity of single-walled carbon nanotubes. *Physical Review B* 59:R2514–R2516.
38. Benedict, L. X., S. G. Louie, and M. L. Cohen. 1996. Heat capacity of carbon nanotubes. *Solid State Communications* 100:177–180.
39. Kim, P., L. Shi, A. Majumder, and P. L. McEuen. 2001. Thermal transport measurements of individual multiwalled nanotubes. *Physical Review Letters* 87:215502.

40. Pop, E., D. Mann, Q. Wang, K. Goodson, and H. Dai. 2006. Thermal conductance of an individual single wall carbon nanotube above room temperature. *Nano Letters* 6:96–100.
41. Aliev, A. E., M. H. Lima, E. M. Silverman, and R. H. Baughman. 2010. Thermal conductivity of multi-walled carbon nanotube sheets: Radiation losses and quenching of phonon modes. *Nanotechnology* 21:035709.
42. Berber, S., Y-K. Kwon, and D. Tomanek. 2000. Unusually high thermal conductivity of carbon nanotubes. *Physical Review Letters* 84:4613–4616.
43. Balandin, A., S. Ghosh, W. Bao, I. Calizo, D. Teweldebrhan, F. Miao, and C. N. Lau. 2008. Superior thermal conductivity of single-layer graphene. *Nano Letters* 8: 902–907.
44. Ghosh S., I. Calizo, D. Teweldebrhan, E. P. Pokatilov, D. L. Nika, A. A. Balandin, W. Bao, F. Miao, and C. N. Lau. 2008. Extremely high thermal conductivity of graphene: Prospects for thermal management applications in nanoelectronic circuits. *Applied Physics Letters* 92:151911.
45. Calizo, I., F. Miao, W. Bao, C. N. Lau, and A. A. Balandin. 2007. Variable temperature Raman microscopy as a nanometrology tool for graphene layers and graphene-based devices. *Applied Physics Letters* 91:071913.
46. Calizo, I., A. A. Balandin, W. Bao, F. Miao, and C. N. Lau. 2007. Temperature dependence of the Raman spectra of graphene and graphene multilayers. *Nano Letters* 7:2645–2649.
47. Nika, D. L., E. P. Pokatilov, A. S. Askerov, and A. A. Balandin. 2009. Phonon thermal conduction in graphene: Role of Umklapp and edge roughness scattering. *Physical Review B* 79:155413.
48. Nika, D. L., S. Ghosh, E. P. Pokatilov, and A. A. Balandin. 2009. Lattice thermal conductivity of graphene flakes: Comparison with bulk graphite. *Applied Physics Letters* 94:203103.
49. Ghosh, S., W. Bao, D. L. Nika, S. Subrina, E. P. Pokatilov, C. N. Lau, and A. A. Balandin. 2010. Dimensional crossover of thermal transport in few-layer graphene. *Nature Materials* 9:555–558.
50. Jiang, J. W., J. S. Wang, and B. Li. 2009. Thermal conductance of graphene and dimerite. *Physical Review B* 79:205418.
51. Jauregui, L. A., Y. Yue, A. N. Sidorov, J. Hu, Q. Yu, G. Lopez, R. Jalilian, D. K. Benjamin, et al. 2010. Thermal transport in graphene nanostructures: Experiments and simulations. *Electrochemical Society Transactions* 28:73–83.
52. Cai, W., A. L. Moore, Y. Zhu, X. Li, S. Chen, L. Shi, and R. S. Ruoff. 2010. Thermal transport in suspended and supported monolayer graphene grown by chemical vapor deposition. *Nano Letters* 10:1645–1651.
53. Chen, S., A. L. Moore, W. Cai, J. W. Suk, J. An, C. Mishra, C. Amos, C. W. Magnuson, J. Kang, L. Shi, and R. S. Ruoff. 2011. Raman measurements of thermal transport in suspended monolayer graphene of variable sizes in vacuum and gaseous environments. *ACS Nano* 5:321–328.
54. Slack, G. A. 1962. Anisotropic thermal conductivity of pyrolytic graphite. *Physical Review* 127:697–701.
55. Taylor, R. 1966. The thermal conductivity of pyrolytic graphite. *Philosophical Magazine* 13:157–166.
56. Faugeras, C., B. Faugeras, M. Orlita, M. Potemski, R. R. Nair, and A. K. Geim. 2010. Thermal conductivity of graphene in corbino membrane geometry. *ACS Nano* 4:1889–1992.
57. Seol, J. H., I. Jo, A. R. Moore, L. Lindsay, Z. H. Aitken, M. T. Pettes, X. Li, Z. Yao, R. Huang, D. Broido, N. Mingo, and R. S. Ruoff. 2010. Two-dimensional phonon transport in supported graphene. *Science* 328:213–216.
58. Koh, Y. K., M-H. Bae, D. G. Cahill, and E. Pop. 2010. Heat conduction across monolayer and few-layer graphenes. *Nano Letters* 10:4363–4368.

59. Naeemi, A., and J. D. Meindl. 2007. Conductance modeling graphene nanoribbon (GNR) interconnects. *IEEE Electron Device Letters* 28:428–431.
60. Lan, J., J. S. Wang, C. K. Gan, and S. K. Chin. 2009. Edge effects on quantum thermal transport in graphene nanoribbons: Tight-binding calculations. *Physical Review B* 79:115401.
61. Murali, R., Y. Yang, K. Brenner, T. Beck, and J. D. Meindl. 2009. Breakdown current density of graphene nanoribbons. *Applied Physics Letters* 94:243114.
62. Hu, J., X. Ruan, and Y. P. Chen. 2009. Thermal conductivity and thermal rectification in graphene nanoribbons: A molecular dynamics study. *Nano Letters* 9:2730–2735.
63. Guo, Z., D. Zhang, and X-G. Gong. 2009. Thermal conductivity of graphene nanoribbons. *Applied Physics Letters* 95:163103.
64. Evans, W. J., L. Hu, and P. Keblinski. 2010. Thermal conductivity of graphene ribbons from equilibrium molecular dynamics: Effect of ribbon width, edge roughness, and hydrogen termination. *Applied Physics Letters* 96:203103.
65. Löfwander, T., and M. Fogelström. 2007. Impurity scattering and Mott's formula in graphene. *Physical Review B* 76:193401.
66. Stauber, T., N. M. R. Peres, and F. Guinea. 2007. Electronic transport in graphene: A semiclassical approach including midgap states. *Physical Review B* 76:205423.
67. Cutler, M., and N. F. Mott. 1969. Observation of Anderson localization in an electron gas. *Physical Review* 181:1336–1340.
68. Zuev, Y. M., W. Chang, and P. Kim. 2009. Thermoelectric and magnetothermoelectric transport measurements of graphene. *Physical Review Letters* 102:096807.
69. Wei, P., W. Bao, Y. Pu, C. N. Lau, and J. Shi. 2009. Anomalous thermoelectric transport of Dirac particles in graphene. *Physical Review Letters* 102:166808.
70. Checkelsky, J. G., and N. P. Ong. 2009. Thermopower and Nernst effect in graphene in a magnetic field. *Physical Review B* 80:081413.
71. Wang, C-R., W-S. Lu, and W-L. Lee. 2010. Transverse thermoelectric conductivity of bilayer graphene in quantum Hall regime. *Physical Review B* 82:121406R.
72. Hwang, E. H., E. Rossi, and S. Das Sharma. 2009. Theory of thermopower in two-dimensional graphene. *Physical Review B* 80:235415.
73. Ferrari, A. C., J. C. Meyer, V. Scardaci, C. Casiraghi, M. Lazzeri, F. Mauri, S. Piscanec, D. Jiang, et al. 2006. Raman spectrum of graphene and graphene layers. *Physical Review Letters* 97:187401.
74. Calizo, I., W. Bao, F. Miao, C. N. Lau, and A. A. Balandin. 2007. The effect of substrates on the Raman spectrum of graphene: Graphene-on-sapphire and graphene-on-glass. *Applied Physics Letters* 91:201904.
75. Dawlaty, J. M., S. Shivaraman, M. Chandrashekhar, F. Rana, and M. G. Spencer. 2008. Measurement of ultrafast carrier dynamics in epitaxial graphene. *Applied Physics Letters* 92: 042116.
76. Sun, D., Z-K. Wu, C. Divin, X. Li, C. Berger, W. A. de Heer, P. N. First, and T. B. Norris. 2008. Ultrafast relaxation of excited Dirac fermions in epitaxial graphene using optical differential transmission spectroscopy. *Physical Review Letters* 101:157402.
77. Bolotin, K. I., K. J. Sikes, J. Hone, H. L. Stormer, and P. Kim. 2008. Temperature-dependent transport in suspended graphene. *Physical Review Letters* 101:096802.
78. Krauss, B., T. Lohmann, D-H. Chai, M. Haluska, K-V. Klitzing, and J. H. Smet. 2009. Laser-induced disassembly of a graphene single crystal into a nanocrystalline network. *Physical Review B* 79:165428.
79. Nair, R. R., P. Blake, A. N. Grigorenko, K. S. Novoselov, T. J. Booth, T. Stauber, N. M. R. Peres, and A. K. Geim. 2008. Fine structure constant defines visual transparency of graphene. *Science* 320:1308.

80. Kim, K. S., Y. Zhao, H. Jang, S. Y. Lee, J. M. Kim, K. S. Kim, J-H. Ahn, P. Kim, et al. 2009. Large-scale pattern growth of graphene films for stretchable transparent electrodes. *Nature* 457:706–710.
81. Mak, K. F., M. Y. Sfeir, Y. Wu, C. H. Lui, J. A. Misewich, and T. F. Heinz. 2008. Measurement of the optical conductivity of graphene. *Physical Review Letters* 101:196405.
82. Ghosh, S., D. L. Nika, E. P. Pokatilov, and A. A. Balandin. 2009. Heat conduction in graphene: Experimental study and theoretical interpretation. *New Journal of Physics* 11: 095012.
83. Srivastava, G. P. 1990. *The physics of phonons*, 99. Philadelphia, PA: IOP.
84. Klemens, P. G. 2000. Theory of the a-plane thermal conductivity of graphite. *Journal of Wide Bandgap Materials* 7:332–339.
85. Klemens, P. G. 2001. Theory of thermal conduction in thin ceramic films. *International Journal of Thermophysics* 22:265–275.
86. Mounet, N., and N. Marzari. 2005. First-principles determination of the structural, vibrational and thermodynamic properties of diamond graphite, and derivatives. *Physical Review B* 71:205214.
87. Subrina, S., and D. Kotchekov. 2008. Simulation of heat conduction in suspended graphene flakes of variable shapes. *Journal of Nanoelectronics and Optoelectronics* 3:249–269.
88. Shamsa, M., W. L. Liu, A. A. Balandin, C. Casiraghi, W. I. Milne, and A. C. Ferrari. 2006. Thermal conductivity of diamond-like carbon films. *Applied Physics Letters* 89:161921.
89. Liu, W. L., M. Shamsa, I. Calizo, A. A. Balandin, V. Ralchenko, A. Popovich, and A. Saveliev. 2006. Thermal conduction in nanocrystalline diamond films: Effects of the grain boundary scattering and nitrogen doping. *Applied Physics Letters* 89:171915.
90. Balandin, A. A., M. Shamsa, W. L. Liu, C. Casiraghi, and A. C. Ferrari. 2008. Thermal conductivity of ultrathin tetrahedral amorphous carbon films. *Applied Physics Letters* 93:043115.
91. Shamsa, M., S. Ghosh, I. Calizo, V. G. Ralchenko, A. Popovich, and A. A. Balandin. 2008. Thermal conductivity of nitrogenated ultrananocrystalline diamond films on silicon. *Journal Applied Physics* 103:083538.
92. Sukhadolau, A. V., E. V. Ivakin, V. G. Ralchenko, A. V. Khomich, A. V. Vlasov, and A. F. Popovich. 2005. Thermal conductivity of CVD diamond at elevated temperatures. *Diamond and Related Materials* 14:589–593.
93. Worner, E., C. Wild, W. Muller-Sebert, R. Locher, and P. Koidl. 1996. Thermal conductivity of CVD diamond films: High-precision, temperature-resolved measurements. *Diamond and Related Materials* 5:688–692.
94. Ko, G., and J. Kim. 2009. Thermal modeling of graphene layer on the peak channel temperature of AlGaN/GaN high electron mobility transistors. *Electrochemical and Solid-State Letters* 12:H29–H31.
95. Prasher, R. 2010. Graphene spreads the heat. *Science* 328:185–186.
96. Shao, Q., G. Liu, D. Teweldebrhan, A. A. Balandin, S. Rumyantsev, M. Shur, and D. Yan. 2009. Flicker noise in bilayer graphene transistors. *IEEE Electron Device Letters* 30:288–290.
97. Shao, Q., G. Liu, D. Teweldebrhan, and A. A. Balandin. 2008. High-temperature quenching of electrical resistance in graphene interconnects. *Applied Physics Letters* 92:202108.

Index

C

F

G

H

N

O

S

Milton Keynes UK
Ingram Content Group UK Ltd.
UKHW051013071024
449327UK00012B/232